Meta-Heuristics: Advances and Trends in Local Search Paradigms for Optimization

Meta-Heuristics: Advances and Trends in Local Search Paradigms for Optimization

Edited by
Stefan Voß
Technical University of Braunschweig, Germany
Silvano Martello
University of Bologna, Italy
Ibrahim H. Osman
University of Kent, United Kingdom
Catherine Roucairol
University of Versailles, France

INRIA

Kluwer Academic Publishers
Boston/Dordrecht/London

Distributors for North, Central and South America:
Kluwer Academic Publishers
101 Philip Drive
Assinippi Park
Norwell, Massachusetts 02061 USA
Telephone (781) 871-6600
Fax (781) 871-6528
E-Mail <kluwer@wkap.com>

Distributors for all other countries:
Kluwer Academic Publishers Group
Distribution Centre
Post Office Box 322
3300 AH Dordrecht, THE NETHERLANDS
Telephone 31 78 6392 392
Fax 31 78 6546 474
E-Mail <orderdept@wkap.nl>

Electronic Services <http://www.wkap.nl>

Library of Congress Cataloging-in-Publication Data

A C.I.P. Catalogue record for this book is available
from the Library of Congress.

Printed on acid-free paper.

Printed in the United States of America

Contents

Part VI: Further Meta-Heuristics

Preface

In recent years, there have been significant advances in the theory and application of meta-heuristics to the approximate solutions of hard optimization problems. A meta-heuristic is an iterative master process that guides and modifies the operations of subordinate heuristics to efficiently produce high-quality solutions. It may manipulate a complete (or incomplete) single solution or a collection of solutions at each iteration. The subordinate heuristics may be high (or low) level procedures, or a simple local search, or just a construction method. The family of meta-heuristics includes, but is not limited to, adaptive memory procedures, tabu search, ant systems, greedy randomized adaptive search, variable neighborhood search, evolutionary methods, genetic algorithms, scatter search, neural networks, simulated annealing, and their hybrids.

Meta-heuristics provide decision-making managers with robust tools that obtain high-quality solutions, in reasonable time horizons, to important applications in business, engineering, economics and science. Finding exact solutions in these applications still poses a real challenge despite the impact of recent advances in computer technology and the great interactions between computer science, management science / operations research and mathematics.

This book grew out of the second Meta-Heuristics International Conference (MIC-97) which was successfully organized by the University of Versailles and the Institut National de Recherche en Informatique et en Automatique (INRIA), at the INRIA center in Sophia-Antipolis, France, between the 21st and 24th of July 1997.

The book is not the conference proceedings, but it comprises a carefully refereed selection of extended versions of the best papers presented at the conference. In this respect it may be referred to as a follow-up of the MIC-95 book [1]. The selected articles describe the most recent developments in theory and applications of meta-heuristics, heuristics for specific problems, and comparative case studies. The book is divided into six parts containing homogeneous articles grouped mainly by the techniques considered.

The extensive first part with twelve papers covers tabu search and its application to a great variety of well-known combinatorial optimization problems (including the resource-constrained project scheduling problem and vehicle routing problems). Real-world applications include the consideration of actuator selection for the control of noise in aircraft interiors, mail collecting for the French postal services, and crew scheduling for a Portuguese metro company.

In the second part we find one paper where tabu search and simulated annealing are investigated comparatively and two papers which consider hybrid methods combining tabu search with genetic algorithms. The third part has four papers on genetic and evolutionary algorithms. While two of these papers aim at making these algorithms (self-) adaptive we arrive at another paradigm within meta-heuristics, i.e., the ant-system (part four). In three papers it is argued that the pure ant system paradigm is a stimulating concept which may be improved when incorporating other ideas such as local search.

The four papers of the fifth part study the behavior of parallel local search algorithms mainly from a tabu search perspective. Applications are again found for well-known combinatorial optimization problems such as the job-shop problem. The final part consists of eight papers examining a great variety of additional meta-heuristics topics. This includes neural networks, variable neighbourhood search as well as guided local search. Furthermore, the integration of meta-heuristics with the branch-and-bound paradigm is investigated.

MIC-97 program committee members

Fred Glover, General Chair, University of Colorado, Boulder, U.S.A.
Ibrahim H. Osman, Co-Chair, University of Kent, Canterbury, UK, and Kuwait University, Safat, Kuwait.
Catherine Roucairol, Co-Chair, University of Versailles, Versailles, France.
Silvano Martello, University of Bologna, Bologna, Italy.
Stefan Voß, Technical University Braunschweig, Braunschweig, Germany.

Acknowledgements

Of course such an ambituous project like organizing the conference and publishing a high quality book to reflect its outcome would not have been possible without the most valuable input of a large number of individuals. We wish to thank all the authors for their contributions, the referees for their input and the attendees for their participation and fruitful discussion. We are grateful to the INRIA staff and the Versailles University staff for their help and financial supports. In particular, we are indebted to Claudie Thenault for her enthusiasm and help when running the conference and throughout the project, to Van-Dat Cung for his help at various stages up to the conference, and to the team at the Technical University Braunschweig, Kai Gutenschwager, Kathryn Dawson-Townsend and Andreas Fink, who helped in putting this book together, and to Gary Folven at Kluwer Academic Publishers for his help and encouragement.

Referees

Emile H.L. Aarts, Eindhoven
Ramon Alvares-Valdes, Valencia
Anita Amberg, Darmstadt
Mohammed Amini, Memphis
Roberto Battiti, Trento
James C. Bean, Ann Arbor
Foued Ben Abdelaziz, Le Bardo
Christian Bierwirth, Bremen
Jose Brandao, Braga
Peter Brucker, Osnabrück
Bernd Bullnheimer, Vienna
Jaques Carlier, Compiegne
Yves Caseau, St. Quentin
Daniel Costa, Neuchatel
Teodor G. Crainic, Montreal
Van-Dat Cung, Versailles
Mauro DellAmico, Modena
Jaques Desrosiers, Montreal
Marco Dorigo, Brussels
Raphael Dorne, Nimes
Emanuel Falkenauer, Brussels
Andreas Fink, Braunschweig
Mark A. Fleischer, Norfolk
Paulo M. Franca, Campinas
Xavier Gandibleux, Valenciennes
Jean-Marc Geib, Villeneuve
Michel Gendreau, Montreal
Hassan Ghaziri, Beirut
Fred Glover, Boulder
Luis Gouveia, Lisbon
Peter Greistorfer, Graz
Tore Grünert, Aachen
Said Hanafi, Valenciennes
Jin-Kao Hao, Nimes
Alain Hertz, Lausanne
Khalil Hindi, Uxbridge
Johann L. Hurink, Twente
Steve Hurley, Cardiff
Toshihide Ibaraki, Kyoto
David S. Johnson, Murray Hill
Martin Kabath, Köln
Jim Kelly, Boulder
Rex K. Kincaid, Williamsburg
Robert Klein, Darmstadt
Gary Kochenberger, Denver
Werner Krabs, Darmstadt
Manuel Laguna, Boulder
Gilbert Laporte, Montreal
V. Jorge Leon, College Station
Arne Lokketangen, Molde
William G. Macready, Santa Fe
Nelson Maculan, Rio de Janeiro
Vittorio Maniezzo, Bologna
Silvano Martello, Bologna
Olivier Martin, Orsay
Joseph B. Mazzola, Durham
Nenad Mladenovic, Montreal
Rolf H. Möhring, Berlin
Heinz Mühlenbein, St. Augustin
Eugeniusz Nowicki, Wroclaw
Luiz S. Ochi, Niteroi
Ibrahim H. Osman, Canterbury
Francois Pellegrini, Bordeaux
Erwin Pesch, Bonn
Gerard Plateau, Paris
Chris Potts, Southhampton
Jean-Yves Potvin, Montreal
Patrick Prosser, Glasgow
Vic Rayward-Smith, Norwich
Colin Reeves, Coventry
Andreas Reinholz, St. Augustin
Mauricio Resende, Florham Park
Celso Ribeiro, Rio de Janeiro
Erik Rolland, Columbus
Jean Roman, Bordeaux
Catherine Roucairol, Versailles
Gabriele Schneidereit, Braunschweig
Czeslaw Smutnicki, Wroclaw
Lutz Sondergeld, Braunschweig
Patrick Soriano, Montreal
Nelson Stephens, London
Thomas Stützle, Darmstadt
Eric Taillard, Lugano
El-Ghazali Talbi, Villeneuve
Huub M.M. ten Eikelder, Eindhoven
Michel Toulouse, Norman
Daniel Tuyttens, Mons
Vicente Valls, Valencia
Alex van Breedam, Antwerp
Stefan Voß, Braunschweig
David L. Woodruff, Davis
Takeshi Yamada, Kyoto

References

[1] I.H. Osman and J.P. Kelly, *Meta-Heuristics: Theory & Applications*, Kluwer Academic Publishers, Boston, 1996.

The Editors:

Stefan Voß
Silvano Martello
Ibrahim H. Osman
Catherine Roucairol

1 TABU SEARCH ALGORITHMS AND LOWER BOUNDS FOR THE RESOURCE-CONSTRAINED PROJECT SCHEDULING PROBLEM

Tonius Baar, Peter Brucker and Sigrid Knust

Universität Osnabrück, Department of Mathematics/Informatics
49069 Osnabrück, Germany.
peter@mathematik.uni-osnabrueck.de, sigrid@mathematik.uni-osnabrueck.de

Abstract: We present two tabu search algorithms for the resource-constrained project scheduling problem. Given are n activities which have to be processed without preemptions. During the processing period of an activity constant amounts of renewable resources are needed where the available capacity of each resource type is limited. Furthermore, finish-start precedence relations between the activities are given. The objective is to determine a schedule with minimal makespan. The first tabu search approach relies on elimination of critical arcs and list-scheduling techniques. The second neighborhood is based on schedule schemes, where neighbors are generated by placing activities in parallel or deleting a parallelity relation. Furthermore, a column-generation approach for a linear programming-based lower bound is presented and computational results are reported.

1.1 INTRODUCTION

The **resource-constrained project scheduling problem (RCPSP)** may be formulated as follows:

Given are n activities $1, \dots, n$ and r renewable resources. A constant amount of R_k units of resource k is available at any time. Activity i must be processed for p_i time units; preemption is not allowed. During this time period a constant amount of r_{ik} units of resource k is occupied. The values R_k, p_i, and r_{ik} are supposed to be non-negative integers. Furthermore, there are precedence relations defined between the activities. The objective is to determine starting times s_i for the activities $i = 1, \dots, n$ in such a way that

- at each time t the total resource demand is less than or equal to the resource availability for each resource type,
- the given precedence constraints are fulfilled, and
- the makespan $C_{\max} = \max_{i=1}^{n} c_i$, where $c_i := s_i + p_i$, is minimized.

The problem may be presented by an activity-on-the-node network with dummy starting and finishing nodes 0 and $n+1$, respectively, and $p_0 = p_{n+1} = 0$. We assume $s_0 = 0$ and identify s_{n+1} with the $C_{\max}$-value.

As a generalization of the job-shop scheduling problem the RCPSP is $\mathcal{NP}$-hard in the strong sense [3]. Therefore, besides exact algorithms like branch-and-bound procedures many authors have developed heuristics for the RCPSP.

The first heuristics are methods based on priority rules which have been obtained in [9]. Later on many different priority rules based on characteristics of the activities, precedence constraints, resources or critical paths have been suggested and tested [1]. None of these are superior to others for all instances. Therefore, recently more complicated algorithms have been proposed. A procedure in [14] uses sampling techniques and disturbed priorities. Problem-space based neighborhoods are defined in [18, 20]. Local search procedures for the RCPSP can be found in [17, 21, 22]. In [2, 5] simulated annealing algorithms are presented and in [11] parallel genetic algorithms are considered.

In this paper we present two tabu search algorithms. The first generalizes neighborhoods known for the job-shop problem. The neighborhood is based on a list of the activities and the elimination of critical arcs. The second tabu search method relies on so-called schedule schemes which have been introduced in [4] in connection with a branch-and-bound algorithm. These schemes represent sets of feasible schedules which are defined by conjunctions, disjunctions, and relations placing activities in parallel.

Furthermore, we describe a column-generation approach for a linear programming lower bound introduced in [19]. This lower bound is used to obtain information about the quality of the calculated solutions.

This paper is organized as follows. In Section 1.2 we describe the tabu search algorithm based on critical arcs. In Section 1.3 we state the main ideas of the tabu search method based on parallelity concepts. Section 1.4 is devoted to the calculation of lower bounds. Computational results for various problem instances are presented in Section 1.5. Finally, in Section 1.6 some concluding remarks can be found.

1.2 TABU SEARCH BASED ON CRITICAL ARCS

The tabu search method presented in this section generalizes critical path concepts developed for the job-shop problem. Contrary to the job-shop problem where the disjunctive graph model is used to define schedules (and corresponding critical paths) we consider list-schedules. The reason for this is that for the

RCPSP not all incompatibilities can be represented by pairs of activities. We start with a list L and associate with L a feasible schedule s. Using properties of s a new list L' and a corresponding schedule s' are created. This process is repeated.

Schedules s and s' are derived from L and L' by a list-scheduling procedure which is described in Section 1.2.1. L' is chosen from a neighborhood $N(L)$ of L which is introduced in Section 1.2.2. The main characteristics of a tabu search procedure based on these concepts are listed in Section 1.2.3.

In the following we denote by C a set containing all original precedence constraints and further precedences induced by transitivity and immediate selection. The relations in C are denoted by $i \to j$.

1.2.1 *A List-Scheduling Heuristic*

Let L be a list of all activities $0, 1, \ldots, n, n+1$. For two activities i, j we write $i \prec j$ if in L activity i is listed before j. All lists considered are assumed to be **compatible** with the precedences $i \to j \in C$, i.e. we always have $i \prec j$ if $i \to j \in C$.

The following Algorithm List-Schedule calculates a feasible schedule $s = H_1(L)$ using the list L. This algorithm is also known as serial scheduling procedure [1, 13]. In this algorithm PS is the set of finishing times of scheduled activities.

Algorithm List-Schedule (L)

1. $PS := \{0\}$;
2. WHILE $L \neq \emptyset$ DO
   BEGIN
3. Eliminate the next activity j from L;
4. $h_j := \max\{s_i + p_i \mid i \to j \in C\}$;
5. Calculate the earliest starting time $t \in PS$, $t \geq h_j$ such that activity j can be scheduled without violating the resource constraints;
6. Schedule j at time $s_j := t$;
7. $PS := PS \cup \{s_j + p_j\}$
   END.

During the algorithm for all $t \in PS$ we keep track of the amounts $r_k(t)$ of resource k used by all activities being processed or starting at time t. Using these $r_k(t)$-values it can be checked whether a new activity can be started at time $t \in PS$ without violating the resource constraints. The complexity of this algorithm is $O(n^2 r)$. It is not difficult to show that there always exists a list L^* such that $s^* = H_1(L^*)$ is optimal.

1.2.2 *The Neighborhood*

Let L be a list and s be the corresponding list-schedule. Consider the directed graph $G_s = (V, A_s)$ where V is the set of all activities and

$$(i, j) \in A_s \text{ if and only if } s_i + p_i = s_j.$$

A **critical path** with respect to s is a path in G_s from 0 to $n+1$. There always exists at least one critical path. An arc (i,j) belonging to a critical path is called a **critical arc** if (i,j) does not belong to C. To improve the schedule s at least one critical arc must become noncritical. Hence we construct neighbor lists of list L in such a way that critical arcs in s may become noncritical. Next, we describe how to construct such a neighborhood.

First, we calculate a critical path $CP(s)$ with a minimal number of critical arcs.

Then the neighborhood $NBH(L)$ is defined by three types of operators. Each depends on the critical path $CP(s)$. The idea of these operators is to cancel a critical arc (i,j).

1. shift_{ij} is a **shift-operator** which is defined for a critical arc $(i,j) \in CP(s)$ with $i \prec j$ in L. It reverses the ordering of activities i and j in L by moving i together with all activities u satisfying

$$i \rightarrow u \in C \text{ and } i \prec u \prec j$$

   immediately after j. This can be illustrated by

$$L = \ldots i \ldots u \ldots j \ldots \Rightarrow L' = \ldots j\, i\, u \ldots$$

2. bshift_{ij} is a **backshift-operator** which is defined for a critical arc $(i,j) \in CP(s)$ with $j \prec i$ in L. It moves the first activity u which satisfies the conditions

$$i \prec u \text{ and } (i,u) \notin C$$

   immediately before i. Such a backshift increases in L the distance between activities j and i and can be illustrated by

$$L = \ldots j \ldots i \ldots l \ldots u \ldots \Rightarrow L' = \ldots j \ldots u\, i \ldots l \ldots$$

   Note that due to the transitivity of C there is no l with $i \prec l$ and $l \rightarrow u$.

3. Symmetrically, the **frontshift-operator** fshift_{ij} is defined. It can be illustrated by

$$L = \ldots u \ldots l \ldots j \ldots i \ldots \Rightarrow L' = \ldots l \ldots ju \ldots i \ldots$$

Finally, we set

$$\begin{aligned} NBH(L) &= \{\text{shift}_{ij}(L) \mid (i,j) \in CP(s) \text{ is critical}, i \prec j \text{ in } L\} \\ &\cup \{\text{bshift}_{ij}(L) \mid (i,j) \in CP(s) \text{ is critical}, j \prec i \text{ in } L\} \\ &\cup \{\text{fshift}_{ij}(L) \mid (i,j) \in CP(s) \text{ is critical}, j \prec i \text{ in } L\}. \end{aligned}$$

Note that $|NBH(L)| = O(n)$ and the effort to calculate all neighbors is $O(n^2)$.

1.2.3 Implementation

In each iteration we have a list L and the corresponding list-schedule $s = H_1(L)$. The next list is chosen from the set $NBH(L)$ which depends on a critical path for s with a minimal number of critical arcs.

Other features of the tabu search procedure are:

- After applying a shift$_{ij}$, bshift$_{ij}$ or fshift$_{ij}$-operator, we store the pair (i, j) in a tabu-list. Then all operators which change the order of j and i are tabu.

- After each iteration which improves the current upper bound, immediate selection, see [4], is applied. This method introduces new precedence constraints C and disjunctions D for an instance which have to be satisfied if the current upper bound is improved. Thus, the set of feasible solutions and the neighborhood become smaller. Each time the precedences C are changed by immediate selection we check if the current list has become infeasible and eliminate all incompatibilities in this case.

- An aspiration criterion is used: If an operator which is tabu leads to a solution with a better $C_{\max}$-value than the best solution value found so far, then the tabu-status of the operation is cancelled.

- The tabu search is stopped if in the last 250 iterations no solution is found which improves the best solution calculated so far.

- We allowed two restarts, i.e., when stopping the tabu search after 250 iterations we start the search again with the best solution found so far and choose a new neighbor randomly.

- To get a good starting solution we applied the serial scheduling procedure with different priority rules (MTS, LPF, GRPW, LST, LFT, RSM, CUMRED, cf. [1]) and selected the best solution found.

- Experiments have shown that dynamic tabu-lists of minimal length 12 and maximal length 24 gave the best results. The length l of the tabu-list is controlled as suggested in [7] for the job-shop problem:

  - if we are in an improving phase and $l > 12$, then l is decreased by one,
  - if we are not in an improving phase and $l < 24$, then l is increased by one,
  - if we have improved the current upper bound in the last step, then we delete the tabu-list completely and set $l = 1$.

1.3 TABU SEARCH BASED ON PARALLELITY

The tabu search method we introduce in this section is based on so-called schedule schemes. A schedule scheme consists of sets of relations which define a set of feasible schedules. The iterative process of going from one solution to the next may be described as follows. We start with a scheduling scheme S and associate with S a feasible schedule s. Using properties of s a new schedule scheme S' and a corresponding feasible schedule s' are created. This process is repeated.

Scheduling schemes are introduced in Section 1.3.1. s and s' are derived from S and S' by a procedure which is described in Section 1.3.2. There are several possibilities of moving from S to S'. These moves define a neighborhood on the set of all schedule schemes which is discussed in Section 1.3.3. In Section 1.3.4 details of a tabu search procedure based on these concepts are given.

1.3.1 Schedule Schemes

Schedule schemes have been introduced in connection with a branch-and-bound algorithm for the RCPSP in [4]. A schedule scheme represents a (possibly empty) set of feasible schedules. It can be defined as follows.

Let V be the set of all activities $0, 1, \ldots, n, n+1$ and define

$$A := \{(i,j) \mid i,j \in V, i \neq j\}.$$

A **schedule scheme** (C, D, N, F) consists of four disjoint relations $C, D, N, F \subseteq A$. It has the following properties:

- C is antisymmetric (i.e. $(i,j) \in C$ implies $(j,i) \notin C$),
- D, N, F are symmetric (i.e. if (i,j) belongs to one of these sets, then (j,i) belongs to the same set), and
- for each $(i,j) \in A$ we have either $(i,j) \in D \cup N \cup F$ or one of the conditions $(i,j) \in C$ or $(j,i) \in C$ holds.

As in the last section, the relations in C are original and induced precedences $i \to j$ which we call **conjunctions**. The set D represents **disjunctions** denoted by $i - j$. The relations in N are called **parallelity relations** and, finally, F contains **flexibility relations**. They are denoted by $i \parallel j$ and $i \sim j$, respectively. Activities i, j with $i \parallel j$ must be processed in parallel for at least one time unit, whereas there is no restriction on activities with $i \sim j$. Due to symmetry we identify $i - j$ with $j - i$, $i \parallel j$ with $j \parallel i$, and $i \sim j$ with $j \sim i$. Two activities i and j are called **incompatible** if $i - j$, or $i \to j$, or $j \to i$ holds.

The schedule scheme (C, D, N, F) **represents** the set $\mathcal{S}(C, D, N, F)$ of all schedules with

- $c_i \leq s_j$ (activity j is scheduled after i) for all $i \to j \in C$,
- $c_i \leq s_j$ or $c_j \leq s_i$ for all $i - j \in D$ (the processing intervals of activities i and j do not overlap), and
- $s_i < c_j$ and $s_j < c_i$ (the processing intervals of activities i and j intersect) if $i \parallel j \in N$.

Note that a schedule scheme may represent infeasible as well as feasible schedules because the resource constraints may be violated. The subset of feasible schedules that are represented is denoted by $\mathcal{S}_f(C, D, N, F)$.

Let C_0 be the set of original precedences and let D_0 be an initial set of disjunctions that is induced by the resource constraints for pairs of activities (i.e. we set $i - j \in D_0$ if and only if $r_{ik} + r_{jk} > R_k$ for some $k = 1, \ldots, r$). Then $(C_0, D_0, \emptyset, F_0)$ with $F_0 = \{i \sim j \mid i \to j \notin C_0, j \to i \notin C_0,\ i - j \notin D_0\}$ represents all feasible schedules.

1.3.2 *A Parallelity Heuristic*

The problem of deciding whether there exists a schedule which belongs to $\mathcal{S}_f(C, D, N, F)$ has been shown to be $\mathcal{NP}$-hard in [16]. Thus, it is unlikely to find such a schedule in polynomial time. For these reasons, in this section we will develop a heuristic procedure H_2 which, for a given schedule scheme $S = (C, D, N, F)$, calculates a schedule $s = H_2(S)$. This schedule

- satisfies all conjunctions C and disjunctions D,
- satisfies a "large" number of parallelity relations N,
- has a "good" $C_{\max}$-value.

The basis of this heuristic is the concept of parallelity components. A **parallelity component** is a connectivity component in the undirected graph induced by the set N of parallelity relations. Let P_1 and P_2 be two different parallelity components. Then a precedence relation $P_1 \to P_2$ is introduced if and only if activities $i \in P_1$ and $j \in P_2$ exist such that $i \to j \in C$ holds (cf. Figure 1.1). These parallelity components and the precedences between components have a high impact on the priority of an activity.

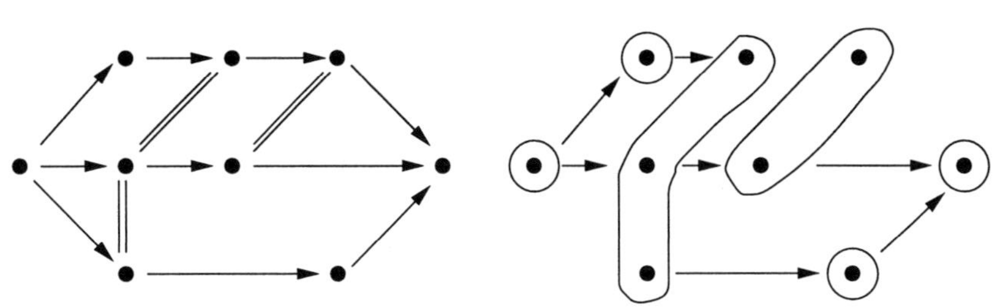

Figure 1.1 Example for parallelity components (○) for a graph with vertices (•)

The heuristic constructs a feasible schedule by adding one activity to the current partial schedule in each step. It can be described as follows:

(1) Calculate the parallelity components and the corresponding precedence constraints between these components.

(2) If no parallelity component without predecessor components exists, then go to (5). Otherwise, one of the parallelity components without predecessor is chosen which contains a job with the smallest p_i+q_i-value, where q_i is the tail of i (i.e. a lower bound for the difference between the makespan and the finishing time of activity i). The component P is now scheduled until it is empty in the following way:

(3) Choose an activity $i \in P$ according to a specific priority rule and calculate an earliest starting time r_i at which i can be started if it is added to the partial schedule. More precisely, let

$$h_i := \max \{\max_{j \to i}\{s_j+p_j \mid j \text{ planned}\}, \min_{k \| i}\{s_k-(p_i-1) \mid k \text{ planned}\}\}.$$

Calculate the earliest starting time $r_i \geq h_i$ such that activity i can be scheduled without violating the disjunctions and resource constraints and schedule i at time r_i. Eliminate i from P, update the resource profile of the partial schedule and repeat Step (3) until P is empty.

(4) When P becomes empty, eliminate this component with its precedences and go to (2).

(5) A cycle between some parallelity components exists. Choose an activity i without an unfinished conjunctive predecessor using the same criterion as in (3). Schedule i and update the parallelity components with its precedences. If a component without predecessors now exists, go to (2). Otherwise, repeat Step (5).

This procedure can be implemented in $O(n^3r)$.

As we will see in the next section, the schedule $s = H_2(S)$ may be used to define different neighbor schemes $S' = (C', D', N', F')$ for the schedule scheme S.

1.3.3 *The Neighborhood*

Contrary to the list-scheduling procedure in Section 1.2, heuristic H_2 generally constructs schedules $s = H_2(S)$ which are not active, i.e. activities can be shifted to the left in s without violating feasibility. The reason for this is that H_2 tries to satisfy a large number of parallelity relations. Therefore, we may not have a critical path corresponding to s. But the ideas of a critical path can be transferred to the parallelity concept in the following way.

A sequence $0 = i_0, i_1, \ldots, i_l = n+1$ of activities is called a **parallel critical path** $CP_{\|}(s)$ if for $i_\nu, i_{\nu+1}$, $(\nu = 0, \ldots, n)$ one of the following conditions holds:

- activity $i_{\nu+1}$ starts when activity i_ν finishes, or
- activity i_ν starts earlier than $i_{\nu+1}$ and both activities are processed in parallel for at least one time unit.

The makespan of a schedule s can only be shortened if more of the activities on $CP_{\|}(s)$ are processed in parallel.

Therefore, we introduce four different types of operators which are defined on a schedule scheme $S = (C, D, N, F)$ and try to place more activities in parallel.

The first operator move $(i \sim j)$ is defined for $i \sim j \in F$.

- move $(i \sim j)(S)$ transforms $i \sim j \in F$ into a parallelity relation $i \parallel j$, i.e.

$$\text{move } (i \sim j)(S) := (C, D, N \cup \{i \parallel j\}, F \backslash \{i \sim j\}).$$

The next two operators are defined for $i \parallel j \in N$ and the parallelity component P to which i, j belong. These operators eliminate parallelity relations with the objective that the parallelity relation $i \parallel j$ may be satisfied by the heuristic H_2.

- move $(P\backslash i, j)$ eliminates all parallelity relations $k \parallel i$ with $k \neq j$, i.e.

$$\text{move } (P\backslash i, j)(S) := (C, D, N\backslash\{k \parallel i \mid k \neq j\}, F \cup \{k \sim i \mid k \parallel i, k \neq j\})$$

- move $(P\backslash i\backslash j)$ is the concatenation of move $(P\backslash i, j)$ and move $(P\backslash j, i)$, i.e. it eliminates all parallelity relations $k \parallel i$ and $k \parallel j$ except $i \parallel j$.

Finally, we have an operator which cancels all parallelity relations in a parallelity component. Then it may be possible that H_2 can satisfy other parallelity relations.

- move (P) eliminates all parallelity relations in a parallelity component P, i.e.

$$\text{move } (P)(S) := (C, D, N\backslash\{k \parallel l \mid k \parallel l \in P\}, F \cup \{k \sim l \mid k \parallel l \in P\}).$$

Neighborhood $NBH_{\parallel}$ is defined by $NBH_{\parallel}(S) := N_1 \cup N_2 \cup N_3 \cup N_4$ with

$$\begin{aligned}
N_1 &:= \{\text{move } (i \sim j)(S) \mid i, j \in CP_{\parallel}(s), i \sim j \in F\}, \\
N_2 &:= \{\text{move } (P\backslash i, j)(S) \mid i, j \in CP_{\parallel}(s), i \parallel j \in N\}, \\
N_3 &:= \{\text{move } (P\backslash i\backslash j)(S) \mid i, j \in CP_{\parallel}(s), i \parallel j \in N\} \text{ and} \\
N_4 &:= \{\text{move } (P_u)(S) \mid P_u \text{ is a parallelity component with at} \\
&\qquad \text{least two activities}\}.
\end{aligned}$$

To keep the parallelity neighborhood $NBH_{\parallel}$ small we choose a parallel critical path $CP_{\parallel}(s)$ where the number of relations $i \parallel j$ and $i \sim j$ with $i, j \in CP_{\parallel}(s)$ is small.

Practical tests with this $O(n^2)$-neighborhood have shown that it is still too large. Therefore we considered a subneighborhood $NBH_{\parallel}^{CritRes}$ of $NBH_{\parallel}$ which is constructed as follows.

Instead of choosing all pairs $i, j \in CP_{\parallel}(s)$ for move-operators of the four types described above, we only consider a limited number of operators in each set. These operators are chosen in such a way that we get some operators that only disturb the current solution slightly, some change it intensively and others which cause changes in the medium range. We estimate the magnitude of these changes as follows.

For $S' =$ move $(i \sim j)(S)$, $S' =$ move $(P\backslash i, j)(S)$ or $S' =$ move $(P\backslash i\backslash j)(S)$ we define Δ_{ij} to be the minimal number of time units we have to move activity i such that i and j are processed in parallel, i.e.

$$\Delta_{ij} = \max\{s_j - c_i + 1, s_i - c_j + 1, 0\},$$

where s_j is the starting time of activity j and c_i is the finishing time of activity i in the schedule $s = H_2(S)$. Let $\Delta_{\max}(\Delta_{\min})$ be the maximal (minimal) value of these numbers. For $S' =$ move $(P_u)(S)$ we consider the number of parallelity relations in P_u. Let $\alpha_{\max}(\alpha_{\min})$ be the maximal (minimal) value of these numbers.

We split $NBH_{\|}(S)$ into nine disjoint sets $M_1, \ldots, M_9$ as follows. We divide both intervals $[\Delta_{\min}, \Delta_{\max}]$ and $[\alpha_{\min}, \alpha_{\max}]$ into three parts of equal length. According to this partition we divide N_1 into sets M_1, M_2, M_3 with small, medium, and large Δ_{ij}-values. In the same way, $N_2 \cup N_3$ is partitioned into sets M_4, M_5, M_6. N_4 is partitioned into sets M_7, M_8, M_9 with respect to the α-values.

Now in each of the nine sets all schedule-schemes are chosen which have the lowest priority value. For schedule schemes $S' \in M_1 \cup \ldots \cup M_6$ this priority value is given by $CritRes(S') = -\frac{r_{ik}+r_{jk}}{R_k}$ where k is the index of a critical resource. A **critical resource** is a resource k which induces the largest number of resource conflicts $u - v$ due to $r_{uk} + r_{vk} > R_k$.

For $S' =$ move $(P_u)(S)$ the priority value is equal to the number of parallelity relations in P_u.

The neighborhood $NBH_{\|}^{CritRes}(S)$ consists of all $S' \in M_i$ $(i = 1, \ldots, 9)$ which have the lowest priority value:

$$NBH_{\|}^{CritRes}(S) = \bigcup_{i=1}^{9}\{S' \mid S' \in M_i \text{ has the lowest priority}\}.$$

1.3.4 Implementation

In this section the main strategies of the second tabu search procedure are presented.

We use the neighborhood $NBH_{\|}^{CritRes}$. In each iteration step we have a schedule scheme S and calculate the corresponding heuristic solution $s = H_2(S)$. The next schedule scheme S' is chosen from $NBH_{\|}^{CritRes}(S)$ which is defined on the basis of a parallel critical path $CP_{\|}(s)$ of the schedule s.

There are situations where an easy test shows that an operator move $(i \sim j)$ leads to a schedule scheme $S' =$ move $(i \sim j)(S)$ with no feasible solution. This situation occurs if in $S = (C, D, N, F)$ we have for some activities k, l the relations $i \to k$, $k \parallel l$, and $l \to j$. These relations imply $i \to j$. Thus, move $(i \sim j)$ leads to a schedule scheme with only infeasible schedules. In order to apply move $(i \sim j)$, all parallelity relations $k \parallel l$ which induce $i \rightarrow j$ or $j \rightarrow i$ are eliminated.

Other features of the tabu search procedure are:

- The tabu-status is marked by storing the relations $i \parallel j$ which are added or eliminated. If $i \parallel j$ is in the tabu-list, then all operators involving $i \parallel j$ are tabu.
- In Step (3) of the heuristic H_2 we used the following priority rule. For each activity $i \in P$ an earliest starting time r_i is calculated as described and an $i \in P$ with smallest r_i-value is chosen. In the case of ties, an activity with the largest $\left(\sum_{k=1}^{r} \frac{r_{ik}}{R_k}\right)$-value is determined.
- Experiments have shown that dynamic tabu-lists of minimal length 3 and maximal length 6 gave the best results. The length of the tabu-list is controlled as described in Section 1.2.3.
- As in the tabu search based on critical arcs, immediate selection, stopping criteria, aspiration criteria, restarts, and several priority based starting heuristics are used.

1.4 CALCULATION OF LOWER BOUNDS

Several lower bounds for the RCPSP are available in the literature [10]. We used lower bounds to obtain bounds for the quality of the calculated solutions. Some tests have shown that the lower bound LB_2 from [19] in connection with immediate selection techniques, as presented in [4], gives very good results.

The lower bound is calculated after the tabu search algorithm. With the calculated solution value TS immediate selection is applied. If immediate selection proves that TS is optimal, i.e. an upper bound $TS-1$ induces infeasibility, we may set $LB = TS$ and do not have to calculate another bound.
Otherwise we calculate $LB = LB_2$ which is the optimal value of a linear program partially relaxing the precedence constraints and allowing preemption. The columns of this LP correspond to so-called nondominated feasible subsets. Feasible sets X are sets of activities which may be processed simultaneously, i.e. there are no conjunctions or disjunctions between any pair of activities $i, j \in X$ and all resource constraints are respected. Such a set is called nondominated if it is not a proper subset $X \subset Y$ of some feasible set Y.

We denote all nondominated feasible sets by $X_1, X_2, \ldots, X_q$. With each set X_j we associate an incidence vector $a_j \in \{0,1\}^n$ defined by

$$a_{ij} := \begin{cases} 1 & \text{if} \quad i \in X_j \\ 0 & \text{otherwise.} \end{cases}$$

Furthermore, let x_j be a variable denoting the number of time units where all activities in X_j are processed simultaneously. Then a LP-formulation of the relaxation may be written as follows:

$$\min \sum_{j=1}^{q} x_j \tag{1.1}$$

$$s.t. \quad \sum_{j=1}^{q} a_{ij} x_j \geq p_i \quad (i = 1, \ldots, n) \tag{1.2}$$

$$x_j \geq 0 \quad (j = 1, \ldots, q). \tag{1.3}$$

Unfortunately, the cardinality of all nondominated feasible sets grows exponentially with the number n of activities. For $n = 60$ we have approximately 300 000 columns for $n = 90$ even 8 000 000.

However, it is not necessary to generate and store all these columns. We can work with only a few of them at a time and only generate new feasible sets when they are really needed. Such an approach which has been suggested in [8] for the cutting-stock problem is called the delayed column-generation technique. In the following we describe the details of our approach. We work iteratively with a set of columns which contains a basis and the n slack-variables which correspond to (1.2).

After solving the corresponding LP to optimality we search for feasible columns which are able to improve the objective value when entering the basis. If y_i $(i = 1, \ldots, n)$ are the values of the dual-variables associated with the current basic solution, we have to find a column $a \in \{0,1\}^n$ with

$$\sum_{i=1}^{n} y_i \cdot a_i > 1 \tag{1.4}$$

$$\sum_{i=1}^{n} r_{ik} \cdot a_i \leq R_k \quad (k = 1, \ldots, r) \tag{1.5}$$

$$a_i \cdot a_j = 0 \quad (i - j \in D, i \to j \in C \text{ or } j \to i \in C). \tag{1.6}$$

Conditions (1.5) and (1.6) ensure that a column a corresponds to a feasible subset (respecting the resource constraints, the disjunctions D, and the conjunctions C). From duality it is known that if (1.4) holds for a column a, then an optimality condition is violated and the current objective value is improved when a enters the basis.

Columns which satisfy conditions (1.4) - (1.6) are generated by a branch-and-bound procedure. If no such columns are found, the current basic solution is an optimal solution for the LP (1.1) - (1.3) and the process is stopped.

Our branch & bound procedure works as follows: Since we have solved the current LP to optimality and the slack-variables are kept in the working set, we have for all dual-variables $y_i \geq 0$. Because of (1.4) and (1.5) we only have to consider activities i with $y_i > 0$ $(i = 1, \ldots, n)$.

At first these activities are sorted according to nonincreasing dual-variables $y_1 \geq y_2 \geq \ldots \geq y_{n_0} > 0$, $y_{n_0+1} = \ldots = y_n = 0$. To calculate incidence vectors a which correspond to nondominated feasible sets we define a branching tree in which the vertices represent partial strings $(a_1, \ldots, a_l)(1 \leq l \leq n_0)$. Such a string has at the most two sons. $(a_1, \ldots, a_l, 0)$ is the right son. The left son $(a_1, \ldots, a_l, 1)$ only exists if the set which corresponds to $(a_1, \ldots, a_l, 1, 0, \ldots, 0)$ is feasible.

We do a depth-first search in the branching tree where leaves correspond to feasible subsets. The sets are generated in lexicographic order. We generate a subset of nondominated feasible columns which satisfy (1.4) - (1.6). Each time a column a^* which satisfies these conditions has been found, we only search for further columns a with

$$\sum_{i=1}^{n_0} y_i a_i > \sum_{i=1}^{n_0} y_i a_i^*. \tag{1.7}$$

This ensures that all sets are nondominated for the following reason: An incidence vector a is dominated if at least one component a_k can be changed from 0 to 1 without violating feasibility. But then a dominating column a' has already been considered (either generated as a leaf or eliminated due to its objective value), since we scan the columns in lexicographic order. Let a^* be the best column found when we consider a. Since a' has been considered before, we have

$$\sum_{i=1}^{n_0} y_i a_i^* \geq \sum_{i=1}^{n_0} y_i a_i' \geq \sum_{i=1}^{n_0} y_i a_i + \underbrace{y_k}_{>0} > \sum_{i=1}^{n_0} y_i a_i,$$

i.e. column a is not generated by our procedure.

When we have found a suitable leaf a^*, we fill this column successively with all activities $n_0 + 1, \ldots, n$ which do not violate feasibility. Then the corresponding column is also nondominated and we backtrack to depth n_0.

We included the following simple bounding rule into our branch-and-bound procedure: If we set $a_k = 0$ in step k and

$$\sum_{\lambda=1}^{k-1} y_\lambda \cdot a_\lambda + \sum_{\lambda=k+1}^{n_0} y_\lambda \leq \sum_{i=1}^{n_0} y_i \cdot a_i^*$$

holds, the best column a^* found so far cannot be improved and therefore we can backtrack.

The search process stops if one of the following conditions holds:

- we have found a given number of columns,
- we have found at least one column and the quotient of the number of generated leaves with the number of leaves satisfying (1.7) exceeds a certain limit.

Then the generated columns are added to the working set of columns and the LP is resolved to optimality. If the number of columns in the working set exceeds a given size, some arbitrary columns are deleted.

1.5 COMPUTATIONAL RESULTS

In this section we present some computational results for the tabu search methods developed in the previous sections. We investigated the performance of our algorithm with respect to various benchmark instances generated by the PROGEN generator [12, 15].

For each of the problem sizes $n = 30, 60$, and 90 with $r = 4$ resources 480 benchmark instances were generated which differ in various parameter constellations.

The results presented in this section were obtained using a Sun Ultra 2 workstation with operating system Solaris 2.5 and 320 MB general storage (167 MHz). The algorithms were coded in C. We used CPLEX as an LP-solver to provide the lower bounds.

The computational results obtained by our implementations of the algorithms for the 480 PROGEN-instances with $n = 30$ and $r = 4$ are presented in Table 1.1. The optimal values are known for all these instances [6]. The number of problems which could be solved to optimality is indicated. For the remaining instances the number of problems is given for which the heuristic starting solution S_0 could be improved by the tabu search algorithms. The number of unsolved problems from 480 is given in parentheses. Furthermore we calculated the average percentage deviations $\Delta(OPT) = \frac{TS-OPT}{OPT}$ and $\Delta^0(OPT) = \frac{S_0-OPT}{OPT}$ of the tabu search value TS and the starting solution S_0 from the optimal value OPT.

Note that the presented computational times include the times needed for calculating the starting heuristics and for immediate selection. The computational results for the 480 PROGEN-instances with $n = 60$ and $r = 4$ are presented in Table 1.2. The optimal values are not known for all these instances. Therefore we used the lower bound described in Section 1.4 to obtain bounds for the quality of the calculated solutions. The number of problems for which the value could be verified to be optimal is indicated. Furthermore we calculated the average percentage deviations $\Delta(LB) = \frac{TS-LB}{LB}$ and $\Delta^0(LB) = \frac{S_0-LB}{LB}$ from the lower bound LB.

The computational results for the 480 PROGEN-instances with $n = 90$ and $r = 4$ are presented in Table 1.3. Some tests showed that for these instances

	Tabu search 1 (critical arcs)	Tabu search 2 (parallelity)
average (maximal) comput. time (sec)	2 (21)	15 (108)
problems solved from 480	347	389
improvements of S_0	88 (133)	75 (91)
average $\Delta^0(OPT)$	2.8%	2.8%
average (maximal) $\Delta(OPT)$	0.7% (7.5%)	0.4% (6.9%)
average number of iterations	350	368

Table 1.1 PROGEN-instances with $n = 30$ and $r = 4$

	Tabu search 1 (critical arcs)	Tabu search 2 (parallelity)
average (maximal) comput. time (sec)	3 (34)	41 (306)
problems verified from 480	286	309
improvements of S_0	147 (194)	134 (171)
average $\Delta^0(OPT)$	7.5%	7.5%
average (maximal) $\Delta(LB)$	5.7% (37.5%)	5.2% (33.9%)
average number of iterations	300	359

Table 1.2 PROGEN-instances with $n = 60$ and $r = 4$

	Tabu search 1 (critical arcs)	Tabu search 2 (parallelity)
average (maximal) comput. time (sec)	4 (48)	61 (480)
problems verified from 480	290	319
improvements of S_0	134 (190)	104 (161)
average $\Delta^0(OPT)$	7.2%	7.2%
average (maximal) $\Delta(LB)$	6.1% (38.4%)	5.7% (36.9%)
average number of iterations	283	313

Table 1.3 PROGEN-instances with $n = 90$ and $r = 4$

the option of restarts in the tabu search method based on parallelity does not lead to much better results than without them. If we omit the restarts, the average (maximal) computational times can be shortened to 19 (170) seconds, 318 instances could be verified to be optimal, and the average value of $\Delta(LB)$ increases only to 5.9%.

The lower bounds were calculated after the tabu search algorithm and immediate selection procedures. If immediate selection proved that TS is optimal we did not calculate the bounds. The average (maximal) computational times for computing the lower bounds with column-generation for all instances where immediate selection could not prove optimality were 4 (25) seconds for $n = 60$ (without the times used for the tabu search and immediate selection). Eliminating one instance which needed more than 2 hours, the computational times for $n = 90$ were 39 (644) seconds.

Tables 1.1-1.3 show that the quality of the tabu search methods is quite good. At the same time they show that the lower bounds must be of good quality. Furthermore, the computational times are small. With respect to computational times, the first method based on critical arcs is superior to the second tabu search. But with respect to quality, the second tabu search based on parallelity is superior to the first method in most cases.[1]

[1]Detailed computational results for all groups of the PROGEN-instances can be found at the location `http://www.mathematik.uni-osnabrueck.de/research/OR/software.html`.

1.6 CONCLUDING REMARKS

Two tabu search algorithms have been developed for the resource-constrained project scheduling problem.

The first method generalizes neighborhoods known for the job-shop problem which rely on the elimination of critical arcs. The concept of schedule schemes which has also been applied successfully in a branch-and-bound algorithm ([4]) leads to very good results in the second tabu search method.

While the second approach calculates high-quality solutions, the first is faster and provides solutions of reasonable quality.

Due to the fact that we have good lower bounds, we were able to prove optimality for 81% ($n = 30$), 64% ($n = 60$), and 67% ($n = 90$) of the benchmark problems.

Acknowledgement

This work was supported by the Deutsche Forschungsgemeinschaft, Project "Komplexe Maschinen-Schedulingprobleme", which is gratefully acknowledged. The authors also appreciate the helpful comments of the referee.

References

[1] R. Alvarez-Valdes, J.M. Tamarit (1989) Heuristic algorithms for resource-constrained project scheduling: a review and an empirical analysis; *in*: Slowinski, R., Weglarz, J. (eds.): Advances in project scheduling, pp. 113-134, Elsevier Science Publ., Amsterdam.

[2] F.F. Boctor (1996) Resource-constrained project scheduling by simulated annealing, Int. J. Prod. Res. 34, 2335-2351.

[3] J. Blażewicz, J.K. Lenstra, A.H.G. Rinnooy Kan (1983) Scheduling projects subject to resource constraints: classification and complexity, Discr. Appl. Math. 5, 11-24.

[4] P. Brucker, S. Knust, A. Schoo, O. Thiele (1998) A branch & bound algorithm for the resource-constrained project scheduling problem, European J. Oper. Res. 107, 272-288.

[5] J.-H. Cho, Y.-D. Kim (1997) A simulated annealing algorithm for resource-constrained project scheduling problems, J. Oper. Res. Soc. 48, 736-744.

[6] E. Demeulemeester, W. Herroelen (1992) A branch and bound procedure for the multiple resource-constrained project scheduling problem, Management Sci. 38, 1803-1818.

[7] M. Dell'Amico, M. Trubian (1993) Applying tabu search to the job-shop scheduling problem, Annals Oper. Res. 41, 231-252.

[8] P.C. Gilmore, R.E. Gomory (1961) A linear programming approach to the cutting-stock problem, Oper. Res. 9, 849-859.

[9] J.E. Kelley jr. (1963) The critical path method: Resources planning and scheduling, in: J.F. Muth, G.L. Thompson (Eds): Industrial scheduling, Prentice Hall, New Jersey, pp. 347-365.

[10] R. Klein, A. Scholl (1997) Computing lower bounds by destructive improvement - an application to resource-constrained project scheduling, Schriften zur Quant. BWL, TH Darmstadt, to appear in European J. Oper. Res.

[11] U. Kohlmorgen, H. Schmeck, K. Haase (1996) Experiences with fine-grained parallel genetic algorithms, to appear in Annals Oper. Res.

[12] R. Kolisch, A. Sprecher, A. Drexl (1995) Characterization and generation of a general class of resource-constrained project scheduling problems, Management Sci. 41, 1693-1703.

[13] R. Kolisch (1996) Serial and parallel resource-constrained project scheduling methods revisited: Theory and computation, European J. Oper. Res. 90, 320-333.

[14] R. Kolisch, A. Drexl (1996) Adaptive search for solving hard project scheduling problems, Naval Res. Log. 43, 23-40.

[15] R. Kolisch, A. Sprecher (1997) PSPLIB - A project scheduling library, European J. Oper. Res. 96, 205-216.

[16] A. Krämer (1995) Scheduling multiprocessor tasks on dedicated processors, PhD thesis, Department of Mathematics/Informatics, Universität Osnabrück.

[17] J.-K. Lee, Y.-D. Kim (1997) Search heuristics for resource-constrained project scheduling, J. Oper. Res. Soc. 47, 678-689.

[18] V.J. Leon, R. Balakrishnan (1995) Strength and adaptability of problem-space based neighborhoods for resource constrained scheduling, Oper. Res. Spekt. 17, 173-182.

[19] A. Mingozzi, V. Maniezzo, S. Ricciardelli, L. Bianco (1994) An exact algorithm for project scheduling with resource constraints based on a new mathematical formulation, Technical Report No.32, Department of Mathematics, Università degli Studi di Bologna, to appear in Management Sci.

[20] K.S. Naphade, S.D. Wu, R.H. Storer (1995) Problem space search algorithms for the resource-constrained project scheduling problem, Working Paper, Leigh University, Department of Industrial and Manufacturing Engineering.

[21] E. Pinson, C. Prins, F. Rullier (1994) Using tabu search for solving the resource-constrained project scheduling problem, Working Paper, Université Catholique de l'Ouest, Angers.

[22] S.E. Sampson, E.N. Weiss (1993) Local search techniques for the generalized RCPSP, Naval Res. Log. Quart. 40, 665-675.

2 METAHEURISTIC FOR THE VEHICLE ROUTING PROBLEM WITH TIME WINDOWS

José Brandão

Gestão e Administração Pública, EEG, Universidade do Minho
Largo do Paço, 4709 Braga codex, Portugal.
sbrandao@ci.uminho.pt

Abstract: The objective of this paper is to describe a tabu search metaheuristic for the vehicle routing problem with time windows, and to compare it with some of the best known algorithms for the same problem developed by other authors. From this comparison it can be concluded that the present algorithm performs rather well, outperforming the latter algorithms in several test problems. Another aim of this paper is to show the enormous influence on the results of the type of objective function used, in spite of the travelling time between customers being numerically equal to the distance.

2.1 INTRODUCTION

The *vehicle routing problem with time windows* (VRPTW) consists of defining a number of routes, in order to serve a given number of dispersed customers whose demand, geographical location and time window for the delivery are known. Each route is travelled by one vehicle (driver) that starts the journey at the depot, visits the customers belonging to the route and returns to it. The sum of the customers' demand in the route must not exceed vehicle capacity, and the routing time must not exceed the drivers working day. All the vehicles have identical characteristics, namely the capacity and maximum travelling time per day. The objective function may be to minimise the total length of the routes, the routing time, the number of the routes, the variable cost (i.e. cost resulting directly from travelling, such as fuel consumption, maintenance, etc.), or even a combination of these objectives with different weights and priorities. Lenstra and Rinnooy Kan [5] had proven that the VRPTW is a

NP-hard problem. This is the reason why there is still a lot of research going on in order to find increasingly better algorithms, in terms of solution quality and computing time, to solve the VRPTW.

The algorithm described and tested in this paper is a tabu search algorithm named TSVRPTW. The best known approximate algorithms for the VRPTW are of the tabu search type, and the comparisons made in this paper are with three of them. The TSVRPTW is derived from another algorithm described in Brandão and Mercer [1], with several modifications dictated by the research made thereafter and by differences in the type of problem studied.

Above all, the distribution companies are concerned with costs. Therefore the main objective function of any VRPTW should be cost minimisation. Nevertheless, they are not easy to calculate accurately and very often the necessary data are not available. The best proxy for the variable costs is the routing time because most of them are direct or indirectly related to it. The objective of this paper is to present and prove the quality of the TSVRPTW algorithm and in addition to achieve the following goals: i) to show the influence of the objective function of the algorithm in the results obtained; ii) to take general conclusions about the application of the tabu search concepts with the VRPTW, which may be applied in the development of other algorithms.

This paper is organised as follows. First the notation and the formulae are presented, then the TSVRPTW algorithm and its initial solution are described. Subsequently, the objective functions used in this paper are described in Section 2.5. In Section 2.6, the computational experiments are presented, the results are shown and conclusions are drawn. In the last section, directions for further research are indicated.

2.2 NOTATION AND FORMULAE

The customers and the depot are represented by a set of vertices $V = \{0, 1, ..., N\}$, where 0 is the depot and N is the total number of customers. The set of arcs that connect the V vertices is called A.

u_i: time of unloading at customer $i, i \in V$.

t_{ij}: travel time between i and j, plus u_i, $(i, j) \in A$.

e_i: earliest start time of the delivery at customer $i, i \in V$.

l_i: latest arrival time for the delivery at customer $i, i \in V$.

a_i: actual arrival time of the vehicle at customer $i, i \in V$.

d_i: departure time of the vehicle from customer $i, i \in V$.

Route: set constituted by the depot and by one or more customers in a given sequence, which is called the sequence of routing or travelling. The route always begins and ends at the same depot. Each route r is represented as $r = [0, 1, ..., n, n+1]$, where the origin, 0, and the destination, $n+1$, are always the depot, and the numbers, $1, ..., n$, represent the customers, which are visited in this sequence. Obviously, a VRPTW solution may have several routes and the value of n may be different in each of them.

Nr: total number of routes in the solution.

W_i: waiting time at customer $i, i \in V$. This is the time which the driver must wait if he arrives at the customer before the earliest start of the delivery.

Sf_i: forward slack of the customer $i, i \in V$. This variable measures for each customer i, in the current route, the amount of time that the arrival of the vehicle at i can be delayed and still complying with the delivery time window at i.

L_i: lateness in arrival to the customer $i, i \in V$.

L: total lateness in a route. If it is necessary to distinguish between routes, it is written $L(r)$ for route r.

p: parameter that defines the search neighbourhood inside an algorithm developed by Gendreau et al. [4] for the travelling salesman problem (TSP), called GENI.

The behaviour of the GENI algorithm depends very much on parameter p that specifies the number of nearest neighbours of each vertex that is considered to define the exchange of positions in the Hamiltonian circuit. According to the authors, the best compromise between speed and quality seems to be $p = 5$. As each route of the VRPTW is a TSP with time windows (TSPW), GENI has been adapted by the author for the TSPW but the same name has been retained, since the main structure is still the same. This algorithm is used inside the TSVRPTW to move one customer from one route to another.

T: maximum daily driving time of a driver.

$D(r)$: distance travelled in route r.

$Ot(r)$: overtime used by route r.

K: is a constant that penalises overtime (the value used is $K = 1$). In the test problems used in this paper no overtime is allowed but, since in phase 1 of the TSVRPTW algorithm infeasible solutions are also accepted, these routes may have overtime which is penalised with this constant.

The waiting time at customer i is given by equation (2.1). If a vehicle arrives late at the customer, i.e., if $a_i > l_i$, it is supposed that $W_i = 0$.

$$W_i = max\{0, e_i - a_i\}. \tag{2.1}$$

The time $t(r)$ to travel route r (or routing time) is given by the equivalent expressions (2.2) and (2.3), this being more suitable for programming purposes.

$$t(r) = \sum_{i=0}^{n} [t_{i,i+1} + W_i] \tag{2.2}$$

$$t(r) = a_{n+1} - a_0. \tag{2.3}$$

For each route r, a_0 is the arrival time of the vehicle at the depot and it coincides with the departure time d_0, and a_{n+1} is the returning time at the depot after servicing the customers. The value d_0 is calculated through (2.4), assuming that it is undesirable to have waiting time at the first customer of the route.

$$d_0 = max\{e_0, e_1 - t_{01}\}. \tag{2.4}$$

The *route schedule*, that establishes both the sequence and the servicing time of each customer of the route, can be entirely determined employing the formulae from (2.5) to (2.11), which are valid for any route r. A route schedule is infeasible if any customer (or the depot) has a negative forward slack (Sf).

$$Ot(r) = max\{0, t(r) - T\}. \tag{2.5}$$

The departure time from a customer i can be calculated either from (2.6) or (2.7) which are mathematically equivalent, but (2.6) has the advantage of avoiding to calculate the waiting time, which is never needed in the computer program for other purposes. It is worth remembering that the unloading time (u_i) has been added to the travelling time between i and j. This means that in practice the departure from customer i is not really at d_i, but u_i units of time later. However, the mathematical model remains valid to describe the real situation if this detail is taken into account when defining the upper limit of the delivery time window of the customer.

$$d_i = max\{a_i, e_i\} \qquad i = 0, 1, ..., n+1 \tag{2.6}$$

$$d_i = a_i + W_i \qquad i = 0, 1, ..., n+1 \tag{2.7}$$

$$a_{i+1} = d_i + t_{i,i+1} \qquad i = 0, 1, ..., n \tag{2.8}$$

$$Sf_i = l_i - a_i \qquad i = 0, 1, ..., n+1 \tag{2.9}$$

$$e_i \leq d_i \leq l_i \qquad i = 0, 1, ..., n+1 \tag{2.10}$$

$$L_i = max\{0, d_i - l_i\} \qquad i = 0, 1, ..., n+1 \tag{2.11}$$

If $L_i > 0$ then $Sf_i < 0$ and (2.10) is not satisfied. The lateness of a route is given by (2.12).

$$L = \sum_{i=1}^{n} (d_i - l_i) \tag{2.12}$$

2.3 INITIAL SOLUTION

The execution of the TSVRPTW algorithm is started with the production of an initial solution by an algorithm called NENBVRP, which mixes two classical approaches: nearest neighbour and insertion. Each route is started with the unrouted customer farthest from the depot. In a first stage, the routes are created in a sequential constructive manner, with the customer introduced in the route being chosen based on its vicinity to all the customers already included. Besides, that customer is selected from a set that contains only the δ nearest neighbours of each of the customers already in the route. The value of δ is the same for every customer and is based on the expected number of customers per route. As a consequence of the restriction imposed by δ some customers may remain unrouted, so they are introduced in the second stage. Among those unrouted customers is chosen the admissible one which increases the route length least. This route length is calculated assuming that the new customer must be inserted immediately before or after the customer to whose δ-neighbourhood it belongs. Obviously, a customer can be inserted near another only if it belongs to its δ-neighbourhood. If no customer is admissible to enter the route, a new route is started. This stage is finished if there are no more unrouted customers or if the maximum number of routes initially defined has been attained. In the second stage, each of the unrouted customers is inserted in one of the existing routes. This is a parallel procedure, because all the routes compete for the same unrouted customers. The customer chosen, is the admissible one with least insertion distance, considering all the routes.

The maximum number of routes is defined before starting the first stage through an estimate based on a ratio of the sum of customers' demand to the capacity of each vehicle. If this number of routes (vehicles) is not enough for all the customers, those remaining unrouted are inserted in other new routes that are created by applying the insertion procedure only. Each of these routes is started as before with the unrouted customer farthest from the depot.

In the TSVRPTW three versions of the NENBVRP were tried:

- NENBVRP1 - The solution is feasible relative to all the routing constraints of the problem but it may be infeasible with regard to the time windows.
- NENBVRP2 - Equal to the previous version except that the solution is feasible in relation to all the VRPTW constraints.
- NENBVRP3 - The objective function and the neighbourhood is defined in terms of time instead of distance as in the two previous versions. In order to define this time based neighbourhood, the time between two customers (i, j) considers not only the travelling time plus the unloading time but also a factor that is an estimate of the waiting time if j is served immediately after i, which is calculated before starting to construct the routes.

2.4 COMPONENTS OF THE TSVRPTW ALGORITHM

The execution of the TSVRPTW algorithm consists of making a given number of iterations whose upper limit is defined in the beginning. Each iteration consists of discovering another solution that in general is different from the previous ones and that hopefully is better or will originate better solutions in future iterations. In each iteration a set of trial moves is generated from which is selected and executed the one that gives the largest gain in the objective function without violating the tabu rules of the algorithm. The TSVRPTW is composed of two phases, phase 1 and phase 2, that are applied sequentially, and inside each phase there are two different procedures designated by part 1 and part 2. Those two phases are identical except that in phase 1 infeasible solutions are allowed while in phase 2 only the feasible ones are accepted. The initial solution for phase 1 is the solution found by the NENBVRP while the initial solution for phase 2 is the best feasible solution found in phase 1. In phase 2, only minor improvements are expected, mainly because the solution passed on from phase 1 is already good. Hence the maximum number of iterations allowed in phase 2 is substantially less than in phase 1. The presence of two phases and two parts inside of them are ways of producing diversification, and the restart of part 1 with the best known solution is an important element of intensification. This algorithm contains several parameters that can be adjusted by the user when executing the program, but there is also a set of predefined values that are used by default. These are the ones that are presented in this paper. In order to find the most appropriate combination of values for the parameters, it is necessary to perform an enormous amount of computational tests. This has not been done; only a few experiments were made, those values being thus defined mainly by intuition and a matter of convenience, such as keeping the computing time under reasonable limits. Besides, one's opinion is that the results will be better if the value of the parameters is proportional to the size of the problem, principally, the number of customers and average number of customers per route, instead of using the same fixed size for every problem. Consequently, the algorithm's performance will be better in real situations where the size and other characteristics of the problems vary a great deal.

The subsections that follow will clarify the contents, the meaning and the interrelation of all the components of the algorithm just described.

2.4.1 *Trial moves*

The solution at each iteration of the tabu search consists of a set of routes Nr. A set of customers C_r with cardinality $\#C_r = \lceil \frac{n}{c} \rceil$ is chosen randomly from each route r.

We used $c = 2$ in the experiments reported in this paper. Joining the set of customers of each route produces the set of candidates C, given as follows:

$$C = \bigcup_{r=1}^{Nr} C_r.$$

There are two kinds of trial moves in TSVRPTW: *insert* and *swap* moves.

The insert move consists of removing one customer from one route (origin) and putting it into another route (destination). The destination route can be an existing route or a new route.

A trial insert of a customer $x_i \in r_i$ and $x_i \in C$, into r_j $(r_i \neq r_j)$ exists if and only if: (i) at least one customer of r_j belongs to the δ-neighbourhood (δ nearest neighbours) of x_i, or r_j is a new route; (ii) the insertion of x_i in r_j does not cause an excess of load in the vehicle that contains r_j.

A *swap* move consists of exchanging two calls belonging to two different routes. A trial swap of a customer $x_i \in r_i$ and $x_i \in C$, with $x_j \in r_j$ and $x_j \in C$ $(r_i \neq r_j)$ exists if and only if: (i) at least one customer of r_j belongs to the δ-neighbourhood of x_i and at least one customer of r_i belongs to the δ-neighbourhood of x_j; (ii) the exchange of x_i and x_j does not cause an excess of load either in r_i or r_j; (iii) at least one of the routes r_i or r_j contains more than one customer.

In phase 2, only feasible solutions can be generated so that, besides the previous conditions, it is also required that the move respects the time window constraints and that the routing time of the route does not exceed T (plus the legal overtime when this is allowed). In phase 1, as soon as a route becomes feasible in relation to the routing time, it is not allowed to become infeasible again.

The parameter δ is a way of limiting the number of potential moves and it is unlikely to have an appreciable negative effect on the quality of the final solutions, if any. This neighbourhood is defined in terms of distance. In spite of this reduction the number of potential swap moves is still very large. Therefore, the swap moves are only tried every seven iterations or in those iterations for which none of the potential insert moves can be executed. The computation of the trial moves is accelerated by the use of a memory structure that keeps much of the information from one iteration to another, avoiding thus its recalculation. The same scheme is used for the swap moves, but less information remains unchanged in each iteration.

The equations below are used to calculate $F1$ and $F2$, which represent the cost of a feasible and an infeasible solution, respectively. The value of $F1$ is based on the travelling time and the overtime. Note that a solution to be feasible the maximum overtime allowed cannot be exceeded. However, in some circumstances as it happens with the test problems used in this article, that maximum is zero. The same factors as in $F1$ are included in $F2$ plus one to evaluate the infeasibility. Since in phase 2 only feasible solutions are accepted, $F2$ is unnecessary.

$$F1 = \sum_{i=1}^{Nr} t(i) + K \times Ot(i) \tag{2.13}$$

$$F1 = \sum_{i=1}^{Nr} t(i) + K \times Ot(i) + P \times L(i) \tag{2.14}$$

The penalty P is initially taken to be one. P is multiplied by two if all the solutions have been infeasible during α consecutive iterations. Conversely, P is divided by two if all the solutions have been feasible during the last α consecutive iterations. A good starting value is $\alpha = 10$.

The procedures *part 1* plus *part 2* existing inside each phase constitute what is called a cycle. This cycle is executed twice starting with *part 1.* These procedures are described in the following.

Part 1 In this part insert and swap moves are executed as stated above. This part is made up of two subcycles after which part 2 starts. The first subcycle is executed with $p = 5$ and $\delta = 5$ and the second with $p = 6$ and $\delta = 10$. A subcycle ends when a given number of iterations ($limit = \lceil 4.5\text{N} \rceil$) has been executed without improving either $F1*$ or $F2*$, where $F1*$ is the best known value of $F1$ from equation (2.13) and $F2*$ is the best known value of $F2$ from equation (2.14). The initial solution of each subcycle of part 1 is the solution corresponding to $F1*$ or $F2*$, if no feasible solution is yet known. In phase 2, *limit* $= 3N$, $p = 6$ and $\delta = 10$, and inside each cycle *part 1* is executed only once.

Part 2 Rule (i) stated above for the trial moves may prevent a customer x_i from leaving his present route r_i if this route has more than δ calls, because the δ nearest neighbours of x_i may be all in r_i, thus impeding the search for better alternatives. In order to take this into account, the procedure *part 2* was created.

During the execution a table of frequencies is created to count the number of times that each customer has been moved from one route to another. Every time that the end of *part 1* is reached, the average frequency for all the calls is calculated. During *part 2*, C is *fixed* and consists of all the calls whose frequency is below the average. Additionally, in the insert moves of $x_i \in r_i$, $\delta = max\{5, n\}$, where n is the number of customers in r_i.

In *part 2*, no swap moves are executed. This decision was taken only in order to avoid spending computing time because the trial swap moves are the ones that take more time. However, it is worth testing in future work whether this has been the most convenient decision.

The initial solution of *part 2* is the best feasible solution from equation (2.13). If no feasible solution is available, the initial solution is the best known infeasible solution given by equation (2.14). The two important benefits derived from *part 2* are that the effect of an inappropriate choice of δ is corrected and the search is diversified through the use of a *long term* memory.

Part 2 finishes after *limit1* $= \lceil 1.5N \rceil$ iterations without improving $F1*$ or $F2*$. The reason for using a limit that is a third of that of *part 1* is that there is a trend for having no allowable moves after a given number of iterations, since the set of candidates is fixed.

2.4.2 Evaluation of the moves and aspiration criteria

In the *insert* moves, if the customer $x_i \in r_i$, with $r_i = [0, ..., x_h, x_i, x_j, ..., 0]$, is moved to the route r_k, the following change is made. The new route r_i is r_i' $= [0, ..., x_h, x_j, ..., 0]$ and the new route r_k is the one generated by the familiar GENI algorithm, with $p = 5$.

In the *swap* moves, if the calls $x_i \in r_i$ and $x_j \in r_j$ are exchanged, the new routes r_i and r_j are constructed in two steps. First, x_i is removed from r_i and x_j from r_j as in the insert move. Second, x_i is inserted in r_j and x_j is inserted in r_i, both with the GENI algorithm.

In phase 1, the cost of a move is obtained from equation (2.14). Therefore, the *allowable* move with the lowest solution cost given by that equation is made. The solution value of a *feasible solution* is defined by equation (2.13), being the minimum value $F1*$. If no feasible solution is obtained, $F1*$ is considered to be infinite.

The two *aspiration criteria* for a move in phase 1 are either $F2 < F2*$, or the solution is feasible and $F1 < F1*$. When at least one of these conditions is satisfied, the tabu status of a move is overridden. The only *aspiration criterion* in phase 2 is that $F1 < F1*$.

2.4.3 *The tabu list size and tabu attributes*

The set of trial moves in the TSVRPTW is dependent primarily on the set of candidates, so the tabu list size is based on it. The tabu list size, q, is taken to be a random number in the interval $\theta \in [\theta_{min}, \theta_{max}]$, where $\theta_{min} = \lceil N/3c_1 \rceil$, and $\theta_{max} = 2\theta_{min}$. The parameter c_1 may be defined by the user, since the best values of θ for each type of problem and tabu search method have to be found empirically. Values defined by taking $c_1 = 2$ were used in this application.

The attribute used either in the insert move of k or in the swap move of the pair (i, j) is the route to which the calls (k or i or j) belonged before the move. The restriction imposed on both types of move is that a customer leaving a route cannot return to the same route during the next θ iterations.

2.5 THE OBJECTIVE FUNCTION

One of the aims of this research is to show the importance and the influence of the objective function (OF) in the final solutions. The decision maker (distribution planner) should choose an OF according to the objectives of the company, which depend on several and diverse circumstances, and the data available. Generally, the main objective of the companies is to minimise the distribution costs. The classical division of these costs is into fixed and variable. To calculate these costs is not an easy task and furthermore a lot of data is needed which frequently is not available. There are also other types of costs that are still more difficult to estimate such as the eventual loss of good will by the customer if he is not served as and when required.

The fixed costs arise from strategic decisions taken by the company and therefore are irrelevant for the daily distribution planning. For example, once a fleet of vehicles has been acquired, in general, there is no benefit for the company if as result of a particular plan some of the vehicles stay unused. The salary of the company's drivers is a fixed cost, but it should be considered as a variable cost if the idle drivers can do another job. Also the overtime payment is a variable cost.

For daily planning, the most important costs are the variable ones. These costs are a function of time and distance, time being the most influential factor. These two variables are interrelated, but they are very different since in practice the shortest path is very different from the fastest path. Furthermore, in the absence of new roads, the shortest path between two customers is always the same, while the fastest path, as well as the time to travel it, may vary across the day, the week, and the year, and with the vehicle used. In the VRPTW, because of the time windows, the waiting time may contribute largely to the routing time. Hence, even assuming that the travelling time among the customers is numerically equal to the distance, the solution obtained depends very much on the OF of the algorithm. If the OF is to minimise the distance travelled, the waiting time has no influence on the search of the best solution, while if the OF is to minimise the routing time, the waiting time is directly taken into account in driving the search across the solution space.

The most common OF used either in practice or in theoretical studies is the minimisation of the distance, immediately followed by the minimisation of the routing time. Nevertheless, some authors state that the most important objective is to minimise the number of routes, assuming that this number is equivalent to the number of vehicles. With small variations, this is the point of view taken by several authors like [8, 7, 9, 10]. They justify this with the fact that the vehicles are the most expensive resource. Such reasoning may be true but it may also contain two fallacies. The first one is that the number of routes is the same as the number of vehicles. This is not true if each vehicle makes more than a route per working day as happens very frequently in practice. This problem of a vehicle travelling multiple routes in a day has not been much studied, but it is now receiving more attention as it deserves.

The second fallacy results from a confusion between fixed and variable costs. As explained earlier, the vehicles and drivers are very expensive but there is only an additional cost if the routes generated require the hiring of more vehicles or drivers, or the buying of new vehicles.

Those authors just mentioned, in spite of giving primary importance to the number of routes, never use them directly in the OF of their algorithms, as can be concluded from the description below.

Solomon [8] states that the number of vehicles is the most important followed by the routing time and then the distance. However, the OF that he uses combines distance and time.

Thangiah et al. [10] state that a solution with fewer vehicles and a greater travelling distance is preferable to one with larger number of vehicles and smaller distance. However, the OF used by them does not include the number of routes, but the distance and also some weight is given to time. Only in the end, when choosing the best solution, among several generated for each testing problem, they select the solution with the fewest number of routes.

Rochat and Taillard [7] defined as a first objective to minimise the number of routes and as a second objective, subject to the minimisation of the first, the minimisation of the distance, but they do not clarify how this is made in the algorithm. Like the previous authors, from a set of solutions for the same problem they choose the one with the fewest routes as being the best.

Taillard et al. [9] use as their objective function the distance travelled plus a lateness penalty component, which obviously is zero if the solution is feasible. So, one can say that the objective is to minimise the distance.

Desrochers et al. [3] present an algorithm for the exact resolution of the VRPTW, defining that the OF is to minimise the distance travelled.

The opinion expressed and defended in this paper is that the planner can only decide which is the primary objective after knowing the specific conditions of the company, such as the number of vehicles and drivers available, the haste for the delivery of some orders, etc. Nevertheless, a priori, the most important are the variable costs, but if these are difficult or impossible to calculate, the best substitute for them is the routing time.

The TSVRPTW was executed in four versions, each one corresponding to a different OF:

i) T_{min} - The OF is defined by (2.13) and (2.14). The initial solution is given by NENBVRP3.

ii) D_{min} - The OF is defined by (2.13) and (2.14), substituting in these equations $t(i)$ by $D(i)$. The initial solution is given by NENBVRP2.

iii) TD_{min} - $F1$ is defined by (2.13), replacing $t(i)$ by $D(i)$, and $F2$ is defined by (2.14). Therefore, in phase 1, the minimisation of distance is combined with the minimisation of the routing time. The initial solution is given by NENBVRP2.

iv) R_{min} - The objective is to minimise the number of routes. This has been implemented by adding to $F2$ a factor Kr that penalises the existence of each route in the solution. This factor is given by (2.15), thus it is related to the number of routes in the solution. If, as result of a move, a new route is created, Kr is added to $F2$, and if a route disappears Kr is subtracted.

$$Kr = F1/(\beta Nr). \tag{2.15}$$

The value of $F1$ and the number of routes Nr are those of the initial solution, and the value of β used was five. In R_{min}, besides the OF there is another modification in phase 1 and phase 2, which affects the choice of the best feasible solution. $F1$ is used only to compare two feasible solutions with the same number of routes. Otherwise, when comparing two feasible solutions with different number of routes, the one with the fewest routes is the best, independently of their $F1$ values. It should be noted that each of the three previous versions of TSVRPTW can give rise to one version of R_{min}.

As was said in Section 2.4.2, in each trial move, the insertion of a customer in a route is made with the GENI algorithm. The objective function for this algorithm is the routing time with the versions T_{min} and TD_{min}, and is the distance with the version D_{min}. The neighbourhood associated with the parameter p is always defined in terms of distance.

2.6 EVALUATION OF THE PERFORMANCE OF THE TSVRPTW

In order to evaluate the quality of the TSVRPTW algorithm and to show the influence of the objective function in the final result, benchmark problems from the literature were used. The results are then compared among them and with the results given by three algorithms for the VRPTW, one due to Thangiah et al. [10] (called GenSAT), another due to Rochat and Taillard [7] and the third is presented by Taillard et al. [9], which so far as it is known are the best published.

Solomon [8], based on data for the VRP from Christofides et al. [2], created a set of 56 VRPTW's, each one with 100 customers. All these problems are Euclidean and it is assumed that the travelling time between each pair of customers is numerically equal to the corresponding distance, which has been calculated with a precision of one digit by the TSVRPTW as follows. First the distances are calculated in double precision then they are multiplied by ten, after which they are truncated and subsequently there are no rounding errors, because all the calculations use integers. In the end the total distance (time) is divided by 10 and the ratio is rounded to the nearest integer. The precision used is equal to that used by Desrochers et al. [3] to determine the optimal solution of some test problems. However, in the other works taken for comparison [10, 7, 9] a different precision was used: real floating point or, in the case of Taillard et al. [9], integers were used resulting from multiplying the distances (in double precision) by 1000 and then rounding. This means that the comparisons are not strictly valid, but likely the differences are almost insignificant.

These problems are divided into six groups: R1, C1, RC1, R2, C2 and RC2. In the sets R1 and R2 the customers are located according to an uniform distribution, while in the sets C1 and C2 they form clusters, and in the sets RC1 and RC2 there are customers uniformly distributed and others in clusters. The time windows for the groups R1, R2, RC1 and RC2 were generated randomly, while for the sets C1 and C2 the customers were first grouped in routes and afterwards the windows were created.

As the TSVRPTW algorithm contains certain stochastic parameters, namely the set of candidates and the tabu list size, the solutions produced in each run may be different. However, for the experiments carried out with the four versions of the objective functions, the average percentage difference between the maximum and the minimum distance of solutions with the same number of routes never exceeded 4.1%, i.e., considering each version separately and taking the average for all the six types of test problems. Each of the three first versions defined in the previous section was executed twice taking the initial solution defined there and twice taking the initial solution given by NENBVRP1. The version R_{min} was executed once with the objective function of D_{min} and TD_{min} and twice with the objective function of T_{min}. The values presented in Table 2.1 result from the selection of the best solution for each problem, which is defined in the following: in T_{min} the solution with minimum routing time was chosen; in D_{min} and TD_{min} the minimum distance solution; in R_{min} the solution with the fewest routes and the ties are broken choosing the solution with minimum routing time. These rules were followed very strictly, otherwise other solutions more favourable in relation to the number of routes could be given. For instance, a different solution found by T_{min} for problem R206 has a routing time of 1051 and three routes, thus this number of routes is much better than the solution with five routes in Table 2.1 and the routing time is only 0.67% above the value in the table. Many other cases like this were discovered with versions T_{min} and D_{min}. The column $SD*$ contains the best solution in terms of distance among the previous four. The column $SR*$ contains the best solution found by the TSVRPTW algorithm in terms of number routes and then in terms of distance.

The best solution found by the GenSAT, Rochat and Taillard [7], and the best known solution as is given by Taillard et al. [9] is also shown in Table 2.1. It is worth saying that there are a few problems for which there is a small discrepancy between the best solution reported by these authors and the best solution reported by other authors. In the following are described three cases that were discovered: for problem C201 Thangiah et al. [10] present $D = 590$ (solution given by Potvin et al. [6]) as being the best solution; for problem C205, these same authors write two different values in different places of the same article ($D = 589$ and $D = 588$); and in the last case the difference concerns the number of routes reported for problem R209, while Thangiah et al. [10, 11] say they are two, Taillard et al. [9] write three. Except for this last case, the differences are so small that to know the exact value is not of major importance. Therefore, it was decided to use the values given by Taillard et al. [9], with the exception of problem R209.

The GenSAT is a system constituted of four algorithms, each one initialised with two different solutions of high quality given by two different heuristics - one equal to Solomon's [8] insertion algorithm and another a genetic heuristic; from the 8 solutions generated, the best one in terms of distance is presented in Table 2.1. Rochat and Taillard [7] present only one solution that is the best resulting from the execution of the algorithm five times with five different initial solutions.

The comparative performance of the four versions of the TSVRPTW algorithm can be observed in Table 2.2 which is constructed with the mean values, for each set of problems, taken from Table 2.1. In Table 2.2, is given the percentage difference among the four versions in relation to the travelling time as well as the distance and the number of routes. This difference is given by $100(x - y)/y$, where x is the value for the version read in the column and y is the value of the version for the row entry.

Observing Table 2.2, it can be concluded that the version that gives the least routing time is T_{min}; the version with the least distance is D_{min}, and R_{min} is the best version in relation to the number of routes. The version TD_{min} lies in the middle with respect to any of the three objectives. The version R_{min} may be considered the most equilibrated because it is not only the best in terms of number of routes, but the routing time and the distance are also good. All these results were to be expected and they show clearly that the decision maker must define unequivocally which objective is the most important for the company, or even make a ranking of relative importance, because there is a conflict among them. If the vehicles are allowed to make multiple trips a day, in general, the objective should be the minimisation of the routing time, which is better accomplished by T_{min}.

In Table 2.3, the best solutions in terms of distance (S_D*) and the best solutions in terms of number of routes (S_R*) are compared with the solutions obtained by the other algorithms in Table 2.1, and also with the results given by the algorithm of Taillard et al. [9] (indicated in the table simply as Taillard), considering the minimum value reported by them. These results do not appear

Probl.	T_{min}			D_{min}			TD_{min}			R_{min}			S^*_D		S^*_R		GenSAT		Rochat		Best	
	T	D	R	T	D	R	T	D	R	T	D	R	D	R	D	R	D	R	D	R	D	R
R101	2037	1772	19	2381	1653	20	2165	1703	20	2278	1653	19	1653	19	1653	19	1600	20	1656	19	1608	18
R102	1714	1649	17	2135	1490	19	1809	1541	18	2051	1526	17	1490	19	1526	17	1493	17	1477	18	1434	17
R103	1319	1271	14	1963	1228	15	1489	1240	14	1658	1255	14	1228	15	1240	14	1207	13	1223	14	1207	13
R104	1046	1039	10	1234	987	11	1104	1012	10	1139	989	10	987	11	989	10	1032	11	1013	10	982	10
R105	1472	1434	14	1551	1362	15	1465	1384	15	1518	1436	14	1362	15	1434	14	1407	15	1405	14	1377	14
R106	1290	1285	13	1417	1247	13	1394	1290	13	1342	1297	12	1247	13	1297	12	1326	14	1294	12	1252	12
R107	1107	1101	11	1210	1082	11	1120	1109	11	1174	1089	11	1082	11	1082	11	1140	12	1086	11	1127	10
R108	977	977	10	1125	960	11	972	968	10	1111	951	10	951	10	951	10	989	10	965	10	969	9
R109	1209	1203	13	1222	1154	13	1229	1179	13	1225	1157	12	1154	13	1157	12	1218	12	1186	12	1215	11
R110	1116	1112	12	1249	1078	12	1117	1109	11	1165	1094	11	1078	12	1094	11	1105	11	1108	11	1080	11
R111	1101	1095	12	1182	1057	12	1086	1071	12	1181	1072	11	1057	12	1072	11	1120	12	1071	11	1105	10
R112	973	969	10	1015	970	10	987	980	10	999	960	10	960	10	960	10	992	10	966	10	954	10
Mean	1280	1242	12.9	1474	1189	13.5	1328	1216	13.1	1403	1207	12.6	1187	13.3	1205	12.6	1219	13.1	1204	12.7	1193	12.1
C101	827	827	10	827	827	10	827	827	10	827	827	10	827	10	827	10	829	10	829	10	827	10
C102	829	829	10	827	827	10	907	839	10	827	827	10	827	10	827	10	829	10	829	10	827	10
C103	846	846	10	1029	834	10	850	845	10	1072	859	10	834	10	834	10	835	10	828	10	828	10
C104	845	845	10	1014	834	10	839	839	10	842	841	10	834	10	834	10	840	10	842	10	825	10
C105	829	829	10	827	827	10	829	829	10	827	827	10	827	10	827	10	829	10	829	10	829	10
C106	827	827	10	827	827	10	827	827	10	834	834	10	827	10	827	10	829	10	829	10	827	10
C107	827	827	10	827	827	10	827	827	10	827	827	10	827	10	827	10	829	10	829	10	827	10
C108	829	829	10	836	836	10	836	832	10	832	832	10	829	10	829	10	829	10	829	10	827	10
C109	829	829	10	831	831	10	882	830	10	830	830	10	829	10	829	10	829	10	829	10	829	10
Mean	834	833	10.0	872	830	10.0	847	833	10.0	858	834	10.0	829	10	829	10	831	10.0	830	10.0	827	10
RC101	1733	1682	15	1909	1647	15	1740	1674	15	1948	1681	15	1647	15	1647	15	1669	14	1737	15	1669	14
RC102	1558	1545	14	1880	1485	15	1664	1509	14	1799	1543	13	1480	14	1543	13	1458	15	1481	13	1555	12
RC103	1296	1286	11	1541	1274	12	1343	1321	11	1378	1287	11	1274	12	1286	11	1110	11	1264	11	1110	11
RC104	1155	1152	10	1211	1110	10	1235	1170	10	1250	1156	10	1110	10	1110	10	1213	11	1157	10	1136	10
RC105	1626	1582	15	1910	1548	16	1768	1592	15	1688	1617	14	1548	16	1617	14	1602	14	1543	15	1643	13
RC106	1421	1415	12	1560	1386	13	1437	1409	13	1545	1480	12	1386	13	1415	12	1420	13	1416	12	1448	11
RC107	1279	1263	12	1316	1212	12	1273	1234	11	1327	1257	11	1212	12	1234	11	1264	12	1262	11	1231	11
RC108	1165	1161	11	1167	1118	11	1215	1135	11	1240	1143	11	1118	11	1118	11	1169	11	1150	11	1140	10
Mean	1404	1386	12.5	1562	1348	13.0	1459	1381	12.5	1522	1400	12.1	1347	13	1371	12.1	1363	12.6	1376	12.3	1367	11.5
R201	1716	1694	5	4583	1161	8	2638	1533	5	2405	1423	4	1161	8	1423	4	1354	4	1485	4	1255	4
R202	1363	1357	6	3213	1060	5	2112	1342	4	1583	1525	4	1060	5	1342	4	1176	4	1101	4	1214	3
R203	1167	1165	4	3747	918	6	1989	1117	4	1769	988	3	918	6	988	3	1126	3	913	4	949	3
R204	883	879	4	2180	753	4	1635	839	3	1686	765	3	753	4	765	3	803	3	825	3	869	2
R205	1179	1174	4	2571	981	5	1295	1103	5	1437	1072	3	981	5	1072	3	1128	3	1206	3	1039	3
R206	1044	1039	5	3070	902	6	1485	1003	3	1475	955	3	902	6	955	3	833	3	956	3	833	3
R207	928	927	3	1772	868	4	1580	942	4	1501	856	3	856	3	856	3	904	3	815	3	815	3
R208	769	768	3	1298	727	3	1432	785	4	1198	710	3	710	3	710	3	823	2	709	3	739	2
R209	1033	1023	3	2526	883	6	1691	997	5	1286	976	3	883	6	976	3	855	2	902	4	855	2
R210	1111	1086	4	3171	920	6	1903	1052	5	1073	1071	3	920	6	1071	3	1052	3	1087	3	968	3
R211	811	811	4	1486	777	4	841	838	4	1061	782	3	777	4	782	3	816	3	794	3	924	2
Mean	1091	1084	4.1	2693	905	5.2	1691	1050	4.2	1498	1011	3.2	902	5.1	995	3.2	988	3.0	981	3.4	951	2.7
C201	589	589	3	589	589	3	589	589	3	589	589	3	589	3	589	3	591	3	592	3	592	3
C202	599	599	3	589	589	3	1239	618	4	589	589	3	589	3	589	3	707	3	592	3	592	3
C203	652	642	3	626	599	3	778	667	3	639	612	3	599	3	599	3	791	3	591	3	591	3
C204	630	614	3	649	611	3	621	609	3	624	598	3	598	3	598	3	685	3	598	3	591	3
C205	607	607	3	590	590	3	596	596	3	589	589	3	589	3	589	3	589	3	589	3	589	3
C206	595	595	3	595	595	3	595	595	3	592	592	3	592	3	592	3	588	3	588	3	588	3
C207	651	615	3	587	587	3	707	633	3	643	609	3	587	3	587	3	588	3	588	3	588	3
C208	586	586	3	749	599	3	609	609	3	670	628	3	586	3	586	3	588	3	588	3	588	3
Mean	614	606	3.0	622	595	3.0	717	615	3.1	617	601	3.0	591	3	591	3	641	3.0	591	3	590	3.0
RC201	1923	1852	4	4467	1283	9	3624	1474	6	2324	1804	4	1283	9	1804	4	1294	4	1470	5	1249	4
RC202	1639	1624	4	4283	1106	8	2178	1454	4	2405	1260	4	1106	8	1260	4	1291	4	1444	4	1164	4
RC203	1252	1251	4	2277	1005	4	2125	1170	4	1286	1271	3	1005	4	1266	3	1203	3	1014	4	1080	3
RC204	963	962	5	1963	805	4	2114	901	6	1536	828	3	805	4	828	3	897	3	843	3	807	3
RC205	1725	1675	5	5123	1180	9	2976	1450	5	2426	1622	4	1180	9	1622	4	1389	4	1287	5	1328	4
RC206	1316	1308	4	2902	1072	6	2057	1279	4	1503	1299	3	1072	6	1299	3	1213	3	1208	4	1159	3
RC207	1135	1133	6	3079	984	7	2965	956	5	1909	1034	4	956	5	1034	4	1181	3	1079	4	1082	3
RC208	856	855	5	1619	786	5	1176	828	5	1038	884	3	786	5	884	3	919	3	920	3	834	3
Mean	1351	1333	4.6	3214	1028	6.5	2402	1189	4.9	1803	1250	3.5	1024	6.3	1250	3.5	1173	3.4	1158	4.0	1088	3.4

Table 2.1 Best solutions given by the TSVRPTW, based on four different objective functions. Legend: T - total travelling time excluding the time to serve the customer; D - total distance; R - number of routes; $\mathbf{S^*_D}$ - best solution in terms of distance given by the TSVRPTW, among the four presented (T_{min}, D_{min}, TD_{min}, R_{min}); $\mathbf{S^*_R}$ - best solution in terms of number of routes given by the TSVRPTW; GenSAT - best solution in terms of distance found by Thangiah et al. [10, 6]; Rochat - best solution presented by Rochat and Taillard [7]; Best - the best known solution in terms of number of routes as presented by Taillard et al. [9]).

		D_{min}			TD_{min}			R_{min}		
		T(%)	D(%)	R(%)	T(%)	D(%)	R(%)	T(%)	D(%)	R(%)
T_{min}	R1	15.2	-4.3	4.5	3.7	-2.2	1.3	9.6	-2.9	-2.6
	C1	4.5	-0.4	0.0	1.6	0.0	0.0	2.8	0.1	0.0
	R2	146.7	-16.5	26.7	55.0	-3.1	2.2	37.3	-6.7	-22.0
	C2	1.3	-1.8	0.0	16.8	1.4	4.2	0.5	-0.8	0.0
	RC2	137.9	-22.9	41.3	77.8	-10.8	5.4	33.5	-6.2	-23.9
	Mean	52.8	-8.1	12.8	26.5	-2.5	2.2	15.4	-2.6	-8.6
D_{min}	R1				-9.9	2.2	-3.1	-4.8	1.5	-6.8
	C1				-2.8	0.3	0.0	-1.6	0.5	0.0
	RC1				-6.6	2.4	-3.8	-2.6	3.9	-6.9
	R2				-37.2	16.1	-19.3	-44.4	11.7	-38.5
	C2				15.3	3.3	4.2	-0.8	1.0	0.0
	RC2				-25.3	15.7	-24.6	-43.9	21.6	-46.2
	Mean				-11.1	6.7	-7.8	-16.4	6.7	-16.4
TD_{min}	R1							5.7	-0.7	-3.8
	C1							1.2	0.1	0.0
	RC1							4.3	1.4	-3.2
	R2							-11.4	-3.7	-23.8
	C2							-13.9	-2.2	-4.0
	RC2							-24.9	5.1	-28.6
	Mean							-6.5	0.0	-10.6

Table 2.2 Comparison of the four versions of TSVRPTW.

		R1(%)		C1(%)		RC1(%)		R2(%)		C2(%)		RC2(%)		Mean(%)		B	E	W
		D	R	D	R	D	R	D	R	D	R	D	R	D	R			
GenSAT	SD*	2.7	-1.9	0.2	0.0	1.2	-1.9	9.6	-41.1	8.4	0.0	14.4	-46.0	6.1	-15.2	46	3	7
	SR*	1.2	4.0	0.2	0.0	-0.7	4.1	-1.6	-5.7	8.4	0.0	-8.5	-6.9	0.4	-0.1	36	3	17
Rochat	SD*	1.4	-5.0	0.2	0.0	2.2	-4.9	8.8	-33.9	-0.1	0.0	13.1	-36.5	4.3	-13.4	42	4	10
	SR*	0.0	0.7	0.2	0.0	0.4	1.0	-1.3	5.7	-0.1	0.0	-7.4	14.3	-1.4	3.6	32	4	20
Taillard	SD*	2.5	-8.1	-0.1	0.0	1.5	-7.7	10.4	-41.1	-0.1	0.0	13.8	-46.3	4.7	-17.2	-	-	-
	SR*	1.0	-2.6	-0.1	0.0	-0.3	-2.0	0.1	-5.7	-0.1	0.0	-6.8	-3.4	-1.0	-2.3	-	-	-
Best	SD*	0.4	-9.4	-0.2	0	1.5	-10.7	5.4	-46.4	-0.2	0	6.2	-46.0	2.2	-18.8	34	6	16
	SR*	-1.0	-4.0	-0.2	0	-0.3	-5.2	-4.4	-14.3	-0.2	0	-13.0	-2.9	-3.2	-4.4	19	6	31
CPU-TSVRPTW(min)		21		3		15		95		27		65						
CPU-Rochat(min)		60		71		64		217		160		173						
CPU-Taillard(min)		230		244		187		337		273		193						

Table 2.3 Comparison of TSVRPTW's average solution values with the solutions of other algorithms and the best known solutions. Legend: B - number of TSVRPTW solutions better than the respective algorithm; E - number of equal solutions; W - number of TSVRPTW solutions worse; CPU - average time in minutes for executing each problem, TSVRPTW on a Pentium Pro at 200 MHz, Rochat on 100 MHz Silicon Graphics Indigo workstation, and Taillard on a Sun Sparc 10 workstation.

in Table 2.1 because the authors only present the individual best solution of each problem found "during the course of multiple experiments" (Taillard et al. [9]). However, to compare this algorithm with TSVRPTW what is more important is the "normal" execution of the algorithm, and for this the authors only give the average for each type of problem. In Table 2.3 the average percentage difference is shown, which is given by $100(x-y)/y$, where x and y are, respectively, the average values obtained by the comparing algorithm and by TSVRPTW.

Table 2.3 shows that $S_D{}^*$ is on average distance better than any of the three algorithms and also than the best known solutions, but it is worse on the average number of routes. On the contrary, $S_R{}^*$ is better in terms of average distance than GenSAT and worse than the other two algorithms, but in relation to the number of the routes is better than the algorithm of Rochat and Taillard [7] and worse than the other two, though the difference is very small, specially in relation to GenSAT. In terms of distance for the 56 problems, 34 new best solutions were found, five that are equal and 17 worse. For 12 problems the TSVRPTW solution is equal or better, simultaneously, in relation to the distance and number of routes, when compared with the best known solutions.

The computing times are difficult to compare because the algorithms have been executed in different kinds of computers. However, it seems that GenSAT is faster than TSVRPTW and this is faster than the algorithms of Rochat and Taillard [7] and Taillard et al. [9]. These last two algorithms provide very good solutions but the computation times are too long to be applied in the daily planning, especially if the planner wishes to solve the same problem several times in order to create different scenarios.

The influence of the several TSVRPTW parameters in its performance was not studied enough to take definitive conclusions, and their value is more based on intuition than on experimentation. However, the few experiments made suggest that TSVRPTW is not very sensitive to small variations of most of the parameters, except p. Although part 2 is applied when the solution is already good, most of the times there is an improvement inside this part which means that it is valuable for the algorithm. It is also very frequent that the best solution from phase 1 is improved in phase 2, but this improvement is usually small.

2.7 DIRECTIONS FOR FURTHER IMPROVEMENT

In comparison with other algorithms for the VRPTW, the TSVRPTW algorithm performs rather well, but this is a very difficult problem and further improvements are always desirable. In the following, the aspects on which the future research is going to be centered are described.

During the execution of TSVRPTW it was observed that in spite of the existence of swap moves there are many iterations on which no move is allowed. This represents a loss of computing time and may be overcome by enlarging the candidate set just in the iterations where this is verified.

Even in phase 1, the trial solutions that are infeasible in relation to the driving time are not accepted. It is worth trying to relax this constraint, because this may have a positive influence by itself and at the same time may contribute to overcoming the difficulty referred to in the previous paragraph.

As was explained previously, the insertion of a customer in a route is made with the GENI algorithm, the neighbourhood being defined in terms of distance. One possible improvement is to define a different kind of neighbourhood, for example, similar to the one used with NENBVRP3. Another way is to use the classical insertion procedure instead of GENI. This diminishes very much the computation time per iteration, but the experiments made so far suggest the quality of the solution is substantially reduced. However, this effect may be smoothed by alternating the two types of insertion.

The last suggestion that is worth testing is to finish an iteration as soon as a new best solution is discovered as defined by equation (2.14).

Acknowledgement

This research was partially supported by Junta Nacional de Investigação Científica e Tecnológica under the program PRAXIS XXI, project n° 3/3.1/ CEG/ 2661/ 95 and by CEEG research centre.

References

[1] Brandão, J. and Mercer, A. A tabu search algorithm for the multi-trip vehicle routing and scheduling problem. European Journal of Operational Research, 100, 180-191, 1997.

[2] Christofides, N., Mingozzi, A., and Toth, P. The Vehicle Routing Problem. In N. Christofides, A. Mingozzi, P. Toth and C. Sandi, editors, Combinatorial Optimization, Willey, Chichester, 1979.

[3] Desrochers, M., Desrosiers, J., and Solomon, M. A New Optimization Algorithm for the Vehicle Routing Problem with Time Windows. Operations Research, 40, 342-354, 1992.

[4] Gendreau, M., Hertz, A., and Laporte, G. New Insertion and Post-Optimization Procedures For the Traveling Salesman Problem. Operations Research, 40, 1086-1094, 1992.

[5] Lenstra, J.K. and Rinnooy Kan, A.H.G. Complexity of Vehicle Routing and Scheduling Problems. Networks, 11, 221-227, 1981.

[6] Potvin, J.-Y. and Rousseau, J.-M. A Tabu Search Heuristic for the Vehicle Routing Problem with Time Windows. Working Paper CRT - 953, Centre de Recherche sur les Transports, Université de Montréal, Canada, 1993.

[7] Rochat, Y. and Taillard, E.D. Probabilistic Diversification and Intensification in Local Search for Vehicle Routing. Working Paper CRT - 95 -13, Centre de Recherche sur les Transports, Université de Montréal, 1995.

[8] Solomon, M.M. Algorithms for the Vehicle Routing and Scheduling Problems with Time Window Constraints. Operations Research, 35, 254-265, 1987.

[9] Taillard, E.D., Badeau, F., Gendreau, M., Guertin, F., and Potvin, J.-Y. A tabu search heuristic for the vehicle routing problem with soft time windows. Transportation Science, 31, 170-186, 1997.

[10] Thangiah, S.R., Osman, I.H., and Sun, T. Metaheuristics for Vehicle Routing Problems with Time Windows", Research Report No: UKC/IMS/OR94/8, Institute of Mathematics and Statistics, University of Kent, UK, 1995.

[11] Thangiah, S.R., Osman, I.H., and Sun, T. Hybrid Genetic Algorithms, Simulated Annealing and Tabu Search Methods for Vehicle Routing Problems with Time Windows. Working Paper UKC/IMS/OR94/4, Institute of Mathematics and Statistics, University of Kent, UK, 1994.

3 NEW HEURISTIC ALGORITHMS FOR THE CREW SCHEDULING PROBLEM

Luis Cavique[1], César Rego[2] and Isabel Themido[3]

[1]Instituto Politécnico de Setúbal
Rua do Vale de Chaves, Estefanilha, 2900 Setubal, Portugal.
lcavique@bocage.ips.pt

[2] Faculdade de Ciências, Universidade de Lisboa
Departamento de Estatística e Investigação Operacional - DEIO
Campo Grande C/2, 1700 Lisboa, Portugal.
cesar@fc.ul.pt

[3]Instituto Superior Técnico - IST/CESUR, Av. Rovisco Pais 1096 Lisboa, Portugal.
ithemido@civil18.civil.ist.utl.pt

Abstract: In this paper we describe the implementation of two Tabu search based algorithms for the Crew Scheduling Problem (CSP). The problem is a standard CSP, for an urban transport problem, formulated as the minimal number of duties necessary to cover a predefined timetable under contractual rules which excludes the possibility of resorting to extra-time. The initial solution is obtained with a run-cutting algorithm. Two algorithms are proposed to improve that initial solution. In the first algorithm the run-cutting procedure is applied again to a subset of the problem in a Tabu search framework guided by a strategic oscillation procedure. The second algorithm uses block partition and matching algorithms embedded in a Tabu subgraph ejection chain to explore the neighborhood. Computational results are presented showing the superior performance of the latter algorithm for the case study.

3.1 INTRODUCTION

In this paper we describe two heuristic algorithms for the Crew Scheduling Problem (CSP). The problem is formulated as the minimum number of duties to cover a predefined timetable under a set of contractual rules, and we consider the case of the Lisbon Metro Company (ML).

As there is no uniformity in the terms used in the literature, it is useful to define the terms used in this paper. The terms trip and block are closely related to the train timetable, while the terms piece of work (or piece) and duty (or run) are related to the crew schedule. A trip is a one-way movement of a vehicle between two terminal points. A block is a sequence of trips assigned to the same vehicle. A piece of work is part of a duty that covers a set of consecutive trips in a block, under certain contractual rules. In the ML case there are two kinds of feasible duties: duties with two pieces of work separated by a meal break and the duties that only have one piece of work, leaving the remaining time free to act as reserve crews. For a glossary of terms in crew scheduling refer to Hartley [14].

Contractual rules can be organized under the following headings:

- *relief points* - must be the same at the beginning and end of the duties and meal breaks. When this does not happen join-up time must be allowed for the crew to travel between relief points.
- *pieces of work* - maximum of three and half hour and minimum of one and half hour.
- *duty* - total durations of nine hours, of which eight are working hours and one is a meal break.
- *meal break* - the meal break is one hour and it must coincide with the period of functioning of the canteen (11:00-15:00 and 17:00-22:00), except for the duties which terminate between 14:00-14:30 and 21:00-21:30.

A number of exact and heuristic algorithms have been developed for the urban transit crew scheduling problem. These methods have been described in detail in the proceedings of six workshops on computer-aided transit scheduling (Bodin and Bergman [1], Wren [22], Rousseau [21], Daduna and Wren [7], Desrochers and Rousseau [8], Daduna, Branco and Paixão [6]).

There are three main approaches to the Crew Scheduling Problem (see Bodin et al. [2] and Wren and Rousseau [23]): the run-cutting heuristic, the matching algorithm and the set covering formulation. The run-cutting algorithm was used in the 1970's, in the American RUCUS package and on the Parker and Smith heuristics for the British TRACS system; this is a constructive algorithm used by manual schedulers. The matching method is used in RUCUS-II and in an initial version of HASTUS; the method is divided into three parts: block partition of the timetable, graph generation and duty achievement. The set covering formulation is used in the package IMPACS of the BUSMAN system and in the Crew-Opt package; the latter belongs to an early version of HASTUS. The set covering method in a first stage generates a large set of feasible duties and in a second stage finds the minimal covering set.

The most widely used decision support systems at present are probably those of HOT from Germany, BUSMAN from the United Kingdom and HASTUS from Canada. We must also include in this list unpublished systems such as Teleride-Sage and UMA.

The aim of this paper is to describe two heuristic algorithms for the CSP as the dimension of the problem recommends a heuristic approach (Falkner and Ryan [9]).

The first algorithm considers two procedures based on a run cutting approach developed so that the generation and feasibility testing of a duty are done separately in order to increase the capacity of the algorithm to cope with different problems. In an initial stage the algorithm determines a feasible solution which is thereafter improved by performing some local perturbations in the schedule using a tabu search framework.

The second algorithm considers the initial schedule given by the run-cutting procedure and improves this schedule by efficiently exploiting the solution space using neighborhoods based on ejection chains and a matching algorithm.

3.2 TABU SEARCH BASED ON A RUN-CUTTING APPROACH: THE TABU-CREW ALGORITHM

In this section we present a tabu search algorithm based on a run-cutting approach for the crew scheduling problem defined in the above section. The algorithm is structured into two phases: construction and improvement.

3.2.1 The run-cutting algorithm

In the construction phase, the algorithm determines an initial feasible schedule using a run-cutting approach for creating runs (duties) with one or two pieces of work.

The run-cutting method is a traditional procedure used in crew scheduling. The method is inspired by manual schedule procedures and can be implemented in different ways to take advantage of the structure of the particular problem. The algorithm generates a feasible duty (or run) by covering (or cutting) uncovered trips from a block of the timetable; this process is repeated until all trips are covered. In other words the trips of the timetable are successively cover by the duties' pieces of work, until the cover is complete.

The quality of a piece of work is evaluated in terms of the number of trips covered and in terms of the room left for subsequent adjacent pieces. A covering piece that does not leave adjacent trips uncovered is not penalized. An inadequate covering piece leaves a few adjacent trips, insufficient to complete a new piece. All feasible pieces must allow enough room for other pieces or else they could compromise the cover of the total train timetable.

In the ML case each duty has a unit cost, so duties with only one piece of work are worse than duties with two pieces, that cover more trips for the same cost. So, the run-cutting heuristic chooses the less costly at each duty stage, allowing a good feasible solution for the CSP. As a second criterion to evaluate alternative pieces a penalty function attributes the inverse of the number of adjacent trips left by a covering piece.

Generating feasible duties

A fundamental drawback of the run-cutting algorithms is their lack of flexibility. In order to overcome these difficulties, we consider a *generate-and-test* strategy (see Rich and Knight [20]) based on two independent procedures: a generator of combinatorial objects and a checker of contractual rules. With this architecture the algorithm can easily be adapted to incorporate additional constraints.

Another drawback of the run-cutting algorithms is their inadequate adaptability to problems where break times must be specified in the runs (Wren and Rousseau [23]). To overcome this we consider combinatorial objects generated in a systematic form. By establishing an analogy with a lexicographic search (Reingold, Nievergelt and Deo [19]), the generation of a piece of work may be viewed as the creation of a subword, of a large word representing a sequence of all the trips in the timetable.

3.2.2 Guiding the run-cutting method using tabu search

The improvement phase is an iterative procedure which attempts to reduce the total number of duties using Tabu search (see Glover [10, 11, 12] for a comprehensive description of the method, and Glover [13] for a survey on Tabu search applications). As the optimal timetable cover presents a number of expensive duties (with a single piece of work), improvements may be obtained by reducing the number of expensive duties. In order to open space for new configurations, expensive duties as well as their adjacent duties are destroyed.

The procedure is an extension of the run-cutting method described above and implemented in a tabu search framework. Starting with the solution obtained in the construction phase, a new schedule is obtained, at each iteration, cutting a certain number of runs (duties) from the current solution and re-running the subproblem using the run-cutting process.

A tabu list is used to avoid the creation of duties (runs) with pieces previously destroyed. As a dynamic tabu list size is used, pieces in expensive duties can be recombined with others in further iterations of the method. Such a procedure, which destroys part of a solution and reconstructs a new solution, step by step, passing through a sequence of incomplete (or non-feasible) solutions until all the trips are covered, represents a form of strategic oscillation. Strategic oscillation is a very suitable technique to explore the solution space in highly constrained problems.

3.3 IMPROVING THE SEARCH USING EJECTION CHAINS: THE RUN-EJECTION ALGORITHM

In this section we will describe a subgraph ejection chain method which when incorporated into a tabu search guiding framework provides an interesting algorithm for the CSP. The method is based on the formulation of the problem as the Maximum Cardinality Matching Problem (MCMP) of a non-bipartite graph $G = (X, A)$. Each piece of work $P(x_i, x_j)$ corresponds to a node of the graph and for each feasible duty $D(x_i, x_j, y_i, y_j)$ we define an edge $A(x_{ij}, y_{ij})$ on G.

The difficulty of this approach is that there are a large number of possible graphs G depending on the way that each block $B(x_1, x_2, x_3, x_4, ..., x_k)$ is partitioned in pieces, where x_1 represents the first trip of the block and x_k the last trip of the block. Moreover, for each graph there are several possibilities of matching pieces for creating duties.

In order to explore this enormous number of alternatives we consider a tabu search framework using compound neighborhoods defined as follows.

Neighborhood structures

The neighborhood search is based on an *embedded neighborhood structure* $\mathcal{N}$ composed by two substructures $\mathcal{N}_1$ and $\mathcal{N}_2$ which separate the neighborhood space into two subsets. $\mathcal{N}_1$ is an intermediate structure responsible for generating sets of new pieces of work which will replace pieces of the current graph $G_k = (X_k, A_k)$ transforming it into another graph $G_{k+1} = (X_{k+1}, A_{k+1})$.

The final neighborhood structure $\mathcal{N}$ is obtained by choosing a set of edges in G_{k+1} in order to obtain a feasible solution.

Notice that $\mathcal{N}_1$ is an incomplete neighborhood because it does not define regions of complete (or feasible) solutions but rather changes the neighborhood space in which a new solution will be searched with $\mathcal{N}_2$. In other words, $\mathcal{N}_1$ defines a set of directions (arcs) on the *solution space graph* $\mathcal{G}$ to be searched from the current solution S_k. These directions, which restrict the neighborhood space, are represented by the graph G_{k+1} and a new neighboring solution S_{k+1} is determined by solving the MCMP on G_{k+1}.

In the algorithm, the neighborhood structure $\mathcal{N}_2$ is constructed using a *greedy* heuristic procedure. At each iteration the procedure searches the piece with the smallest number of links (i.e. the node of G_{k+1} with lower degree) and selects one alternative linking it to another piece.

We define $\mathcal{N} = \mathcal{N}_1 \oplus \mathcal{N}_2$ as the combination of the two neighborhood structures and use the classical notation, introduced by Hertz and de Werra [15] to denote $S_{k+1} = S_k \oplus m$, the transition from a solution S_k to a new solution S_{k+1} by the application of a move m. Let X be the set of feasible solutions, M_1 and M_2 the set of moves defined respectively in $\mathcal{N}_1$ and $\mathcal{N}_2$.

The neighborhood considered in the algorithm can be defined as $N(S_k) = \{S_{k+1} \mid \exists m_1 \in M_1, m_2 \in M_2 \text{ with } S_{k+1} = (S_k \oplus m_1) \oplus m_2, S_{k+1} \in X\}$.

We now describe the various components of the neighborhood structure considered in this tabu search procedure.

As is typical in practical crew scheduling problems, the reduction of the number of duties, even for a randomly generated schedule is hard to realize because problems are very constrained and usually a small change in one part of the schedule requires propagation of changes throughout the entire schedule to produce an improvement (Clement and Wren [5]).

Generating new pieces of work

Based on this assumption we have conceived an *ejection chain method* in which at each level l ($l = 1, 2, \cdots, L$) of the chain a subgraph $\overline{G}_k^l$ of G_k^l ($G_k^1 = G_k$) is deleted (ejected) and replaced by another subgraph $\tilde{G}_k^l$ which is added to the current graph transforming G_k^l into $G_k^{l+1} = G_k^l \setminus \overline{G}_k^l \cup \tilde{G}_k^l$. The process is defined by successive delete/add ejection moves, where each delete operation of a subgraph in order to add a new subgraph leads to the generation of another move of the same type. The process ends when no other pair of these operations is carried out.

In the ejection chain process, delete/add moves can be carried out by any of the elementary operations defined as follows:

- a *shift operation*, which shifts x_j the extreme of a piece to the right or to the left transferring a trip between adjacent pieces $P_1(x_i, x_j)$ and $P_2(x_{j+1}, x_k)$ to obtain partitions $P_1(x_i, x_p)$ and $P_2(x_{p+1}, x_k)$ with $p = j+1$ or $p = j-1$.
- a *cut operation*, which cuts one piece $P_1(x_i, x_j)$ into two pieces $P_1(x_i, x_k)$ and $P_2(x_{k+1}, x_j)$,
- a *merge operation*, which searches for adjacent pieces $P_1(x_i, x_j)$ and $P_2(x_{j+1}, x_k)$ which can be combined into a single piece $P_1(x_i, x_k)$.

The move m_1 mentioned above is composed by combinations of these operations or submoves (one at each level of the chain), $m_1 = m_{11} \oplus m_{12} \oplus \cdots \oplus m_{1L}$, which defines the ejection chain process generated by $\mathcal{N}_1$. However, such a procedure only defines sequences of *ejection moves* which create sets of new pieces and their corresponding links to existent pieces, modifying G_k at each level of the chain.

Creating new duties

To obtain a *trial solution* it is necessary to define a *trial move* to evaluate the complete transition from one solution to another. A trial move may be done for each level l of the chain by solving the corresponding MCMP on the current graph G_k^l. Of course, this would be extremely expensive from the computational point of view.

To overcome this difficulty an auxiliary objective function

$$F'(S_k) = \mid A_k^{l+1} \mid - \mid A_k^l \mid$$

was introduced to evaluate the transition from $G_k^l = (X_k^l, A_k^l)$ to $G_k^{l+1} = (X_k^{l+1}, A_k^{l+1})$. The objective is to maximize $F'(S_k)$, the criterion for the selection of a move carried out by one of the three operations defined above. Thus, the graph of the problem is successively enriched with new edges allowing a greater number of variables (duties) to be evaluated at each iteration k of the algorithm. Once the chain ends, the final move m_2 is carried out by applying $\mathcal{N}_2$ on the graph G_k^{l*}, where l^* represents the level of the chain where the best value of F' was found.

The quality of a solution S is given by the objective function $F(S)$ which measures the corresponding number of duties.

It is clear that if no condition is imposed for the selection of operations, the cut operation will be frequently chosen for the move according to the F' criterion. In order to maintain a good balance of operations and introduce a suitable diversification of the search, operations are randomly selected throughout the ejection chain.

In the tabu search algorithm, the tabu list is used with two purposes: 1) guiding moves in the ejection chain process; 2) avoiding the search to fall into solutions previously visited. For the first point, a tabu tenure of a single iteration is considered forbiding a piece of work to be modified again, up to the end of the ejection chain in order to propagate the search throughout the entire schedule. This also satisfies the second point; nevertheless to prevent cycling, reverse operations of moves are put into the tabu list for a certain number θ of iterations, which is randomly generated within an interval $[\underline{\theta}, \overline{\theta}]$. Reverse operations are shift-left/shift-right and cut/merge.

In addition to the strategic oscillation implicit in the ejection chain process, a diversification strategy was implemented by introducing an addition term in the move evaluation function in order to favour the insertion of edges linking nodes (pieces) which have frequently represented duties with a single piece of work. To do so we consider a vector $\phi = (p_1, p_2, \cdots, p_b)$ to record frequencies of pieces identified by their positions, where b is an upper bound for the number of pieces. The frequency values indicate the number of times that each piece was left isolated in the solution of a matching problem. More formally, the move evaluation function can be described as follows:

$$F'(S_k) = \mid A_k^{l+1} \mid - \mid A_k^l \mid + \sum_{(x_{ij}, y_{ij}) \notin G_k^l} \max\{\phi(x_{ij}), \phi(y_{ij})\}.$$

3.4 COMPUTATIONAL EXPERIENCE

The performance of Tabu-Crew (TC) and Run-Ejection (RE) algorithms was evaluated for a set of eight timetables represented in Table 3.1. Results are also presented for the Run-Cutting (RC) algorithm which provides the initial schedule for TC and RE algorithms. Comparisons were made between the results obtained by our algorithms and those of the Metro of Lisbon (ML).

The computational results for the algorithms are reported in Table 3.2. The computations were performed on a HP9000/712 workstation and algorithms were coded in Pascal.

The quality of the solutions is evaluated on the basis of three performance measures: the relative improvements to the number of duties considered by the ML; the matching ratio, that is the percentage of the duties with two pieces of work; and the average of the driving time (in hours) per duty. We can see that all algorithms produce a number of duties which is on average inferior

Problem	Number of Terminus	Driving Hours	Number of Blocks	Number of Trips
H62 (week-days)	3	341	26	718
H63 (week-end)	3	306	22	654
H65 (Saturdays)	3	349	25	658
H66 (Sundays)	3	280	21	528
H68 (week-days)	3	366	26	685
H65a (Saturdays)	2	349	25	658
H66a (Sundays)	2	280	21	528
H68a (week-days)	2	366	26	685

Table 3.1 Characteristics of test problems

	Nb. duties/Driv. time				% Improvement			% Matching			CPU time		
Prob.	ML	RC	TC	RE	RC	TC	RE	RC	TC	RE	RC	TC	RE
H62	76/ 4.5	79/ 4.3	76/ 4.5	75/ 4.5	-3.9	0.0	1.3	59	63	65	1.7	1642	134
H63	66/ 4.6	68/ 4.5	65/ 4.7	64/ 4.8	-3.0	1.5	3.0	70	72	75	1.2	639	87
H65	80/ 4.4	83/ 4.2	80/ 4.4	78/ 4.5	-3.8	0.0	2.5	63	66	69	1.3	863	92
H66	80/ 3.5	68/ 4.1	63/ 4.4	61/ 4.6	15.0	21.3	23.8	68	72	74	1.4	173	68
H68	90/ 4.1	92/ 4.0	87/ 4.2	86/ 4.3	-2.2	3.3	4.4	66	70	71	1.9	1254	153
H65a	80/ 4.4	82/ 4.3	79/ 4.4	76/ 4.6	-2.5	1.3	5.0	71	75	77	1.2	298	97
H66a	80/ 3.5	65/ 4.3	61/ 4.6	59/ 4.7	18.8	23.8	26.3	71	78	80	0.8	915	72
H68a	90/ 4.1	91/ 4.0	86/ 4.3	85/ 4.3	-1.1	4.4	5.6	69	74	76	1.5	743	85
Avg. Avg.	80.3/ 4.1	78.5/ 4.2	74.6/ 4.4	73.0/ 4.5	2.1	6.9	9.0	67.1	71.3	73.4	1.4	816	99

Table 3.2 Performance measures and computational times (in seconds).

to the manual solutions implemented by the Lisbon Metro Company. However, the average improvement is not very meaningful for the evaluation of the Run-Cutting algorithm, as manual results for problems H66 and H66a are conditioned by the rostering scheme of the ML which calls for an identical number of duties on Saturdays and Sundays. Nevertheless, we can consider that RC algorithm produces reasonable feasible solutions in a very short time span which makes the algorithm useful for determining starting solutions for improving methods. Moreover, the results obtained by the Tabu-Crew algorithm, which reduced the number of duties of the initial schedule by about 5%, show the efficiency of the method when incorporated into a tabu search procedure.

Finally, the RE algorithm reduced the number of duties of the initial schedule by about 7% representing an improvement of 9% on the manual solutions. Also, the Run-Ejection algorithm produces duties of better quality than Tabu-Crew with respect to the matching criterion and achieves the solutions in a much shorter time. The Run-Ejection algorithm is thus very suitable for obtaining high quality solution for planning purposes.

We have verified that feasible solutions are difficult to find in timetables with small blocks as occurs in problems H62 and H68. This suggests that the computation time depends on the size of blocks more than on the number of trips.

Another important feature for the ML is the total driving time which, on average, should be longer than 4.5 hours per duty. This objective is reached by our algorithms for all the problems tested.

3.5 CONCLUSION

We have presented two heuristic algorithms for a real crew scheduling problem faced by the Lisbon Metro Company. The proposed algorithms are part of a decision support system for crew scheduling management.

In a first stage we have developed an efficient run-cutting algorithm which provides rapidely an initial feasible solution for two improvement algorithms based on tabu search. The Run-Cutting algorithm considers duties generated in a systematic form, based on the notion of lexicographic search; the feasibility of a duty is tested by an independent procedure. This makes it easy to incorporate additional rules in the algorithm and introduces a straightforward manner for implementations of run-cutting algorithms.

As the run-cutting approach constructs feasible solutions from scratch using a greedy criterion, it is reasonable to suspect that better solutions may be found in a restricted neighborhood of this solution. So, the algorithm was adapted to consider local improvements of the initial schedule using a tabu search framework. The implementation of the Tabu-Crew algorithm has shown the importance of using strategic oscillation for searching new crew scheduling solutions, and a candidate list from which the best run is selected for improving algorithm performance (Cavique [3], Cavique and Themido [4]).

Ejection chains have produced interesting results in an number of scheduling related problems. Motivated by successful experiments obtained from node based ejection chains (Rego and Roucairol [18]) and subpath ejection chain methods (Rego [16, 17]) for vehicle routing and traveling salesman problems, we have introduced in this paper a special type of ejection chain based on the ejection of subgraphs for the crew scheduling problem. Again, the method proved quite effective for finding good solutions quickly. As the degree of modification of a solution can be easily measured at each level of an ejection chain, this algorithm is thus very suitable for a future real-time decision support system required by ML for daily operation management.

References

[1] L. Bodin and D. Berman, editors. *Workshop on Automatic Techniques for Scheduling of Vehicle Operator for Urban Public Transportation Services*, 1975.

[2] L.D. Bodin, B.L. Golden, A.A. Assad, and M.O. Ball. Routing and scheduling of vehicles and crews. The state of the art. *Computers & Operations Research*, 10:63–212, 1983.

[3] L. Cavique. *Sequenciamento de pessoal tripulante.* M.Sc. Thesis, Universidade Técnica de Lisboa - Instituto Superior Técnico, Lisbon, 1994.

[4] L. Cavique and I.H. Themido. Sequenciamento de serviços de pessoal tripulante: uma abordagem baseada num conjunto de heuristicas. *Investigação Operacional*, 15:123–141, 1995.

[5] R. Clement and A. Wren. Greedy genetic algorithms, optimizing mutations, and bus driver scheduling. *Computer-Aided Transit Scheduling*, volume 430 of *Lecture Notes in Economics and Mathematical Systems.* J.R. Daduna, I. Branco, and J.M.P. Paixão (eds.), Springer Verlag, Berlin Heidelberg, 213–235, 1995.

[6] J.R. Daduna, I. Branco, and J.M.P. Paixão. *Computer-Aided Transit Scheduling*, volume 430 of *Lecture Notes in Economics and Mathematical Systems.* Springer Verlag, Berlin Heidelberg, 1995.

[7] J.R. Daduna and A. Wren. *Computer-Aided Transit Scheduling*, volume 308 of *Lecture Notes in Economics and Mathematical Systems.* Springer Verlag, Berlin Heidelberg, 1988.

[8] M. Desrochers and J.-M. Rousseau. *Computer-Aided Transit Scheduling*, volume 386 of *Lecture Notes in Economics and Mathematical Systems.* Springer Verlag, Berlin Heidelberg, 1992.

[9] J.C. Falkner and D.M. Ryan. Express: Set partitioning for bus crew scheduling in christchurch. Proceedings of the Fifth International Workshop on Computer-Aided Scheduling of Public Transport. Montréal, Canada, 1990.

[10] F. Glover. Tabu search - part I. *ORSA Journal on Computing*, 1:190–206, 1989.

[11] F. Glover. Tabu search - part II. *ORSA Journal on Computing*, 2:4–32, 1990.

[12] F. Glover. Tabu search fundamentals and uses. Technical report, Graduate School of Business and Administration, University of Colorado at Boulder, 1995.

[13] F. Glover. Tabu search and adaptive memory programming: Advances, applications and challenges. Technical report, Graduate School of Business and Administration, University of Colorado at Boulder, 1996.

[14] T. Hartley. A glossary of terms in bus and crew scheduling. Proceedings of the Second International Workshop on Computer Scheduling of Public Transport. Leeds, 1981.

[15] A. Hertz and D. de Werra. Using tabu search techniques for graph coloring. *Computing*, 39:345–351, 1987.

[16] C. Rego. Relaxed tours and path ejections for the traveling salesman problem. *European Journal of Operational Research*, 106:522–538, 1998.

[17] C. Rego. A subpath ejection method for the vehicle routing problem. *Management Science*, 1996. Forthcoming.

[18] C. Rego and C. Roucairol. *Parallel Tabu Search Algorithm based on Ejection Chains for the Vehicle Routing Problem.* Methaheuristics: Theory & Applications. I.H. Osman and J.P. Kelly (eds), Kluwer Academic Publishers, 661–675, 1996.

[19] E.M. Reingold, J. Nievergelt, and N. Deo. *Combinatorial Algorithms.* Prentice-Hall, 1977.

[20] E. Rich and K. Knight. *Artificial Intelligence*, volume 386. McGraw-Hill, 1991.

[21] J.-M. Rousseau. *Computer Scheduling of Public Transport 2.* North-Holland, Amsterdam, 1985.

[22] A. Wren. *Computer Scheduling of Public Transport Urban Passenger Vehicle and Crew Scheduling.* North-Holland, Amsterdam, 1981.

[23] A. Wren and J-M. Rousseau. Bus driver scheduling: an overview. *Computer-Aided Transit Scheduling*, volume 430 of *Lecture Notes in Economics and Mathematical Systems.* J.R. Daduna, I. Branco, and J.M.P. Paixão (eds), Springer Verlag, Berlin Heidelberg, 173–187, 1995.

4 ENHANCED CONTINUOUS TABU SEARCH: AN ALGORITHM FOR OPTIMIZING MULTIMINIMA FUNCTIONS

Rachid Chelouah and Patrick Siarry

LGI2P, Institut Universitaire de Technologie de Cergy-Pontoise
Département de Génie Electrique et Informatique Industrielle
Groupe de Recherche Systèmes Complexes
Allée des Chênes Pourpres 95014 Cergy Cedex, France.
chelouah@u-cergy.fr, siarry@u-cergy.fr

Abstract: An algorithm called Enhanced Continuous Tabu Search $ECTS$ is proposed for the global optimization of multiminima functions. It results from an adaptation of combinatorial Tabu Search which aims to follow, as close as possible, Glover's basic approach. In order to cover a wide domain of possible solutions, our algorithm first performs diversification: it locates the most promising areas, by fitting the size of the neighborhood structure to the objective function and its definition domain. When the most promising areas are located, the algorithm continues the search by intensification within one promising area of the solution space. The efficiency of $ECTS$ is thoroughly tested by using a set of benchmark multimodal functions, of which global and local minima are known. $ECTS$ is compared to other published versions of continuous Tabu Search and to some alternative algorithms like Simulated Annealing.

4.1 INTRODUCTION

Tabu Search TS is a metaheuristic originally developed by Glover [6, 7], that has been successfully applied to a variety of directly inspired from combinatorial optimization problems. However, very few works deal with its application to the global minimization of functions depending on continuous variables. Up to now, we are aware of only works [2, 4, 8, 11] related to the subject. In this paper we propose an adaptation of TS to continuous optimization problems, called Enhanced Continuous Tabu Search $ECTS$, which is directly inspired from Glover's approach. We produce an efficient algorithm by improving the simplistic approach proposed by Siarry and Berthiau in [11]. To achieve this, we introduce, diversification and intensification concepts emphasized by Glover, and implemented by Battiti and Tecchiolli [1] in ECTS.

First, we will summarize how Glover's idea is adapted to the continuous case in [11]. The algorithm starts from a randomly selected initial solution s. From this current solution s, a set of neighbors, called s', is generated (details at the beginning of Section 4.2). To avoid the danger of the appearance of a cycle, the neighbors of the current solution, which belong to a subsequently defined "tabu list", are systematically eliminated. The objective function to be minimized is evaluated for each generated solution s', and the best neighbor of s becomes the new current solution, even if it is worse than s. After each "move", s is put into a circular list of tabu solutions: when the tabu list is full, it is updated by removing the first solution entered. Then a new "iteration" is performed: the previous procedure is repeated by starting from the new current point, until some stopping condition is reached. Usually, the algorithm stops after a given number of iterations without any improvement of the objective function value.

This rudimentary algorithm, directly inspired from the simple combinatorial Tabu Search algorithm, is tested through classical multiminima functions of which global and local minima are known [11]. The results obtained are encouraging, but not wholly satisfactory, mainly due to the excessive simplicity of the algorithm. To improve the approach, it is necessary to carry out diversification (detection of promising areas) allowing the scanning of a wide solution space, and intensification (search inside the most promising area) for a more accurate result. These two operations are described in the current paper.

In order to compare *ECTS* to other algorithms we have implemented one set of test functions and various algorithms using the object-oriented language C^{++}.

The paper is organized as follows. In Section 4.2, we present the general setting of the method. In Section 4.3, Enhanced Continuous Tabu Search is explained in detail. Experimental results are discussed in Section 4.4 and Section 4.5 provides the conclusion.

4.2 GENERAL SETTING OF THE METHOD

First we state the way of discretizing the solution space. We have adapted the method described in detail in [11]. The neighborhood is defined in [11] by using the concept of "ball". A ball *B(s,r)* is centered on s with radius *r;* it contains all points s' such that: $||s' - s|| \leq r$ (the symbol $||.||$ is used to denote the Euclidean norm). To obtain homogeneous exploration of the space, we consider a set of balls centered on the current solution s, with radius h_0 , h_1, .., h_k. Hence the space is partitioned into concentric "crowns" C_i (s, h_{i-1}, h_i), such that:

$$C_i(s, h_{i-1}, h_i) = \{s'/h_{i-1} \leq ||s' - s|| \leq h_i\}.$$

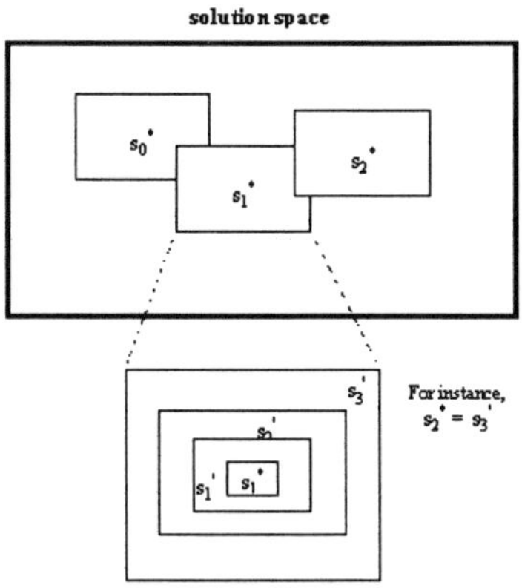

Figure 4.1 Partition of current solution neighborhood and diversification strategy.

The k neighbors of s were obtained by random selection of one point inside each crown C_i, for i varying from *1* to k. Finally, we select the best neighbor of s among these k neighbors, even if it was worse than s. In $ECTS$, we replace the balls by hyperrectangles for the partition of the current solution neighborhood (see Figure 4.1), and we generate neighbors in the same way. The reason for using a hyperrectangular neighborhood instead of crown "balls" is the following: it is mathematically much easier to select a point inside a specified hyperrectangular zone than to select a point inside a specified crown ball. Therefore in the first case, we only have to compare the coordinates of the randomly selected point with the bounds that define the hyperrectangular zone at hand.

The initial size of the hyperrectangular neighborhood (the more external hyperrectangle) is obtained by dividing the size of the search domain by a factor ρ_{neigh}, called the neighborhood reduction parameter.

$ECTS$ consists composed of the five stages: initialization, diversification, search for the most promising area, intensification, and output of the best point found. We show the principle of the three central stages. Like the Continuous Tabu Search proposed in [11], $ECTS$ accepts one newly generated neighbor if it is not tabu. $ECTS$ checks that each newly generated neighbor is not inside one of the tabu balls, the centers of which are saved in the tabu list. In order to handle a wide domain of possible solutions, our algorithm performs diversification for locating the most "promising areas". A new promising area is detected whenever an "unacceptable" deterioration (in the sense defined below) of the objective function occurs: if all neighbors at hand are worse than the current solution, this solution is a local optimum. If all related deteriorations of the value of the objective function are higher than a given threshold, we can then consider that this local optimum is at the center of a hyperrectangular zone called promising area. The centers of successively obtained promising areas are stored in a list, called the "promising list". To avoid returning to already visited areas, we check that each newly detected "promising solution" is not inside one of the promising balls, the centers of which are saved in the promising list. This restriction forces the search toward solutions far enough away from the solutions previously obtained.

After a specified number of iterations without any detection of a new promising area, diversification stops and $ECTS$ determines the "most promising area" of the solution space among the promising areas saved in the promising list. The most promising area is not necessarily the area associated with the best element of the promising list. This particular point is explained in the following section.

$ECTS$ continues by intensifying the search within the selected area. The center of this area becomes the new current solution, and a classical tabu search (with iterative computation of neighbors, selection and update) is performed restricted to the specified region of the solution space. The algorithm stops after a given number of iterations without any improvement of the objective function value.

4.3 DETAILED PRESENTATION OF ENHANCED CONTINUOUS TABU SEARCH

The general flow chart of $ECTS$ is shown in Figure 4.2. In this chart, the main stages of $ECTS$ (initialization; diversification; search for the most promising area; intensification; and output of the best point found) are evident. It can be seen that diversification and intensification use the same routines: generating neighbors, selecting the best neighbor, updating various lists and adjusting parameters. The modular organization of the software in C^{++} takes full advantage of this feature. Therfore the same software compounds are used in the different steps of the search.

4.3.1 Initialization

At this stage, we define the following parameters: the length N_t of the tabu list, the length N_p of the promising list, the parameters ρ_t and ρ_p which determine, the radius ϵ_t and ϵ_p of the tabu and promising balls (see details below) respectively, and the neighborhood reduction parameter ρ_{neigh}. These parameters are fixed empirically, by taking the characteristics of the objective function into account. We set d to the length of the smallest edge of the hyperrectangular search domain. The values of ϵ_t and ϵ_p are equal to d/ρ_t and d/ρ_p respectively.

In order to create the promising list, the algorithm randomly draws a point. This point is accepted as the center of an initial promising ball, if it does not belong to an already generated ball. So the algorithm generates N_p sample points which are uniformly dispersed in the whole space solution S. This sample provides also the initial best point found: we detect the best point among the

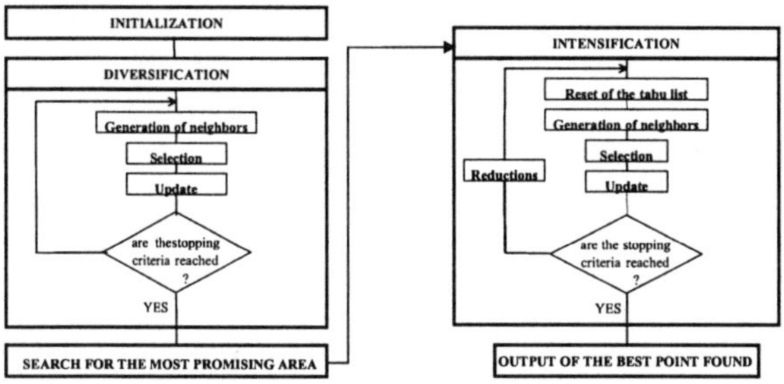

Figure 4.2 General flow chart of Enhanced Continuous Tabu Search.

previous N_p sample points. We calculate the threshold, equal to the average of the objective function values over this sample. Finally we randomly draw the initial solution s^* inside the hyperrectangular search domain.

4.3.2 *Diversification*

At this stage, the process starts with the initial solution. *ECTS* generates a specified number of neighbors: one point is selected inside each hyperrectangular zone around the current solution, as explained in Section 4.2. Each neighbor is accepted only if it does not belong to the tabu list. The best of these neighbors becomes the new current solution, even if it is worse than the previous one. A new promising solution is detected whenever an unacceptable deterioration of the objective function occurs inside the neighborhood area centered around the current solution. This promising solution defines a new promising area if it does not yet belong to a promising ball. If a new promising area is accepted, the worst area of the promising list is replaced by the newly accepted promising area. The use of the promising and tabu lists stimulates the search for solutions far from the current one and the identified promising areas. Diversification process stops after a given number (usually five) of successive iterations without any detection of a new promising area. Then the algorithm determines the most promising area among those present in the promising list.

4.3.3 *Search for the most promising area*

In order to determine the most promising area, we proceed in three steps. First, we calculate the average value of the objective function over all the solutions present in the promising list. Second, we eliminate all the solutions for which the function value is higher than this average value. Third, we deal with the thus reduced list in the following way. We halve the radius of the tabu balls and the size of the hyperrectangular neighborhood. For each remaining promising solution, we perform the generation of neighbors and selection of the best. We replace the promising solution by the best neighbor located, yet only if this neighbor is better than that solution. After having scanned the whole promising list, the algorithm removes the least promising solution. This process is reiterated after halving again the two parameters started above. It stops when just one promising area remains.

4.3.4 *Intensification*

The first step of intensification stage is the resetting of the tabu list. The remaining promising area allows the definition of a new search domain. The center of this area is taken as the current point, and the tabu search stars again: generation of neighbors not belonging to the tabu list, selection of the best, and insertion into the tabu list. This selected neighbor becomes the new current solution, even if it is worse than the previous one. A tuning of the tabu balls' radius and of the size of the hyperrectangular neighborhood is performed (division by two after a specified number of successive iterations). To stop the algorithm, we use a specified number of successive iterations without any improvement of the objective function value and a maximum specified number of iterations. These two specified numbers are fixed automatically by taking the size of the search domain and the number of variables into account, both of which are characteristics of the objective function.

4.3.5 *Tuning of the parameters influencing the algorithm convergence*

Some parameters of $ECTS$ are automatically computed according to the characteristics of the objective function at hand: for instance, d, ϵ_t and ϵ_p are calculated as explained above. Other parameters that characterize $ECTS$ are the following:

- the number of neighbors generated around the current solution;
- the length of the tabu list;
- the length of the promising list;
- the parameters ρ_t, ρ_p, and ρ_{neigh};
- the maximum number of iterations without any improvement in the value of the objective function;
- the maximum number of iterations without any detection of a promising area.

After a series of trials we fix the value of some of the above-mentioned parameters. The fixed parameters are the length of the tabu list (set to seven), the length of the promising list (set to ten) and the parameters ρ_t, ρ_p and ρ_{neigh} (set to 100, 50 and 5 respectively). The other parameters are automatically tuned, by taking the function's characteristics into account. The maximum number of successive iterations without any improvement in the value of the objective function is equal to five times the number of variables, and the maximum number of successive iterations without any detection of a promising area is equal to twice the number of variables. After a predetermined number of iterations without any improvement in the value of the objective function, the size of the hyperrectangular neighborhood and the radius of tabu balls are halved.

4.4 EXPERIMENTAL RESULTS

The efficiency of $ECTS$ was tested using a set of benchmark functions (two to 100 variables), which are listed in the Appendix. To avoid any misinterpretation of the optimization results, relating to the choice of particular starting points, we performed each test 100 times, starting from various randomly selected points in the hyperrectangular search domain given in the usual literature [12, 3, 5, 9, 10]. Therefore, the $ECTS$ tests are systematically performed with different initial seeds for the pseudo-random number generator, and hence different starting points.

$ECTS$ is tested and compared against other Tabu Search implementations and methods. For functions of two to six variables, we compare our results with those previously published. We accept these published results as valid ones, and we do not ourselves program the corresponding algorithms. For functions of more than six variables, there are few available results: consequently $ECTS$ results are only compared to those produced by Siarry et al. [11, 12].

The results of $ECTS$ tests performed on 20 functions are shown in Table 4.1. To evaluate the efficiency of the algorithm, we retain the following criteria summarizing results from 100 minimizations per function: the rate of successful minimizations, the average of the objective function evaluation numbers and the average error. These criteria are defined precisely below.

To determine how long $ECTS$ takes to run, we give an execution time for one function on a $Pentium$133 personal computer: the optimization of the Goldstein and Price function takes about three seconds at an average.
When at least one of the stopping tests is verified, $ECTS$ stops and provides the coordinates of a located point, and the objective function value "$FOBJ_{ECTS}$" at this point. We compare this result with the known analytical minimum "$FOBJ_{ANAL}$;" and we considere this result to be "successful" if the following inequality held: $|FOBJ_{ECTS} - FOBJ_{ANAL}| \leq \epsilon_{rel} * FOBJ_{ANAL} + \epsilon_{abs}$, where $\epsilon_{rel} = 10^{-4}$ and $\epsilon_{abs} = 10^{-6}$.

The average of the objective function evaluation numbers is evaluated in relation to only the successful minimizations. The average error is defined as the average of $FOBJ$ gaps between the best successful point found and the known global optimum.

The results presented in Table 4.1 deserve the following comments. For functions of two variables, the average of the objective function evaluation numbers does not exceed 550 with an appreciable accuracy, except for the Easom function, however, the global minimum of the Easom function laws in a very narrow "hole" of the solution space. Furthermore, for the Shekel

Test Function	*Rate of successful minimizations %*	*Average of objective function evaluation numbers*	*Average error*
RC	100	245	0.05
ES	100	1284	0.01
GP	100	231	2E-3
SH	100	370	1E-3
R_2	100	480	0.02
Z_2	100	195	2E-7
DJ	100	338	3E-8
$H_{3,4}$	100	548	0.09
$S_{4,5}$	75	825	0.01
$S_{4,7}$	80	910	0.01
$S_{4,10}$	75	898	0.01
R_5	100	2142	0.08
Z_5	100	2254	4E-6
$H_{6,4}$	100	1520	0.05
R_{10}	85	15720	0.02
Z_{10}	100	4630	2E-7
R_{50}	75	63210	0.02
Z_{50}	100	63970	2E-7
R_{100}	75	162532	0.05
Z_{100}	100	152030	1E-3

Table 4.1 Results of Enhanced Continuous Tabu Search for 20 classical test functions.

Method	**Reference**
Enhanced Continuous Tabu Search (ECTS)	this work
Continuous Tabu Search (CTS)	Siarry and Berthiau [11]
Continuous Reactive Tabu Search (CRTS min)	Battiti and Tecchiolli [2]
Continuous Reactive Tabu Search (CRTS ave)	Battiti and Tecchiolli [2]
Taboo Search (TS)	Cvijovic and Klinowski [4]
Enhanced Simulated Annealing (ESA)	Siarry and Berthiau [12]
INTEROPT (INT)	Bilbro and Snyder [3]

Table 4.2 Listing of various methods used in the comparison.

functions, $ECTS$ is sometimes trapped in a local minimum. We can avoid this by increasing the number of iterations without any detection of a promising area or by introducing an aspiration criteria to $ECTS$. In the case where the problem dimension is between five and nine, the rate of successful minimization is equal to 100 %, and the average of objective function evaluation numbers does not exceed 2300. Finally, we test two functions of more than ten variables: the Zakharov and Rosenbrock functions. The results obtained for the first one are better than the ones obtained for the second. We explain this difference by the dissymetry existing between the variables of the Rosenbrock function.

The performance of $ECTS$ is then compared to four other published versions of continuous TS and alternative algorithms, in particular Simulated Annealing. These various methods are listed in Table 4.2. The difference between $CRTS_{min}$ and $CRTS_{ave}$ is as follows: in $CRTS_{min}$, the algorithm uses the minimum $FOBJ$ value among the neighbors under consideration, whereas, in $CRTS_{ave}$, the algorithm uses the average $FOBJ$ value.

Meth *Fct*	*ECTS*	*CTS*	*CRTS* *min*	*CRTS* *ave*	*TS*	*ESA*	*INT*
RC	245	668	41	38	492	-	4172
GP	231	1636	171	248	486	783	6375
SH	370	1123	-	-	727	-	-
R_2	480	1616	-	-	-	796	-
Z_2	195	689	-	-	-	15820	-
$H_{3,4}$	548	628	609	513	508	698	1113
$S_{4,5}$	825 (.7)	4889 (.5)	664	812	-	1137 (.5)	3700 (.4)
$S_{4,7}$	910 (.8)	5110 (.6)	871	960	-	1223 (.5)	2426 (.6)
$S_{4,10}$	898 (.7)	5232 (.6)	693	921	-	1189 (.5)	3463 (.5)
R_5	2142	52733	-	-	-	5364	-
Z_5	2254	27659	-	-	-	69799	-
$H_{6,4}$	1520	2250	1245	750	2845	2638	17262

Table 4.3 Average number of objective function evaluations used by 7 methods to optimize 12 functions of less than 10 variables. The numbers in parentheses are the ratios of runs for which the algorithm found the global minimum rather than being trapped into a local minimum.

The experimental results obtained for twelve test functions of less than ten variables, using the seveb different methods, are given in Table 4.3. For each function, we give the average number of function evaluations for 100 runs. Some results are not available for some methods. The numbers in parentheses are the ratios of runs for which the algorithm found the global minimum rather than being trapped into a local minimum.

The results from $ECTS$ are far better than results from CTS, our previous version of continuous Tabu Search. $ECTS$ results are satisfactory for all of the functions. For instance, the average computing cost of $ECTS$ is equal to 231 function evaluations for the Goldstein-Price function, GP, and to 548 for the Hartmann function, $H_{3,4}$. These results are close to the best available results, obtained by Battiti and Tecchiolli [2]: for GP, 171 evaluations with $CRTS_{min}$ and 248 evaluations with $CRTS_{ave}$; and for $H_{3,4}$, 609 evaluations with $CRTS_{min}$ and 513 evaluations with $CRTS_{ave}$.

All the methods, except CRTS, occasionally failed to detect the global minimum of the Shekel functions; although the rate of successful minimizations is best for $ECTS$.

Table 4.4 shows the comparison of $ECTS$ to Continuous Tabu Search [11] and Enhanced Simulated Annealing [12] for functions of 10 to 100 variables (no data is available for other published methods for this type of function), and it appears that our algorithm provides the best results.

4.5 CONCLUSION

In this paper, we show that Tabu Search can be efficiently applied to the optimization of continuous multiminima functions. Our main contribution is the adaptation to the continuous case of two concepts widely used in combinatorial Tabu Search, namely diversification and intensification. We propose diversifica-

Method / *Function*	*ECTS*	*CTS*	*ESA*
R_{10}	15720 (.85)	263299 (.7)	12403
Z_{10}	4630	295820 (.5)	15820
R_{50}	63210 (.75)	$\geq 10E+6$	78224
Z_{50}	63970	$\geq 10E+6$	195726
R_{100}	162532 (.75)	$\geq 10E+6$	188227
Z_{100}	152030	$\geq 10E+6$	789718

Table 4.4 Average number of objective function evaluations used by 3 methods to optimize 6 functions of 10 to 100 variables. The numbers in parentheses are the ratios of runs for which the algorithm found the global minimum rather than being trapped into a local minimum.

tion for the detection of promising areas, and intensification as a search within the most promising area.

The results are satisfactory, and for functions having less than ten variables, we obtain similar or better results than the ones provided by other methods or other versions of continuous Tabu Search. We avoid performing *CPU* time costly partitions of the solution space and using approaches too sophisticated for the neighborhood structure. So, *ECTS* can be applied to functions having a large number of variables without the prohibitive increase of *CPU* time. Furthermore, that *ECTS* is a lot simpler because it is naturally derived from classical combinatorial Tabu Search.

With regard to the future, we believe that the management of the promising list and the strategy for detection of promising areas could be further improved. *CPU* time could be drastically reduced by using a parallel version of Enhanced Continuous Tabu Search. Promising areas, e.g., could be detected by several processors working simultaneously.

Appendix: List of test functions

Branin RCOS (RC) (2 variables)

$$RC(x_1,x_2) = (x_2-(5/(4.\pi^2)).x_1^2+(5/\pi).x_1-6)^2+10(1-(1/(8.\pi))).cos(x_1)+10$$

search domain : $-5 \leq x_1 \leq 10,\ 0 \leq x_2 \leq 15$
3 global minima : $(x_1,\ x_2)^* = (-\pi,\ 12.275),\ (\pi,\ 2.275),\ (9.42478,\ 2.475)$
$RC((x_1,\ x_2)^*) = 0.397887$.

Easom (ES) (2 variables)

$$ES(x_1,x_2) = -cos(x_1).cos(x_2).exp(-((x_1-\pi)^2+(x_2-\pi)^2))$$

search domain : $100 \leq x_j \leq 100, j = 1,\ 2$
1 global minimum : $(x_1,\ x_2)^* = (\pi,\ \pi)\ \ ES((x_1,\ x_2)^*) = -1$.

Goldstein and Price (GP) (2 variables)

$GP(x_1,\ x_2) = [1 + (x_1 + x_2 + 1)^2 * (19 - 14x_1 + 13x_1^2 - 14x_2 + 6x_1x_2 + 3x_2^2]*$
$[30 + (2x_1 - 3x_2)^2 * (18 - 32x_1 + 12x_1^2 - 48x_2 - 36x_1x_2 + 27x_2^2]$
search domain : $-2 \leq x_j \leq 2, j = 1,\ 2$
4 local minima;
1 global minimum : $(x_1,\ x_2)^* = (-1,\ 0)\ \ GP((x_1,\ x_2)^*) = 3.$

Shubert (SH) (2 variables)

$$SH(x_1, x_2) = \{\sum_{j=1}^{5} j cos[(j+1)x_1 + j]\} * \{\sum_{j=1}^{5} j cos[(j+1)x_2 + j]\}$$

search domain : $-10 \leq x_j \leq 10, j = 1,\ 2$
760 local minima 18 global minima : $SH((x_1,\ x_2)^*) = -186.7309.$

De Joung (DJ) (3 variables)

$$DJ(x_1, x_2, x_3) = x_1^2 + x_2^2 + x_3^2$$

search domain : $-5.12 \leq x_j \leq 5.12, j = 1, 3$
1 minimum : $(x_1,\ x_2,\ x_3)^* = (0,\ 0,\ 0)\ \ DJ((x_1,\ x_2,\ x_3)^*) = 0.$

Hartman ($H_{3,4}$) (3 variables)

$$H_{3,4}(\mathbf{x}) = -\sum_{i=1}^{4} c_i\ exp\ [-\sum_{j=1}^{3} a_{ij}\ (x_j - p_{ij})^2]$$

search domain : $0 \leq x_j \leq 1, j = 1,\ 3$
4 local minima : $\mathbf{p_i} = (p_{i1}, p_{i2}, p_{i3}) = i^{th}$ local minimum approximation;
$f((\mathbf{p_i})) \cong -c_i$.
1 global minimum : $\mathbf{x}^* = (0.11,\ 0.555,\ 0.855);\ H_{3,4}(x^*) = -3.86278.$

i	a_{ij}			c_i	p_{ij}		
1	3.0	10.	30.	1.0	0.3689	0.1170	0.2673
2	0.1	10.	35.	1.2	0.4699	0.4387	0.7470
3	3.0	10.	30.	3.0	0.1091	0.8732	0.5547
4	0.1	10.	35.	3..2	0.0381	0.5743	0.8828

Shekel ($S_{4,n}$) (4 variables)

$$S_{4,n}(\mathbf{x}) = -\sum_{i=1}^{n} [\ (\mathbf{x} - \mathbf{a_i})^T(\mathbf{x} - \mathbf{a_i}) + c_i]^{-1}$$

$\mathbf{x} = (x_1,\ x_2,\ x_3,\ x_4)^T$
$\mathbf{a_i} = (a_i^1,\ a_i^2,\ a_i^3,\ a_i^4)^T$
3 functions $S_{4,n}$ were considered : $S_{4,5},\ S_{4,7},\ S_{4,10}$
search domain: $0 \le x_j \le 10, j = 1,\ 4$
10 local minima : $\mathbf{a_i}^T = i^{th}$ local minimum approximation;
$S_{4,n}((\mathbf{a_i}^T)) \cong -1/c_i$
1 global minimum : $\mathbf{x}^* = (4,\ 4,\ 4,\ 4); S_{4,n}(\mathbf{x}^*) = -10.40294$.

i	$\mathbf{a_i}^T$				c_i
1	4.	4.	4.	4.	.1
2	1.	1.	1.	1.	.2
3	8.	8.	8.	8.	.2
4	6.	6.	6.	6.	.4
5	3.	7.	3.	7.	.4
6	2.	9.	2.	9.	.6
7	5.	5.	3.	3.	.3
8	8.	1.	8.	1.	.7
9	6.	2.	6.	2.	.5
10	7.	3.6	7.	3.6	.5

Hartman ($H_{6,4}$) (6 variables)

$$H_{6,4}(\mathbf{x}) = -\sum_{i=1}^{4} c_i\ exp\ [-\sum_{j=1}^{6} a_{ij}\ (x_j - p_{ij})^2]$$

search domain : $0 \le x_j \le 1, j = 1,\ 6$
4 local minima: $\mathbf{p_i} = (p_{i1}, \ldots, p_{i6}) = i^{th}$ local minimum approximation;
$f((\mathbf{p_i})) \cong -c_i$.

i	a_{ij}						c_i
1	10.0	3.00	17.0	3.50	1.70	8.00	1.0
2	0.05	10.0	17.0	0.10	8.00	14.0	1.2
3	3.00	3.50	1.70	10.0	17.0	8.00	3.0
4	17.0	8.00	0.05	10.0	0.10	14.0	3.2

i	p_{ij}					
1	0.1312	0.1696	0.5569	0.0124	0.8283	0.5886
2	0.2329	0.4135	0.8307	0.3736	0.1004	0.9991
3	0.2348	0.1451	0.3522	0.2883	0.3047	0.6650
4	0.4047	0.8828	0.8732	0.5743	0.1091	0.0381

Rosenbrock (R_n) (n variables)

$$R_n(\mathbf{x}) = \sum_{j=1}^{n-1}[100(x_j^2 - x_{j+1})^2 + (x_j - 1)^2]$$

5 functions were considered : $R_2,\ R_5,\ R_{10},\ R_{50}$ *and* R_{100}
search domain : $-5 \leq x_j \leq 10,\ j = 1, ..., n$
global minimum $\mathbf{x}^* = (1, ..., 1)\ R_n(\mathbf{x}^*) = 0.$

Zakharov (Z_n) (n variables)

$$Z_n(\mathbf{x}) = (\sum_{j=1}^{n} x_j^2) + (\sum_{j=1}^{n} 0.5jx_j)^2 + (\sum_{j=1}^{n} 0.5jx_j)^4$$

5 functions were considered : $Z_2,\ Z_5,\ Z_{10},\ Z_{50}$ *and* Z_{100}
search domain : $-5 \leq x_j \leq 10,\ j = 1, ..., n$
global minimum $\mathbf{x}^* = (0, ..., 0)\ Z_n(\mathbf{x}^*) = 0.$

References

[1] R. Battiti and G. Tecchiolli, The reactive tabu search. *ORSA J. Comput. 6, 126-140*, 1994.

[2] R. Battiti and G. Tecchiolli, The continuous reactive tabu search: Blending Combinatorial Optimization and Stochastic Search for Global Optimization. *Ann. Oper. Res. 63, 53-188*, 1996.

[3] G.L. Bilbro and W.E. Snyder, Optimization of Functions with Many Minima. *IEEE Trans. Syst. Man, and Cybernetics 21, 840-849*, 1991.

[4] D. Cvijovic and J. Klinowski, Taboo search. An approach to the Multiple Minima Problem. *Science 667, 664 - 666*, 1995.

[5] L.C.W. Dixon and G.P. Szeg, Towards Global Optimization 2. *North-Holland, Amsterdam*, 1978.

[6] F. Glover, Tabu search, Part I. *ORSA J. Comput. 1, 190-206*, 1989.

[7] F. Glover, Tabu search, Part II. *ORSA J. Comput. 2, 4-32*, 1990.

[8] N. Hu, Tabu search method with random moves for globally optimal design. *Int. J. Numer. Methods in Eng. 35, 1055-1070*, 1992.

[9] K. Schittkowski and W. Hock, Test examples for nonlinear programming codes. *Lecture Notes in Economics Math. Syst. 187, Springer*, 1981.

[10] K. Schittkowski and W. Hock, More test examples for nonlinear programming codes. *Lecture Notes in Economics Math. Syst. 282, Springer*, 1987.

[11] P. Siarry and G. Berthiau, Fitting of tabu search to optimize functions of continuous variables. *Int. J. Numer. Methods Eng. 40, 2449-2457*, 1997.

[12] P. Siarry, G. Berthiau, F. Durbin and J. Haussy, Enhanced simulated annealing for globally minimizing functions of many continuous variables. *ACM Trans. Math. Software 23, 209-228*, 1997.

5 LOCAL SEARCH IN CONSTRAINT PROGRAMMING: EXPERIMENTS WITH TABU SEARCH ON THE VEHICLE ROUTING PROBLEM

Bruno De Backer and Vincent Furnon

ILOG
9 rue de Verdun, BP85 94253 Gentilly Cedex, France.
debacker@ilog.fr, furnon@ilog.fr

Abstract: Constraint Programming has proved to be useful for solving many combinatorial problems. In particular, it provides a language allowing to easily model complex problems, and to describe a way to solve them. However, it has seldom been used to solve Vehicle Routing Problems (VRP). Indeed, traditional Constraint Programming systems explore the search space in a depth-first fashion. On the other hand, local search methods have proved to be efficient for solving VRP. In this article, a constraint programming model for the VRP is described, along with a method allowing to perform local non-monotonic search within a Constraint Programming framework. This framework has been used to implement Tabu Search meta-heuristics. The implementation choices are described, as well as the results of experiments performed on problems from the literature.

5.1 INTRODUCTION

Problems involving the routing of vehicles, whether they are represented mathematically as Traveling Salesperson Problems, Vehicle Routing Problems or Pickup and Delivery Problems are very common in the operations of many industries. The mathematical models of real-life problems actually include a number of important side constraints, including precedence constraints, compatibility constraints, multiple capacity constraints or complex time windows.

Both the economic importance of such problems and the complexity of their expressions make it necessary to design a tool generic enough to describe the problems in their entirety. Constraint Programming has proved to be a powerful framework for solving complex combinatorial problems. Moreover, the rich set of constraints already available can be extended to meet the specifications of the real problem.

It is therefore natural to try to use Constraint Programming for solving VRPs. However, this raises many problems. Search in Constraint Programming is usually based on depth-first search (although some work has already been carried out on integrating local search in Constraint Programming [9]). This means that the domains of each variable are monotonically reduced by propagation during the search. Although this approach can be useful for finding a first solution to a VRP, it is not practicable when an optimized solution is sought. This is the reason why much research has been devoted to the design and the implementation of local search techniques in the context of routing problems.

In this article basic principles for implementing local search techniques and meta-heuristics in Constraint Programming are presented. These principles have been applied to Tabu Search, with experimental results on various symmetric TSPs and various VRPs (with capacities, time windows, deadlines).

This article is organized as follows: After describing the types of VRP considered here, a general Constraint Programming model for the VRP is given in Section 5.3 that allows the description of various routing problems, ranging from TSPs to PDPs. In particular, this model allows to take into account various side constraints expressed as standard Constraint Programming constraints. The solving method is then described in Section 5.4, which consists in an adaptation of the standard two-phase approach. A feasible solution is first computed using a basic construction heuristic. It is subsquently improved using local search. A way to implement local search in Constraint Programming is thus given, along with some algorithmic improvements specific to the case of the VRP. Sections 5.5 and 5.6 are devoted to the main choices that were made with respect to the Tabu Search metaheuristic. Finally, experimental results are given in Section 5.7 that show the performance of the proposed approach.

5.2 VEHICLE ROUTING PROBLEMS

Vehicle Routing Problems (VRP) are an extension to the well-known Traveling Salesperson Problem. A number of visits being given, the goal is to perform these visits with vehicles, using a set of minimal-cost tours, each of which must start and end at the same position.[1] The VRP is, like the TSP, an NP-hard problem.

If no extra constraints on the capacity of vehicles, a maximum driving time or tour length are given, the solution to a VRP would be a single tour. However, real-life VRP come precisely with these kinds of constraints, or even more complex ones. For example, multiple capacity constraints can be expressed in

[1] For the sake of clarity, the multiple-depot VRP is not considered in this article.

several units or *dimensions* (weight, volume, length, number of pallets, *etc.*). Some problems involve constraints where the total capacity of the vehicle cannot be used; instead, the loading of the vehicle must follow specific rules or legislation. This is for example the case in Europe with oil tanks.

Time is also an important aspect. Customers may request to be served during specific time windows. There may be one time window or many disjoint time windows for each customer. Depots, parcels, goods, roads, drivers, and vehicles can also have time windows during which they are available. Finally, many kinds of compatibility constraints exist: some vehicles cannot access some customer sites, some parcels have to be delivered by a certain kind of vehicle, *etc.*

Expressing such constraints can be tedious, and yield problems with huge models, especially in the case of traditional linear programming (LP) models, or make programs solving VRP very complex and difficult to maintain. The next section shows how to design a model that circumvents these two problems.

5.3 A CONSTRAINT PROGRAMMING MODEL FOR THE VRP

5.3.1 Decision variables

A set of decision variables of a model is a minimal subset of the variables in the model such that knowing the values of these decision variables enables to reconstruct the whole solution to the problem.

In standard LP models (see for example [13]), decision variables usually belong to a set of Boolean variables x_{ij}^k which take the value 1 if vehicle k is used to travel from visit i to visit j. Therefore, these models use $O(mn^2)$ decision variables, where m is the number of vehicles and n is the number of visits to perform.

It is actually possible to describe a VRP with a number of variables that is linear (instead of quadratic) with respect to the number of visits. For this purpose, each visit i is associated two *finite-domain* variables $next_i$ and veh_i, representing respectively the possible visits following i and the vehicle serving visit i. $next_i = j$ if visit j immediately follows visit i in the tour, and $veh_i = veh_j$ if $next_i = j$. Two visits per vehicle are introduced to represent visits to the depot. Let N denote the set of indices corresponding to (regular) visits, S denote the set of indices of starting visits (*i.e.* the ones corresponding to departures from the depot), and E denote the set of indices of ending visits (returns to the depot). $next_i = j$ is possible iff $i \in N \cup S$ and $j \in N \cup E$. This ensures the conservation of the flow of vehicles through each customer and the depot, and, together with $(next_i = j \wedge veh_i = k) \Rightarrow veh_j = k$, is equivalent to the set of linear constraints $\forall i \in V\ \forall k \in M \sum_{j \in V} x_{ij}^k = \sum_{j \in V} x_{ji}^k = y_{ik}$, where $V = N \cup E \cup S$. This shows the compactness of the representation, both in terms of number of variables and number of constraints.

5.3.2 *Path Constraints*

After performing the choice of decision variables, it is necessary to devise a generic way of taking into account constraints on dimensions that can be as diverse as weight, time, or volume. For this purpose, the notion of *path constraint* [7] is introduced. Path constraints are able to propagate accumulated quantities such as time and weight along a vehicle tour. This constraint, denoted as $path(N, S, E, \sigma, f)$ can be formally described as: $\forall (i, j) \in (S \cup N) \times (N \cup E)) next_i = j \Rightarrow \sigma_i + f(i, j) \leq \sigma_j$, where the σ_i are variables taking their values over a real interval denoting the accumulated quantity at visit i.

5.3.3 *Using Path Constraints for capacitated VRP with time windows*

This expression, although simple, is very versatile, since the function $f(i, j)$ can take different forms:

- In the case of *capacity constraints*, a constraint of the form $path(N, S, E, w, Q)$ with $Q(i, j) = q_i$, (it depends only on i) can be used. In this case q_i denotes the quantity (expressed in terms of weight, volume, *etc.*) to be delivered at visit i and w_i corresponds to the load at visit i. Moreover, simply adding bounds on the w_i variables $\forall i \in N \cup S \cup E \; 0 \leq w_i \leq C$, will ensure that the capacity is not exceeded.

- In the case of time, a constraint of the form $path(N, S, E, \tau, T)$ with $T(i, j) = s_i + t_{i,j}$ can be used. Here, s_i represents the service time at i and $t(i, j)$ the travel time between i and j. τ_i then represents the arrival time at i. The path constraint therefore accumulates the arrival times along the tours. A simple time window can then be written as $e_i \leq \tau_i \leq l_i$. Defining disjoint time windows is also very easy. For example, ILOG Solver allows constraints such as $(e_i \leq \tau_i \leq l_i) \vee (e'_i \leq \tau_i \leq l'_i)$ to be written without introducing extra variables.

- The special case $path(N, S, E, c, U)$ where $\forall i \; \forall j \; U(i, j) = 1$ is also useful. Here, c_i corresponds to the number of visits performed before visit i. Constraining these c_i allows to state easily precedence constraints. Another interesting use of this version of the path constraint is to forbid cycles, by stating $\forall i \in E \; c_i = 0$ and $\forall i \in N \; c_i <= |N| + 1$.

5.3.4 *Cost functions*

In many VRPs, the goal is to minimize $\sum_{i \in E} \tau_i$ (the total duration of the tours). More complicated cost functions can also be taken into account, such as ones including the number of vehicles.

5.4 LOCAL SEARCH IN CONSTRAINT PROGRAMMING

Some work has already been devoted to the solving of TSP and VRP is Constraint Programming. [1] proposed a way to solve quickly small TSPs using a complete approach based on implicit enumeration and Branch and Bound. However, such an approach will not work for problems with more than 30 visits. Our goal is rather to be able to find good solutions to problems with 1000 or more visits.

5.4.1 A two-phase approach

A classical two-phase solving mechanism was therefore chosen. A first solution is first generated using classical first solution generation heuristics such as the savings heuristic [2] or the nearest insertion heuristic. The description of the implementation of these heuristics is beyond the scope of this article. This solution, whose quality is often quite good, is subsequently optimized using local search techniques guided with metaheuristics.

This scheme, although common in the OR community, has not been used a lot in the Constraint Programming community. Indeed Constraint Programming relies on depth-first search. This means that decisions can only be reconsidered in reverse order, and this makes it difficult to implement a local search approach where any decision (i.e. the values of a variable) can be changed at any time.

5.4.2 Decision Variables and Heuristics

The idea to implement local search in Constraint Programming uses the fact that a solution can be completely described by the values of the decision variables of the problem. Therefore a simple process to implement Local Search can be the following:

- generate a new solution using a move operator
- set the decision variables affected by the move operator to their new values
- set the remaining decision variables to their saved values
- let the constraints propagate
- if the propagation fails, the new solution is refused. Otherwise, the solution is accepted, and the decision variables of the new solution are saved
- the domains of variables are restored to their initial values on backtracking

5.4.3 The case of the VRP

In the VRP model, the decision variables are the variables $next_i$ associated to each visit i (knowing the value of the $next$ variable of the visit associated to the departure of a vehicle allows to propagate the veh_i variables along the tour). The move operators modify the position of visits in a tour or move them from one tour to another. To modify the solution, the $next$ variables are set to their saved values except for those that the move operator modifies, which are set to their new values. If the new solution is feasible and satisfies the criteria of the metaheuristic, it is accepted (the decision variables are saved). The solver then backtracks regardless of whether the new solution was accepted or rejected.

In the implementation described here, both intra-route move operators (2-opt, Or-opt) and inter-route move operators (cross, exchange, relocate) are used; note that inter-route operators can also be used intra-route. Actually, generalized versions of the inter-route operators to clusters of nodes have been implemented.

5.4.4 Reducing the Number of Constraint Checks

The difficulty with the generic approach described above is that all the *next* variables have to be set at each step. For the VRP, this time-consuming behaviour can be improved. When a visit is moved from one position to another, the feasibility of the new solution can be tested by only looking at the one or two tours involved in the modification.[2]

Another improvement can be applied which should be generally applicable to local search within any application. Before testing if the move is feasible, the variation of cost can be computed. In the case where the cost of a solution for the VRP is its total length, the variation of cost between two solutions is the difference between the sum of the lengths of the added arcs and the sum of the lengths of the arcs removed. In the case of a descent approach, if the cost value increases, the move will not be accepted and the checking phase is not necessary.

Knowing time or capacity windows can also be used to filter out infeasible solutions. It is possible to use the results of propagation to derive filters similar to those described in [14], the main difference lying in the fact that *all* the constraints (including side constraints) are taken into account to tighten the bounds of the time variables, thanks to the generic propagation scheme provided by Constraint Programming.

Once a solution is set (all *next* variables are bound, i.e. their values are fixed), the lower and the upper bounds of the σ variables are known (due mainly to path constraints). In the case of the capacitated VRP with time windows, σ_i can either be the arrival time or the load of the vehicle at visit i. These bounds are saved after each successful move to be used in testing the feasibility of subsequent moves.

As an example, such a filter for a move operator which inserts visit k between visit i and visit j works as follows, supposing a capacitated vehicle routing problem with time windows but with no particular side-constraint:

- σ_i is the arrival time at visit i
  Let $tmin_i$ be the saved propagated lower bound of σ_i and $tmax_j$ denote the saved propagated upper bound of σ_j. Let $Tmin_k$ and $Tmax_k$ be the

[2]This is true as long as there are no side constraints that act between a tour involved in the move and one not involved. When this is not the case, all tours which may be affected by the move (via propagation on the side constraints) should be examined.

Visits	No. of Full Constraint Tests			Time (sec)		
	basic	+ cost filter	+ insertion filter	basic	+ cost filter	+ insertion filter
100	451442	8875	6926	112.7	5.0	4.2
500	3634385	74978	35007	N/A	98.5	56.0

Table 5.1 Application of performance enhancement techniques to search

initial bounds of σ_k. The new solution is not feasible if $tmin_i + s_k + t_{ik} > Tmax_k$ or $Max(tmin_i + s_k + t_{ik}, Tmin_k) + s_j + t_{kj} > tmax_j$.

- σ_i is the load of the vehicle at visit i
  Let $wmin_i$ and $wmax_i$ be the saved propagated bounds of σ_i. In the case where only pickups or only deliveries are considered, the new solution is infeasible if $q_k > wmax_i - wmin_i$. ($\forall(i,j) wmax_i - wmin_i = wmax_j - wmin_j$ represents the slack at visit i.)

Similar filters have also been implemented in the case of an exchange of visits or parts of tours.

Table 5.1 shows results that were obtained when applying the set of techniques described above to two capacitated VRPs with time windows, one with 100 visits, the other with 500 visits. There are three cases to consider. The basic case performs full constraint testing after each attempted neighbourhood move. The case with a cost filter performs a full constraint test if the cost is reduced. The final case performs a cost check and a constant time insertion test which both must succeed before performing a full constraint test. From these results, it appears that checking the cost improvement before performing any constraint check provides a very important speedup, but most importantly that the constant-time insertion test becomes more interesting for larger problem sizes.

5.5 TABU SEARCH

Two-phase procedures as presented in this article have the following drawback: If the improvement procedures used in the second phase only consider moves which improve a solution, either by choosing moves which improve the solution the most (best descent) or by randomly accepting improving moves (random descent), it eventually gets stuck in a local minimum.

To circumvent this problem, many methods, known as meta-heuristics, have been devised that help the search to climb out of such minima. The following paragraphs focus on Tabu Search [5, 6]. Tabu Search is indeed very appealing to our approach for two main reasons. It has proved to be one of the most effective metaheuristics when dealing with vehicle routing problems. Moreover, it is easy to implement.

Tabu Search can be described as shown in Figure 5.1.

Here, M($\mathcal{S}$) denotes the set of moves applicable to $\mathcal{S}$ and m($\mathcal{S}$) denotes the application of a certain move m to $\mathcal{S}$. Therefore, the neighborhood of $\mathcal{S}$ is the

```
S := InitialSolution()
TabuList := ∅
while not StoppingCondition() do
    Choose m ∈ M(S) such that
        m ∉ TabuList and m(S) has the best cost
    Stop if there are no such moves
    Insert m in TabuList for T iterations
    Update TabuList
    S:= m(S)
end
```

Figure 5.1 Tabu Search algorithm

set of the solutions attainable by applying the m in M($\mathcal{S}$) to $\mathcal{S}$. InitialSolution() is a provided function that computes an initial solution to the problem, while StoppingCondition() will return true if the number of iterations has been exhausted, for example.

Tabu Search moves out of local minima by preventing the search from choosing moves which lead back to solutions that have been explored in a recent past. These unauthorized moves are called *Tabu* moves. A move remains tabu during a certain number of iterations. To tell whether a move is tabu or not requires a fixed-size queue, known as the *Tabu list*, in which the tabu moves are kept for a given amount of time.

Tabu Search is usually enhanced with diversification and intensification procedures [5, 6] which try to focus the attention of the search on the interesting parts of a problem. To prevent the search from staying in the same region of the search space, diversification tries to pull the search to other variables (or region). An example of diversification can be to penalize the features of a problem that has not changed for a while so that the search starts modifying them. Intensification has an opposite goal: it prevents the search from skipping by interesting variables without fully exploring the solutions they lead to.

5.6 TABU SEARCH FOR THE VRP

Many papers have already dealt with the VRP and Tabu Search, for example in [16, 4, 17]. The method described in this paper derives from the one of [11]. Tabu search is well suited for the VRP because the features of a solution are easy to identify: they are the arcs present in the solution. Therefore moves can be characterized by a set of added arcs AA and a set of removed arcs RA. For example, the Relocate move operator removes three arcs and adds three new arcs. Two Tabu Lists can then be created, one for storing the added arcs (denoted as ATL) and the other for storing the removed arcs (denoted as RTL). Their size corresponds to the number of arcs they contain and not the number of moves. For example, Relocate will add three arcs to both lists. A move is considered tabu if $card(AA \cap RTL)) + card(AR \cap ATL)) > threshold$, where *threshold* is a parameter. The smaller *threshold* is, the more moves will be tabu.

One can notice that this threshold has a diversification effect on the search; moves that have never been used can be tabu just because they have arcs in common with the ones in the lists. Intuitively, this pulls the search out of certain regions that have been explored intensively.

For diversification, we used a scheme derived from Guided Local Search [8]. The idea is to penalize the cost of the most commonly used arcs in a VRP solution, and to perform the search on this penalized cost. More precisely, an integer $p(a)$, initially equal to zero is associated to each arc a. Let $c(a)$ and $c^*(a)$ denote the cost and modified cost of a, respectively. A priority $\pi(a)$ is then defined as:

$$\pi(a) = \frac{c(a)}{1 + p(a)}$$

Let also S denote the set of arcs present in a given solution. Then, at each iteration:

$$\forall a \in S \; s.t. \; \pi(a) = \max_{a \in S}\{\pi(a)\} \tag{5.1}$$

$$\text{increment } p(a) \tag{5.2}$$

$$c^* = c(1 + \gamma p(a)) \tag{5.3}$$

Where γ is a parameter, usually equal to 0.15 or 0.20. Therefore, each time a costly and seldom penalized arc appears in a solution, its cost gets penalized by 15% or 20%, thus making it less attrative. This tends to remove this arc from the solution in favor of other arc whose unmodified cost might be greater. This results in diversifying the search.

It may happen, that a tabu move yields a solution whose cost is better than the best cost found so far. In this case, a classical aspiration criterion is used, that accepts such a move, even if it is considered as tabu.

Finally, it is worthwhile to notice that the proposed approach does not consider unfeasible solutions (ones that do not respect all the constraints) like it is done in [4], but it could be extended in that direction in order to reach unconnected regions in the search space.

5.7 EXPERIMENTAL RESULTS

5.7.1 Experimental conditions

The proposed approach was implemented within ILOG Dispatcher, an add-on module to ILOG Solver dedicated to routing problems. Its versatility was tested by solving several benchmark problems with the same basic code. The length of the tabu lists is 30 arcs. All results given are found after generating a first solution with a Savings heuristic derived from [2], improving it with a greedy local search (in order to obtain a good solution quickly), then performing 3000 iterations of Tabu Search.

Five move operators (2-opt, Or-opt, Relocate, Exchange and Cross) have been used to explore the neighborhood of a solution, in the greedy descent, while Tabu Search used only Relocate, Exchange and Cross. Table 5.2 gives the thresholds used for the different move operators. The first column is the name of the move operator, the second one the number of arcs added and removed by this operator and the third the threshold chosen. Tuning these thresholds of course influences the quality of the results and the speed at which they are achieved. Notice also that the number of arcs modified in 2-opt is variable because the inverted part of the tour may be of any length.

Operator	arcs	threshold
cross	4	3
exchange	8	6
relocate	6	5
2-opt	*	3
Or-opt	6	5

Table 5.2 Thresholds for different move operators

It is also very important to notice that all results given thereafter were found using double-precision floating-point numbers.

5.7.2 The Capacitated Vehicle Routing Problem with Time Windows

The Solomon [15] benchmarks include both time windows and capacity constraints. There are three sets of problems in these benchmarks: in C1 and C2 the visits are clustered, in R1 and R2 the visits are randomly distributed in the plane and in RC1 and RC2 the situation is in between. Problems of type 1 have a shorter planning horizon, and need more vehicles than problems of type 2. The goal of the Solomon benchmarks is first to minimize the number of vehicles, then to minimize the number of vehicles. To achieve that, we ran the programs by giving it the best number of vehicles found in the literature. The savings heuristic, however, uses more vehicles than this number. To allow the use of the savings heuristic, a number of fake vehicles was added. When one of these fake vehicles serves a visit, the cost is penalized by 1000. Because of this high cost, Tabu Search will tend to have all the visits served using standard vehicles, and thus to minimize the number of vehicles. This approach also allows the solving process to go through unfeasible solutions.

Tables 5.3 and 5.4 show the results compared to the best results published. The colmun TS shows the results obtained with a simple Tabu Search with no diversification, while the column GTS shows the results obtained using Tabu Search, diversified by penalizing the cost function in the manner of Guided Local Search. The first number indicates the number of vehicles, while the second one indicates the length of the tour.

The best solutions use 410 vehicles in total, for a total length of 56611.6. At 462 vehicles and a length of 59667.97, simple Tabu Searh is far from the best result. On the other hand, the diversification provided in GTS performed very well since it allowed to reduce the number of vehicles to 421 and the total length to 57267.45. Moreover, several best results (particularly in the C class) were matched or approached (marked with an ^{m}) and five new best results were found (marked with a ^{b}).

As far as speed is concerned, 3000 iterations of GTS are performed in about 50 minutes on a 200 MHz Pentium Pro PC. For all classes of problems, except RC1, solutions within 6% of the best solution found by our method are found within 5 minutes. For RC1, the running time for getting good solutions is about half an hour.

Pb	Reference	Best	TS	GTS	
C101	[3]	10/827.3	10/828.94	10/828.94	^{m}
C102	[3]	10/827.3	10/834.64	10/829.86	
C103	[12]	10/828.06	10/872.25	10/828.99	
C104	[12]	10/824.78	10/865.5	10/825.87	
C105	[10]	10/828.94	10/828.94	10/828.94	^{m}
C106	[3]	10/827.3	10/828.94	10/828.94	^{m}
C107	[3]	10/827.3	10/828.94	10/828.94	^{m}
C108	[3]	10/827.3	10/828.94	10/828.94	^{m}
C109	[10]	10/828.94	10/828.94	10/828.94	^{m}
R101	[3]	18/1607.7	21/1707.53	19/1652.18	
R102	[3]	17/1434	19/1502.58	17/1498.88	
R103	[18]	13/1207	15/1261.84	14/1235.02	
R104	[12]	10/982.01	11/1016.93	10/992.46	
R105	[12]	14/1377.11	16/1378.67	14/1431.31	
R106	[12]	12/1252.03	14/1284.71	12/1259.46	
R107	[17]	10/1126.69	13/1174.3	11/1094.36	
R108	[17]	9/968.59	11/994.68	9/971.98	
R109	[17]	11/1214.54	13/1190.42	12/1163.95	
R110	[12]	11/1080.36	12/1155.63	11/1100.36	
R111	[17]	10/1104.83	12/1131.81	11/1086.48	
R112	[12]	10/953.63	11/1045.83	10/985.91	
RC101	[18]	14/1669	18/1766.54	15/1718.14	
RC102	[17]	12/1554.75	16/1636.1	13/1537.73	
RC103	[18]	11/1110	13/1394.23	11/1303.05	
RC104	[12]	10/1135.83	11/1187.47	10/1144.86	
RC105	[17]	13/1643.38	18/1670.73	15/1570.05	
RC106	[17]	12/1448.26	14/1483.10	12/1425.60	^{b}
RC107	[17]	11/1230.54	13/1381.05	11/1238.7	
RC108	[17]	10/1139.82	11/1226.87	10/1218.04	

Table 5.3 Results for the Solomon benchmarks of type 1

5.8 CONCLUDING REMARKS

We have proposed a simple framework for local search in Constraint Programming. This framework is both simple and general. All that is necessary is to identify the decision variables of a given problem. Moreover, it allows to implement many different meta-heuristics easily. For example the implementation

Pb	Reference	Best	TS	GTS	
C201	[10]	3/591.56	3/591.56	3/591.56	m
C202	[10]	3/591.56	3/591.56	3/591.56	m
C203	[12]	3/591.17	3/617.62	3/591.17	m
C204	[10]	3/590.6	3/778.95	3/590.6	m
C205	[10]	3/589	3/588.88	3/588.88	m
C206	[10]	3/588	4/615.52	3/588.49	m
C207	[10]	3/588	3/588.29	3/588.29	m
C208	[10]	3/588	3/588.32	3/588.32	m
R201	[17]	4/1254.8	4/1298.84	4/1253.23	b
R202	[17]	3/1214.28	4/1165.69	4/1101.91	
R203	[12]	3/948.74	3/1169.88	3/956.16	
R204	[12]	2/869.29	3/796.92	2/880.51	
R205	[17]	3/1038.72	3/1111.58	3/1019.4	b
R206	[18]	3/833	3/1070.41	3/940.27	
R207	[12]	3/814.78	3/866.19	3/824.95	
R208	[12]	2/738.55	3/759.60	2/740.20	
R209	[18]	3/855	3/963.36	3/926.96	
R210	[12]	3/967.6	4/1011.4	3/982.79	
R211	[17]	2/923.8	3/820.97	3/803.52	
RC201	[18]	4/1294	4/1469.37	4/1433.76	
RC202	[17]	4/1164.25	4/1329.58	4/1161.61	b
RC203	[12]	3/1079.57	3/1246.29	3/1093.73	
RC204	[12]	3/806.75	3/931.31	3/812.96	
RC205	[17]	4/1328.21	4/1426.73	4/1301.49	b
RC206	[17]	3/1158.81	4/1193.26	3/1228.92	
RC207	[17]	3/1082.32	4/1023.44	3/1122	
RC208	[12]	3/833.97	3/916.45	3/873.29	

Table 5.4 Results for the Solomon benchmarks of type 2

of Tabu Search in ILOG Dispatcher takes less than 150 lines of C++ code, including the code managing the Tabu Lists. It has also been shown that, in order to make this framework usable, it is necessary to include some knowledge of the problem at hand, such as constant-time insertion in the case of the VRP.

A Constraint Programming model for the VRP was also given, that allows to describe many different kinds of routing problems along with their side constraints. This model has proved usable for describing many benchmarks in the literature as well as real-world applications.

This yielded an implementation of Tabu Search with a Constraint Programming tool for solving VRPs. The advantages of Constraint Programming in terms of ease of description of a problem are kept, and the approach also proved to be efficient and to give promising results on a number of problems. Dispatcher is currently used on real-world problems, including simple capacitated VRP, VRP with Time Windows and various other constraintss.

Obviously, there is room for improvement in this approach. This include simple things such as tuning parameters, better first-solution generation heuristics, and improvement of the Tabu approach to reduce the number of vehicles. This may result both in a gain in performance, by reducing the number of iterations necessary to reach a solution, and in a gain on the quality of the solutions.

Acknowledgement

The production of this paper was supported by the GreenTrip project, a research and development undertaking partially funded by the ESPRIT Programme of the Commission of the European Union as project number 20603. The partners in this project are Pirelli (I), ILOG (F), SINTEF (N), Tollpost-Globe (N), and University of Strathclyde (UK).
WWW—`http://www.si.sintef.no/GreenTrip`

References

[1] Y. Caseau and F. Laburthe. Solving small TSPs with constraints. To appear in ICLP-97, MIT Press., 1997.

[2] G. Clarke and G. W. Wright. Scheduling of vehicles from a central depot to a number of delivery points. *Operations Research*, 12:568–581, 1964.

[3] M. Desrochers, J. Desrosiers, and M. M. Solomon. A new optimization algorithm for the vehicle routing problem with time windows. *Operations Research*, 40:342–354, 1992.

[4] M. Gendreau, A. Hertz, and G. Laporte. A tabu search heuristic for the vehicle routing problem. *Management Science*, 40:1276–1290, 1994.

[5] F. Glover. Tabu search – part I. *ORSA Journal on Computing*, 1:190–206, 1989.

[6] F. Glover. Tabu search – part II. *ORSA Journal on Computing*, 2:4–32, 1990.

[7] ILOG. Solver 4.2 manuals. 1997.

[8] P. Kilby, P. Prosser and P. Shaw. Guided local search for the vehicle routing problem with time windows. This volume.

[9] G. Pesant and M. Gendreau. A view of local search in constraint programming. In *CP-96*. Springer-Verlag, 1996.

[10] J.-Y. Potvin and S. Bengio. The vehicel routing problem with time windows - part ii: Genetic search. *INFORMS J. Comput.*, 8:165–172, 1996.

[11] V.M. Pureza and P.M. Franca. Vehicle routing problems via tabu search metaheuristic. Technical Report CRT-747, Centre de Recherche sur les Transports, University of Montreal, 1991.

[12] Y. Rochat and E.D. Taillard. Probabilistic diversification and intensification in local search for vehicle routing. *Journal of Heuristics*, 1:147–167, 1995.

[13] M.W.P. Savelsbergh. Computer aided routing. Centrum voor Wiskunde en Informatica, Amsterdam, 1988.

[14] M.W.P. Savelsbergh. An efficient implementation of local search algorithm for constrained routing problems. *European Journal of Operational Research*, 47:75–85, 1990.

[15] M.M. Solomon. Algorithms for the vehicle routing and scheduling problem with time window constraints. *Operations Research*, 35:254–265, 1987.

[16] E. Taillard. Parallel iterative search methods for vehicle routing problems. *Networks*, 23:661–676, 1993.

[17] E.D. Taillard, P. Badeau, M. Gendreau, F. Guertin, and J.-Y. Potvin. A tabu search heuristic for the vehicle routing problem with soft time windows. *Transportation Science*, 31:170–186, 1997.

[18] S.R. Thangiah, I.H. Osman, and T. Sun. Algorithms, simulated annealing and tabu search methods for vehicle routing problems with time windows. Technical Report UKC/OR94/4, Institute of Mathematics & Statistics, University of Kent, Canterbury, UK, 1994.

6 TABU SEARCH FOR GRAPH COLORING, T-COLORINGS AND SET T-COLORINGS

Raphaël Dorne and Jin-Kao Hao

LGI2P, EMA-EERIE, Parc Scientifique Georges Besse, F-30000 Nimes, France.

dorne@eerie.fr, hao@eerie.fr

Abstract: In this paper, a generic tabu search is presented for three coloring problems: graph coloring, T-colorings and set T-colorings. This algorithm integrates important features such as greedy initialization, solution re-generation, dynamic tabu tenure, incremental evaluation of solutions and constraint handling techniques. Empirical comparisons show that this algorithm approaches the best coloring algorithms and outperforms some hybrid algorithms on a wide range of benchmarks. Experiments on large random instances of T-colorings and set T-colorings show encouraging results.

6.1 INTRODUCTION

The graph coloring problem is one of the most studied *NP-hard* problems and can be defined informally as follows. Given an undirected graph, one wishes to color the nodes of the graph with a minimal number of colors in such a way that two colors assigned to two adjacent nodes must be different, i.e., have a minimal distance greater than zero. Graph coloring has many practical applications such as timetabling and resource assignment. Given the NP-completeness of the coloring problem, it is natural to design heuristic methods. Indeed many heuristic methods have been developed: constructive methods in the 1960s and 1970s [1, 23], local search meta-heuristics in the 1980s and 1990s [15, 3, 18, 10] and genetic or population-based local search methods in the 1990s [9, 5, 24, 6]. Moreover, there are a large number of well-known benchmarks for evaluating and comparing different algorithms.

Several extensions of this classical graph coloring problem exist allowing more applications to be embraced. T-colorings and set T-colorings are two important extensions allowing one to model the frequency assignment problem. In T-colorings, the forbidden separation distance for colors assigned to adjacent nodes is no longer limited to the singleton $\{0\}$, but may be any set of positive integers. In set T-colorings, a node may receive several colors verifying some forbidden separation distances. There is extensive literature on the application side, i.e., the frequency assignment problem, including studies on specific heuristic algorithms [13, 12, 22], meta-heuristic algorithms [20, 8, 2, 21, 14] and lower bounds [11, 16]. On the contrary, studies on heuristic methods for the general problems of T-colorings and set T-colorings are much limited. Costa experimented with some known methods such as *Dsatur*, tabu search, and simulated annealing for T-colorings [4]. Jiang studied various methods including greedy, dynamic ordering, and tabu search as well as some combinations of these methods for set T-colorings [17]. Until now, there are very few well-established benchmarks available for evaluating algorithms for T-colorings and set T-colorings.

In this paper, we present a generic tabu search algorithm for these three coloring problems. To evaluate its performance, we use well-known benchmarks for graph coloring and introduce a set of random instances for T-colorings and set T-colorings. Moreover, we make these instances (and our instance generator) available to other researchers in the hope that these instances may serve as benchmarks for further studies.

The paper is organized as follows. Section 6.2 defines the three families of coloring problems. Section 6.3 introduces our random instance generator for the T-colorings and set T-colorings problems. Section 6.4 presents our generic tabu search algorithm for these coloring problems. Section 6.5 shows experimental results on a wide range of random instances. Section 6.6 gives some conclusions.

6.2 COLORING PROBLEMS

6.2.1 Graph coloring

Given an undirected graph $G{=}(V,E)$ with $V{=}\{v_1,...,v_N\}$ the set of nodes and $E{=}\{e_{ij}|\ \exists$ an edge between v_i and $v_j\}$ the set of edges. The graph coloring problem is to determine a partition of V with a *minimum* number of color classes $C_1, C_2, ..., C_k$ such that for each edge $e_{ij} \in E$, v_i and v_j are not in the same color class [25]. Given $c(v_i)$ which is the color (a positive integer) assigned to the node v_i, a proper coloring must meet the following *color constraint*:

$$\forall e_{ij} \in E, \quad |c(v_i) - c(v_j)| \neq 0 \tag{6.1}$$

The *chromatic number* $\chi(G)$ corresponds to the smallest value of k such that G is *k-colorable*.

6.2.2 T-colorings

Given an undirected graph $G{=}(V,E)$ as above, a collection of sets $T{=}\{T_{ij} \in \mathbb{N}|$ for each $e_{ij} \in E\}$ is now defined to determine for each edge e_{ij} the color separations which are not allowed between the nodes v_i and v_j. Each T_{ij} is a set of unsigned integers such as $\{0, 2, 4, 7\}$, and the *color constraint* to be met is:

$$\forall e_{ij} \in E, \quad |c(v_i) - c(v_j)| \notin T_{ij} \tag{6.2}$$

The separation of colors assigned to two adjacent nodes v_i and v_j must be different from those of T_{ij}. A T-coloring of a graph is a partition of V in different color classes $C_1, C_2, ..., C_k$ such that the property (6.2) is verified for each edge of G. The *chromatic number* $\chi_T(G)$ corresponds to the minimum number of different color values used to color G. The *span* of a T-coloring is the difference between the smallest and the highest color values needed to obtain the T-coloring of G. The *T-colorings problem* is to determine the *minimum span* $sp_T(G)$ for all the possible colorings of G [13].

If each $T_{ij} \in T$ is a set of consecutive integers of the form $T_{ij} = \{0,1,2,...,$ t_{ij} -1$\}$, the *restricted T-colorings problem* can be defined where constraint (6.2) becomes:

$$\forall e_{ij} \in E, \quad |c(v_i) - c(v_j)| \geq t_{ij} \tag{6.3}$$

It is easy to see that the graph coloring problem is a special case of the T-colorings problem where all $T_{ij} = \{0\}$.

$sp(G)$=$\chi(G) - 1$.

6.2.3 Set T-colorings

Given an undirected graph G=(V, E) and a collection of sets T as above, a set of *demand* values D=$\{d_1, ..., d_N | \ d_i \in I\!N\}$ is now added for each node v_i corresponding to the number of colors required by the node that is:

$$\forall v_i \in V \quad (d_i = l \implies c(v_i) = \{c_{i,1}, ..., c_{i,l}\}) \tag{6.4}$$

And for each set of colors assigned to the same node, a set of forbidden separations called *co-node constraints*, is also defined:

$$\forall c_{i,m}, c_{i,n} \in c(v_i), m \neq n \quad |c_{i,m} - c_{i,n}| \notin T_{ii} \tag{6.5}$$

Finally, the color constraint between two adjacent nodes can be stated as:

$$\forall e_{ij} \in E, \forall c_{i,m} \in c(v_i), \forall c_{j,n} \in c(v_j) \quad |c_{i,m} - c_{j,n}| \notin T_{ij} \tag{6.6}$$

The *problem of set T-colorings* of G consists of finding a coloring such that the properties (6.4), (6.5), (6.6) are satisfied for the graph. The *chromatic number* $\chi_T^D(G)$ corresponds to the minimum number of different color values used to color G. The *span* is the difference between the smallest and the highest color values needed to obtain such a coloring of G. The *problem of set T-colorings* is the optimization problem of finding the *minimum span* $sp_T^D(G)$

of the graph G [17]. As stated above, if each T_{ij} is a set of consecutive integers, the *restricted set T-colorings* problem can be defined by replacing constraints (6.5) and (6.6) with:

$$\forall c_{i,m}, c_{i,n} \in c(v_i), m \neq n \quad |c_{i,m} - c_{i,n}| \geq t_{ii} \tag{6.7}$$

$$\forall e_{ij} \in E, \forall c_{i,m} \in c(v_i), \forall c_{j,n} \in c(v_j) \quad |c_{i,m} - c_{j,n}| \geq t_{ij} \tag{6.8}$$

It is easy to see that the T-colorings problem is a special case of the set T-colorings problem where each node v_i has the same demand d_i=1. Note that a graph of set T-colorings G=(V, E) can be transformed into a graph of T-colorings G'=(V', E') by creating a node for each demand ($|V'|$=$\sum_{v_i \in V} d_i$). A co-node constraint becomes a set of edges forming a clique of d_i nodes with a separation equal to T_{ii} for each edge. And for each color separation e_{ij}, each new node of v_i is connected to each new node v_j with a color value separation equal to T_{ij} (see Fig. 6.1).

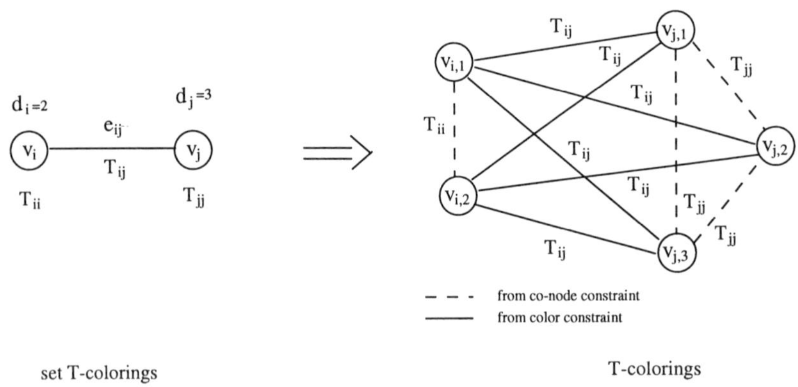

Figure 6.1 Transformation of a graph of set T-colorings into a graph of T-colorings

6.3 BENCHMARKS

6.3.1 Graph coloring

For graph coloring, there exists a large number of well-known benchmarks [15, 18][1]. Instances used in this study will be introduced in Section 6.5 when experimental results are presented.

[1]Most of these benchmarks can be downloaded via ftp from: dimacs.rutgers.edu/pub/challenge/graph/benchmarks/

6.3.2 T-colorings and Set T-colorings

Definitions

For these two classes of problems, no benchmark is available. Therefore, we developed a random instance generator for *restricted T-colorings* and *restricted set T-colorings*, following the same principles defined for generating random graphs for graph coloring [18].

For restricted T-colorings, each instance is defined by three parameters: N, the number of nodes; $d \in [0,1]$, the edge density; and Sep, the largest color separation. To generate such an instance, we first build a graph of N nodes with $d(N(N\text{-}1))/2$ edges uniformally distributed on these nodes. Then, a uniform random value from $[1, Sep]$ is assigned to each edge.

For *restricted set T-colorings*, two more parameters are necessary: D, the maximum number of colors required by a node and $CoSep$, the maximum separation required between two colors of the same node. As above, to generate a restricted set T-colorings instance, we make for each node two uniform random choices from $[1, D]$ and $[1, CoSep]$.

Note that from an instance for restricted set T-colorings, we can get an instance for restricted T-colorings (with an identical edge topology) by setting for each node v_i its separation d_i to 1. In the same way, we can obtain a random *graph coloring* instance with an identical edge topology by setting for each edge e_{ij} the separation T_{ij} to $\{0\}$.

Probabilistic estimations for the minimum span

Unlike graph coloring, there is no estimation available for the minimum span for T-colorings. There are some theoretical lower and upper bounds, but these bounds are of little use in our case because they are often far from the real minimum span. Establishing a good estimation for T-colorings is very helpful for us, but seems to be difficult for the general T-colorings. We thus limit our attention to restricted T-colorings with a color separation uniformly chosen in $[1..Sep]$. So we can define the following estimation:

$$\widetilde{sp}_T(G_{n,p}) = (\widetilde{\chi}(G_{n,p}) - 1) * Sep_{ave} \tag{6.9}$$

where Sep_{ave} is the average of color separations over the graph. This estimation considers that from an edge topology which allows a k-coloring, if each edge has a color separation equal to Sep_{ave} on average, then the minimum span of T-colorings is equal to the difference between the highest color value $1 + (k - 1) * Sep_{ave}$ and the lowest color value, i.e., 1. For the instances used in this study, Sep_{ave} is equal to 3 because the color separation is uniformly chosen between 1 and 5 (see next subsection).

As for T-colorings, we try to establish an estimation for the minimum span for random set T-colorings graphs. In order to do this, we need to transform the graph of set T-colorings into a graph of T-colorings. As explained in Section 6.3, a co-node constraint becomes a clique of d_i nodes with a separation equal to t_{ii} on each edge. And for each color separation e_{ij}, each new node of v_i is connected to each new node of v_j with a color value separation equal to t_{ij}. Since the color separations for co-nodes and for adjacent nodes are very close in our case, the influence of the demand of each node, thus the estimation, depends mainly on the edge density. The lower the edge density is, the higher the influence of the demand is on the *minimum span* $sp_T^D(G_{n,p})$ of G.

Given D_{ave} the average demand per node, and Sep_{ave} the average color separation over the graph, we can define two estimations: one for low edge density (6.10) and one for high edge density (6.11).

$$\widetilde{sp}_T^D(G_{n,p}) = ((\widetilde{\chi}(G_{n,p}) - 1) * Sep_{ave}) * D_{ave} \tag{6.10}$$

The first estimation considers that the graph is divided into D_{ave} similar and strongly connected subgraphs.

$$\widetilde{sp}_T^D(G_{n,p}) = (\widetilde{\chi}(G_{D_{ave}*n,p}) - 1) * Sep_{ave} \tag{6.11}$$

In the second case, because we have a high edge density, only the number of nodes in the transformed graph of T-colorings is modified and the edge density before and after the transformation remains nearly the same.

For the instances used in this study, the demand for each node and the co-node separation are on average equal to 3 (demands and co-node separations are uniformly chosen between 1 and 5) (see next subsection).

Random instances

For the purpose of this study, we generate 15 random instances of restricted set T-colorings with the following possible values: $N \in \{30, 100, 300, 500, 1000\}$, $d \in \{0.1, 0.5, 0.9\}$, *Sep*=*D*=*coSep*=5. Each instance is defined by a name of the form $N.D_{sum}.d.STcol$, where $STcol$ means set T-colorings and D_{sum} is the sum of all the demands ($D_{sum} = \sum_{i \in [1,N]} d_i$).

From these instances, we built 15 random instances of T-colorings denoted by $N.d.Tcol$ (resp. graph coloring instances denoted by $N.d.col$) by assigning $\forall i \in [1, N], d_i = 1$ (resp. by assigning $\forall i \in [1, N], d_i = 1$ and $\forall T_{ik} \in T, T_{ik} = \{0\}$ for coloring). So we have the same graph topology (edge structure) of different densities for the three families of problems. These 45 instances[2] enable us to study the differences among them, in terms of separation distance, demands and edges density.

6.4 GENERIC TABU SEARCH FOR COLORING PROBLEMS

Given the fact that graph coloring and T-colorings are special cases of set T-colorings, the following presentation is oriented to set T-colorings. It should be clear that some components of the algorithm are not necessary for graph

[2] These instances and the generator are available from the authors via email and via internet at http://www.eerie.fr/~dorne

coloring or T-colorings.

Configuration and search space

Given a graph involving N nodes, with d_i (i $\in$ {1..N}) demand per node and NC available colors numbered from 1 to NC, a configuration $s = < c_{1,1} ... c_{1,d_1} ... c_{i,1} ... c_{i,d_i} ... c_{N,1}, c_{N,d_N} >$ is a complete coloring satisfying the following condition:

$$\forall c_{i,m}, c_{i,n} \in c(v_i), m \neq n \quad |c_{i,m} - c_{i,n}| \notin T_{ii}$$

The search space S is thus composed of all possible configurations meeting the *co-node constraints.* Integrating co-node constraints into configurations is an important factor in improving the search efficiency for the set T-colorings problem. This point is developed in [7] for the frequency assignment problem.

Cost function

For each configuration $s \in S$, f(s) is simply the total number of unsatisfied color constraints.

$$f(s) = \sum_{e_{ij} \in E} \sum_{\substack{c_{i,m} \in c(v_i) \\ c_{j,n} \in c(v_j)}} q(c_{i,m}, c_{j,n})$$

where $q(c_{i,m}, c_{j,n}) = \begin{cases} 1 & \text{if } |c_{i,m} - c_{j,n}| \in T_{ij} \\ 0 & \text{otherwise} \end{cases}$

Neighborhood and candidate list

Given $s \in S$, let $s(i, m)$ be equal to the value of the mth color of the node v_i in s. Then the neighborhood N is defined as follows. $s' \in N(s)$ if and only if the following condition is verified:

$\exists$! i $\in$ {1..N}, $\exists$! m $\in$ {1..d_i} such that $s(i, m) \neq s'(i, m)$

A neighbor of s can thus be obtained by changing the value of a color of a node in s in such a way that the new value always satisfies the co-node constraint. A move is thus characterized by a triplet $< i, m, v >$, i, m and v being a node, a demand of the node, and a color value, respectively. Note that the number of neighbors of a configuration may be very high for large graphs.

In this work, the following strategy is used to define the candidate list V^* = {$s' \in N(s)$ such that s' and s are different only at the color of a *conflicting* demand of a node in s}. A demand of a node is said to be conflicting if its color value violates some color constraints. Let CD be the set of conflicting demands of the nodes in s, then $|V^*|$=$|CD| * (NC - 1)$. Clearly, $|V^*|$ varies during the search according to the number of conflicting demands. This candidate list strategy reduces the number of neighbors to be considered at each iteration;

more importantly, it helps the search to concentrate on influential moves.

Incremental evaluation and neighborhood examination

To evaluate the configurations, we use an approach inspired by a technique proposed in [10]. The main idea consists of maintaining incrementally δ, the *move value* or *cost variation* for each possible move from the current configuration s, where W $= \sum_{i=1}^{N} d_i$ in a $NC * W$ matrix. Each time a move is carried out, the elements affected by the move are updated accordingly. Initializing δ takes time $O(NC * W^2)$. The matrix can be updated in time $O(NC * W)$ in the worst case. Now a best neighbor in V^* can be evaluated in time $O(|V^*|)$. Thus each iteration takes time $O(NC * W + |V^*|)$.

Tabu list and tabu tenure

When the color of a demand m of a node v_i in a configuration s is changed to a new value, the triplet $< v_i, m, old\_value >$ is classified tabu for l (tabu tenure) iterations. That is, the old value will not be allowed to be re-assigned to v_i during this period. The tabu tenure l is dynamically adjusted by a function defined over the number of conflicting demands of the nodes. More precisely, $l = \alpha * |CD| + random(g)$ where the function $random(g)$ returns an integer value uniformly chosen in $[1..g]$. The values of α and g are empirically determined and l is limited by $|V^*|$. Since CD varies during the search, so does l for any fixed α.

Aspiration criteria

A very simple aspiration criterion is used to override the tabu restriction. The tabu status of a move will be cancelled if the move leads to a better configuration than the best configuration s^* encountered so far.

Generic Tabu Search (GTS)

Our Generic Tabu Search algorithm is composed of three parts: greedy construction of initial coloring, configuration re-generation and searching for proper coloring (see Fig. 6.2).

- Initial configuration: GTS uses a *Dsatur-based* greedy algorithm [1] to generate initial configurations. This greedy step is fast and provides, with the tabu algorithm, a far better initial configuration than a random approach.

- Configuration re-generation: the re-generation aims at producing quickly a k-1 coloring with a minimum of color conflicts. It proceeds as follows: given a coloring with k colors, the nodes colored with the kth color are given a new color from $[1..k-1]$ in such a way that the new color minimizes the color conflicts over the graph (break ties randomly).

```
Input: G, a graph
Output: NC, the minimum number of color used (minimum span for
T-colorings and set T-colorings is equal to NC-1)
%Variables
% f,f*: objective function and its best value encountered so far
% s,s*: current configuration and the best configuration encountered so far
% V*,l,MAX: candidate list, size of tabu list, limit of the iterations
% NC: the minimum number of color used
% colors(s): returns the highest color value
begin
    M = 0 /* to initialize tabu matrix */
    NB = 0 /* to initialize iteration counter */
    s = generate() /* initial configuration generated with a greedy algorithm */
    NC = colors(s)-1
    s = re-generate(s,NC) /* re-generation from s with NC colors */
    while (NB < MAX) do
        s* = s
        f* = f(s*)
        while ( f* > 0 and NB < MAX) do
            if ( f(best_neighbor(s)) < f* ) then
                s = best_neighbor(s) /* Aspiration criterion, s(i,m) = new_v */
            else
                s = best_non_tabu_neighbor(s) /* s(i,m) = new_v */
            M[i,m,old_v] = NB + l; /* < v_i, m, old_v > becomes tabu */
            l = α * |CD| + random(g);
            if (f(s) < f*) then
                s* = s
                f* = f(s*)
            NB = NB + 1
        if (f* = 0) then
            NC = NC - 1
            s = re-generate(s,NC)
            NB = 0
    return NC+1
end
```

Figure 6.2 Algorithm: Generic Tabu Search (GTS)

- Searching for proper coloring: beginning from such an improper coloring, the tabu algorithm tries to reduce the number of color conflicts to zero. If this happens, the algorithm finds a proper coloring and proceeds to re-generate a new improper coloring with one less color.

The algorithm stops when an optimal or a best known coloring is obtained or when MAX iterations have been carried out without finding a conflict-free (proper) coloring for the current number of colors k. The lowest number of colors used to find a conflict-free coloring is returned.

6.5 EXPERIMENTAL RESULTS

6.5.1 Settings for experiments

In this section, we report experimental results on various instances for the three coloring problems. All the tests have been performed on an Ultra Sparc station with a 143 MHz processor and 128 MB of memory. The GTS algorithm was implemented in C++ and compiled by CC compiler with option -O5. The computing time reported corresponds to the average running time of the entire algorithm including the above three steps.

For graph coloring, GTS runs with a dynamic tabu tenure determined by $2 * |CN| + random(10)$ for all the instances except for the three flat1000... graphs where the chosen value is $4 * |CN| + random(10)$ (CN corresponds to the set of conflicting nodes). For T-colorings and set T-colorings, tabu tenure is set to $l{=}4 * |CD| + random(10)$ for all instances (CD is the set of conflicting demands). The maximum number of iterations for an attempt of finding a proper coloring is fixed at 10,000,000 (for some large or hard graph coloring instances this value is increased to 20,000,000).

6.5.2 Graph coloring

For graph coloring, we used benchmarks from 2nd Dimacs Challenge, Hertz and De Werra [15], and Johnson et al. [18]. Results are compared with the best ones published in the literature:

1. *Fleurent and Ferland*, a tabu algorithm [10] (denoted by 1a in the tables) and a genetic tabu algorithm (denoted by 1b) [9]. These algorithms use an efficient pre-processing technique of [15], which reduces the initial graphs by removing a large number of independent sets. Coloring algorithms are then used to color the residual graphs. This technique is applied to graphs larger than 300 nodes and systematically used by many existing coloring algorithms.

2. *Costa*, an evolutionary hybrid algorithm *EDM* (denoted by 2) with the above pre-processing technique [5].

3. *Morgenstern*, distributed local search algorithms (denoted by 3) based on a particular neighborhood and initialized by a parallelized version of *Johnson et al's XRLF* algorithm [24].

4. *Johnson et al.*, the Successive Augmentation Method *XRLF* (4a) and a set of methods based on simulated annealing: *Penalty Function Annealing* (4b), *Kempe Chain Annealing* (4c), and *Fixed-K Annealing* (4d) [18].

problems	Best Known			Generic Tabu Search			
	k	Method	Time(sec.)	*runs(fail.)*	*k*	Iterations	Time(sec.)
R125.1.col	5	3,1a	1	10(0)	5	1	1
R125.5.col	35	1a	1,380	10(0)	36	147,000	65
R125.1c.col	46	3,1a	1	10(0)	46	1	1
R250.1.col	8	3,1a	1	10(0)	8	1	1
R250.5.col*	65	3	181	5(2)	66	7,800	6
R250.1c.col	64	3,1a	60	5(1)	64	462	1
R1000.1.col	20	3	18	3(0)	20	1	1
R1000.5.col*	241	3	2,078	3(0)	242	6,027,000	18,758
R1000.1c.col*	98	3	1,240	3(0)	98	1,623,000	4,500
flat300_20_0.col	20	3	1	5(0)	20	33,000	17
flat300_26_0.col	26	3	22	5(0)	26	1,723,000	850
flat300_28_0.col*	31	3	4,214	2(0)	31	17,000,000	9,200
flat1000_50_0.col*	50	3	1	10(0)	50	1,664,000	3,020
flat1000_60_0.col*	60	3	1	5(0)	60	5,548,000	10,579
flat1000_76_0.col*	84	1b	14,520	10(5)	89	5,410,000	8,015

Table 6.1 2nd Dimacs Challenge instances

Tables 6.1-6.3 give comparative results for Dimacs, Hertz and De Werra, and Johnson et al. graphs. Table 6.1 shows our results on Dimacs instances with the best-known results given in the above papers. The best known results are summarized in columns 2-4: the smallest number of colors ever obtained, methods which produced such a coloring and the best computing time required. For example, the third line indicates that two methods find a coloring with 46 colors for the problem R125.1c.col, and the best method requires 1 second. Note that information about computing time is only for indicative purpose because these methods have been run on different machines. The last four columns report results obtained by our Tabu algorithm. We give the number of total runs with the number of failures (unsuccessful runs) in parentheses (5th column), the smallest number of colors obtained (6th column), the number of iterations and computing time averaged over successful runs (7th and 8th columns).

From Table 6.1, we see that the results of GTS are very competitive on these instances. Indeed, except for four instances, GTS manages to produce the best-known result. This is remarkable if we compare these results with those of Fleurent and Ferland (1a): GTS gives better colorings on a wide range of instances (all these instances are marked with a star "*") and in particular we find an optimal coloring for the problems flat1000_50_0.col and flat1000_60_0.col while the method 1a produces a solution with 90 colors.

Table 6.2 shows a comparison on Hertz and De Werra instances between GTS and the methods 1a, 1b and 2. These instances belong to four classes of 100, 300, 500 and 1000 nodes, respectively composed of 20 (g1-g20), 10 (gg1-gg10), 5 (ggg1-ggg5[3]), and 2 instances (gggg1-gggg2[4]). k^* corresponds to

[3] ggg5 is identical to DSJC500.5.col

[4] gggg2 is identical to DSJC1000.5.col

problems	Best Known				Generic Tabu Search				
	k^*	k_{moy}	Method	Time	$runs$ $(fail.)$	k^*	Iterations	Time	k_{moy}
g1-20	15	14.95	1b	9.5 sec.	10(0)	15	8,000	2,5 sec	14.95
gg1-10	34	33.3	2	11,000 sec.	5(0)	33	1,302,000	635.9 sec	32.90
ggg1-5	49	49	1b,3	0 sec.	3(0)	50	4,037,000	3,500 sec	49.75
gggg1-2	84	84	1b	41 hours	2(0)	90	12,355,000	19,514 sec	90.0
residual	23	23	1b	5 hours	5(0)	23	834,000	382	23.0 sec

Table 6.2 Hertz and De Werra instances

the smallest number of colors used for all the instances of one class and k_{moy} is the average. On small instances (gg1-gg10), GTS gives better results and brings down the value of k^* to 33. For large instances, we get worse results. However, remember that these methods (1a, 1b and 2) use the "independent sets removing" pre-processing, this makes it impossible to compare these results directly. To get a fair comparison, we run GTS on the residual graph of gggg2 (or DSJC1000.5.col). This residual graph is obtained after having removed 61 independent sets and is given in [9]. GTS finds a 23 coloring leading to an 84 coloring to the initial graph. GTS took 382 seconds and 834,000 iterations to reach a 23-coloring, while 1a needs more than 19 hours and about 50,000,000 iterations, and 1b more than five hours. Note also that Costa's algorithm obtains only 85-colorings.

problems	Best Known			Generic Tabu Search			
	k	Method	Time	$runs(fail.)$	k	Iterations	Time(sec.)
DSJC125.1.col	5	4a	1	10(0)	5	5,000	<1
DSJC125.5.col	17	3	14	10(0)	17	348,000	136
DSJC125.9.col	44	4d	1,080	10(0)	44	9,000	5
DSJC250.1.col	8	4d	9,360	5(0)	8	168,000	32
DSJC250.5.col*	28	3	591	5(0)	28	3,604,000	1,716
DSJC250.9.col	72	4c	72,000	5(0)	72	720,000	591
DSJC500.1.col	12	3	5,452	3(0)	13	16,000	5
DSJC500.5.col*	48	3	49,000	3(0)	50	3,078,000	2,327
DSJC500.9.col	126	3	158,400	3(0)	127	4,211,000	6,150
DSJC1000.1.col	21	3	210	3(0)	21	290,000	154
DSJC1000.5.col*	84	3,1b	118,000	3(1)	90	11,211,000	16,799
DSJC1000.9.col	226	3	65,500	2(0)	226	13,283,000	39,554

Table 6.3 Johnson et al. instances

Table 6.3 shows the results on random instances from Johnson et al. [18] with different edge densities of 10, 50, and 90 %. GTS remains competitive on instances for graphs of low or high edge density. On large instances with 50 % of edges, our method has some difficulties and certainly needs an independent set pre-processing to improve its results.

6.5.3 T-colorings and set T-colorings

Before giving our results on the 15 random T-colorings and set T-colorings instances (Section 6.3), Table 6.4 shows the results on the instances considered as graph coloring. The chromatic number $\widetilde{\chi}$ is estimated with the Johri and

Matula probabilistic method [19]. Note that the results obtained by GTS are similar to those of the Table 6.3.

problems	$\tilde{\chi}$	Generic Tabu Search				
		$runs(fail.)$	k_{best}	k_{ave}	Iterations	Time(sec.)
30.1.col	4	3(0)	3	3.0	2	<1
30.5.col	8	3(0)	7	7.0	14	<1
30.9.col	15	3(0)	15	15.0	10	<1
100.1.col	6	3(0)	5	5.0	81	<1
100.5.col	16	3(1)	14	14.33	2,514,000	1,875
100.9.col	36	3(0)	37	37.0	29,000	32
300.1.col	11	3(0)	9	9.0	15,597	6
300.5.col	35	3(2)	32	32.67	4,746,000	2,262
300.9.col	84	3(0)	84	84.0	790,000	376
500.1.col	14	3(0)	13	13.0	12,000	9
500.5.col	50	3(2)	49	49.67	10,951,000	8,279
500.9.col	124	2(1)	127	127.5	2,908,000	4,247
1000.1.col	22	1(0)	21	21.0	2,256,000	1,198
1000.5.col	85	1(0)	91	91.0	4,013,000	6,013
1000.9.col	222	1(0)	226	226.0	16,484,000	49,085

Table 6.4 Results for graph coloring

To evaluate the GTS algorithm on the T-colorings instances, the estimation of the minimum span, defined in Section 6.3.2, allows us to have a rough idea about the performance of the GTS algorithm. In order to further evaluate the performance of GTS, we adapted *Dsatur* algorithm for T-colorings and set T-colorings during the constructive process, when a color c is assigned to a node v_i the values included in $[c-(t_{ij}-1), c+(t_{ij}-1)]$ are now forbidden for any adjacent node v_j. This is the only difference from *Dsatur* for coloring. Table 6.5 gives comparative results on the 15 T-colorings instances.

From Table 6.5, several remarks may be made. First, GTS gives far better results than *Dsatur*. Indeed, GTS requires much fewer colors (up to -231 colors) for these graphs. Second, the estimation given in the equation (6.9) seems reasonably good. Finally, the computing time necessary to get good T-colorings is high for large and dense graphs.

Table 6.6 gives comparative results for the 15 set T-colorings instances. From the data, we may make similar remarks as for T-coloring. In particular, we see that GTS outperforms *Dsatur* on all the instances, especially on large

problems	$\tilde{sp}_T$	Dsatur		Generic Tabu Search				
		sp_{best}	sp_{ave}	$runs(fail.)$	sp_{best}	sp_{ave}	Iterations	Time(sec.)
30.1.Tcol	9	9	9.0	3(0)	9	9.0	1	<1
30.5.Tcol	21	23	25.2	3(0)	17	17.0	55,000	21
30.9.Tcol	42	40	44.9	3(0)	31	31.0	293,000	114
100.1.Tcol	15	23	23.0	3(0)	14	14.0	154,000	82
100.5.Tcol	45	63	66.1	3(0)	43	43.67	936,000	365
100.9.Tcol	105	124	125.0	3(1)	82	82.33	4,327,000	2,317
300.1.Tcol	30	40	43.8	3(0)	29	29.0	4,720,000	899
300.5.Tcol	102	159	161.3	3(2)	110	111.0	11,967,000	6,285
300.9.Tcol	249	304	318.9	3(2)	216	217.0	32,734,000	26,869
500.1.Tcol	39	61	62.5	3(0)	43	43.0	2,363,000	711
500.5.Tcol	147	240	247.5	3(0)	175	175.67	6,900,000	5,216
500.9.Tcol	369	483	493.1	2(0)	351	351.0	14,523,000	20,811
1000.1.Tcol	63	104	105.1	1(0)	77	77.0	12,537,000	6,657
1000.5.Tcol	252	436	441.4	1(0)	328	328.0	31,554,000	47,281
1000.9.Tcol	663	896	904.2	1(0)	665	665.0	18,949,000	56,426

Table 6.5 Comparative results for T-colorings

problems	$\widetilde{sp}_T^D$		Dsatur		Generic Tabu Search				
	low	high	sp_{best}	sp_{ave}	$runs$ $(fail.)$	sp_{best}	sp_{ave}	Iterations	Time (sec.)
30.86.1	27	15	28	33.6	2(0)	26	26.0	193,000	101
30.95.5	63	42	74	76.1	2(1)	61	61.50	2,468,000	2,721
30.90.9	126	99	145	153.7	3(1)	102	102.33	2,004,000	2,050
100.275.1	45	30	62	62.0	2(1)	45	45.50	1,563,000	1,294
100.304.5	135	102	170	176.2	2(1)	114	114.50	7,332,000	17,974
100.299.9	315	249	320	340.3	3(1)	224	224.33	2,690,000	6,020
300.937.1	90	57	112	114.6	2(1)	80	80.50	11,526,000	20,890
300.905.5	306	234	431	443.9	2(1)	267	268.50	17,169,000	129,291
300.940.9	747	609	902	921.2	3(2)	572	573.33	19,350,000	96,800
500.1507.1	117	84	157	166.0	2(1)	112	113.0	19,300,000	20,308
500.1484.5	441	351	677	677.0	1(0)	402	402.0	14,750,000	133,673
500.1536.9	1,107	915	1,409	1,432.3	1(0)	931	931.0	3,593,000	42,938
1000.3049.1	189	138	281	285.7	2(1)	178	178.50	27,879,861	126,854
1000.3024.5	756	615	1,220	1,226.3	1(0)	846	846.0	19,179,000	382,898
1000.2975.9	1,989	1,665	2,522	2,541.0	1(0)	1,723	1,723.0	9,000,000	670,000

Table 6.6 Comparative results for set T-colorings.

instances with a difference of about 500 colors. The computing times to color these instances are high because the search space is huge (up to about 3,000 variables and 4,000,000 constraints). Indeed, each node requires several colors and the number of colors needed for the graph is very high (up to about 2,000).

Finally, we mention that in addition to these tests, a variant of GTS was also applied to solve frequency assignment instances in mobile-radio networks coming from French National Center for Telecommunications [14]. GTS was compared with methods based on constraint programming, simulated annealing, Gamst's constructive method and evolutionary algorithms. Experimental results showed that GTS finds the best results for all the tested instances.

6.6 CONCLUSIONS

In this paper, a robust and effective tabu search algorithm has been presented for three coloring problems: graph coloring, T-colorings and set T-colorings. The algorithm integrates some important features such as greedy construction of initial configurations, re-generating configurations, dynamic tabu tenure, and co-node constraint handling. Compared with many best known algorithms, this algorithm remains simpler and easier to tune.

The performance of the algorithm was evaluated on a wide range of random graphs. For graph coloring, the algorithm produces highly competitive results compared with some well-known and more complicated algorithms. The "independent sets" pre-processing technique should improve the performance of the algorithm. For T-colorings and set T-colorings, experimental results show that GTS outperforms largely a *Dsatur* algorithm adapted to these problems for all the instances tested. Given that there are few benchmarks for T-colorings and set T-colorings, the random generator and instances used in this study may help to improve this situation.

A last remark to conclude the paper: the independent sets extracting is a widely-used technique for the graph coloring problem and has proven to be important for all well-known algorithms to color large graphs. Unfortunately, no equivalent technique is available for T-colorings or set T-colorings. It will certainly be interesting and important to develop such techniques in the future.

Acknowledgements

We would like to thank the referees for their useful comments which helped to improve the paper.

References

[1] D. Brélaz. New methods to color vertices of a graph. *Communications of ACM*, 22:251–256, 1979.

[2] D.J. Castelino, S. Hurley, and N.M. Stephens. A tabu search algorithm for frequency assignment. *Annals of Operations Research*, 63:301–320, 1996.

[3] M. Chams, A. Hertz, and D. De Werra. Some experiments with simulated annealing for coloring graphs. *European Journal of Operational Research*, 32:260–266, 1987.

[4] D. Costa. On the use of some known methods for T-colorings of graphs. *Annals of Operations Research*, 41:343–358, 1993.

[5] D. Costa, A. Hertz, and O. Dubuis. Embedding of a sequential procedure within an evolutionary algorithm for coloring problems in graphs. *Journal of Heuristics*, 1:105–128, 1995.

[6] R. Dorne and J.K. Hao. A new genetic local search algorithm for graph coloring. To appear in the Proceedings of PPSN'98, Lecture Notes in Computer Science, 1998.

[7] R. Dorne and J.K. Hao. Constraint handling in evolutionary search: A case study of the frequency assignment. In *Intl. Conf. on Parallel Problem Solving From Nature*, volume 1141 of *Lectures Notes in Computer Science*, pp 801–810, Berlin, Germany, 1996.

[8] M. Duque-Anton, D. Kunz, and B. Rüber. Channel assignment for cellular radio using simulated annealing. *IEEE Transactions on Vehicular Technology*, 42:14–21, 1993.

[9] C. Fleurent and J.A. Ferland. Genetic and hybrid algorithms for graph coloring. *Annals of Operations Research*, 63:437–461, 1995.

[10] C. Fleurent and J.A. Ferland. Object-Oriented Implementation of Heuristic Search Methods for Graph Coloring, Maximum Clique, and Satisfiability. Volume 26 of *Discrete Mathematics and Theoretical Computer Science*, pp 619–652. American Mathematical Society, 1996.

[11] A. Gamst. Some lower bounds for a class of frequency assignment problems. *IEEE Transactions on Vehicular Technology*, 35:8–14, 1986.

[12] A. Gamst. A ressource allocation technique for FDMA systems. *Alta Frequenza*, 57(2):89–96, 1988.

[13] W.K. Hale. Frequency assignment: Theory and applications. *IEEE Transactions on Vehicular Technology*, 68(12):1497–1514, 1980.

[14] J.K. Hao, R. Dorne, and P. Galinier. Tabu search for frequency assignment in mobile radio networks. *Journal of Heuristics*, 4:47–62, 1998.

[15] A. Hertz and D. De Werra. Using tabu search techniques for graph coloring. *Computing*, 39:345–351, 1987.

[16] S. Hurley and D.H. Smith. Bounds for the frequency assignment problem. *Discrete Mathematics*, 167-168:571–582, 1997.

[17] M. Jiang. Méthodes heuristiques pour le problème du T-coloriage avec intervalles. *RAIRO Operational Research*, 32:75–99, 1997.

[18] D.S. Johnson, C.R. Aragon, L.A. McGeoch, and C. Schevon. Optimization by simulated annealing: An experimental evaluation; part II, graph coloring and number partitioning. *Operations Research*, 39:378–406, 1991.

[19] A. Johri and D.W. Matula. Probabilistic bounds and heuristic algorithms for coloring large random graphs. Technical Report 82-CSE-06, Southern Methodist University, Department of Computing Science, Dallas, 1982.

[20] D. Kunz. Channel assignment for cellular radio using neural networks. *IEEE Transactions on Vehicular Technology*, 40:188–193, 1991.

[21] W.K. Lai and G.G. Coghill. Channel assignment through evolutionary optimization. *IEEE Transactions on Vehicular Technology*, 45:91–95, 1996.

[22] R.A. Leese. Tiling methods for channel assignment in radio communication networks. *Zeitschrift für Angewandte Mathematik und Mechanik*, 76:303–306, 1996.

[23] F.T. Leighton. A graph coloring algorithm for large scheduling problems. *Journal of Research of the National Bureau Standard*, 84:79–100, 1979.

[24] C. Morgenstern. Distributed Coloration Neighborhood Search. Volume 26 of *Discrete Mathematics and Theoretical Computer Science*, pp 335–358. American Mathematical Society, 1996.

[25] C.H. Papadimitriou and K. Steiglitz. Combinatorial optimization - algorithms and complexity. *Prentice Hall*, 1982.

7 TABU SEARCH WITH CRITICAL EVENT MEMORY: AN ENHANCED APPLICATION FOR BINARY QUADRATIC PROGRAMS

Fred Glover[1], Gary Kochenberger[2], Bahram Alidaee[3] and Mohammed Amini[4]

[1]Graduate School of Business, University of Colorado
Campus Box 419, Boulder, CO 80309, USA.
fred.glover@colorado.edu

[2]College of Business, University of Colorado, Denver, CO 80217, USA.
gkochenberge@castle.cudenver.edu

[3]College of Business, University of Mississippi, University, MS 38677, USA.
alidaee@bus.olemiss.edu

[4]College of Business, University of Memphis, Memphis, TN 38152, USA.
mamini@msuvx2.memphis.edu

Abstract: A recently developed tabu search approach for solving binary quadratic programs has been shown to be highly effective in finding optimal solutions to known difficult problems. Organized chiefly to employ strategic oscillation, based on critical event memory, the method has not only quickly found optimal solutions in those cases where they are known, but has also found high quality (perhaps optimal) solutions to new test problems that are larger than previously reported in the literature.

This paper extends our previous work by introducing a simple but effective scheme for accelerating the evaluation of candidates' moves (for transforming one solution into another), and for updating associated problem information. Accompanying this we also introduce methods for generating advanced starting solutions and additional trial solutions at critical events. Our preliminary computational experiments on a Pentium 200 PC platform show that the method finds optimal solutions (where they are known) in two to six seconds of computation time. We also include tests for problems containing 1000 variables, which are substantially larger than any that appear in the literature prior to this work. Our enhanced approach obtains high quality solutions to these problems within one to four minutes of computation time.

7.1 INTRODUCTION

Recently, a tabu search (TS) method incorporating strategic oscillation has proved highly successful for solving 0-1 Quadratic Programming Problems [6]. A key feature of this method is the use of a *critical event memory* to guide the oscillation process. We examine enhancements of this approach designed to improve its efficiency and to obtain higher quality solutions. As a foundation for this, we introduce a simple but effective procedure for accelerating the evaluation of moves made by the method, and for updating associated problem information. In addition, we introduce methods for generating advanced starting solutions and for creating additional trial solutions at critical events. These developments make it possible to find best-known solutions to benchmark problems more effectively than before. We also include tests on problems that are substantially larger than those previously examined in the literature. Our study of these problems is motivated by the wide range of discrete optimization problems that can be equivalently formulated as a 0-1 Quadratic Program (see, e.g., [1, 2, 4, 9]).

We organize the remainder of the paper as follows. Section 7.2 introduces basic notation and sketches the specialized tabu search approach on which our study is based. Section 7.3 provides the ideas and algorithms that underlie our proposed enhancements. Section 7.4 elaborates our approach for generating advanced starting solutions and trial solutions at critical events. Then, Section 7.5 presents our computational results. Finally, we present our assessments and conclusions.

7.2 NOTATION AND OVERVIEW

We formulate the binary quadratic programming problem (QP) as follows:

$$QP : \max x_0 = xCx \text{ , where } x \text{ is a zero-one vector.}$$

The tabu search approach we employ to solve QP relies chiefly on the strategic oscillation component of TS. The strategic oscillations alternate between constructive phases (progressively setting variables to 1) and destructive phases (progressively setting variables to 0). We refer to the process of setting variables to 1 during a constructive phase as "adding" variables and the process of setting variables to 0 in a destructive phase as "dropping" variables.

To control this form of oscillation we use a memory structure that is updated at *critical events.* For the binary quadratic programs considered here, we define a critical event as when the next move (either an add or a drop) causes the objective function to decrease, hence to worsen in the maximization context.

The corresponding solution is called a *critical solution.* We utilize a parameter, *span*, to indicate the amplitude of oscillation about a critical event, measured by the number of moves made beyond the critical event. We begin with span equal to 1 and gradually increase it to some limiting value ($MaxSpan$). For each value of *span*, a series of alternating constructive and destructive phases is executed before progressing to the next value. At the limiting point, we begin to gradually decrease *span*, allowing again for a series of alternating constructive and destructive phases. When *span* reaches a value of 1, a complete span cycle is concluded and once again a gradual increase in span is initiated, starting the next span cycle.

Information stored at critical events is used to influence the search process in the following manner. We record (in a circular list) the t most recent critical solutions found. From these solutions, we maintain a short term recency vector ($TabuR$) that is the sum of these t solutions. By summing the associated solutions, this memory is not only a recency memory but a frequency memory, since each element indicates the number of times that the associated variable receives a value of 1 in the last t iterations. In a like manner, we maintain a long term frequency vector ($TabuF$) which is the sum of all critical solutions found to date. These arrays are used to influence the search process to head in new directions in pursuit of new critical solutions. This is accomplished by applying the memory to affect the choice of variables to receive new value assignments at the *turn around* points of oscillation, where the method changes from a constructive phase to a destructive phase, or vice versa.

We emphasize that this variant of tabu search embodies only a restricted subset of its ideas, but a subset that has proved effective in the present context. Other elements of tabu search are elaborated, e.g., in [7].

To illustrate the rationale underlying our current approach, suppose we have just switched from the destructive phase to the constructive phase. Other things being equal, we would like to choose the variable (to add next) that contributes the largest net increase to the objective function value x_o. To facilitate a search that appropriately balances intensification and diversification, however, we also would like to choose variables that have not appeared at a value of 1 in recent critical solutions. Thus, we seek to impose the condition that the first variable added back, x_j, will have $TabuR(j) = 0$. That is, we can conceive the condition $TabuR(j) > 0$ as implying that variable x_j is tabu. More generally, we seek to require that the first p variables added (after turn around) to have $TabuR(j) = 0$. Such a requirement might not be strictly possible. Consequently, we attach a positive penalty weight, $PenR$, to $TabuR(j)$ and create a penalized evaluation of x_j by subtracting the penalty term $PenR \cdot TabuR(j)$ from the normal evaluation of variable x_j. Long term information, via $TabuF(j)$, is included in a similar fashion but with a much smaller penalty coefficient.

By a corresponding rationale, when we switch from the constructive phase to the destructive phase, we seek to influence the choice of drops made after this turn-around by using the tabu vectors to create inducements to drop variables appearing in recent critical solutions at a level of 1. Thus, we proceed as previously described except that the penalty terms are added (rather that subtracted) to determine the evaluation of a variable.

The parameter $NewAssign$, which identifies the number of tabu-influenced adds or drops made immediately after a turn-around, is managed in a fashion that fosters additional diversity in the search process. We start with $NewAssign = 1$, and after a number of iterations (e.g., $2t$), we set $NewAssign = NewAssign + 1$. We continue in this fashion until NewAssign reaches a limit ($MaxNewAssign$) at which point we set $NewAssign$ back to 1 and the process repeats.

Between the constructive and destructive phases is the transfer phase which orchestrates the oscillations, manages key parameters, and does various bookkeeping chores. We refer to an execution of the Transfer Phase as an iteration of our heuristic. How the various parameters are managed can make a difference in the performance of the heuristic on certain problems. Given that our goal in this paper was to test the usefulness of ideas for accelerating the method and for generating additional trial solutions (rather than trying to optimize parameters settings), we utilized the following parameter values for all of the computational testing: $t = 3$, $MaxSpan = 7$, $MaxNewAssign = 4$.

7.3 FOUNDATION FOR GENERATING TRIAL SOLUTIONS

7.3.1 Efficient Move Evaluations

Both in the basic operation of our TS method, and in the procedures subsequently described for generating advanced starting solution and trial solutions, we require variables to be evaluated in terms of the net increase in the objective function associated with flipping their value. To facilitate the efficiency of such processes, we employ the following design. Let $N = \{1, ..., n\}$ denote the index set for components of x. Throughout this paper we use the convention that x' and x'' denote two successive trial solutions, where x'' is obtained from x' by identifying a specific index $k \in N$, for which x''_k differs from x'_k while $x''_i = x'_i$ for all $i \in N$, $i \neq k$. Let E denote an *evaluation matrix* given by

$$E_{i,i} = C_{i,i} \text{ for } i \in N$$
$$E_{i,j} = C_{i,j} + C_{j,i} \text{ for } i, j \in N, i \neq j$$

and define $e_i = E_i x, i \in N$, where E_i denotes the ith row of E. We understand e'_i and e''_i to be the form of e_i when $x = x'$ and $x = x''$, respectively.

The evaluation matrix E and the associated quantity e_i provide a basis for conveniently identifying the change in the objective function when x' is replaced by x''. In order to identify the element x'_k of x' that will be changed to yield x'', we will make reference to the change δ_k in x_o defined by $\delta_k = x''_o - x'_o$ (where x''_o and x'_o depend on the choice of k as indicated). Computing this change directly from the definition of δ_k (i.e., $\delta_k = x''Cx'' - x'C'$) requires $O(n^2)$ effort. The following result relies on memory of E and allows δ_k to be identified for all $k \in N$ with $O(n)$ effort, including the effort to update for e'_i to e''_i all $i \in N$ once the decision is made to replace x' with x''.

Proposition Let x' and x'' be two n-vectors such that $x'_i = x''_i$ for $i \in N, i \neq k$. Then

$$\delta_k = (e'_k + E_{k,k}x''_k)(x''_k - x'_k) \tag{7.1}$$

and the updated value e''_i for e'_i when x'' replaces x' becomes

$$e''_i = e'_i + E_{i,k}(x''_k - x'_k) \text{ for all } i \in N \tag{7.2}$$

Proof First we determine e''_i. From the relation between x' and x'' we have $E_i x'' = E_i x' - E_{k,k}x'_k + E_{k,k}x''_k$. Since $E_i x' = e'_i$ and $E_i x'' = e''_i$ this establishes the updated e''_i value as specified in (7.2). Next, by direct expansion from the definition of δ_k and from the definitions of e'_i and e''_i we can determine that $\delta_k = x''_k e''_k - x'_k e'_k$. Substituting the value for e''_i established in (7.2), for $i = k$, yields $= x''_k(e'_k + E_{k.k}(x''_k - x'_k)) - x'_k e'_k$. Collecting and rearranging terms then yields the expression specified in (7.1).

Under the situation where one of x' and x'' has the value 0 and the other has the value 1, the expressions (7.1) and (7.2) simplify to provide particularly fast determinations of δ_k and e''_i for $i \in N$.

7.3.2 Trial Solution Generation

To extend the scope of our investigation, we will make use of our approaches for generating trial solutions as a basis for generating starting solutions as well. The utility of good starting points for combinatorial optimization problems has been suggested by a variety of studies. For the binary quadratic problem we are considering here, Pardalos and Rodgers [12] used a simple gradient ascent method to identify a starting point to initialize their branch and bound algorithm. Based on computational testing over a wide variety of problems, they comment that their starting point was in fact optimal for roughly 60% of the problems and consistently within 3% of optimum when not optimal.

The straightforward ascent methods we use to generate trial solutions are structured to operate on any *seed solution* as an input, and thus can be used to generate starting solutions by selecting various simple seeds, such as setting all variables to 1 or all variables to 0. The new solution produced by our algorithms differs from the seed solution whenever the seed solution is not locally optimal. We continue to assume, for any given $k \in N$, that x'' is the solution that complements (flips) the value of x'_k while leaving x'_i for $i \neq k$ unchanged.

Method P (Partial Ascent) Let x' be a chosen 0-1 seed solution, and create a 0-1 trial solution x'' by the following rule. For every $k \in N$: If $\delta_k > 0$, let $x^*_k = x''_k$. Otherwise, let $x^*_k = x'_k$.

We call Method P a "Partial" Ascent Method because it operates by identifying the flips that improve the solution x', and making all such flips at once.

The outcome does not guarantee that the trial solution x^* will improve x', since moves that are improving when considered independently may not be improving in combination.

It is possible to show that algorithm 4.2 given by Pardalos and Rodgers [12] (P&R), which is based on gradient arguments, is a special case of Method P when $x' = (1, 1, 1, 1, ..., 1)$ (although the P&R method does not make reference to a seed solution). This connection provides a basis for interesting comparisons, described later.

Method A (Ascent)

1. Let x' be a chosen 0-1 seed solution.
2. Let J be the subset of N defined by $J = \{j \in N : \delta_j > 0\}$.
3. If J is empty, denote the final x' obtained by x^* and stop. Otherwise, choose an element k from J and redefine x' to be the solution x''. Then return to Step 2.

Method A is a classical ascent method, which we implement by taking advantage of expressions (7.1) and (7.2) of the Proposition. For our present purposes we choose k in Step 3 by the "steepest ascent" rule, to give $k = \arg\max(\delta_j : j \in N)$. Finally, we join Method P and Method A to give the following procedure.

Method PA

1. Apply Method P, starting with a chosen seed solution x'.
2. Redefine $x' = x^*$ where x^* is the solution obtained from Method P. Then apply Method A.

7.4 EXPLOITING STARTING SOLUTIONS AND TRIAL SOLUTIONS

There are several possible ways to take advantage of the procedures P, A and PA for generating trial solutions from various seed solutions. We first apply the algorithms P, A and PA to simple initial seed solutions as a basis for generating starting solutions for our method. We are motivated to do this particularly for the PA approach, because in the special case where the seed is given by $x' = (1, 1, ..., 1)$ the resulting solution can be shown to be the same one created by Pardalos and Rodgers (although their method is not defined in terms of seed solutions).

We also use the ascent methods to generate trial solutions at critical events of the strategic oscillation process, where the critical solutions produced at such points are treated as seed solutions. Such an approach can be used either to replace the seed solutions by the resulting trial solutions, or to use the trial solutions as supplementary solutions which do not alter the search process. In our present preliminary study, to provide a more direct comparison with our previous strategic oscillation approach for QP, we rely only on the use of trial solutions to supplement the search, without altering its progression. The hypothesis to test is that this will make it possible to identify optimal solutions in fewer iterations than by our previous approach, while also providing the ability to find better solutions than previously found in cases where optimal solutions are unknown and may not have been previously obtained.

7.5 COMPUTATIONAL EXPERIENCE

The computational experience reported in this paper is based on the three sets of test problems described in Table 7.1. All problems were obtained using the generator given by Pardalos and Rodgers [11] and all have characteristics designed to make them challenging. (This generator has also been used in the recent heuristic QP study by Chardaire and Sutter [3].) The first 25 problems (the "a", "b" and "c" problems) are small to moderate sized problems which are also reported on by Pardalos and Rodgers. Problem set 2 contains new test problems of intermediate size while problem set 3 contains new test problems much larger than any reported in the literature to date. Optimal solutions for the problems of sets 2 and 3 have yet to be proven.

The branch and bound (B&B) algorithm of Pardalos and Rodgers successfully and rapidly solved the problems of set 1 except for the last three "c" problems. However, the problems of sets 2 and 3 ("d", "e", "f", and "g") have combinations of complexity and dimension that put them beyond the capability of the B&B approach.

The 1000 variable "g" problems are new test problems that are significantly larger than those previously considered in the literature. In fact, the problems of sets 2 and 3 all extend the test bed of problems examined in previous studies. Sets 1 and 2 were used to test the quality of the starting point methods proposed. Then, all three sets were used to test the performance of our TS heuristic enhanced by advanced starts and/or additional trial solutions at critical points. All runs on all problems are carried out with the same set of parameters in our TS heuristic, with no attempt at fine tuning. In every case considered, the problems were run until reaching a limit on the number of complete span cycles. The cycle limit for problems sets 1, 2 and 3 were 20, 40 and 50 cycles, respectively.

Note, that the solutions given for problems "a", "b", and "c" are known to be optimal [12]. Best known solutions to other problems appear later in the paper. All problems were developed using the Pardalos/Rodgers generator [11] with element ranges as given in Table 7.2. The alternative start and trial solution combinations tested are listed in Table 7.3. The seed solutions, the initial x' vectors, used to generate starting solutions in this study are $seed_1 : x' = (1,1,1,1,..)$, $seed_2 : x' = (0,0,0,0,..)$, $seed_3 : x' = (1,0,1,0,..)$, $seed_4 : x' = (0,1,0,1,..)$.

Set 1				Set 2			Set 3		
Problem	Size	Den	Solution	Problem	Size	Den	Problem	Size	Den
1a	50	0.1	3414	1d	100	0.1	1g	1000	0.1
2a	60	0.1	6063	2d	100	0.2	2g	1000	0.2
3a	70	0.1	6037	3d	100	0.3	3g	1000	0.3
4a	80	0.1	8589	4d	100	0.4	4g	1000	0.4
5a	50	0.2	5737	5d	100	0.5	5g	1000	0.5
6a	30	0.4	3980	6d	100	0.6	6g	1000	0.6
7a	30	0.5	4541	7d	100	0.7	7g	1000	0.7
8a	100	0.0625	11109	8d	100	0.8	8g	1000	0.8
1b	20	1	133	9d	100	0.9	9g	1000	0.9
2b	30	1	121	10d	100	1	10g	1000	1
3b	40	1	118	1e	200	0.1	11g	1000	0.1
4b	50	1	129	2e	200	0.2	12g	1000	0.2
5b	60	1	150	3e	200	0.3	13g	1000	0.3
6b	70	1	146	4e	200	0.4	14g	1000	0.4
7b	80	1	160	5e	200	0.5	15g	1000	0.5
8b	90	1	145	1f	500	0.1			
9b	100	1	137	2f	500	0.25			
10b	125	1	154	3f	500	0.5			
1c	40	0.8	5058	4f	500	0.75			
2c	50	0.6	6213	5f	500	1			
3c	60	0.4	6665						
4c	70	0.3	7398						
5c	80	0.2	7362						
6c	90	0.1	5824						
7c	100	0.1	7225						

Table 7.1 Problem Sets

Problem	main diagonal	off diagonal
a	$+/-100$	$+/-100$
b	$-63to0$	$0to100$
c	$+/-100$	$+/-50$
d	$+/-75$	$+/-50$
e	$+/-100$	$+/-50$
f	$+/-75$	$+/-50$
lg-10g	$+/-75$	$+/-50$
11g-15g	$+/-100$	$+/-50$

Table 7.2 Element ranges for the Pardalos/Rodgers generator

Code	Start Method(s)	Trial Solution Method(s)
TS-0	default (zero)	none
TS-1	P, A, PA	none
TS-2	default (zero)	P, A
TS-2	P, A, PA	P

Table 7.3 Alternatives tested, Conditions

7.5.1 *Advanced Starting Points*

The three starting point methods were tested with the four initial seed solutions specified above. The results are presented in Tables 7.4 and 7.5.

As expected, Method P is dominated by Methods A and PA in terms of the quality of starting solutions. The results from Method P are listed in Table 7.4 to give the reader some indication of the method's performance, but they are omitted from Table 7.5 since they are inferior outcomes. Out of the 180 cases reported in Tables 7.4 and 7.5 (25 problems in Table 7.4, 20 problems in Table 7.5, and 4 seed solutions for each), Method PA gives the best starting solution 89 times while Method A gives the best solution 65 times. The two methods tie 26 times. Method PA produces better starting solutions more often than Method A for three of the four seed solutions, but performs slightly worse than Method A for the remaining seed solution (seed solution 2).

Of the 200 starting points generated by Methods A and PA (100 points generated by each) for the relatively easy problems that are considered in Table 7.4, 70 are in fact optimal. Of these, Method PA gives an optimal solution 41 times and Method A 29 times. Conditioning for initial seed solutions, out of the 25 possibilities (in Table 7.4) for each seed, either or both (A and PA) is optimal 19 times for $seed_1$, 14 times for $seed_2$, 24 times for

Table 7.5, which involves a harder set of problems, paints a different picture of starting solution quality. Of the 160 cases presented there, the best starting point generated is equal to the best known solution only three times. While the easier problems of set 1 give rise to optimal starting solutions much more readily than the problems of set 2, even there the percentage of optimal starting solutions is nowhere near the 60% figure referred to by Pardalos and Rodgers [12]. As previously noted, Method PA initialized with $seed_1$ is precisely the starting solution used by Pardalos and Rodgers [12]. Out of the corresponding 25 problems of Table 7.4, the starting points produced are optimal only 11 times. ($seed_3$ gives substantially better results for these easier problems.) And, for the corresponding cases in Table 7.5 for $seed_1$, none of the 25 starting points produced are best known solutions.

7.5.2 *Performance of Advanced Starts and Trial Solutions*

While the quality of various starting points is of interest, our main concern is to gain insight into the extent to which the performance of our TS heuristic may be improved by employing the ideas of Section 7.4, including the use of advanced starts and additional trial solutions. To this end, we solved the problems of all three problem sets with each of the alternatives given in Table 7.3. We refer to our previous TS approach as TS-O, which we have streamlined to take advantage of the efficient updating provided by the Proposition. TS-1 augments TS-0 by including an advanced start, which is always the better of the two starts produced by Method A and PA, using seed 3. TS-2, in contrast, does not employ an advanced start, but rather generates additional trial solutions at critical points using both Methods A and P. Alternative TS-3 is included

	Method			x'_o : $seed_2$ Method			x'_o : $seed_3$ Method			x'_o : $seed_4$ Method		
Prob	P	A	PA	P	A	PA	P	A	PA	P	A	PA
1a	2353	3388	3414*	1599	3404	3414*	2979	3414*	3414*	1360	3355	3414*
2a	4790	6063*	6063*	2246	5967	5989	2943	6063*	6063*	3379	5916	5989
3a	4403	6035	6035	1044	6026	6037*	4025	5949	6033*	3081	5960	6037*
4a	8067	8461	8589	2386	7066	8598*	3178	8598*	8598*	4986	8495	8535
5a	4488	5666	5353	2649	5704	5476	2249	5476	5712	2436	5300	5230
6a	3436	3980*	3980*	1032	3404	3980*	2142	3980*	3846	2304	3980*	3980*
7a	4404	4541*	4541*	1125	4541*	4263	2163	4541*	4541*	3729	4263	4541
8a	8800	11109*	11109*	4245	11031	10965	8075	11109*	11109*	7087	11101	10869
1b	0	98	133*	< 0	133*	98	0	85	133*	0	98	133*
2b	0	121*	91	< 0	91	121*	0	113	91	0	115	91
3b	0	56	102	< 0	102	56	0	88	102	0	118*	102
4b	0	86	101	< 0	101	86	0	84	101	0	118	101
5b	0	70	150*	< 0	150*	70	0	150*	150*	0	54	150*
6b	0	87	113	< 0	113	87	0	105	113	0	106	113
7b	0	90	160*	< 0	160*	160*	0	160*	160*	0	86	160*
8b	0	87	117	< 0	117	87	0	145*	117	0	63	117
9b	0	96	127	< 0	127	99	0	99	127	0	88	127
10b	0	69	121	< 0	121	100	0	128	121	0	111	121
1c	4319	5058*	5058*	430	5058*	5858*	2457	5058*	5058*	2962	5058*	5058*
2c	4943	6213*	6213*	2274	6151	6151	3672	6119	6151	4112	6102	6213*
3c	5941	6665*	6665*	2052	6649	6665*	3922	6665*	6582	3281	6649	6649
4c	5951	7398*	7277	3187	7200	7398*	5316	7398*	7398*	5458	7366	6940
5c	6818	7336	7326	3279	7301	7336	4497	7362*	7326	4320	7272	7272
6c	4981	5762	5707	3699	5718	5745	3926	5694	5762	4040	5697	5763
7c	5916	7124	7211	4431	7015	7173	4418	7147	7087	4493	7225*	7160
Opt	0	8	11	0	5	9	0	12	12	0	4	9

Table 7.4 Starting Point Results: Problem Set 1. All values marked with an asterisk are optimum x_o values. All computation times are less than 1 second.

	x'_o : $seed_1$ Method		x'_o : $seed_2$ Method		x'_o : $seed_3$ Method		x'_o : $seed_4$ Method	
Prob	P	PA	P	PA	P	PA	P	PA
1d	6310	6272	6275	6054	6248	6232	6272	6333*
1d	6534	6381	6217	6050	5722	6267	6040	6673
3d	9205	9108	8647	9056	9193	8969	9085	9140
4d	10660	10548	10629	10714	10659	10461	9893	9974
5d	11421	11491	11604	11477	11410	11565	11290	11474
6d	14024	13981	13946	13762	14207*	14207*	13581	14206
7d	13686	13950	13806	13800	13349	13762	14384	14311
8d	16238	16143	16136	16093	15942	16216	16149	16128
9d	15089	15584	15588	15588	15417	15458	15557	15333
10d	19039	18901	18962	18198	18497	18538	19018	18961
1e	16379	16403	16289	16419	15895	16013	16284	16362
2e	23179	23312	23133	23031	22405	23245	22982	23203
3e	24775	24793	24445	24312	24060	24270	23129	23501
4e	35373	35373	35578	35424	34532	35431	35317	35366
5e	34387	35039	34827	34046	34114	34601	34403	34895
1f	60427	60005	60047	59953	59879	60276	59949	60441
2f	99150	99759	97193	97489	98003	98424	99310	97062
3f	135935	136678	135456	134903	133702	136520	134564	134457
4f	170028	171184	169719	168263	170763	170519	169806	170930
5f	187803	189291	186672	187760	186737	188117	186505	185695
best	0	0	0	0	1	1	0	1

Table 7.5 Starting Point Results: Problem Set 2. All values marked with an asterisk are optimum x_o values. All computation times are less than 1 second.

to give some indication of the usefulness of allowing for both advanced starts and additional trial solutions. It uses only the weakest (although fastest) of the methods to generate trial solutions, Method P.

The results of our runs are shown in Table 7.6. These tables indicate, for each alternative tested, the best objective function value found in the cycle limit allowed and the iteration count at which the best value was found. The bottom of each table also shows average run times for the various problems. For Alternatives TS-2 and TS-3, the tables also indicate (next to the iteration count) the procedure for generating trial solutions that produced the best result. For example, for problem 3a, TS-2 gave the optimal solution at iteration 74 via Method A, while TS-3 gave the optimal solution at iteration 110 via Method P. When no trial method is shown, the associated trial procedure did not produce the best result. The average runtimes are given in Table 7.7.

Results for Problem Set 1 Referring to Table 7.6, we see that all four alternatives produced the optimal solution to each of the 25 problems of Set 1. Comparing the results we see general improvement (in iteration count) by the alternatives to TS-0 in certain cases but degradation of performance in others. These problems were very easy for our TS-0 procedure and, aside from the fact that some of the problems were immediately solved due to optimal starting points, little improvement was anticipated to be possible on the other problems. Solution times are quite modest in all cases.

Compared to the base case results of TS-0, method TS-2 (which generates trial solutions) generally gave somewhat reduced iteration counts. For those problems for which the starting point was not optimal, TS-1 and TS-3 often required more iterations than TS-0. In several cases where the starting point was not optimal, initializing with an advanced starting solution actually impeded the search process. For example, the optimal solution of problem 3a is 6037. The best starting point for this problem yields an objective function value of 6033. Yet, initializing TS-1 at this starting point causes the TS heuristic to run for 603 iterations before finding an optimal solution. In contrast, the TS-0 heuristic found the optimal solution at iteration 127. Table 7.6 displays other examples of this behavior as well.

Results for Problem Set 2 As shown in Table 7.6, an advanced starting solution matched the "best known" solution only once (problem 6d) for Set 2. While the various methods generally produced the same "best" results, there are some differences. Taking TS-0 as the reference, we see that TS-1 (which uses an advanced start) produced the same results 18 times, an inferior result 1 time, and a better result 1 time. TS-2 (which employs trial solutions) gave the same results as TS-0 for all 20 problems, but often found these solutions in significantly fewer iterations. Finally, in terms of solution quality, TS-3 produced the same result as TS-0 19 times and a superior result 1 time.

	TS-0 (Base case)		TS-1		TS-2		TS-4	
Problem	best	iter	best	iter	best	iter	best	iter
				Problem Set 1				
1a	3414	25	3414	0	3414	4P	3414	0
2a	6063	5	6063	0	6063	4A	6063	0
3a	6037	127	6037	603	6037	74A	6037	110P
4a	8598	99	8598	0	8598	25A	8498	0
5a	5737	107	5737	41	5737	1A	5737	8P
6a	3980	12	3980	0	3980	3A	3980	0
7a	4541	2	4541	0	4541	1P	4541	0
8a	11109	30	11109	0	11109	1A	11109	0
1b	133	1	133	0	133	1	133	0
2b	121	60	121	612	121	8A	121	612
3b	118	51	118	51	118	12A	118	51
4b	129	2	129	2	129	1A	129	2
5b	150	1	150	0	150	1	150	0
6b	146	28	146	54	146	27A	146	54
7b	160	1	160	0	160	1	160	0
8b	145	61	145	0	145	61	145	0
9b	137	1686	137	2666	137	682A	137	2665P
10b	154	55	154	30	154	10A	154	30
1c	5058	1	5058	0	5058	1	5058	0
2c	6213	3	6213	13	6213	3	6213	13
3c	6665	36	6665	0	6665	1A	6665	0
4c	7398	77	7398	0	7398	9A	7398	0
5c	7362	289	7362	0	7362	6A	7362	0
6c	5824	293	5824	79	5824	9A	5824	79
7c	7225	27	7225	217	7225	6A	7225	12P
				Problem Set 2				
1d	6333	604	6333	26	633	15A	6333	26
2d	6579	50	6579	320	6579	26A	6579	157P
3d	9261	26	9261	26	9261	25A	9261	26
4d	10727	77	10727	29	10727	37A	10727	29
5d	11626	29	11626	529	11626	4A	11626	529
6d	14207	44	14207	0	14207	43A	14207	0
7d	14476	89	14476	33	14476	89	14476	13
8d	16352	98	16352	26	16352	11A	16352	26
9d	15656	4	15656	30	15656	3A	15656	30
10d	19102	5	19102	30	19102	1A	19102	30
1e	16464	914	16458°	1265	16464	203A	16464	615P
2e	23395	2043	23395	532	23395	109A	23395	202P
3e	25243	63	25244	75	25243	46A	25243	75
4e	35594	82	35594	87	35594	3A	35594	36P
5e	35154	1443	35154	1204	35154	396A	35154	244P
1f	61194	434	61194	347	61194	433A	61194	347
2f	100161	779	100161	3592	100161	42A	100161	3592
3f	138035	126	138035	679	138035	126	138035	679
4f	172618	1811	172771*	1254	172618	1188A	172771*	1254
5f	190507	297	190507	854	190507	297	190507	339P
				Problem Set 3				
1g	131456	2863	131448°	1471	131456	1992A	131456	1198P
2g	172788	4566	172788	5584	172788	2244A	172788	5584
3g	192565	199	192565	201	192565	195A	192565	201
4g	215010	6511	215010	1269	215010	166A	215010	1269
5g	242367	1249	242367	4852	242367	1248A	242367	4852
6g	243393	5663	240569°	1469	243293	5663	240569°	1332P
7g	252664	4264	252466°	1439	252664	4213A	252466°	1438P
8g	264061	3314	263637°	1707	264061	3312A	263637°	1707
9g	261792	3053	262658*	2026	261792	2856A	262658*	2026
10g	273434	3337	273795*	5993	273434	3325A	273795*	5993
11g	131623	1883	131615°	1932	131631*	1379A	131631*	1387P
12g	173615	900	173523°	4039	173615	900	173529°	3690P
13g	192854	218	192864	462	192864	217A	192864	462
14g	215611	6029	215852*	4537	215612*	1090A	215852*	4443P
15g	242832	2573	242832	631	242832	2572P	242832	631

Table 7.6 Starting Point Results: Problem Set 2, Set 2 and Set 3. An iteration count of zero means the starting point was optimal. Each TS variant above found an optimal solution to every problem of Problem Set 1. Here, * denotes solution better than base case, while ^{o} denotes solution worse than base case.

Problems	TS-0	TS-1	TS-2	TS-3
"a"	2	2	4	3
"b"	2	2	5	2
"c"	2	2	6	3
"d"	4	4	15	10
"e"	8	8	43	24
"f"	21	21	320	133
"g"	65	67	1126	424

Table 7.7 Average time for 20 cycles for Problem Set 1, 40 cycles for Problem Set 2 and 50 cycles for Problem Set 3 (in seconds on a Pentium 200 PC)

Results for Problem Set 3 For Problem Set 3 (Table 7.6) we see more variability than for Problem Set 2. Again comparing to TS-O, TS-1 gave the same "best" solution 6 times, an inferior solution 6 times, and a better solution 3 times. In contrast, TS-2 gave the same result 14 times and a better result 1 time. Finally, TS-3 gave the same result 7 times, an inferior result 4 times, and a superior result 4 times. Once again we see general improvement in iteration count (over the base case), but isolated cases of enlarged iteration counts were found.

General remarks on the results for the problem sets A glance at Table 7.6 indicates how frequently Method A produced the best solution. For the 60 problems for which Method A was employed to produce trial solutions, it gave the best solution 48 times. Excluding those cases where the starting point was optimal, Method P was employed on 105 problems and produced the best result 20 times.

Looking at our experience with all 60 problems and the various testing carried out, Method A appears to be the preferred approach for generating additional trial solutions. However, the base case TS-0 method, with no advanced start and no additional trial solutions, gave the best overall performance in terms of solution quality and execution time, even though it generally required more iterations. Alternative TS-2 (which included trial solutions) seldom improved upon the solution quality of TS-0 and yet required substantially greater computation times. Alternatives TS-1 and TS-3, while improving upon the TS-0 solution 5 times, were just as likely to produce inferior solutions.

Since the main source of improved solutions in the preceding tests comes from the use of the advanced trial solution, but the outcomes also include results worse than those for the base case TS-O, we were motivated to test the result of using advanced solutions based on all four of our seed solutions. Since this increases the solution time by a factor of 4, we also investigate the option of simply running TS-0 for four times as many span cycles.

The outcomes appear in Table 7.8. The tests are limited only to the 1000 variable problems. We see that running TS-0 four times longer finds two "new

	BQ1		BQ2		
Problem	best	iter	best	iter	seed#
1g	131456	2863	131456#	5217	4
2g	172788	4566	172788	2234	2
3g	192565	199	192565	201	3
4g	215679*	14067	215679*	6475	4
5g	242367	1249	242368	629	4
6g	243293	5663	243293#	365	4
7g	253181*	12558	253181*	1010	2
8g	264061	3314	264061#	2547	4
9g	261792	3053	262658(+)	2026	3
10g	273719#	18226	274264*(+)	2982	1
11g	131631#	8093	131632	6858	1
12g	173615	900	173542(−)	5908	2
13g	192864	218	192865	296	2
14g	215685#	27606	216012*(+)	6073	4
15g	242832	2473	242833	492	2

Table 7.8 Multiple Starts vs. Base Case. Again, * denotes new best solution (over all runs referenced in this paper), # denotes better than previous best for corresponding method, (+) denotes better than base case result and (−) denotes inferior to base case result. The base case results (BQ1) were run for 200 cycles. The average run times (for 200 cycles) was approximately four minutes for each problem.

best" solutions out of the fifteen problems tested - that is, two solutions better than obtained by any of the previous methods and solution attempts. The method also improves the solution of three additional problems by comparison to the version of TS-0 that was run for fewer iterations. (In fact, as shown in the concluding section, we obtain still better results for some of these problems by a simple variation.)

The BQ2 results show the best objective function value found via initializing BQ2 with the four different seeds. For each seed, the heuristic was run for 50 cycles. The combined time for all 200 cycles was approximately four minutes for each problem.

The method TS-1, which restarts from the four different seeds (and uses the same total number of iterations as the corresponding TS-0 from Table 7.8), eliminates all but one of the cases where TS-1 previously was worse than TS-0, and additionally finds four solutions better than the best previously found by any of the methods. (These match the two new best found by TS-1, and include two others.) Overall, TS-1 obtains three solutions that are better than TS-0 and one solution that is worse. This suggests that there may be value in generating a larger collection of appropriately diverse solutions for re-starting. Alternatively, it may indicate the possibility that the use of longer term frequency memory within TS-0, which serves also to create diversity, should be refined to obtain better outcomes. Frequency memory can also take a role in determining new starting solutions. We stress again that no effort has been made to tune any parameters of the current method, which may offer an opportunity for refinement. The three parameter settings used are the same as those used to solve the smaller and easier problems from our earlier study.

7.6 CONCLUSIONS

Our study has provided several findings. Advanced starting solutions, based on the use of simple seeds (which provide methods more general than those previously used to generate advanced starts), have a widely variable and inconsistent effect on the performance of procedures that use these solutions.

None of the seeds tested produced outcomes that dominated those produced by other seeds. A seed that set all variables equal to 1, which is motivated as "best" by gradient arguments, performed notably worse than a seed that alternately set variables to 0 and 1.

Contrary to reports of a B&B study that starting solutions based on gradient arguments produced optimal solutions in 60% of the cases, we found that such solutions were optimal in less than half of the cases - when these cases were restricted to problems that are easy to solve. For harder problems, these starting solutions (and starting solutions generated by alternative seeds) were almost never optimal.

Trial solutions using the same methods employed for generating starting solutions, but using critical solutions provided by the TS method as seeds, often uncovered previously best known solutions in substantially fewer iterations than required by the TS method that did not make use of these trial solutions. However, these trial solutions did not find solutions superior to the best previously found, and the computation time required to generate these solutions significantly increased the overall solution time of the method.

Allowing the base case method TS-0 to run four times longer (which still consumes only about 4 minutes for each of the large 1000 variable problems), generates improved solutions for 1/3 of the 15 problems, and two of these are better than the best previously found by any method. Allowing the advanced start augmentation of TS-0 to run for the same duration, by using advanced starts from the 4 different seed solutions, does even better, finding better solutions than TS-0 in 3 cases and a worse solution in 1 case.

Several avenues are suggested for future research. First, we anticipate the value of identifying an "indicator" that signals which critical solutions are more likely than others to be useful as a basis for generating trial solutions. An appropriate threshold defined in terms of this indicator (which becomes refined as the search yields additional information), may allow trial solutions to be generated only for a special subset of critical events, therefore significantly reducing the computation time of the variants that create such solutions. Second, the development of such an indicator and threshold may lead to an altered design of the search process (for example, by inducing oscillations of a somewhat different pattern than those currently employed). Third, the question arises whether related information may be used simply to specify a progressively changing set of seed solutions for a collection of "re-starts." Conceivably, the structure of the method may beneficially change by employing a special approach to generate diverse seeds, which shifts the method to rely more fully on a restarting component. Since the design we use causes each starting solution (e.g., produced by Method A or PA) to be a local optimum, this approach can potentially be stronger than the type based on the conception of a starting solution that may be far removed from local optimality.

Finally, we note again that no effort has been made to tune any parameters of our basic TS-0 method, on which all our current variants are based. The parameter values are the same as those used in our earlier study that examined smaller and easier problems. Reasonably, the parameters should be a function of problem size and adaptively determined, rather than pre-set, according to successes in finding improved solutions as the values are varied during the search. (E.g., changes in amplitude of the oscillation can be governed in this way.) In addition, our use of longer term memory and its integration with short term memory are very primitive, offering another source of potential improvement. We note that a recent advance that uses partially overlapping ideas in an evolutionary framework is provided by Lodi, Allemand and Liebling [10].

The types of gains that can be achieved simply by an adaptive setting of parameter values is illustrated by two additional runs we made on the "g" problems using TS-0, subsequent to making the runs summarized in Table 7.8. For the first set of runs, the parameters t, $MaxSpan$, and $MaxNewAssign$ were set to 7, 7, and 5. For the second set of runs, they were set to 7, 7, and 7. For each set of runs, each problem was run for a total of 400 cycles.

Each set of runs produced two new "best known" solutions. Run set #1 gave new best solutions for problems g10 and g14 while run set #2 gave best known solutions for problems g7 and g8. Collecting the best results from all methods considered in this paper, we have generated the following "best known" solutions for the "g" problems: (gl, 131456), (g2, 172788), (g3, 192565), (g4, 215679), (g5, 242367), (g6, 243293), (g7, 253590), (g8, 264268), (g9, 262658), (g10, 274375), (g11, 131631), (g12, 173615), (g13, 192864), (g14, 216168), (g15, 242832). The results from these additional runs motivate our interest in conducting a rigorous target analysis in order to better understand which parameter values are best suited for certain types of problems. Such analysis is currently underway.

Although those considerations indicate that latitude exists to improve our basic approach, the significant advances achieved by the current study over previous results may have useful implications in view of the wide range of problems that can be cast in the nonlinear QP formulation. In particular, the ability to obtain high quality solutions for 1000 variable QP problems with only one to four minutes of computing time on a PC opens the door to handling practical problems of these and larger sizes that have been beyond the scope of previous investigations in this area.

References

[1] Alidaee, B., G. Kochenberger, and A. Ahmadian (1994). 0-1 Quadratic Programming Approach for the Optimal Solution of Two Scheduling Problems, Int. J. Syst. Sci., 25, 401-408.

[2] Barahona, F.M. Jünger, and G. Reinelt (1989). Experiments in Quadratic 0-1 Programming, Math. Program. 44, 127-137.

[3] Chardaire, P., and A. Sutter (1995). A Decomposition Method for Quadratic Zero-One Programming, Management Sci. 41, 704-712.

[4] Gallo, G., P. Hammer, and B. Simone (1980). Quadratic Knapsack Problems, Math. Program. 12, 132-149.

[5] Glover, F., and G. Kochenberger (1996). Critical Event Tabu Search for Multidimensional Knapsack Problems. Meta Heuristics: Theory & Applications, Kluwer, 407-427.

[6] Glover, F., G. A. Kochenberger, and B. Alidaee (1998). Adaptive Memory Tabu Search for Binary Quadratic Programs, Management Sci 44, 336-345.

[7] Glover, F., and M. Laguna (1997). Tabu Search, Kluwer.

[8] Glover, F., M. Laguna, E. Taillard, and D. de Werra, eds, (1993). Tabu Search, Special issue of the Ann. Oper. Res. 41, J. C. Baltzer.

[9] Gulati, V.P., S.K. Gupta, and A.K. Mittal (1984). Unconstrained Quadratic Bivalent Programming, European J. Oper. Res. 15, 121-125.

[10] Lodi, A., K. Allemand, and T.M. Liebling (1997). A Genetic Algorithm for Quadratic 0-1 Programming, Working Paper, University of Bologna.

[11] Pardalos, P.M., and G.P. Rodgers (1988). Computational Aspects of a Branch and Bound Algorithm for Quadratic 0-1 Programming, Technical Report CS-88-04, Penn State University.

[12] Pardalos, P.M., and G.P. Rodgers (1989). Computational Aspects of Branch and Bound Algorithm for Quadratic 0-1 Programming, Computing, 45, 131-144.

8 ACTUATOR SELECTION FOR THE CONTROL OF MULTI-FREQUENCY NOISE IN AIRCRAFT INTERIORS

Rex K. Kincaid[1]* and Sharon L. Padula[2]

[1]Department of Mathematics, The College of William and Mary
Williamsburg, Virginia 23187-9795, USA.
rrkinc@math.wm.edu

[2]Multidisciplinary Optimization Branch, NASA Langley Research Center
Hampton, Virginia 23681.

Abstract: In this paper we study the selection of actuator locations for a constrained multi-frequency active noise control optimization problem. Data regarding the effectiveness of each actuator is gathered from the fuselage of a DC-9 aircraft. A constraint is placed limiting the force that can be applied to each actuator. Tabu search is used to select locations for the actuators and a quadratically constrained complex least squares problem must be solved to determine the control forces for the selected actuators.

8.1 INTRODUCTION

Tabu search and, more generally, adaptive memory programming continues to have startling success in efficiently generating high-quality solutions to difficult practical optimization problems. [4] provides forty-two vignettes, each of which describes a different application of tabu search by researchers and practitioners. The applications include standard problems such as the p-median problem, job shop scheduling and the quadratic assignment problem, as well as unusual applications including polymer chemistry, forest management, and the control of flexible space structures.

Here, we study an optimization problem that arises in active noise control of interior noise in a turboprop aircraft. In Section 8.2, we give a brief introduc-

*The author was supported in this research by NASA grant NAG-1-1783.

tion to active noise control as well as a description of the optimization model. Previous research efforts have focused on single frequency models with no force constraints (see [13, 1, 10]). In these models, given a set of actuator sites, an unconstrained complex least squares problem is solved to determine the control forces and noise reduction potential for each set of actuators. Tabu search (see [5]) has been shown to dominate other engineering methods for the selection of actuator sites in the single frequency case. Here we extend the earlier models in two ways—actuator sites that work well for multiple frequencies are sought and a constraint on the maximum control forces applied to the actuators is enforced. Each of these extensions is a strengthening of the model's link with the real problem of dampening in-flight interior noise for turboprop aircraft.

8.2 ACTUATOR SELECTION FOR ACTIVE NOISE CONTROL

Most aircarriers use turboprop aircraft to transport passengers between local or nearby airports and their hubs. Turboprop aircraft are preferred for short hops due to their superior fuel efficiency at low speeds. However, the decreased cost to the aircarrier comes at the cost of increased levels of cabin noise for the customer. Two broad classes of control methods exist to address the reduction of cabin noise - passive and active control. Passive control is synonymous with vibration absorption methods, such as the installation of liners in the shells of aircraft, which attempt to isolate the cabin from the noise and vibration caused by the engines. Passive methods, unfortunately, have proven ineffective in reducing cabin noise for turboprop aircraft [2], particularly in the low frequency range associated with the blade passage frequency of propellers. Active control methods attempt to counteract propeller noise by introducing a secondary (or multiple secondary) sound source which is of the same frequency and amplitude as the propeller noise, but is 180 degrees out of phase. Thus, in theory, the result is a complete silencing of the propeller noise source.

An early example of one-dimensional active noise control [7] will serve to illustrate the salient features of active noise control systems. Consider a single frequency noise source S_1, traveling down a duct as illustrated in Figure 8.1. The sound is detected with a microphone, M, and is passed through a controller, V, to a loudspeaker, L. The controller adjusts the sound emitted by the loudspeaker so that the wave produced, S_2, completely cancels out S_1. The controller finds the appropriate sound for the loudspeaker by means of a transfer function, which establishes the relationship between measurements at M and the best input (phase and amplitude) at L. For more details on the history of active noise control the interested reader is referred to [3].

A variety of active structural acoustic control systems have been proposed for aircraft (see [13]). We consider an acoustic control model in which the control inputs, which are used to reduce interior noise, are applied directly to a vibrating structural acoustic system (e.g., an aircraft fuselage). The feasibility of this approach has been demonstrated by [1] using measured data from the aft cabin of a Douglas DC-9 fuselage.

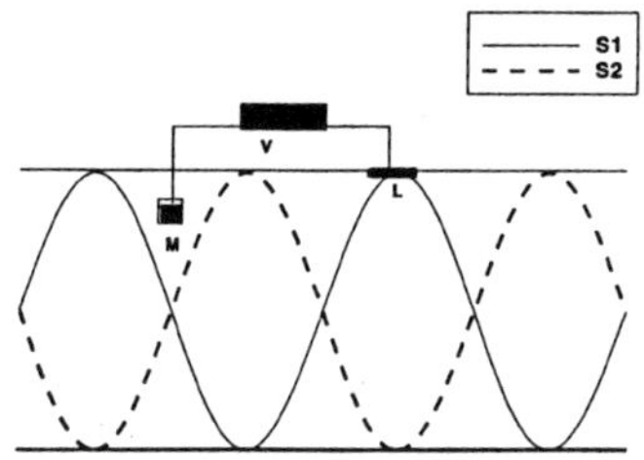

Figure 8.1 Simple Active Noise Control System

8.2.1 Description of test facility

The data for our computational experiments was taken at the McDonnell Douglas Fuselage Acoustic Research Facility (FARF) and is reported [9]. The FARF is a large anechoic chamber that contains the rear section of a DC-9 aircraft (without engines or tail). The interior is complete with seats and trim panels. Two acoustically treated barriers divide the test section from the rest of the aircraft. A large external speaker was used as the primary noise source. To measure the performance of the actuators, 18 microphones (sensors) were mounted at head height, one for each of the 15 seats and three in the aisle.

The data were originally acquired to support broadband jet noise control experiments, but were adapted for our purposes (i.e., propeller noise) by choosing a few discrete frequencies. A typical interior noise spectra in the frequency range of interest due to the loudspeaker is shown in Figure 8.2 for one microphone. The spectra at each of the other 17 microphones is similar. Also shown in Figure 8.2 are five discrete frequencies (228 Hz., 272 Hz., 316 Hz., 360 Hz., and 404 Hz.) that were chosen because they were equidistant from each other (44 Hz.) and near peaks in the sound pressure field for each microphone. This is similar to the harmonics of propeller noise, which would be even multiples of the blade passage frequency and would tend to have the highest noise at the lower frequencies ($<$ 500 Hz.)

The layout of the 64 piezo-electric actuators for the aircraft is shown in Figure 8.3. The figure depicts a 2-dimensional unrolling of the DC-9 fuselage when cut along the bottom of the fuselage. The placement of the actuator sites and the numbers assigned to these locations were done at FARF for the experiment reported in [9]. The left side faces the noise source. The darkened strip in Figure 8.3 (seen also in Figures 8.5 and 8.6) are the window areas along each side of the aircraft fuselage. The actuators were bonded to the interior of the aircraft skin within the bay area formed by the ring frames and longerons.

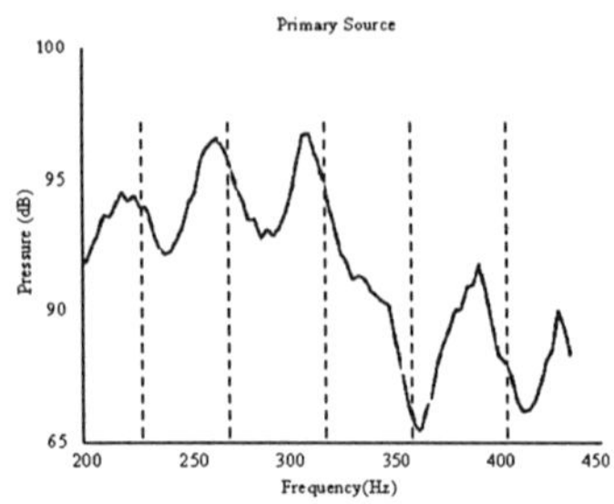

Figure 8.2 Primary Noise Source Response

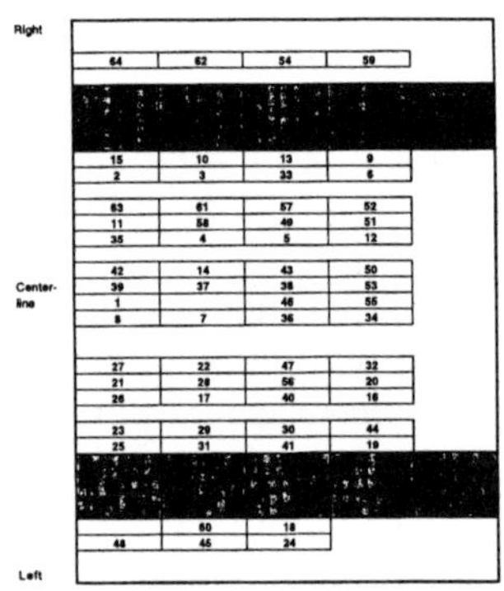

Figure 8.3 Actuator Locations

8.2.2 Measures of performance

The performance measure for a given set of actuator locations for a single frequency is given by the transfer matrix H for that frequency. The entries of H provide the contribution of a unit amplitude input at each of the $n = 64$ actuator sites to interior noise reduction at each of the $m = 18$ microphones (sensors). Thus, each row i of H records the contribution of a unit amplitude input at all 64 actuator sites in reducing the noise at sensor location i. That is, H has one row for each of the 18 sensors and one column for each of the 64 actuators. Given a particular subset K (let $k = |K|$) of columns (or column indices) of H, the submatrix of H induced by these columns, H^K, is the associated transfer matrix needed for computing the forces to drive the k actuators. A complex vector $\mathbf{p}$ of dimension m contains the amplitude and phase of the noise signal as measured at each of the 18 microphones when the controller is turned off. Once a particular set K of actuators sites has been selected a force (amplitude and phase) must be determined for each actuator so that the resultant force cancels out **p** as much as possible. A standard approach to determine a force vector **f** for a single frequency to solve the complex least squares problem can be stated as:

$$\min ||H^K \mathbf{f} + \mathbf{p}||_2. \tag{8.1}$$

The solution to (8.1) (for fixed K) is found by solving

$$H^{K*}H^K\mathbf{f} = -H^{K*}\mathbf{p} \tag{8.2}$$

for **f** ($*$ denotes the complex conjugate transpose). The range of (8.1) is $\mathcal{R}$, the set of real numbers.

Next, we must compute the performance of this set of k columns (or any set of k columns) over all five frequencies. We do this by summing up the single frequency solutions and then applying the standard decibel scale. That is,

$$10 \cdot log_{10} \frac{\sum_{l=1}^{5}(||{H_l}^K \mathbf{f}^{\ l} + \mathbf{p}^{\ l}||_2)^2}{\sum_{l=1}^{5}(||\mathbf{p}^{\ l}||_2)^2} \tag{8.3}$$

The decibel value computed in (8.3) compares the interior noise norm using active control with the norm of the uncontrolled interior noise. Thus, a negative decibel (dB) level signifies a decrease in interior noise due to the control effects of the actuators.

There is one additional assumption. We have assumed that a linear control law appropriately models the active noise control of the aircraft. There are at least two reasons for this assumption. First, we would like to use the elegant closed-form solution given in (8.2). Second, the typical laboratory controllers used to test the quality of our solutions implement a linear control system to adjust the amplitudes and phases of the actuators. That is, when we provide a set of actuator sites the computed control forces associated with this solution are not used. Instead, the laboratory controller selects its own value for the forces via its own optimization procedure, which assumes that the actuators are linearly independent. Hence, as long as the laboratory controllers assume a linear control law, there is no reason for our optimization model to do otherwise.

8.2.3 Constraints on the actuator forces

Earlier efforts to address the actuator selection problem [5] focused on single frequencies and did not consider any constraint on the amount of force applied to an actuator. In practice, the actuator force is limited by many factors, which include the design of the actuator, the way it is mounted, and the actuator power supply. A natural way to constrain the force is to place a bound on the voltage input to each actuator. This is particularly useful here. In the FARF data set, the entries of ${h_{ij}}^l$ record the noise reduction at microphone i due to a one volt input at actuator j. Thus, if $\mathbf{f}_j$ is in polar coordinates, the real part of $\mathbf{f}_j$ is in volts. The sum of these real components for each of the five frequencies must not exceed a specified maximum voltage for actuator j. However, to convert the complex variable problem into a real variable problem, rectangular coordinates must be used. If each component of $\mathbf{f}$ is written in rectangular coordinates and *volt* is the maximum voltage allowed for each actuator, then our constrained optimization problem is:

$$\min \sum_{l=1}^{5} (||H_l{}^K \mathbf{f}^{\,l} + \mathbf{p}^{\,l}||_2)^2 \tag{8.4}$$

$$\begin{array}{rlr} \text{subject to} & & \\ \sum_{l=1}^{5} ||\mathbf{f_j}^{l}||_2 & \leq volt & \text{for all } j = 1, \ldots, k \\ K & \subseteq \text{columns of } H_l & \text{for all } l = 1, \ldots, 5 \\ \mathbf{f}^{\,l} & \in \mathcal{C}^k & \text{for all } l = 1, \ldots, 5 \end{array}$$

Since $\mathbf{f}$ is a vector with complex components, the above (for a given K) is a quadratically constrained complex least squares problem. (Note, that it is possible to write this constraint as a linear constraint if $\mathbf{f}$ is in polar coordinates.) There are a variety of special purpose codes for real variable least squares problems. However, when we convert (8.4) to a real variable problem, we still have the quadratic constraint and to the best of our knowledge all of the special purpose codes require linear constraints or simple bound constraints. The resulting real variable problem is a quadratically constrained least squares problem (for a given K) with twice as many rows and columns as before. When we include the selection of the set of columns of H, we assign a 0/1 variable to each of the n columns of H. The problem is now a mixed integer quadratic program with quadratic constraints. Let $H = (A + iB)$, $f = w + iv$, and $p = s + it$. Our multi-frequency optimization problem can now be written as:

$$\min \sum_{l=1}^{5} \left(|| \begin{pmatrix} A^l & -B^l \\ B^l & A^l \end{pmatrix} \begin{pmatrix} X & X \\ X & X \end{pmatrix} \begin{pmatrix} w^l \\ v^l \end{pmatrix} + \begin{pmatrix} s^l \\ t^l \end{pmatrix} ||_2 \right)^2 \tag{8.5}$$

$$\begin{array}{rlr} \text{subject to} & & \\ \sum_{j=1}^{64} x_{jj} & = k & \\ x_{ij} & = 0 & \text{for all } i \neq j \\ \sum_{l=1}^{5} || \begin{pmatrix} w_j^l \\ v_j^l \end{pmatrix} ||_2 x_{jj} & \leq volt & \text{for all } j = 1 \ldots, 64 \\ \mathbf{w}^l, \mathbf{v}^l & \in \mathcal{R}^{64} & \text{for all } l = 1, \ldots, 5 \\ x_{jj} & \in \{0, 1\} & \text{for all } j = 1, \ldots, 64 \end{array}$$

To find good solutions to the above problem, we use a nested approach. In the outer loop we search over the discrete variables with tabu search, which assigns the 0 or 1 value to each x_{jj}. In the inner loop we solve a quadratically constrained quadratic program (QCQP). We compare two methods for solving QCQP—a Lagrangian relaxation and CONMIN (for a feasible direction search, see [14]). The Lagrangian approach has been used previously to limit actuator forces in Filtered-X LMS noise control systems (see [9]). These efforts have been concerned with increasing the robustness of a control system by penalizing the high actuator forces, which may result from the response at resonance or instabilities introduced by an ill-conditioned system. To implement the penalty, a term dependent on the actuator force is added to the objective function in (8.4). That is,

$$g(\mathbf{f}, \mathbf{r}) = (||H^K \mathbf{f} + \mathbf{p}||_2)^2 + \sum_{j=1}^{K} r_j \ (||f_j||_2 - volt) \tag{8.6}$$

In (8.6) it is easy to see that $\mathbf{r}$ acts as a vector of Lagrange multipliers for the constraints in (8.4). Given a set of values for $\mathbf{r}$ and a set of columns K, we follow the same derivation that led us from (8.1) to (8.2) and we find that a solution to (8.6) is given by

$$\mathbf{f} = -(H^{K*}H^K + R)^{-1} \cdot H^{K*}\mathbf{p} \tag{8.7}$$

where R is an n by n diagonal matrix with $\mathbf{r}$ as the diagonal entries and the off diagonal entries are all 0. To extend this solution procedure to multiple frequencies, we sum together the single frequency solutions. That is, we seek to minimize a performance measure related to $G(\mathbf{f}, \mathbf{r}^{\,1}, \ldots, \mathbf{r}^{\,5}) = \sum_{l=1}^{5} g(\mathbf{f}, \mathbf{r}^{\,l})$.

8.3 DESCRIPTION OF SOLUTION STRATEGIES

In this section we describe two procedures embedded within tabu search for obtaining solutions to QCQP as well as the tabu search methodology. A plain vanilla tabu search (constructive start, attribute based tabu list, and aspiration criterion) controls the selection of the k actuator sites. Given a set of K actuator sites, we solve the QCQP for the control forces $\mathbf{f}$ with a Lagrangian relaxation method or with CONMIN.

8.3.1 Plain vanilla tabu search (PVTS)

The selection of actuator sites (columns of H) is done by a tabu search approach. Each column of H is either in the current solution (ON) or not in the current solution (OFF). Hence, ON is a collection of k column indices of H which has n columns in total. There are a total of $\binom{n}{k}$ possible solutions. Each solution is characterized by an ON vector (or an OFF vector). The search explores the exponentially large solution space by examining the neighbors of ON—all pairwise swaps (a move) between a column of ON with a column of OFF. After completely examining all neighbors of ON, the best admissible move is selected and a new solution is constructed. A simple 2-dimensional tabu list is maintained by updating the tabu age of each move. That is, the upper triangular portion of a 64 by 64 matrix records the tabu age of each move. For example, if column 2 (an element of ON) is to be exchanged with column 29 (an element of OFF) at iteration 50, then entry (2,29) of this matrix would be assigned the value 50+*tabu tenure*. In subsequent iterations the move (29,2) is considered tabu for *tabu tenure* iterations unless the move meets the aspiration criterion (the decibel level is better than any solution yet observed).

An initial set ON is constructed by selecting a set of k columns indices $K \subseteq \{1, \ldots, n\}$ based on the individual properties of each H_l for all $l = 1, \ldots, 5$. We refer the reader again to the single frequency example in Figure 8.1. If there exists a column of H_l that is collinear with $\mathbf{p}^{\ l}$ for some l then, as in Figure 8.1, we would pick this single actuator site, driven at the appropriate amplitude, and total silencing would be achieved at frequency l. Typically, this situation does not exist. However, a likely set of columns for a starting solution are the columns of H_l most nearly collinear with $\mathbf{p}^{\ l}$. To find these columns, we determine the angle between each actuator j site and $\mathbf{p}^{\ l}$ for all $l = 1, \ldots, 5$. This is done by calculating the dot product of each column of H_l matrix, $\mathbf{h}_j{}^l$, with the vector $\mathbf{p}^{\ l}$ and dividing that product by the product of the norms of each vector. That is, the expression

$$\frac{\mathbf{h}_j{}^l \cdot \mathbf{p}^{\ l}}{||\mathbf{h}_j{}^l|| \cdot ||\mathbf{p}^{\ l}||}$$

calculates the cosine of the angle between the two vectors. Since the actuators that would be most effective, theoretically, in silencing the outside noise source would be those either in phase or 180 degrees out of phase, we seek cosine values close to either 1 or -1. Thus, the k initial actuators are chosen that have the absolute value of the cosine closest to 1.

We have made no attempt to include more sophisticated diversification and intensification strategies for the multi-frequency constrained actuator location problem. Embedded within the tabu search is a call to solve the QCQP previously described.

8.3.2 Lagrangian approach to QCQP

The Lagrangian relaxation of the QCQP (see equations (8.6) and (8.7)) is solved as a subproblem within our PVTS described in Section 8.3.1 That is, as tabu search examines all neighbors of a particular solution, a Lagrangian relaxation of the associated QCQP is solved. When $k = 12$, we have 624 relaxations to solve per iteration of tabu search. Clearly, this is the bottleneck operation in the actuator selection problem. There are a variety of approaches to selecting Lagrange multipliers. Standard schemes include multiplier adjustment and subgradient optimization. However, both of these are computationally expensive. Here we use a simple uniform weighting scheme from [12].

In the uniform method the diagonal entries for R_l are found for each frequency by defining a scalar r^l, so that, $R_l = r^l I$ (i.e. $r^l = r_j{}^l$ for all $j = 1, \ldots, K$) and $r^l = (\sigma_{max}||\mathbf{p}^{\ l}||_2/||\mathbf{f}^{\ l}||_2) - \sigma^2{}_{min}$ where σ_{max} and σ_{min} are the minimum and maximum singular values of H_l. A conservative estimate for r^l can be found by substituting *volt* for $||\mathbf{f}^{\ l}||_2$. The main criterion in selecting an estimate for $||\mathbf{f}^{\ l}||_2$ is to find a value that satisfies the constraint so that a feasible $\mathbf{f}$ is found. Figure 8.4 contains two bar charts. There are 14 actuators

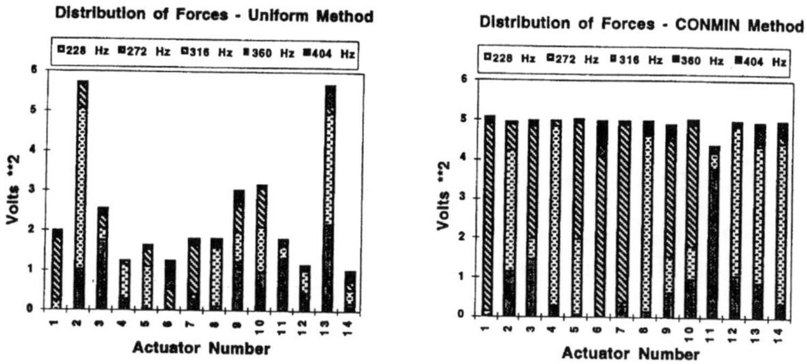

Figure 8.4 Distribution of Actuator Forces for CONMIN and Uniform Method

along the x-axis. The y-axis records the portion of the force, in *volt* $**2$, allocated to each of the 5 frequencies. The left half of Figure 8.4 provides a typical solution pattern for the uniform method when $k = 14$. Notice that only two of the 14 actuators use the maximum amount of volts allowed ($volt^2 = 5$). This can be remedied by a more sophisticated approach to the selection of r^l, such as subgradient optimization or a multiplier adjustment scheme. The right half of Figure 8.4 shows a solution found by using CONMIN to select the r^l values. Here we see that nearly all of the 14 actuators use the maximum amount of force allowed. The inferiority of the uniform method solution is not surprising. However, we will show that although the values found for $\mathbf{f}^{\,l}$ by the uniform method are not particularly good, the actuator sites chosen are quite good.

8.3.3 CONMIN approach to QCQP

CONMIN requires the user to select a variety of parameter values that affect both the execution time of the procedure and the quality of solutions generated. We used as many of the default values as possible. Non-default values specified were the two stopping conditions—maximum number of iterations allowed (200) and number of iterations (10) for which the objective value does not improve. The numbers 200 and 10 were chosen based on the performance of CONMIN in a series of trial runs.

Calling CONMIN to evaluate every neighbor of the current solution within the tabu search proved to be extremely time consuming. We made several adjustments to bring the runtime into the realm of practicality (less than five hours). First, we solved the unconstrained problem equation (8.1) for each frequency and evaluated its performance with equation (8.3). We kept track of the best unconstrained neighbor. Any unconstrained solution that was within 95 percent (90 percent in some cases) of the best unconstrained neighbor was evaluated with CONMIN. In addition, no more than 15 percent of the neighborhood of any solution was allowed to be evaluated with CONMIN. Hence, we have a candidate selection procedure within PVTS. At the end of each neighborhood search the best admissible move was made. Next a random permutation ordering of the neighborhood indices was done so that the neighbors on the subsequent iteration were considered in a different order than before.

8.4 EXPERIMENTAL RESULTS

In this section we present numerical experiments to evaluate the possible solution strategies. In each case we use PVTS to select actuator locations but we use either the uniform method or CONMIN to evaluate the performance of each set of proposed actuators. We seek to discover whether the computationally efficient PVTS with the uniform method yields actuator sites whose performance is comparable to those found by PVTS with CONMIN. We expect the CONMIN approach to be more reliable since it uses the actuator forces as decision variables and directly imposes constraints on those forces (recall equation (8.5) and the right half of Figure 8.5). In addition, we compare results with and without force constraints to determine whether the force constraints are influential in the selection of actuator sites.

We require some method to compare the combinations of actuators chosen by each solution strategy. Ideally, we would test all combinations in the laboratory as in [10]. Since this was not practical we devised a numerical evaluation. Given a fixed set of actuator locations, CONMIN determines values for $\mathbf{r}^{\,l}$ so that $g^l(\mathbf{f}, \mathbf{r}^{\,l})$ is minimized. The associate decibel value is then used to compare competing sets of actuator locations for each solution strategy.

Table 8.1 catalogs the decibel values for the best solutions found under varying conditions and Table 8.2 catalogs the CPU time required for each solution. In Table 8.1, column k designates the number of actuators to be selected. Column 1 records the decibel value for the best solution found by PVTS with the uniform method as the solution procedure for solving the Lagrangian relaxation of the QCQP. Column 3 records the decibel value associated with the best solution found by PVTS when CONMIN solves the QCQP directly (not the Lagrangian relaxation). Column 5 gives the best decibel value when no constraint is placed on the control force at each actuator site. The even numbered columns (2, 4 and 6) record the decibel value obtained when the actuator sites associated with the best solutions in the odd numbered columns are evaluated with the Lagrangian relaxation of QCQP with the Lagrange multipliers selected by CONMIN. That is, we seek to minimize $G(\mathbf{f}, \mathbf{r}^{\,1}, \ldots, \mathbf{r}^{\,5})$ with CONMIN. If we view each of the solution procedures as methods for selecting actuator sites, then the decibel values recorded in the even numbered columns can be used to evaluate the performance of the selected actuator locations.

Several observations can be drawn from the results presented in Table 8.1 First, comparing columns 3 and 4, the noise reduction predictions are quite similar. That is, there is only a small difference between using CONMIN to solve the Lagrangian problem associated with QCQP or CONMIN to solve QCQP directly. Second, comparing columns 2 and 4, the performance of the actuators selected by PVTS with the uniform method is not significantly different from

	PVTS	uniform	PVTS	CONMIN	PVTS	unconst.
k	**1**	**2**	**3**	**4**	**5**	**6**
2	-1.08	-1.34	-1.35	-1.35	-4.21	-0.57
4	-1.55	-1.98	-2.12	-2.03	-5.93	-1.10
6	-1.97	-2.56	-2.56	-2.65	-8.22	-1.59
8	-2.31	-2.99	-3.10	-3.11	-10.13	-1.59
12	-2.94	-4.22	-3.42	-3.48	-14.16	-1.22

Table 8.1 Comparison of Tabu Search Methods (in decibels)

that obtained by PVTS with CONMIN. (A difference of .1 dB is not detectable by the average listener.) However, the PVTS with CONMIN runs take roughly 100 times longer to compute than the PVTS with the uniform method (see Table 8.2) when all computational experiments were conducted on an *SUN ultra2 3D* workstation. Moreover, the filtering technique for PVTS with CONMIN to screen which unconstrained solutions should be solved by CONMIN works well for small number of actuators, but it is not effective for the 12 actuator case (-3.48 CONMIN vs. -4.22 uniform). It appears that for 12 actuators, the top 15 percent of the best unconstrained solutions do not include the best constrained solutions. This degradation can also be seen in the 12 actuator solutions for unconstrained PVTS (columns 5 and 6). Here, the -1.22 for 12 is worse than the -1.59 for 6 and 8 actuators.

For small numbers of actuators (8 or less) there is a strong similarity between the unconstrained solutions and the constrained solutions, but starting at 12 actuators the solutions become quite different. Figures 8.5 and 8.6 show the actuator sites chosen by the three methods—PVTS with the uniform method, PVTS with CONMIN, and PVTS unconstrained—for the 6 actuator problem (Figure 8.5) and the 12 actuator problem (Figure 8.6). The unconstrained solution for 12 actuators uses 7 actuator sites on the right side of the fuselage (i.e., away from the noise source) and none of these 7 actuator sites appeared in any of the uniform or CONMIN solutions from 2 to 12 actuators. However, for the 6 actuator problems, 5 out of the 6 sites chosen for the unconstrained solution are on the half of the fuselage facing the noise source. In addition, 3 out of the 6 sites chosen in the unconstrained case are contained in both the uniform and the CONMIN best solutions for 6 actuators. The results in Table 8.1 for PVTS unconstrained also illustrate that the actuators chosen for the unconstrained solutions are not good solutions for the constrained problem (compare columns 5 and 6).

8.5 DISCUSSION

We have extended previous research efforts on single frequency unconstrained active noise control to include multiple discrete frequencies and constraints on the force needed to control each actuator. The overall noise attenuation

k	Uniform	CONMIN
2	1 sec	52 sec
4	5 sec	1026 sec
6	14 sec	4330 sec
8	32 sec	8199 sec
12	121 sec	17507 sec

Table 8.2 Run-time Comparison: PVTS–uniform versus PVTS–CONMIN

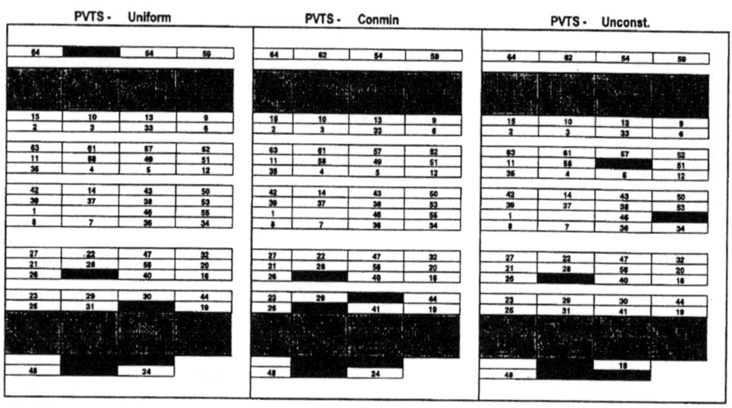

Figure 8.5 Best Locations for 6 Actuators

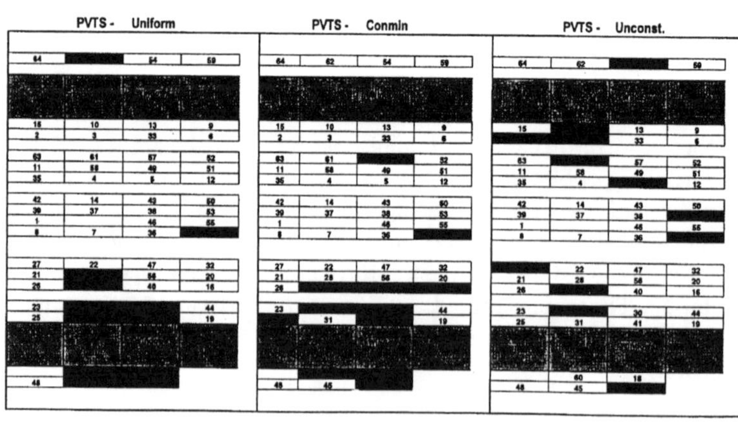

Figure 8.6 Best Locations for 12 Actuators

achieved by 12 (piezo-electric) actuators of -4.2 decibels may encourage researchers to consider other types of actuators with less severe actuator force restrictions. Current experiments at NASA-Langley are examining alternative types of actuators. However, the methods described here are not restricted to a particular type of actuator.

The computational experiments verify that PVTS with the uniform method solving the Lagrangian relaxation of QCQP provides high-quality actuator locations in a reasonable amount of computing time. For example, the 12 actuator case took only 2 minutes to solve on a *SUN ultra2 3D* workstation. We also verified that the uniform heuristic procedure performed as well or better than a more conservative approach. PVTS with CONMIN solved the QCQP directly to evaluate each set of actuator sites. It found slightly better solutions for 4 and 8 actuators but performed poorly when 12 (or more) actuators were considered. Even with our candidate selection strategy this method was still more than 100 times slower than PVTS with the uniform method. Hence, although there are arguably better ways than CONMIN to solve the QCQP, it appears unlikely that any method will find significantly better sets of actuators within the same amount of time as PVTS uniform.

There is an interesting connection to the location/allocation literature that should be explored. Here, the actuators act as the facilities and the sensors act as the customers. However, in the allocation phase instead of assigning specific customers to specific facilities, we assign a portions of the demand for each customer to be met by each facility. Furthermore, we hope that the sensors (customers) are representative of the of the system-wide demand pattern (interior noise). Thinking of the problem in this way gives a window into the associated computational complexity issues. Nearly every location problem of interest is NP-Hard. We strongly suspect that the actuator selection problem described in this manuscript is NP-Hard as well. [6] provide a discussion of complexity issues for an actuator placement problem on large flexible space structures in which there is no allocation phase to the optimization model.

References

[1] Cabell, R.H. and H.C. Lester "Investigation of Force Distributions for Interior Noise Control Using a Neural Network," *Proceedings of the Second Conference on Recent Advances in Active Control of Sound and Vibration*, Blacksburg, VA, April 1993, 55-69.

[2] DeMeis, R. "Quieting Cabin Noise," *Aerospace America*, February 1995, 20-21.

[3] Fuller, C.R and A.H. von Flotow "Active Control of Sound and Vibration," *IEEE Control Systems*, December 1995, 9-19.

[4] Glover, F. "Tabu Search and Adaptive Memory Programming–Advances, Applications and Challenges," in Interfaces in Computer Science and Op-

erations Research, Barr, R.S., Helgason, R.V. and Kennington, J.L. (eds.) Kluwer Academic Publishers, 1997, 1-75.

[5] Kincaid, R.K., K.E. Laba, and S.L. Padula "Quelling Cabin Noise in Turboprop Aircraft via Active Control," *J. of Combinatorial Optimization*, **1**, 1997, 229-250.

[6] Kincaid, R.K., R.T. Berger "The Damper Placement Problem on Space Truss Structures," *Location Science*, **1**, 1993, 219-234.

[7] Leug, P. "Process of Silencing Sound Oscillations," 1936, U.S. Patent No. 2,043,416.

[8] Lyle, K.H. and R.J. Silcox "A Study of Active Trim Panels for Noise Reduction in an Aircraft Fuselage," paper no. 951179 in SAE General, Corporate and Regional Aviation Meeting and Exposition, May 1995.

[9] Mathur, G., B. Tran, and M. Simpson "Broadband Active Structural Acoustic Control of Aircraft Cabin Noise–Lab Tests," Proceedings 3rd AIAA/CEAS Aeroacoustics Conference, AIAA-97-1636, 1997.

[10] Palumbo, D., S.L. Padula, K.H. Lyle, J.H. Cline, and R.H. Cabell "Performance of Optimized Actuator and Sensor Arrays in an Active Noise Control System," presented at 2nd AIAA/CEAS Conference, AIAA 96-1724, 1996.

[11] Palumbo, D. and S.L. Padula "Optimizing an Actuator Array for the Control of Multi-Frequency Noise in Aircraft Interiors," presented at 3rd AIAA/CEAS Aeroacoustics Conference, May (AIAA 97-1615), 1997.

[12] Rosetti, D.J., M.R. Jolly, and S.C. Southward Control Effort Weighting in FeedForward Adaptive Control Systems," *J. Acoustical Society of America*, Vol. 99, No. 5, 1996, 2955-2964.

[13] Silcox, R.J., C.R. Fuller, and H.C. Lester "Mechanisms of Active Control in Cylindrical Fuselage Structures," *AIAA Journal*, Vol. 28, August 1990, 1397-1404.

[14] Vanderplaats, G.N. "CONMIN – A Fortran Program for Constrained Function Minimization User's Manual," NASA Technical Memorandum X-62 1973, 282.

9 NEIGHBORHOOD SEARCH ALGORITHM FOR THE GUILLOTINE NON-ORIENTED TWO-DIMENSIONAL BIN PACKING PROBLEM

Andrea Lodi, Silvano Martello and Daniele Vigo

Dipartimento di Elettronica, Informatica e Sistemistica, University of Bologna
Viale Risorgimento, 2 - I - 40136 Bologna, Italy.
alodi@deis.unibo.it, smartello@deis.unibo.it, dvigo@deis.unibo.it

Abstract: We are given a set of rectangular small pieces which may be rotated by 90°, and an unlimited number of identical rectangular large stock pieces. We consider the problem of determining the guillotine-cuttable patterns that provide all the pieces, such that the total number of required stock pieces is minimized. We show how some classical greedy algorithms given in the literature for the case where no rotation is allowed can be adapted to our problem; in addition, an original heuristic algorithm is presented. We then introduce a tabu search approach to the problem, and analyze its average performance through extensive computational testing.

9.1 INTRODUCTION

In two-dimensional geometrical packing problems we are given a set of n rectangular *pieces*, each characterized by width w_j and height h_j $(j = 1, \ldots, n)$, which must be cut from standardized stock pieces. Two main versions of the problem have been considered in the literature: strip packing and bin packing. In the *Two-Dimensional Strip Packing Problem*, we have a unique stock piece (*strip*) having width W and infinite height: the objective is to determine a cutting pattern that provides all the pieces, such that the height to which the strip is filled is minimized. In the *Two-Dimensional Bin Packing Problem*, we have an unlimited number of identical rectangular stock pieces (*bins*) having width

W and height H: the objective is to determine cutting patterns that provide all the pieces, such that the total number of required bins is minimized.

Both problems are strongly NP-hard, as can be easily seen by transformation from the well-known (one-dimensional) *Bin Packing Problem*, in which a set of n positive values w_j has to be partitioned into the minimum number of subsets so that the total value in each subset does not exceed a given bin capacity W.

Two-dimensional packing problems have a number of important applications in industrial cutting (of glass, wood, textiles, paper, etc.), and in logistics and transportation (packing on floors, shelves, truck beds, etc.). According to specific applications, for both versions of the problem several variants are possible. For example, the pieces may have fixed orientation, i.e., they cannot be rotated, or *rotation* (usually by 90 degrees) may be allowed. Moreover, in cutting problems it is frequently required that the patterns be such that the pieces can be obtained by sequential edge-to-edge cuts parallel to the edges of the bin or of the strip (*guillotine cutting*).

In this paper we consider the two-dimensional bin packing problem, with 90 degrees rotation allowed and a guillotine cutting constraint: the problem will be denoted by 2D-BPRG. A practical situation where this type of problem is relevant arises in the glass industry: the cutting machines can only operate in guillotine mode, and the orientation of the pieces in a pattern is unimportant (if the glass is not decorated). Real world numerical instances of this type of problem are also presented in the computational experiments of Section 9.4.

Two-dimensional packing problems received considerable attention in the literature, although, to our knowledge, no work has been devoted explicitly to 2D-BPRG. For the two-dimensional bin packing problem in the case where no rotation is allowed and the guillotine constraint is not imposed, Martello and Vigo [18] presented lower bounds and an exact branch-and-bound algorithm, while Chung, Garey and Johnson [8] and Frenk and Galambos [14] gave approximation algorithms with worst-case analysis. Berkey and Wang [6] and Lodi, Martello and Vigo [17] proposed heuristic algorithms for the variant in which rotation is not allowed but the guillotine constraint is satisfied. Bengtsson [5] and El-Bouri, Popplewell, Balakrishnan and Alfa [13] gave heuristic algorithms for the case where rotation is allowed but the guillotine constraint is not imposed. Approximation algorithms for the two-dimensional strip packing problem were studied by Baker, Coffman and Rivest [1], Coffman, Garey, Johnson and Tarjan [9], Bartholdi, Vande Vate and Zhang [2], Coffman and Shor [10] and Golan [16]. An annotated bibliography on the cutting and packing area has been presented by Dyckhoff, Scheithauer and Terno [12].

In Section 9.2 we show how some classical greedy algorithms given in the literature for the case where no rotation is allowed can be adapted to our problem; in addition, an original heuristic algorithm is presented. In Section 9.3 we give a tabu search approach to 2D-BPRG, obtained by improving the basic scheme used in [17] and by incorporating original features. An extensive computational testing is presented in Section 9.4, showing the effectiveness of the proposed algorithms.

We assume in the following, without loss of generality, that all input data are positive integers satisfying $\min\{w_j, h_j\} \leq \min\{W, H\}$ and $\max\{w_j, h_j\} \leq \max\{W, H\}$ for $j = 1, \ldots, n$.

9.2 GREEDY HEURISTICS

Berkey and Wang [6] experimentally compared simple and effective algorithms that satisfied the guillotine constraint for the case where no rotation is allowed: good average results were obtained by the algorithms called FFF and FBS. Both algorithms start by sorting the pieces according to non-increasing height, and pack the pieces in rows forming levels (*shelves*). The first shelf is the bottom of a bin. Pieces are packed from left to right with their bottoms at the selected level. When a new level is needed, it is created along the horizontal line which coincides with the top of the first (i.e., tallest) piece packed in the lower level.

In the *Finite First Fit* (FFF) algorithm the bins are directly stratified into shelves. The next piece j is packed left-justified into the lower level of the first bin where it fits. If no shelf can accommodate j, then a new level is created in the first suitable bin. If no bin has enough vertical space for a new level accommodating j, then a new bin is created and initialized with j.

The *Finite Best Strip* (FBS) algorithm works in two phases. First, a strip of width W and infinite height is stratified into shelves according to the following strategy. The next piece j is packed into the feasible shelf having the minimum residual horizontal space; if no shelf can accommodate j, a new shelf is created. Let $h_1, \ldots, h_m$ be the heights of the resulting shelves. In the second phase, the shelves are combined into finite bins using the well-known *Best-Fit Decreasing* algorithm for a one-dimensional bin packing instance consisting of the m values h_j with bin capacity H.

As previously mentioned, FFF and FBS produce patterns satisfying the guillotine constraint: a sequence of horizontal cuts separates the shelves and, for each shelf, a sequence of vertical and horizontal cuts produces the pieces. We have modified such algorithms so as to handle the 90 degrees piece rotations: we call these modified versions FFF_{RG} and FBS_{RG}. We assume, without loss of generality, that $W \geq H$; we will say that a piece is packed *horizontally* (resp. *vertically*) if its longest (resp. shortest) edge is parallel to the bottom of the shelf.

Both modified algorithms start by sorting the pieces according to non-increasing value of the shortest edge. Algorithm FFF_{RG} is like FFF, with the following extensions:

(i) a piece that initializes a shelf is always horizontally packed, in order to keep the vertical occupancy of the corresponding bin as low as possible;

(ii) when a piece is packed into an existing shelf, if both orientations are feasible then the vertical one is selected, in order to keep the horizontal occupancy of the shelf as low as possible.

Analogously, the extension of FBS to FBS_{RG} is obtained by modifying the evaluation of a possible placement: if both orientations of the piece are feasible, then the vertical one is used, in order to maximize the probability of additional packings on the shelf.

We now introduce a new heuristic for 2D-BPRG, which works as FBS_{RG}, in two phases but differs from it in the way the pieces are packed into the shelves. A shelf is initialized by placing a piece with its left edge touching the left edge of the strip, into the lowest admissible position: we call *floor* (resp. *ceiling*) of the shelf the horizontal line touched by the bottom (resp. top) edge of such piece. Algorithms FFF_{RG} and FBS_{RG} pack the pieces exclusively on the floor. Our algorithm *Floor-Ceiling* (FC_{RG}), instead packs the pieces into the shelves either with their bottom edge touching the floor or with their top edge touching the ceiling, in such a way that guillotine cutting of the pieces allocated to the shelves is possible. On the floor the pieces are packed from left to right, while on the ceiling they are packed from right to left in an attempt to fill the space over the lowest pieces packed on the floor. A piece packed on the floor always has its left edge touching the right edge of another piece packed on the floor, while pieces packed on the ceiling may be separated by empty spaces in order to allow guillotine packing.

The *right-end* of the shelf coincides with the right edge of the strip, while its *left-end* is the right edge of the rightmost piece packed on the floor and such that the vertical space between the top edge of the piece and the ceiling is too small to allocate any of the pieces that are not yet packed. Figure 9.1 illustrates the definitions and properties above.

We say that a piece is *floor-feasible* for a given shelf if there is room for packing it (in either orientation) on the floor, at the right of the rightmost piece already packed on the floor. A shelf is called *ceiling-initialized* if there are pieces packed on its ceiling. Ceiling initialization is always performed by packing a piece that is *not* floor-feasible: the piece is packed with its right edge touching the right edge of the strip (see piece number 5 in Figure 9.1), in an orientation chosen according to rules given below.

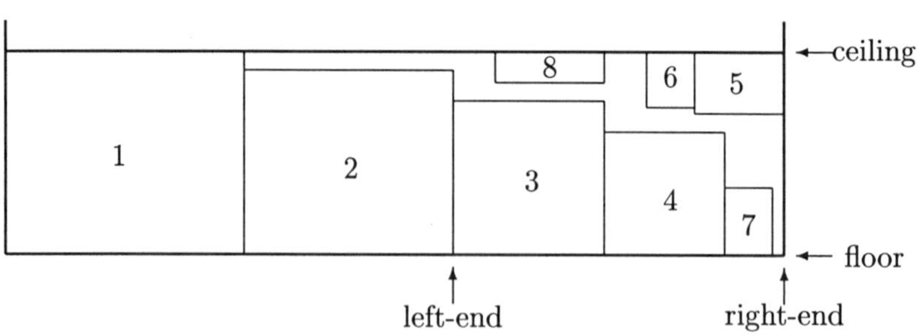

Figure 9.1 A shelf pattern produced by Algorithm FC_{RG}.

A piece is *ceiling-feasible* for a given shelf if either it can initialize the ceiling (see above) or it can be packed on the ceiling, in either orientation, to the left of the pieces already packed on the ceiling and in such a way that guillotine cutting is possible. To this end, first observe that a series of horizontal edge-to-edge cuts along the ceilings separate the shelves. Given a shelf:

(i) a series of vertical cuts obviously separates the pieces packed on the floor;

(ii) when a piece j^* initializes the ceiling, there is a horizontal cut C^* along its lower edge, and extending from the right edge of the strip to the vertical cut $\overline{C}$ that separates (see (i)) the leftmost piece packed on the floor which occupies an horizontal strip portion in common with j^*: hence, no further piece will be packed so as to cross C^*;

(iii) each piece packed on the ceiling to the left of $\overline{C}$ is placed in the rightmost position where it fits without crossing any line extending the vertical edge of a piece packed on the floor (see Figure 9.2).

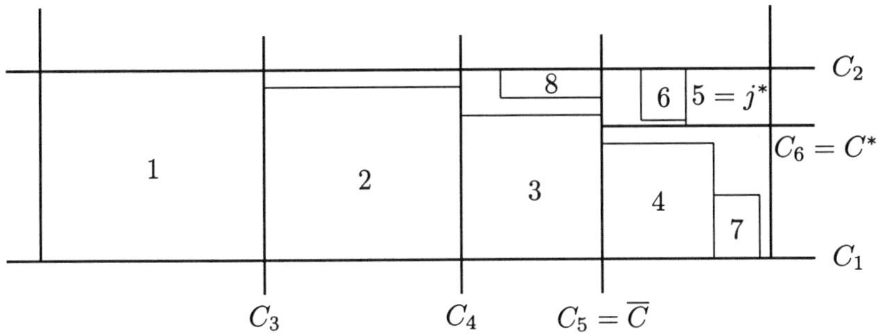

Figure 9.2 Guillotine-feasible cutting pattern for the solution of Figure 9.1.

The algorithm can thus be outlined as follows. (The term "best", used to select the shelf, will be defined later.)

Algorithm FC_{RG}

1. sort and renumber the pieces according to non-increasing value of the shortest edge;

2. **for** $j := 1$ **to** n **do**

  if j is ceiling-feasible for some shelf **then**

    pack j on the ceiling of the best shelf

  else

   if j is floor-feasible for some shelf **then**

    pack j on the floor of the best shelf

   else

    initialize a new shelf by horizontally packing j;

3. use a Best-Fit Decreasing 1-dimensional bin packing algorithm to pack the resulting shelves into finite bins.

(Note that once the ceiling has been initialized, we always prefer ceiling packings, since a good filling of the upper portions of the shelves tends to delay the initialization of new shelves).

When both orientations are feasible, the piece is always packed in the orientation which favors the possibility of future packings within the considered shelf. To this end, for ceiling placements we always select the vertical orientation, since this maximizes the residual horizontal space in the upper part of the shelf. For floor placements, a strategy depending on the current status of the shelf is adopted: if the shelf is ceiling-initialized, the vertical orientation is selected, since this maximizes the residual horizontal space in the lower part of the shelf; otherwise, the horizontal orientation is selected, thus increasing the possibility of future ceiling placements.

In order to decide which is the best shelf, we use a *score* defined by the residual horizontal space that would be left by such placing. More precisely, for a floor (resp. ceiling) placing the score is the distance between the right (resp. left) edge of the piece and the right-end (resp. left-end) of the shelf: the best solution is then the one corresponding to the minimum score, in the spirit of the best-fit rule for the 1-dimensional bin packing problem.

9.3 TABU SEARCH

In this section we present a tabu search approach for 2D-BPRG. The reader is referred to [15] for a general introduction to tabu search and to basic definitions.

Our algorithm combines the strategy adopted in [17] (based on a main loop where, at each iteration, two neighborhoods are explored) and an original restart policy (that can be viewed as a third neighborhood). The general structure of the algorithm is outlined below, while a detailed description of the various components is given in the following sections.

Algorithm TS
1. initialize the tabu lists to empty and determine a starting feasible solution;
2. select a bin b;
  comment: first neighborhood;
  for each non-tabu piece j currently packed into b **do**
    try to move j to a different bin i^*;
    comment: the tabu list stores the recently moved pieces;
  if bin b is empty **then go to** 2
  else
    begin
3.   **comment**: second neighborhood;
      for each piece j currently packed into b **do**
        try to re-combine the contents of two bins h,k ($\neq b$) so that one of them can accommodate j, and associate a score with such move;
      if *restart conditions* do not apply **then**
        perform the non-tabu move with the highest score, store such score

in the second tabu list and **go to** 2
else
if *stopping conditions* do not apply **then**
4. perform a restart action and **go to** 1
end.

In the next sections we describe in detail the various steps of algorithm TS.

9.3.1 *Initialization and basic definitions*

The algorithm starts by applying the heuristics of the previous section and computing the continuous lower bound

$$LB = \left\lceil \frac{\sum_{j=1}^{n} w_j h_j}{WH} \right\rceil \tag{9.1}$$

Let UB denote the value of the best heuristic solution found. If $UB > LB$, the tabu search algorithm is initialized with the solution produced by algorithm FC_{RG}. Let z denote the number of bins used in the current solution. The algorithm accepts moves that either decrease z or re-distribute pieces among the z bins; moves increasing z are never accepted. Restart strategies are provided in order to have diversification and intensification during the search (see below). The search terminates if a feasible solution of value LB is determined, or if a prefixed maximum number of iterations (or a time limit) is reached.

Both neighborhoods consist of moves involving the pieces of a particular bin, which is called the *weakest bin* and is defined as follows. Let S_i denote the set of pieces that, in the current solution, are packed into bin i ($i = 1, \ldots, z$). The weakest bin is then defined as the one that minimizes the quantity

$$\varphi(i) = \alpha \, \frac{\sum_{j \in S_i} w_j h_j}{WH} - \frac{|S_i|}{n} \tag{9.2}$$

that gives an estimate of the difficulty of emptying bin i: a weak bin packs a small fraction of the available area (first term) using a relatively high number of pieces (second term); α is a prefixed nonnegative parameter measuring the relative weight of the two terms. Given a subset of pieces S, let $\mathrm{FC}_{RG}(S)$ denote the number of $W \times H$ bins used by algorithm FC_{RG} to pack all the pieces of S.

9.3.2 *First neighborhood*

Let b be either the weakest bin determined through (9.2), or, in case of restart (see Section 9.3.4), a bin received on input. In the first neighborhood we consider, in turn, each non-tabu piece $j \in S_b$ and we compute $\mathrm{FC}_{RG}(\{j\} \cup S_i)$ for all bins $i \neq b$: whenever a bin i^* is found for which $\mathrm{FC}_{RG}(\{j\} \cup S_{i^*}) = 1$, piece j is moved into bin i^*, it is inserted into tabu list $T1$, and the next piece of b is considered. The following aspiration criterion is adopted: if $|S_b| = 1$ then the search is performed even if piece j is tabu, since the possible move would improve z. When a series of moves empties the weakest bin, z is decreased by one, the new weakest bin is determined and the search continues.

As soon as no move is possible, the search explores the second neighborhood. Tabu list $T1$, which stores the recently moved pieces, is preserved for the next exploration of the first neighborhood.

9.3.3 *Second neighborhood*

In the second neighborhood, for each piece $j \in S_b$ and for each pair of bins h and k $(h, k \neq b)$, we compute $FC = \mathrm{FC}_{RG}(\{j\} \cup S_h \cup S_k)$. Four cases may occur:

- if $FC = 1$ then z may be decreased by one or two units, so the move is immediately performed and the exploration is resumed from the first neighborhood;
- if $FC = 2$ then piece j is removed from bin b and two subcases are considered:
  - if $|S_b| = 0$ then z is decreased by one and the exploration is resumed from the first neighborhood;
  - if $|S_b| > 0$ then the new weakest bin $\bar{b}$ is determined among b, h and k, and the search continues with the second neighborhood if $\bar{b} \equiv b$, or with the first neighborhood, otherwise;
- if $FC = 3$ then let Q be the set of pieces packed by FC_{RG} into the bin with minimum $\varphi(i)$ among the three resulting bins: we compute $\overline{FC} = \mathrm{FC}_{RG}(Q \cup (S_b \setminus \{j\}))$ and consider two subcases:
  - if $\overline{FC} > 1$ then the move is not accepted since it would increase z;
  - if $\overline{FC} = 1$ then we assign to the move a *score* defined by the φ value of the resulting packing of the pieces in $Q \cup (S_b \setminus \{j\})$: if this score is not tabu, the best acceptable move not improving z is possibly updated;
- if $\overline{FC} > 3$ then the move is rejected since it would increase z.

At the end of the exploration the best non-improving acceptable move (if any) is performed, its score is stored in the tabu list $T2$, and the search continues with the first neighborhood. For both tabu lists we used static tabu tenures of length τ_1 and τ_2, respectively.

9.3.4 *Restart (third neighborhood)*

The algorithm performs three different types of restart actions (denoted by R1, R2 and R3) obtained through a combination of diversification and intensification strategies. Some of these actions imply to *fix* a bin, i.e., to continue the search over a reduced instance obtained by removing the pieces currently packed into the fixed bins and assuming that the corresponding patterns constitute an imposed partial solution:

- R1: let b be the current weakest bin, define $\bar{b}$ through $\varphi(\bar{b}) = \min\{\varphi(i) : \varphi(i) \geq \varphi(b), i \neq b\}$ (see 9.2) and restart from the first neighborhood giving $\bar{b}$ on input;
- R2: let P be the set of pieces currently packed into the $\lfloor z/2 \rfloor$ less weak nonfixed bins (i.e., nonfixed bins which, in the current solution, have the highest φ values), re-pack each piece of P into a separate bin and restart from the first neighborhood. In this way the current solution is increased by $|P| - \lfloor z/2 \rfloor$;
- R3: let $\tilde{b}$ be the less weak nonfixed bin, fix it and perform a restart of type R2.

These restart actions are performed in two different situations (called S1 and S2), arising at the end of the exploration of the second neighborhood:

- S1 = no acceptable move was found, i.e., some of the moves are tabu and the others would produce an increase of the value of z;
- S2 = z was not improved during the last μ moves.

In other words, the term "restart conditions" in the **if** statement of Step 3 of algorithm TS means: S1 **or** S2.

The restart action of Step 4 is executed as follows. Let f be the number of currently fixed bins and r be the number of restarts of type R1 recently performed. Initially, f and r take value 0; t denotes the computing time spent so far; R and T are prefixed parameters:

...
4. if S1 **and** $r < R$ **then**
  perform a restart of type R1 and set $r := r + 1$
 else
  begin
   set $r := 0$;
   if $t < T$ **then**
    perform a restart of type R2
   else
    if $f < \lfloor z/2 \rfloor$ **then**
     set $f := f + 1$ and perform a restart of type R3
    else
     reinsert into the instance all pieces packed into the fixed bins,
     set $f := 0$ and perform a restart of type R2
  end;
...

The restart of type R1 can be seen as a combination of intensification and diversification, that of type R2 as a pure diversification and that of type R3 as an evolution of R2 obtained by adding an intensification strategy. In particular, we follow the idea that after T time units the current solution is becoming more and more well-structured, so we try to speed up the convergence to a local optimum.

The fixing strategy adopted in the restart of type R3 can be seen both as the definition of a *third neighborhood* (fixed bins are tabu) and as a *cooling strategy* according to a simulated annealing approach (in some sense we decrease the probability that bad moves are accepted).

9.4 COMPUTATIONAL EXPERIMENTS

We coded the tabu search algorithm in FORTRAN 77 and run it on a Silicon Graphics INDY R10000sc 195Mhz on test instances from the literature. We assigned a time limit L of 100 CPU seconds for the solution of each instance. We selected, through a series of preliminary tests, the following values for the parameters: $\alpha = 5.0$, $\tau_1 = 3$, $\tau_2 = 5$, $R = 5$, $T = L/5$ and $\mu = 50$.

Table 9.1 gives the results obtained on single instances involving up to 164 pieces. Problems `beng1`–`beng8` are the instances of Bengtsson [5]. The remaining problems are instances proposed in the literature for other two-dimensional cutting problems and transformed into 2D-BPRG instances by using the relative bin and piece sizes (if the demand of a piece j is d_j, we consider d_j different pieces with height h_j and width w_j): problems `cgcut1`–`cgcut3` are described in Christofides and Whitlock [7], while problems `gcut1`–`gcut13` and `ngcut1`–`ngcut12` are described in Beasley [3] and [4], respectively. In the last four lines of Table 9.1, we report the results with four new instances, `glass1`–`glass4`, of a real-world 2D-BPRG problem arising in a glass factory (the instances are available on request from the first author).

For each problem the table gives the problem name, the number of pieces, the value of LB, the value of the best solution found by FFF_{RG} and FBS_{RG} (indicated in tables with UB_F), the value of the solution found by FC_{RG}, the value TS found from tabu search and the computing time, expressed in seconds, required by tabu search. An asterisk indicates the instances for which it could be proved that the solution found by tabu search is optimal (by comparison with LB and other lower bounds studied in Dell'Amico, Martello and Vigo [11]).

In Table 9.2 we present the results for six classes of instances randomly generated as in Berkey and Wang [6]. Each class is characterized by a different size of the bins and by the ranges in which the piece dimensions were uniformly randomly generated. The table gives the intervals in which the sizes of the pieces were uniformly randomly generated, the (fixed) bin sizes, the number of pieces, the average ratios of the solutions obtained by greedy heuristics and by TS with respect to the value of LB, the average tabu search computing time, expressed in seconds, and the number of instances for which it can be proved

Name	n	LB	UB_F	FC_{RG}	TS		$time$
beng1	20	3	4	4	4		100.01
beng2	40	6	7	7	7		100.01
beng3	60	9	9	9	9	*	0.01
beng4	80	11	12	12	11	*	3.70
beng5	100	14	14	14	14	*	0.01
beng6	40	2	2	2	2	*	0.01
beng7	80	3	3	3	3	*	0.01
beng8	120	5	5	5	5	*	0.01
cgcut1	16	2	2	2	2	*	0.01
cgcut2	23	2	3	2	2	*	0.01
cgcut3	62	16	23	24	24		100.01
gcut1	10	4	4	4	4	*	100.01
gcut2	20	5	7	7	7		100.01
gcut3	30	7	9	9	8		100.01
gcut4	50	12	15	15	14		100.01
gcut5	10	3	3	3	3	*	0.01
gcut6	20	5	8	8	8		100.01
gcut7	30	9	11	11	11		100.01
gcut8	50	12	14	14	14		100.02
gcut9	10	3	3	3	3	*	0.01
gcut10	20	6	8	8	8		100.01
gcut11	30	7	10	10	9		100.01
gcut12	50	13	17	17	16		100.01
gcut13	32	2	2	2	2	*	0.01
ngcut1	10	2	3	3	3		100.01
ngcut2	17	3	4	4	4		100.01
ngcut3	21	3	4	4	4		100.01
ngcut4	7	2	2	2	2	*	0.01
ngcut5	14	3	4	4	4		100.01
ngcut6	15	2	3	3	3		100.01
ngcut7	8	1	1	1	1	*	0.01
ngcut8	13	2	2	2	2	*	0.01
ngcut9	18	3	4	4	4		100.01
ngcut10	13	2	3	3	3		0.01
ngcut11	15	2	3	3	3		100.01
ngcut12	22	3	4	4	4		100.01
glass1	82	8	9	9	8	*	3.75
glass2	112	11	13	13	12		100.02
glass3	146	15	19	18	17		100.07
glass4	164	9	11	11	11		100.28

Table 9.1 Results on problem instances from the literature and on four new real-world instances arising in glass industry. Time limit of 100 CPU seconds on a Silicon Graphics INDY R10000sc.

pieces	*bins*	n	$\frac{UB_F}{LB}$	$\frac{FC_{RG}}{LB}$	$\frac{TS}{LB}$	*time*	*#opt*
		20	1.06	1.06	1.06	40.00	6
		40	1.12	1.11	1.09	70.41	3
$[1,10] \times [1,10]$	10×10	60	1.10	1.10	1.09	92.76	1
		80	1.10	1.10	1.08	91.08	1
		100	1.08	1.07	1.05	72.31	0
		20	1.00	1.00	1.00	0.01	10
		40	1.10	1.10	1.10	0.01	9
$[1,10] \times [1,10]$	30×30	60	1.10	1.05	1.05	10.00	9
		80	1.07	1.03	1.03	10.00	9
		100	1.03	1.03	1.03	10.00	9
		20	1.25	1.22	1.16	50.02	5
		40	1.21	1.21	1.20	100.01	0
$[1,35] \times [1,35]$	40×40	60	1.18	1.20	1.12	100.02	0
		80	1.16	1.16	1.11	100.03	0
		100	1.16	1.15	1.11	99.29	0
		20	1.00	1.00	1.00	0.01	10
		40	1.10	1.00	1.00	0.01	10
$[1,35] \times [1,35]$	100×100	60	1.10	1.10	1.10	20.01	8
		80	1.10	1.10	1.10	30.00	7
		100	1.07	1.07	1.07	20.01	8
		20	1.12	1.12	1.11	30.00	7
		40	1.18	1.17	1.17	70.01	3
$[1,100] \times [1,100]$	100×100	60	1.16	1.16	1.14	100.02	0
		80	1.15	1.16	1.15	100.03	0
		100	1.13	1.13	1.10	100.04	0
		20	1.00	1.00	1.00	0.01	10
		40	1.40	1.40	1.40	0.01	6
$[1,100] \times [1,100]$	300×300	60	1.05	1.05	1.05	10.00	9
		80	1.00	1.00	1.00	0.01	10
		100	1.07	1.07	1.07	20.01	8

Table 9.2 Results on the random problem instances proposed by Berkey and Wang. Time limit of 100 CPU seconds on a Silicon Graphics INDY R10000sc.

that the solution found by tabu search is optimal. The entries in each row are average values computed over ten random problem instances.

In Table 9.3 we consider randomly generated instances described in [18] having a different mix of pieces. Four types of pieces were considered, each defined by different intervals in which the sizes were uniformly randomly generated (see Table 9.3). Each class is characterized by a different percentage of pieces generated for each type. The table gives, for each pair of size intervals, the percentage of pieces that were uniformly randomly generated in such interval (the bin size is always 100×100), plus the same information as in Table 9.2.

Tables 9.1, 9.2 and 9.3 show a satisfactory behavior both of the new heuristic and of the tabu search algorithm. With only a few exceptions (four times over a total of 540 instances), the solution found by FC_{RG} dominates the best solution found by FFF_{RG} and FBS_{RG}. The tabu search algorithm finds the optimal solution or a solution of value close to that of the lower bound for many cases. For the third class of Table 9.3 the high value of the ratios are only due to a very poor behavior of the continuous lower bound: indeed the solutions found

$h_j \in [1, \frac{H}{2}]$	$[\frac{2}{3}H, H]$	$[\frac{H}{2}, H]$	$[1, \frac{H}{2}]$						
$w_j \in [\frac{2}{3}W, W]$	$[1, \frac{W}{2}]$	$[\frac{W}{2}, W]$	$[1, \frac{W}{2}]$	n	$\frac{UB_F}{LB}$	$\frac{FC_{RG}}{LB}$	$\frac{TS}{LB}$	*time*	*#opt*
				20	1.19	1.19	1.17	80.01	2
				40	1.17	1.17	1.17	100.02	0
70%	10%	10%	10%	60	1.18	1.18	1.16	100.04	0
				80	1.17	1.17	1.16	100.06	0
				100	1.17	1.17	1.16	100.07	0
				20	1.17	1.17	1.17	62.05	3
				40	1.19	1.19	1.19	100.02	0
10%	70%	10%	10%	60	1.18	1.18	1.17	100.03	0
				80	1.16	1.16	1.15	100.07	0
				100	1.17	1.17	1.17	100.07	0
				20	1.52	1.52	1.52	0.01	10
				40	1.52	1.52	1.52	30.01	7
10%	10%	70%	10%	60	1.58	1.58	1.58	30.01	7
				80	1.54	1.54	1.54	50.04	5
				100	1.54	1.54	1.54	50.02	5
				20	1.15	1.15	1.15	40.00	6
				40	1.09	1.09	1.06	40.03	6
10%	10%	10%	70%	60	1.09	1.09	1.07	63.89	4
				80	1.07	1.06	1.06	80.03	2
				100	1.07	1.07	1.06	89.36	1

Table 9.3 Results on the random problem instances proposed by Martello and Vigo. Time limit of 100 CPU seconds on a Silicon Graphics INDY R10000sc.

are often optimal (see the last column) or very close to the optimum (as could be proved by comparison with the improved lower bounds presented in [11]).

The effectiveness of the tabu search approach is due to the efficient combination of the strategies described in Section 9.3. In particular, preliminary experiments to test the relevance of each component of TS showed that the absence of the second or third neighborhood dramatically decreases the performance of the algorithm. The absence of the first neighborhood, instead, is not dramatic since its moves can be performed by the second neighborhood. However, in this case the main effect is a considerable reduction of the speed of the algorithm. We also tried to use the algorithm as a greedy combination of local searches (i.e., with empty tabu lists) but the results were not competitive with those of TS.

The values we adopted for the parameters (see above) turned out to be robust with respect to all our computational experiments. In order to evaluate the sensitivity of the results with respect to changes of these values, we also tried to individually modify the used values by $\pm 20\%$. These experiments showed, e.g., that it is convenient to always have $\tau_1 < \tau_2$ and, in particular, the chosen value for τ_1 ($\tau_1 = 3$) has to be considered as an upper bound in order to guarantee an efficient exploration of the first neighborhood. Anyway, as usual in metaheuristics, the goal of tuning is to provide at each step a sufficiently effective intensification before any diversification. Hence, the more sensitive parameters are those providing the re-start actions. Our suggestion if a problem with different characteristics is given, is to modify the parameters in order to identify such a border between intensification and diversification.

Acknowledgments

We acknowledge Ministero dell'Università e della Ricerca Scientifica e Tecnologica (MURST) and Consiglio Nazionale delle Ricerche (CNR), Italy for the support of this project. We thank Andrea Alberghini, Andrea Bedeschi and Carlo Montagnani for their help with programming. Thanks are also due to an anonymous referee for valuable comments.

References

[1] B.S. Baker, E.G. Coffman, Jr. and R.L. Rivest. Orthogonal Packing in Two Dimensions. *SIAM J. Comput.*, 9(4):846–855, 1980.

[2] J.J. Bartholdi III, J.H. Vande Vate and J. Zhang. Expected Performance of the Shelf Heuristic for the 2-Dimensional Packing. *Oper. Res. Lett.*, 8:11–16, 1989.

[3] J.E. Beasley. Algorithms for Unconstrained Two-Dimensional Guillotine Cutting. *J. Oper. Res. Soc.*, 36:297–306, 1985.

[4] J.E. Beasley. An Exact Two-Dimensional Non-Guillotine Cutting Tree Search Procedure. *Op. Res.*, 33:49–64, 1985.

[5] B.E. Bengtsson. Packing rectangular pieces – a heuristic approach. *The Computer Journal*, 25:353–357, 1982.

[6] J.O. Berkey and P.Y. Wang. Two dimensional finite bin packing algorithms. *J. Oper. Res. Soc.*, 38:423–429, 1987.

[7] N. Christofides and C. Whitlock. An Algorithm for Two-Dimensional Cutting Problems. *Op. Res.*, 25:30–44, 1977.

[8] F.K.R. Chung, M.R. Garey and D.S. Johnson. On packing two-dimensional bins. *SIAM J. Algebraic Discr. Meth.*, 3(1):66–76, 1982.

[9] E.G. Coffman, Jr., M.R. Garey, D.S. Johnson and R.E. Tarjan. Performance Bounds for Level-Oriented Two-Dimensional Packing Algorithms. *SIAM J. Comput.*, 9(4):801–826, 1980.

[10] E.G. Coffman, Jr. and P.W. Shor. Average-Case Analysis of Cutting and Packing in Two Dimensions. *European J. Oper. Res.*, 44:134–144, 1990.

[11] M. Dell'Amico, S. Martello and D. Vigo. An exact algorithm for Non-Oriented Two-Dimensional Bin Packing Problem (in preparation).

[12] H. Dyckhoff, G. Scheithauer, and J. Terno. Cutting and Packing (C&P). In M. Dell'Amico, F. Maffioli, and S. Martello, editors, *Annotated Bibliographies in Combinatorial Optimization*, pages 393–413. John Wiley & Sons, Chichester, 1997.

[13] A. El-Bouri, N. Popplewell, S. Balakrishnan and A. Alfa. A search based heuristic for the two-dimensional bin-packing problem. *INFOR*, 32(4):265–274, 1994.

[14] J.B. Frenk and G.G. Galambos. Hybrid next-fit algorithm for the two-dimensional rectangle bin-packing problem. *Computing*, 39:201–217, 1987.

[15] F. Glover and M. Laguna. *Tabu Search.* Kluwer Academic Publishers, Boston, 1997.

[16] I. Golan. Performance bounds for orthogonal oriented two-dimensional packing algorithms. *SIAM J. Comput.*, 10(3):571–582, 1981.

[17] A. Lodi, S. Martello and D. Vigo. Approximation algorithms for the oriented two-dimensional bin packing problem. *European J. Oper. Res.* (to appear).

[18] S. Martello and D. Vigo. Exact solution of the two-dimensional finite bin packing problem. *Management Sci.* 44(3):388-399, 1998.

10 CANDIDATE LIST AND EXPLORATION STRATEGIES FOR SOLVING 0/1 MIP PROBLEMS USING A PIVOT NEIGHBORHOOD

Arne Løkketangen[1] and Fred Glover[2]

[1]Molde College, Britveien 2, 6400 Molde, Norway.
arne.lokketangen@himolde.no

[2]School of Business,
CB419, University of Colorado, Boulder, CO 80309, USA.
fred.glover@colorado.edu

Abstract: Candidate list strategies play an important role in designing efficient local search heuristics for combinatorial optimization problems. We look at various neighborhood exploration and exploitation strategies for 0/1 MIP problems using a pivoting neighborhood and simple tabu search. This neighborhood is quite costly to evaluate, particularly for large problems. Special attention is also given to the tableau based neighborhood structure. General guidelines based on computational experience are reported.

10.1 INTRODUCTION

Candidate list (CL) strategies are vitally important for developing efficient iterative search methods for combinatorial optimization problems. Such strategies are especially emphasized in tabu search approaches, where they are critical to the focus of Tabu Search (TS) on seeking a best or near best move at each step. In this case, "best" is influenced not only by the problem context, but by short term memory processes embodied in tabu status and aspiration criteria, and by longer term memory processes that underlie diversification and intensification strategies. When the neighborhood is costly to evaluate or explore, or prohibitively large, the use of CL strategies takes on added significance.

We examine the design of effective CL strategies to take advantage of the special pivoting neighborhood used by adjacent extreme point methods for 0/1 mixed integer programming (MIP) problems. This search neighborhood is characterized by several distinct features. The moves are pivots that exchange selected basic and nonbasic variables in a bounded variable simplex method implementation, with the starting point usually given by an optimal solution to the relaxed LP problem. The solutions generated typically include many that are integer infeasible, and we employ a composite move evaluation function, designed for the twin goals of achieving integer feasibility and maintaining a good objective function value. The resulting evaluations can be quite expensive to calculate over a full neighborhood of possible pivot moves, particularly for MIP problems that are relatively large.

We conduct an experimental study of various CL and neighborhood exploration strategies for such pivot neighborhoods, with particular attention to:

- How much (or how little) of the neighborhood must be explored in each iteration to find good solutions.
- To what degree the fixed structure of the tableau influences the exploration strategies.
- The relative merits of randomized versus deterministic exploration.
- How much information should be remembered from iteration to iteration.
- The use of a supplementary Elite CL with its own management rules.

The intended goal of these experiments is to develop more general guidelines for neighborhood exploration and exploitation in this context, and to identify proper associated mechanisms and search parameter values, all in the spirit of [7].

The results of the present work are to be used as part of an enhanced first-level tabu search routine that is to be embedded in a B&B framework as a subproblem solver, see [17, 11]. It thus contains no long term strategies, as these are to be taken care of by the higher levels. For previous work on extreme-point pivoting and associated long-term strategies, see [8, 9].

Our procedure is implemented in a C++ class library specifically designed for heuristic search in combinatorial optimization [16], and incorporates subroutines via an EXPRESS-MP interface [2] to provide the relaxed LP solution and pivoting mechanisms. Experimental results are reported for selected well-known test problems, chosen to reflect the relative merit of the various CL strategies.

For other treatment of candidate list strategies, and for a more general treatment of tabu search, see [3, 4, 5, 6].

The layout of the paper is as follows. This introduction is followed in Section 10.2 by the necessary background and fundamentals. In Section 10.3 and Section 10.4 we examine the various neighborhood exploration and exploitation strategies, respectively, while Section 10.5 deals with elite candidate lists. The conclusions are given in Section 10.6.

10.2 BACKGROUND AND FUNDAMENTALS

The necessary background for the tabu search extreme point pivot search method is outlined here, using the notation and definitions of earlier studies of the direct zero-one tabu search approach. For more details, see [9]. The reader is assumed to be familiar with tabu search (see, e.g., [5]) and with the basic concepts of the bounded variable simplex method.

10.2.1 MIP problem formulation

We represent the MIP problem formulation in the form

$$\begin{array}{llll} \text{Minimize} & z = \sum_{j \in N} c_j x_j & & \\ \text{Subject to} & \sum_{j \in N} A_j x_j \leq b & & \\ & 0 \leq x_j \leq 1 & \text{and } x_j \text{ integer} & j \in I \subseteq N \\ & 0 \leq x_j \leq U_j & & j \in C \subseteq N - I \end{array} \quad (10.1)$$

The first-level Tabu Search heuristic for this problem, TS - MIP, may then be outlined as follows:

Let x^* denote the best MIP feasible solution, and let z^* denote its objective function value.

1. Solve the LP relaxation (i.e., ignoring integer constraints) to obtain an optimal LP basic (extreme point) solution.

2. Consider the feasible pivot moves that lead to adjacent basic LP feasible solutions. If a candidate move would lead to an 0-1 MIP feasible solution x whose associated z value yields $z > z^*$, record x as the new x^* and update z^*.

3. Select the pivot move with the highest move evaluation, applying tabu restrictions and aspiration criteria.

4. Execute the selected pivot, updating the associated tabu search memory and guidance structures. Return to Step 2.

Note that the method may not necessarily visit a best MIP feasible neighbor of the current solution, since the move evaluation of Step 3 depends on factors in addition to the objective function value (see below).

10.2.2 Neighborhood structure

Let $x(0)$ denote the current extreme point solution, let $\{x_j : j \in NB\}$ denote the current set of nonbasic variables, and let $\{x_j : j \in B\}$ denote the set of basic variables ($B = N - NB$). The extreme points adjacent to $x(0)$ have the form

$$x(h) = x(0) - D_h \theta_h \, for \, h \in NB$$

where D_h is a vector associated with the nonbasic variable x_h, and θ_h is the change in the value of x_h that moves the current solution from $x(0)$ to $x(h)$ along their connecting edge.

We thus start the search integer infeasible, and may also spend large parts of the search visiting integer infeasible solution states. The usefulness of moving in infeasible space has been demonstrated, e.g., by [3]. An example of encountering problems when trying to stay feasible all the time is shown in [1], where movement in the solution space is severely restricted for tightly constrained problems.

10.2.3 Move evaluation

The move evaluation function is composite, based on two independent measures. The first measure is the change in *objective function value* when going from $x(0)$ to $x(h)$, and the second measure is the associated change in *integer infeasibility*. Let $near(x_j(h))$ denote the integer nearest to the value $x_j(h)$. Then we define $u(h)$ to be the amount of integer infeasibility for the solution $x(h)$ where

$$u(h) = \sum_{j \in I} | x_j - near(x_j(h)) |$$

Restricting consideration to $h \in NB$, we define

$$\Delta z(h) = z(h) - z(0)$$
$$\Delta u(h) = u(0) - u(h)$$

Note that it is not necessary to execute a pivot to identify $x(h)$ or the values of $u(h)$ and $z(h)$, since only the vector D_h, the scalar θ_h and the current solution $x(0)$ are required to make this determination.

10.2.4 Move types and choice rules

The moves are grouped into four distinct move types, according to the values of $\Delta z(h)$ and $\Delta u(h)$, as shown in Table 10.1.

Move Type	Defining Conditions
I	$\Delta z(h) < 0, \Delta u(h) > 0$
II	$\Delta z(h) > 0, \Delta u(h) < 0$
III	$\Delta z(h) \geq 0, \Delta u(h) \geq 0$
IV	all others

Table 10.1 Move Types

The classification of move types is made according to the degree to which they may be considered to be favorable. Moves of type III are generally the most favorable, improving in both of the measures used in the move evaluation function, while moves of type IV are the least favorable, worsening in both measures. Moves of types I and II are improving in one of the measures, and worsening in the other.

The ranking of moves *within* each move group is shown in Table 10.2, with the best move in the group indicated.

Move Type	Move Ranking
I	$Max(\Delta z(h)/\Delta u(h))$
II	$Max(\Delta u(h)/\Delta z(h))$
III	$Max(\Delta u(h) * \Delta z(h))$
IV	$Min(\Delta u(h) * \Delta z(h))$

Table 10.2 Move classifications and ranking functions

When move evaluations tie within a group, we choose the move with the largest value of the positive component for moves of type I and II. For moves of type III and IV we choose the move that is most balanced in its components, i.e., where the $\Delta z(h)$ and $\Delta u(h)$ components are most equal. The ranking of different move type groups is done by always considering type III moves first, and type IV moves last, with move types I and II grouped together in the middle. To be able to properly rank move types I and II, their values need to be normalized.

We first define the ratio R by

$$R = \frac{\sum abs(z(h))}{w \sum abs(u(h))}$$

where the summations are taken over the currently available moves of type I and II, and where w, the skew factor, can be used to change the emphasis between the two measures making up the move evaluation function, with low values giving a bias towards selecting moves resulting in a lower integer infeasibility measure. The normalized move values are

$$MoveType\, I : R1(h) = (\Delta z(h)/\Delta u(h))/R$$
$$MoveType\, II : R2(h) = (\Delta u(h)/\Delta z(h)) * R$$

The preferred move selection from group I and II is then

$$MoveType\, I\, and\, II : Max(R1(h), R2(h))$$

10.2.5 Tabu status and aspiration criteria

We impose tabu restrictions on the variable that leaves the basis in a move. In other words, a variable that has just left the basis is not allowed to reenter the basis until its tabu tenure has expired.

The aspiration criterion is fulfilled if a new best solution is found. As mentioned, a new integer feasible solution may be encountered during a move evaluation without subsequently selecting the move leading to it. This is due to the process for evaluating and ranking the moves. However, such a solution is picked by the aspiration criterion, and will thus be selected.

For more detailed explanations of the workings of this section, the reader is again referred to [9].

10.3 NEIGHBORHOOD EXPLORATION STRATEGIES

When the neighborhood is too costly or too big to evaluate, the natural avenue of approach would be to only explore parts of it. This can, however, be done in many ways, depending on the structure of the problem at hand and potential knowledge of the solution space. In this paper we will explore the round robin and randomized approaches, but it should be noted that including information about consistent and determined variables could easily augment these basic mechanisms, when this information is available, possibly through the use of an elite candidate list (ECL) (see Section 10.5).

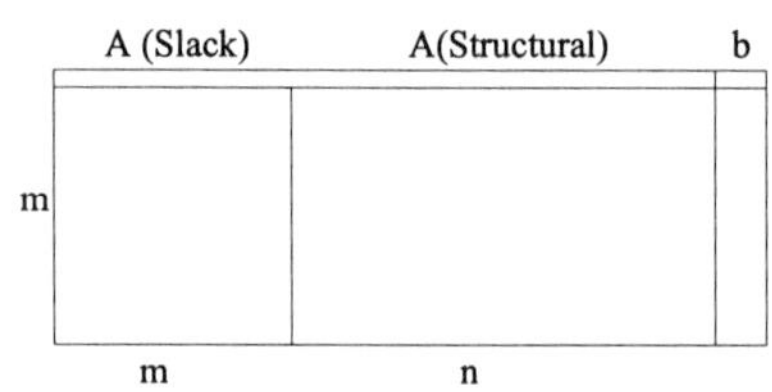

Figure 10.1 Layout of tableau

Our methods work on the vertices of a bounded variable simplex. The actual layout of the tableau, as implemented by XPRESS-MP (see [2]) is as shown in Figure 10.1, for a problem with **m** constraints and **n** variables. The slack variables are grouped together, and, as we will show, special care must be taken to assure that we have a proper mix of slack and structural variables among our candidates. It should also be noted that the total number of variables for the pivot search is **n** + **m**, while the neighborhood size is **n**, consisting of those variables (structural and slack), not present in the basis that represents the current solution. To help in the neighborhood exploration, we keep a list of the current status of all the variables, with the status being either basic or nonbasic (or fixed at one of its bounds when used as a subproblem solver for B&B, see [11], in which case it is not considered).

10.3.1 Test cases and search parameters

We restricted the testing to only a few test cases, to be able to focus on the behavioral aspects of the various candidate list approaches. The problems and their dimensions are shown in Table 10.3. The *Fleisher* and *Wei Shih 30* problems are multiconstraint knapsack problems, selected from the OR library [13], and are also in the test suite used in [8], while *misc05* is from Miplib [12]. The search parameter values follows the same general guidelines as used in [8].

Name	Total No. Variables	Binary Variables	Constraints
Fleisher	20	20	10
Wei Shih 30	90	90	5
misc05	136	74	300

Table 10.3 The test cases

10.3.2 The round robin approach (RR)

Round robin neighborhood exploration is the sequential selection of variables for evaluation as entering pivot, following the general sequence of variables of Figure 10.1. RR usually starts at the first variable, and then selects the next nonbasic variable in sequence, etc., until the desired number of variables have been selected. We also define the search parameter *RR_Inc*, specifying the relative position of the next variable to select, i.e., selecting the variable *RR_Inc* positions to the right in the variable list. All RR is with wrapping when coming to the end of the variable list, and the position of the last variable chosen is remembered from one iteration to the next. We also defined *NB_Expl_Nr*, as the number of candidate moves to generate in the current neighborhood.

Because the *Fleisher* problem is small, and thus tractable, we chose it for the initial tests of neighborhood exploration. In order to see how the fixed block of slack variables influences the efficiency of the RR scheme, we first ran a test with *RR_Inc* $= 1$ and *NB_Expl_Nr* ranging from 20 to 1. The tests were run for a maximum of 1000 iterations, but were stopped when integer feasibility was obtained, with a skew-factor of 0.1. (As the intended use is as a subproblem solver in a B&B setting, it is quite reasonable to have integer feasibility as a goal.)

The results are shown in Table 10.4, in the column with heading **1**, for all the 20 possible values of *NB_Expl_Nr*. The numbers give the number of iterations taken to find an integer feasible solution, with italics indicating that the optimum solution was found, and a dash indicating that no feasible solution was found. As is evident from the table, when the block of variables we try to evaluate decreases to the size of the number of slack variables (or constraints), we experience a dramatic drop in search efficiency, with either no integer feasible solution found, or the one found is inferior or needs many more iterations.

The rest of the columns under the **RR_Inc** heading show the results for all the different sensible values of *RR_Inc*. (Note that in order to visit all the variables, then *RR_Inc* and $\mathbf{n} + \mathbf{m}$ ($= 30$ for this example) should have no common factors.) The results are now far more favorable for smaller values of *NB_Expl_Nr*, both in terms of finding integer feasible solutions, and giving

NB_Expl_Nr	RR_Inc								Rnd	
	1	7	11	13	17	19	23	29	Feas	Avg
20	20	20	20	20	20	20	20	20	10/0/0	20
19	20	21	19	20	20	20	19	20	10/0/0	25
18	34	21	21	41	19	24	20	27	10/0/0	25
17	19	20	23	45	32	19	39	18	10/0/0	26
16	35	22	31	23	20	19	30	19	10/0/0	28
15	29	47	20	45	31	28	20	19	10/0/0	31
14	21	43	24	35	19	37	91	18	8/2/0	34
13	46	63	38	28	20	37	42	34	9/1/0	46
12	23	28	20	85	39	21	32	37	10/0/0	50
11	64	22	75	43	27	22	52	221	10/0/0	42
10	86	22	81	112	38	98	96	31	9/1/0	59
9	237	82	105	25	77	23	59	-	8/2/0	55
8	-	28	60	108	16	53	16	-	9/1/0	116
7	301	92	139	84	92	78	78	-	5/4/1	119
6	-	40	40	73	15	143	39	-	5/5/0	212
5	-	88	57	132	42	115	191	725	6/4/0	285
4	-	146	247	231	30	187	72	-	2/8/0	406
3	-	345	-	161	-	326	-	-	0/9/1	232
2	-	-	225	790	-	-	-	-	0/10/0	443
1	112	-	94	-	199	127	-	24	0/10/0	515

Table 10.4 RR and Rnd neighborhood exploration for the *Fleisher* problem

higher quality solutions for a larger range of values of NB_Expl_Nr. Please note that setting *RR_Inc* = 29 is equivalent to setting it to -1, giving very similar results as for *RR_Inc* = 1. (We also tried to start the RR (on the first iterations) from a different variable, with (as expected) very similar results.)

The same set of tests was tried on the other test cases with similar results. It is thus evident that if RR is to be used as a basic method for neighborhood exploration, then care should be taken to have a value of *RR_Inc* different from 1 (and -1) such that the set of explored variables will have a sufficient diversification so as not to obstruct the search. This also implies that if knowledge of problem structure is available, this knowledge should be used when selecting appropriate values for *RR_Inc*. In general it also seems that setting *RR_Inc* to a *large* fraction of the neighborhood works better, with *large* meaning somewhere in the middle.

10.3.3 The random approach (Rnd)

In this approach we select the entering variables randomly from the available set, until *NB_Expl_Nr* different variables have been selected. This approach is only suitable for partial neighborhood exploration. (It makes no sense to randomly select until the full neighborhood has been selected.) Randomization provides extra diversification, and has been used successfully, e.g., in move selection for the first-level search outlined in Section 10.2 (see also [10]).

We similarly used the *Fleisher* problem for our tests of random neighborhood exploration. For each value of *NB_Expl_Nr* ranging from 20 to 1, we ran 10 tests with different random seed values, and with the same parameter settings as used for the RR tests. The results are shown in Table 10.4, under the **Rnd** heading.

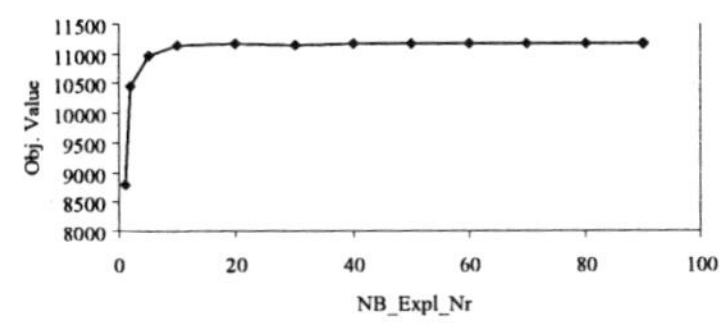

Figure 10.2 Average solution quality for *Wei Shih 30*

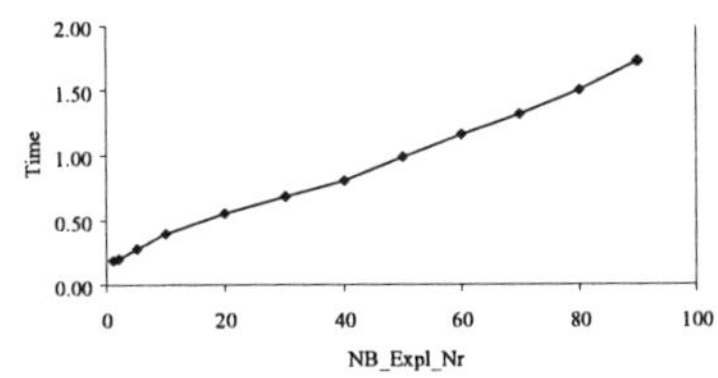

Figure 10.3 Average solution times for *Wei Shih 30*

The three numbers in the **Feas** column show, from left to right, the number of runs yielding the optimal solution, a suboptimal solution, and an integer infeasible solution. The **Avg** column gives the average number of iterations for those runs yielding integer feasible solutions.

As can be seen, only two runs failed to find an integer feasible solution. This indicates that the randomization helps by giving extra diversification to the search when compared to the RR case. However, the randomized approach does suffer from the same degradation in solution quality and increased search time for smaller values of *NB_Expl_Nr* for this instance. It should be noted that the time used per iteration decreases proportionately with decreasing values of *NB_Expl_Nr*. This means, with reference to Table 10.4, that even though it roughly takes twice as many iterations to find an integer feasible solution when only exploring half the neighborhood per iteration, the overall search time remains roughly the same. This is because the cost of the neighborhood evaluation dominates the other search costs.

We ran a similar set of tests for the *Wei Shih 30* problem, which has 90 variables and five constraints. For this instance integer feasibility is very easily obtained, so the tests were instead terminated after 100 iterations. In Figure 10.2 the average solution value for *NB_Expl_Nr* ranging from 1 to 90, with random neighborhood exploration is shown. As can be seen, if at least about 20% of the neighborhood is explored, we get good, nearly indistinguishable, results. Figure 10.3 similarly shows the average computational time for the same set of tests. It is evident that for the full neighborhood exploration, about 90% of the total search time is used for the exploration, while only 10% is used for the rest of the search activities. When comparing Figure 10.2 and Figure 10.3, we see that for this instance great savings in search time can be achieved.

These findings are in agreement with the findings of [15] and [14], who both report great savings with either a fixed (small, see [15]) value or for a range of values (see [14]) of their equivalent of *NB_Expl_Nr*. In [14] is also demonstrated the beneficial use of randomized exploration.

The use of reduced neighborhood exploration induces extra tabu tenure, equal to the ratio between the neighborhood size and *NB_Expl_Nr*. This is most easily seen in the RR case with *RR_Inc* = 1. Here it takes n/NB_Expl_Nr iterations for a variable already considered until it will be considered again, where n is the neighborhood size. The tabu tenure of a variable tends to expire faster, as the original TT will be counted down at a faster rate as compared to the case with full neighborhood exploration, with this rate being the same as the above-mentioned ratio. We have not explored the added effect of this.

10.4 NEIGHBORHOOD EXPLOITATION

In each iteration, after the neighborhood exploration phase is done as outlined in Section 10.3, a collection of candidate moves is gathered. This collection is sorted according to the criteria of Section 10.2, and the best move is selected. This does not mean that all of the other moves are not worthy of consideration, but only that the potentially best move has been selected.

It might therefore be sensible to try to use as pivots the sequence of the top *CL_Ret_Nr* number of moves before reevaluating the neighborhood, with *CL_Ret_Nr* being a relatively small number. This involves some problems, as the neighborhood center that these moves were calculated relative to, is not the center after the first (and presumed best) pivot has been selected. The entering pivot is still valid, but the leaving pivot might not be, even though one must still exist. A reevaluation of the moves considered for extra pivots is therefore necessary in order to determine the (new) leaving pivot. Another problem is that the actual move value might be distorted, and higher accuracy is obtained if it is reevaluated. The move value and the determination of leaving pivot are therefore reevaluated for every move considered for selection after the first (since the last neighborhood exploration).

One crucial question is how much of the neighborhood needs to be explored in order to provide enough potentially good moves to support this exploitation phase. Connected to this is the problem of finding the best value for *CL_Ret_Nr*, or rather, how many moves to consider, and possibly explore, before re-evaluating the neighborhood. It is also important to note that since the moves for the current neighborhood exploration are sorted, it might be prudent to add a threshold to reduce the number of considered moves. In other words, when the next move considered is below the threshold value, re-explore the neighborhood. This is a somewhat imprecise measure, as the true move evaluation changes after each pivot, but we use this threshold on the reevaluated move values. There is thus no guarantee that the next move on the candidate list might not be a good one, but it is prudent to avoid taking potentially bad pivots. The threshold, *MR_Thresh*, defined in terms of the move types defined in Section 10.2, has the following levels:

1. No rejection. Accept all moves.
2. Reject moves of type IV.
3. Reject moves of type II and IV.
4. Reject moves of type I, II and IV.

Based on the results from Section 10.3, we will use random neighborhood exploration when not exploring the full neighborhood.

We chose the test case *misc05*, with 136 columns and 300 rows, to find values for the various search parameters. Again, the tests were run for 1000 iterations or until integer feasibility was obtained, with random neighborhood exploration except when using the full neighborhood. The following search parameters were used:

- For *NB_Expl_Nr* we used the following fractions of the neighborhood: 1, 0.75, 0.5, 0.25. This gives values for *NB_Expl_Nr* of 136, 102, 68 and 34.
- *CL_Ret_Nr*: 1,2,3,4,5,6,7,8,9,10.
- *MR_Thresh*: 1,2,3,4.

All the randomized tests (i.e., the ones with reduced neighborhood exploration) were run 10 times each with different seed values.

CL_Ret_Nr	MR_Thresh			
	1	2	3	4
1	1,00	1,00	1,00	1,00
2	1,23	1,16	1,12	1,05
3	1,61	1,39	1,27	1,10
4	2,14	1,62	1,38	1,13
5	2,75	1,83	1,50	1,15
6	3,45	1,95	1,57	1,16
7	4,21	2,22	1,59	1,15
8	5,12	2,31	1,60	1,16
9	5,78	2,46	1,62	1,16
10	6,57	2,45	1,68	1,16

Table 10.5 Average number of moves taken per iteration

Table 10.5 shows the average number of moves taken per iteration for the selected values of *NB_Ret_Nr* and *MR_Thresh*, over all the runs. The general trend is that the higher the threshold (i.e., value of *MR_Thresh*), the fewer iterations are actually taken per neighborhood exploration, and the higher the value of *CL_Ret_Nr*, the more iterations are taken per exploration. It is also fairly clear that a value of *MR_Thresh* equal to 4 (meaning that only Type III moves are considered after the first iteration,) is too restrictive. This setting has a very small potential effect, and should only be used with *CL_Ret_Nr* up to 3, if at all. Similar values of *CL_Ret_Nr* indicated by Table 10.5 are 5 for $MR\_Thresh = 3$, 7 for $MR\_Thresh = 2$, and possibly 10 for $MR\_Thresh = 1$.

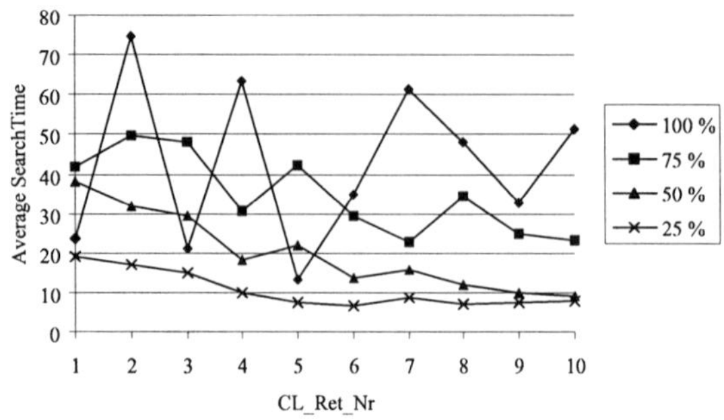

Figure 10.4 Average solution times vs. *CL_Ret_Nr* for *misc05*

When it comes to solution quality, preliminary screening of the data showed a similar trend. It was evident, however, that the best solution quality (in terms of number of integer feasible runs, and optima obtained), was with $MR\_Thresh = 2$, especially for the smaller values of *NB_Expl_Nr*. $MR\_Thresh = 4$ was worst, again more pronounced for the smaller neighborhood explorations. It is also worthy to note that searches with smaller *NB_Expl_Nr* should really have had a larger iteration cutoff limit to make a fairer comparison, and that the iteration cutoff is in terms of pivots executed and not in terms of neighborhood explorations.

The average search times for the tested range of *CL_Ret_Nr* and different values of *NB_Expl_Nr* are shown in Figure 10.4. (The reason for the ragged behavior of the full neighborhood exploration, is that it consists of only one sample per value of *L_Ret_Nr*, as it contains no randomization, but the general trend is still there.) As expected there are substantial savings in search time for the smaller neighborhood explorations. The general solution quality is fairly uniform for all the values of *NB_Expl_Nr*, indicating that substantial savings can be had with a 25% neighborhood exploration, *MR_Thresh* = 2, and a *CL_Ret_Nr* value of, e.g., 6 or 7.

10.5 ELITE CANDIDATE LISTS

In *Elite Candidate Lists* (ECL), potentially good moves are collected, remembered between neighborhood explorations, and used as a supplement to the normal neighborhood selection rules. It is thus only sensible when using the reduced neighborhood exploration.

Our implementation of an ECL consists of moves that were collected in the preceding exploration phase, and passed by the ensuing exploitation iterations. As our exploitation phase is designed to empty the candidate list of good moves (until the cutoff defined by *MR_Thresh* is reached), we do not remember moves that were not considered by the preceding exploitation phase. We thus include potentially good moves that were not taken in an iteration. When generating a new list of candidate moves in the next exploration phase, we first include the moves on the ECL, and then supplement them with the required number of new moves. One necessary search parameter is *ECL_Nr*, which gives the maximum number of moves to remember.

We ran a series of tests on *misc05* with *NB_Expl_Nr* set to 50% and 25% of the neighborhood size, *MR_Thresh* set to 1 or 2, $CL\_Ret\_Nr = 10$, and *ECL_Nr* ranging from 1 to 5. All tests were run with 10 different random seeds. The results were slightly inferior to the similar runs without the use of an ECL. The reasons for this might be that the neighborhood exploitation phase really uses most of the good moves, and thus the ECL moves supplants some of the new moves that could otherwise have been explored.

We believe that to get an efficient ECL, guidance from higher levels of the search is required, and it can then be used in support of diversification, intensification and/or Strategic Oscillation schemes. Problem-specific information would also be helpful, as well as the identification of consistent and/or determined variables.

10.6 CONCLUSIONS

The neighborhood of an adjacent extreme point method for 0/1 MIP problems is very costly to evaluate, and the time spent doing neighborhood explorations thus takes a substantial part of the overall search time when used in an iterative search method like tabu search. We have shown that properly designed neighborhood exploration and exploitation strategies can be used to significantly reduce this time spent in evaluating the neighborhood, and thus speed up the overall search, without significantly reducing the quality of the solutions found.

We similarly predict a good effect of an ECL approach in combination with guidance coming from higher levels of the search, giving support to the higher meta-strategies of intensification, diversification and strategic oscillation.

References

[1] Connolly, D., "General Purpose Simulated Annealing", *J. Oper. Res. Soc.*, 43/5, pp. 495-505, 1992.

[2] Dash Assoc., *XPRESS-MP Reference Manual*, Version 9.32, Binswood Ave, Leamington Spa, Warwickshire, CV32 5TH, UK, 1996.

[3] Gendreau, M., A. Hertz and G. Laporte, "A Tabu Search Heuristic for the Vehicle Routing Problem". *Management Sci.*, Vol 40, No. 10, pp 1276-1290, 1994.

[4] Glover, F., "Tabu Search - Part I", *ORSA J. Comput.*, Vol 1, No 3, pp 190-206, 1989.

[5] Glover, F. and M. Laguna, Tabu Search, Kluwer Academic Publishers, Norwell, MA, 1997.

[6] Glover, F., M. Laguna, E. Taillard and D. de Werra, "A Users Guide to Tabu Search", *Ann. Oper. Res.*, Vol 41, pp 3-28, 1993.

[7] Hooker, J., "Testing Heuristics: We Have It All Wrong", *Journal of Heuristics*, Vol 1, No 1, pp 33 - 42, 1995.

[8] Løkketangen, A., and F. Glover, "Tabu Search for Zero-One Mixed Integer Programming with Advanced Level Strategies and Learning", *International Journal of Operations and Quantitative Management*, Vol 1, No 2, 1995.

[9] Løkketangen, A., and F. Glover, "Solving Zero-One Mixed Integer Programming Problems Using Tabu Search," *European J. Oper. Res., Special Issue on Tabu Search*, Vol 106, pp 624-658.

[10] Løkketangen, A. and F. Glover, "Probabilistic Move Selection in Tabu Search for Zero-One Mixed Integer Programming Problems," in *Meta-heuristics: Theory & Applications*, I.H. Osman and J.P. Kelly, eds., pp 467-487, 1996.

[11] Løkketangen, A. and D. L. Woodruff, "Integrating Pivot Based Search with Branch and Bound for Binary MIPs". Working paper, University of California at Davis, 1997.

[12] Miplib. ftp://softlib.cs.rice.edu/softlib/miplib/

[13] OR library. http://mscmga.ms.ic.ac.uk/info.html

[14] Reeves, C. R., "Improving the Efficiency of Tabu Search for Machine Sequencing Problems", *J. Oper. Res. Soc.*, Vol 44, No 4, pp 375-382, 1993.

[15] Sun, M., "A Tabu Search Heuristic Procedure for Solving the Transportation problem with Exclusionary Side Constraints", Working Paper. University of Texas at San Antonio, 1997.

[16] Woodruff, D. L., "A Class Library for Heuristic Search Optimization: Preliminary Users Guide", Working Paper, University of California at Davis, 1993.

[17] Woodruff, D. L., "A Chunking Based Selection Strategy for Integrating Meta-heuristics with Branch and Bound", this volume.

11 GLOBAL AND LOCAL MOVES IN TABU SEARCH: A REAL-LIFE MAIL COLLECTING APPLICATION

Redouane Mechti[1], Stephane Poujade[1],
Catherine Roucairol[1] and Bernard Lemarié[2]

[1]PRiSM Laboratory, University of Versailles
45 avenue des États-Unis, 78035 Versailles Cedex, France.
mechti@prism.uvsq.fr, pos@prism.uvsq.fr, roucairo@prism.uvsq.fr

[2]Service de Recherche Technique de la Poste, B.P 86334
10 rue de l'Ile Mabon, 44263, Nantes Cedex 02, France.
bernard.lemarie@srt-poste.fr

Abstract: The problem we deal with is the optimization of mail collecting at several customer sites that are scattered around an urban area. It involves the design of a set of minimum cost routes, originating and terminating at a central depot, for a fleet of vehicles that service those customer sites with known demands. We develop a Tabu Search approach where at each iteration the best move is selected among a large variety of possible moves. This new version of the metaheuristic Tabu leads us to determine a good vehicle fleet mix (cheapest cost incorporating routing and fixed vehicle costs) without violating constraints such as time restrictions and capacity.

11.1 INTRODUCTION

The Vehicle Fleet Mix Problem with Strong Time Windows (**VFMP-STW**) is a variant of the standard Vehicle Routing Problem (**VRP**). In its broadest sense, the vehicle routing problem can be described as the problem of designing optimal pick-up routes from one warehouse to a set of geographically scattered customers. The problem we deal with is to organize mail collecting at several customer sites that are scattered around an urban area. It involves the design of a set of minimum cost routes, originating and terminating at a central depot, for a heterogeneous fleet of vehicles that service those customer sites with known demands.

Each customer is serviced exactly once and, furthermore, all the customers must be assigned to vehicles such that the vehicle capacities are not exceeded. The company has three types of vehicles: motorcycles, trucks and vans, located at a central depot. These vehicles have different fixed and routing costs. They are also characterized by their capacities.

In the context of our problem, the service of a customer, involving a pick-up, must begin within the time window defined in the contract between La Poste, the French national mail post office company, and the customer. This time window sets the earliest and the latest times that a customer can be served.

Our problem is composed of very tight time windows, approximately gathered in a same period of time and the mail collect has to be done in a short time. This reduces on one hand the number of feasible solutions and makes them hard to find. But, on the other hand, the fact that many customers can be in very close areas (same street or building) increases the total number of solutions.

After creating an initial feasible solution respecting time windows and capacity constraints, we use at each iteration of our algorithm a variant of tabu search (Variable Move Tabu Search), and an appropriate move is selected from either a **local** move or a **global** move generated from the current solution.

Local moves favour the optimization of only one route, hence we have a local exploration and solutions close to the current solution. Global moves significantly change the solution structure and allow a wide exploration of other possible solutions. We alternate between the two classes through some activating tests that lead us to select the route on which we act and the type of moves to use in order to improve the current solution as much as possible. This approach has been tested on a practical problem occurring in mail collecting.

The main objective of this paper is to present an efficient algorithm for solving the CVFMP-TW with characteristics of mail collecting problems. The paper is organized as follows. Sections 11.2 and 11.3 are devoted to literature review and problem formulation. The two main stages of our algorithm (Transfleet) are briefly described in Section 11.4. As our tabu search approach is based upon two types of moves, we present in Section 11.5 the first type, global moves followed by their activating tests and in Section 11.6 the second type, local moves with their activating tests as well. Section 11.7 describes some hybrid moves while Section 11.8 is devoted to a new way to intensify and diversify the tabu search. Computational results and comments are described in Section 11.9. Finally, we present our conclusions and future research in Section 11.10.

11.2 LITERATURE REVIEW

There are relatively few research articles on the Vehicle Fleet Mix Problem (VFMP). Golden et al. [10] were among the first to address this problem in the early 1980's. The only ones known to us are based on adaptations of the savings algorithm of Golden et al. [10], on tour partitioning and on the generalized assignment based heuristic of Gheysens et al. [8] and on the matching based saving heuristics of Desrochers and Verhoog [4, 5]. We conclude the section by reviewing the more recent algorithms of Osman and Salhi [13].

Savings algorithms for Fleet Size and Mix Vehicle Routing Problem are adaptations of the well known Clarke and Wright savings algorithm [3] which was originally designed for the classical VRP. In this algorithm, each of the n customers is initially served by a single vehicle. Two distinct routes containing customers i and j can be merged if i and j are both adjacent to the depot and the demand of the combined route does not exceed the vehicle capacity. The saving obtained due to a combination of two routes is $S1_{ij} = C_{oi} + C_{jo} - C_{ij}$ if feasible, $-\infty$ otherwise. Since the total costs do not only consist of routing costs but also of fixed costs, the savings in fixed cost should be considered as well. Let $Cf_k(R_i)$ be the fixed cost of the smallest vehicle of type k that can serve a total demand D_i of a route R_i. The total savings, due to a feasible combination of the two routes containing customers i and j, is $S2_{ij} = S1_{ij} + Cf_k(R_i) + Cf_k(R_j) - Cf_k(R_i + R_j)$.

The procedure terminates when $S2_{ij}$ is no positive for all pairs (i, j). It is referred to as the Combined Saving Approach. This approach is the basis for the two remaining savings algorithms which are known as "Opportunity Savings Algorithms".

When merging two routes, a larger vehicle is required for the combined route, hence a higher fixed cost. In many cases, this may mean that the considered fusion is unattractive because of the under-utilization of the new vehicle. One way to deal with this problem is to give a value to the unused portion of new vehicle capacity. Golden et al. [10] have defined opportunity savings as a function of the unused capacity of the larger vehicle serving a combined route and developed several algorithms by varying the exact definition of the opportunity savings.

Giant Tour Partition Algorithms (GTPA), first developed by Gillett and Miller [9] represent an approach to solve the Fleet Size and Mix Vehicle Routing Problem (FSMVRP) significantly different from the savings methods. They are examples of the well-known heuristics "route first-cluster second" of Bodin et al. [2]. Similar approaches can be found in papers by Newton and Thomas [11], Bodin et al. [1] and Stern and Dror [20].

The Single Partition Giant Tour Algorithm (SPGTA), Golden et al. [10], starts with a tour that begins at the central depot, visits each customer exactly once and then returns to the central depot. The total costs is made of total routing costs of a route and the cost of the smallest vehicle that can serve the route. The algorithm finds then the shortest path from the first customer of the route to the last one, the depot. The Multiple Partition Giant Tour Algorithm (MPGTA) of Golden et al. [10] starts with a tour that visits each customer but not the depot. As the latter depot is not included in the giant tour, several starting points can be used that may give multiple partitionings. The auxiliary graph and the total costs are defined as in SPGTA except for the total demand which is set to exceed the largest vehicle type. Both for SPGTA and MPGTA, a single application of one of these algorithms brings unsatisfactory results, the authors introduce a variety by starting from a number of TSP tours. Because of the more powerful partitioning, a sample size of 5 has proved to be effective for MPGTA. They refer to this variant of MPGTA as MGT^5. Other improvements have been carried out. The first one, (MGT+2-opt), is an adaptation of the 2-opt algorithm and the second one, (MGT+OROPT), is an adaptation of the modified 3-opt procedure proposed by Or [12]. Since MGT^5 gives results comparable to (MGT+2-opt), variety is again achieved by starting from several TSP tours (in this case, 5 tours). This leads to $(MGT + 2 - opt)^5$ and $(MGT + OROPT)^5$.

Gheysens et al. [8] have developed a two phase heuristic. The first phase consists of finding the fleet composition using the lower bounding procedure of Golden et al. [10] while the famous Generalized Assignment based heuristic of Fisher and Jaikumar [7] was adapted to the heterogeneous fleet case during the second phase. Excellent results were obtained.

Ferland and Michelon [6] gave a formulation of the problem and showed that the exact methods and heuristics developed for the VRP-TW for a homogeneous fleet can be extended to the heterogeneous case. No computational results were reported.

Desrochers and Verhoog [4] developed a Matching Based Savings Algorithm (MBSA) that globally considers the savings associated with all feasible combinations of two different routes by using a weighted matching problem to select them. This algorithm merges the two routes having the largest savings in the matching solution. The MBSA can also be applied to the FSMVRP by changing the weights for the maximum weighted matching problem. Notice that the latter weights are set, for the classical VRP, equal to the savings in routing costs, originally used in the Clarke and Wright savings methods. For the FSMVRP, the same savings, as in the Savings heuristics of Golden et al. [10] can be used as weight entries in the matching problem. Thus the weights can be defined as the total savings, including savings in routing costs, savings in vehicle costs, or opportunity savings, associated with each feasible combination. Desrochers and Verhoog [5] extend these latter heuristics for the FSMVRP to a new saving heuristic based on successive routes fusion. At each iteration of the algorithm, the best fusion is selected by solving a weighted matching problem.

Another mixed integer formulation with fixed and variable running costs was presented by Salhi et al. [17]. They modified the route first-cluster second and the savings algorithms in order to take into account both fixed and variable costs. Rochat and Taillard [15] and Taillard [21] have proposed very efficient local search heuristics for a VRP with homogeneous fleet. They have also proposed nice heuristics for the heterogeneous case. Several techniques have been proposed by Semet and Taillard [18] and Rochat and Semet [14] to simplify and accelerate the re-assignment of vehicles to tours. Salhi and Rand [16] proposed an interactive procedure denoted RPERT. This procedure builds a VFM solution using a series of seven phases. Each phase uses one perturbation module in order to improve the solution of its predecessor if the new solution is better and feasible in terms of vehicle capacities and maximum travel times. It is then accepted as the current solution and passed into the next phase for further improvements. Otherwise the previous solution is retained and the search continues until the RPERT procedure is terminated at the end of the last phase. Osman and Salhi [13] enlarged the neighbourhood of RPERT by performing moves that were considered infeasible by the old RPERT. This method, denoted MRPERT, Modified Route Perturbation Procedure, and based on tabu search is developed using an appropriate data structure. MRPERT has produced significant improvements over RPERT but required extra computing time. New best known results were found for a VFMP without time windows.

These types of local search methods, which use a large variety of different moves, seem to fit well to our problem. But they do not take into account time window constraints. To our knowledge, strong time window constraints are introduced in a VFMP here for the first time. The time windows are very tight and distributed over a same time period of one day: for example, we have to visit 102 customers in two hours and the time window width of each customer is between 5 and 10 minutes (min.). Furthermore, several customers are located close to each other and could be served within their time window in any order.

11.3 PROBLEM FORMULATION

We first give a mathematical formulation of the problem that we call **Vehicle Fleet Mix Problem with Strong Time Windows (VFMP-STW)** to illustrate the difference with the more classical problems of the VRP-TW type. This formulation corresponds to a single depot problem.

Problem constants:

- K number of vehicle types (K is fixed). Unlike the more classical problem VFMP, we assume that an infinite supply of each vehicle type is available.
- n number of customers; customers are numbered from 1 to n and 0 denotes the central depot,
- Cf_k fixed cost of vehicle type k,
- $C1^k_{ij}$ driving cost, i.e., (driver immobilization cost given in French Francs (FF) per time unit) $\times$ (time used to go from i to j using vehicle type k),
- $C2^k_{ij}$ travel cost, i.e., (driver cost given in FF per distance unit) $\times$ (distance used to go from i to j using vehicle type k),
- $C3_j$ service cost, i.e., (driver immobilization cost given in FF per time unit) $\times$ (service time at customer j); it includes waiting time at customer j,
- Q_k capacity (or volume) of vehicle type k

d_i	demand of customer i,
$[a_i, b_i]$	time window associated with customer i,
t_{ij}^k	travel time from customer i to customer j using vehicle type k including the service time (st_i) at customer i,
α, β, γ	fixed parameters.

Problem variables:

del_i^k	arrival time (early or late) to customer i using vehicle type k,
del_0^k	due time (early or late) to depot,
X_{ijk}	binary variable equals 1 if vehicle type k travels from customer i to customer j, 0 otherwise,
Y_{ij}	non negative variable which gives the total load on the vehicle just after visiting customer i and on the way to visit customer j.

$$\begin{aligned} Min \sum_{k=1}^{K}\sum_{i=0}^{n}\sum_{j=0}^{n}(C1_{ij}^k + C2_{ij}^k + C3_j) \times X_{ijk} + \sum_{k=1}^{K} Cf_k \times \sum_{j=1}^{n} X_{0jk} \\ +\alpha\left(\sum_{k=1}^{K}\max(0, \sum_{i=1}^{n} d_i - Q_k)\right) + \gamma\left(\sum_{k=1}^{K}\max(0, del_0^k - b_0)\right) \\ +\beta\sum_{k=1}^{k}\left(\sum_{i=1}^{n} max(0, del_i^k - b_i) + \sum_{i=1}^{n} max(0, a_i - del_i^k)\right) \end{aligned} \tag{11.1}$$

$$\sum_{k=1}^{K}\sum_{i=0}^{n} X_{ijk} = 1, \qquad j = \overline{1,n} \tag{11.2}$$

$$\sum_{i=0}^{n} X_{ijk} - \sum_{l=0}^{n+1} X_{jlk} = 0, \quad j = \overline{1,n}, k = \overline{1,K} \tag{11.3}$$

$$\sum_{l=0}^{n+1} Y_{jl} - \sum_{i=0}^{n} Y_{ij} = d_j, \qquad j = \overline{1,n} \tag{11.4}$$

$$\sum_{j=1}^{n} Y_{j0} = \sum_{i=1}^{n} d_i, \tag{11.5}$$

$$\sum_{i=1}^{n} Y_{0i} = 0, \tag{11.6}$$

$$Y_{ij} \le \sum_{k=1}^{K}\sum_{i=0}^{n} Q_k \times X_{ijk}, \qquad j = \overline{1,n} \tag{11.7}$$

$$X_{ijk}(del_i^k + t_{ij}^k - del_j^k) \le 0, \; i = \overline{0,n}, \quad j = \overline{1,n}, \; k = \overline{1,K} \tag{11.8}$$

$$X_{ijk} \in \{0,1\}, \; Y_{ij} \ge 0, \; Y_{ii} = 0, i = \overline{0,n}, \quad j = \overline{1,n}, \; k = \overline{1,K}$$

The objective function (11.1) consists of fixed costs, routing costs and penalties in time and capacity with different weight parameters (α, β, γ). Constraints

(11.2) ensure that each customer is visited only once. Constraints (11.3) and (11.4) show the conservation of flow between customers and evaluate the needs between routes and vehicles. Constraint (11.5) shows that the total demand is satisfied. Constraints (11.6) denote that all vehicles are empty when they leave the depot. The capacity constraints (11.7) ensure that a vehicle goes from customer i to customer j if its total load after the service is complete at customer i does not exceed a capacity Q_k of vehicle type k. Finally, constraints (11.8) are subtour elimination constraints which are easily linearized.

11.4 "TRANSFLEET": A VARIABLE MOVE TABU SEARCH

The proposed algorithm **Transfleet** for solving the VFMP-STW is made up of two main phases. The first phase, building route phase, builds routes in an iterative way. The second phase, the improving phase, improves the current solution obtained at the end of the first phase through a series of different moves within Tabu Search (TS) leading to a variant of TS (Variable Move Tabu Search). In this section, we briefly present these two phases before describing the different steps of our tabu search procedure.

11.4.1 Phase 1: a greedy method of priorities

We start by using a heuristic proposed by Ferland and Michelon [6] for building a starting solution but this offers solutions with high cost. This is due to waiting times that are not taken into account within the algorithm. The heuristic seeks to reach the nearest customer (requesting the lowest time to be reached within a neighborhood of selected customers). Its disadvantage is that time window constraints are not considered. Customers can be reached quickly but the postman spends a long time before collecting his mail. Hence this leads to a considerable cost. In order to make up for this problem, we set up the following algorithm:

- Rank the customers in a list L in decreasing order of their priority P_i, calculated as $P_i = a_i + \frac{1}{b_i - a_i} + st_i$ where st_i is the service time at customer i. The tighter the time window of a customer and the longer its service time, the harder it is to insert it in an already constructed route.

- Creating a route:
  - create a route with the first element of L,
  - **while** L is not empty,
    - take the element at the head of L,
    - for each created route, evaluate the cost of inserting this element in that route,
    - select the route where the insertion cost is the smallest.

      if this cost is smaller than the cost resulting from the creation of a new route for that customer **then** insert this customer,

      else create a new route with that customer.

11.4.2 Phase 2: solution improvement by a variable move tabu search

After creating a feasible starting solution respecting capacity and time window constraints, we use a tabu search heuristic in order to get new solutions. As the meta-heuristic used here is of local search type, we explore the solution space by moving from one solution to another. A move could lead to an unfeasible solution (violating one or several constraints). The originality of our tabu search approach is that at each iteration, this move is chosen following some special criteria. Two types of moves are used: **local** and **global moves**. We briefly give our motivation before describing them in detail together with the tests which activate them. The best known heuristics up to now use moves that exchange customers between routes. The efficiency of these heuristics has already been proved; However, we note that they favour the optimization of one route only, hence a local exploration and not the solutions far from the starting solution. We call them **local moves**. In order to widen the exploration of the solutions we propose new moves that change significantly the structure of the solution. They act on the number of routes and can modify the whole solution by disrupting all routes. We call them **global moves**. Those two types of moves are used alternately in our application following some activating tests. Local moves (exchange, reallocation) are used to improve the solution by proceeding by small steps. When they show a weakening of their efficiency, we use global moves (reduction, splitting) to regenerate a solution. The activating tests of those two types of moves are described in the next section.

11.5 GLOBAL MOVES AND THEIR ACTIVATING TESTS

Two global moves are to be distinguished, route reduction and route splitting.

11.5.1 Route reduction

The main goal of this move is to remove one route that seems to be costly or inefficient. Customers are removed from that route and inserted in the best possible place in the other routes. Since it is more significant to insert first the customer having the larger quantity of mail to send in a set of already constructed routes, we establish a list of customers sorted in decreasing order of their assigned loads. For each element of this list, we seek the least costly feasible place in the other routes.

11.5.2 Route splitting

This move is also applied to a route which is too costly. It generates two routes from one by removing the penalized arc of the route. This arc could be the most costly (fixed costs plus routing costs) or the least efficient (long wait time at customers or too few customers on the route). We connect the first "arc initial node" of that penalizing arc to the depot terminating henceforth the route. Likewise, the second "arc terminal node" is connected to the depot setting up the first customer of a new route.

11.5.3 Activating tests

Route selection Global moves have an effect on the number of routes. They add or remove one route from the solution set. The main question is how to specify the target route and the global move type. We concentrate on the most costly route. However, let us first define the word "**costly**". Two routes can have the same total cost but different partial ones. As costs are of different types, we can evaluate separately the value of one route. A route can be costly even if it contains no customers-waiting-time-cost, while another route with a lesser cost can have its customers-waiting-time-cost higher than its own total cost. Hence we see that the first route has a higher intrinsic value (in terms of solution quality) than the second one and that it is preferable to seek to optimize this one.

The example led us to change our route selecting criteria to optimize and define several parameters necessary to make this choice. For each route R where k is the type of vehicle used, we save its total cost, but also its partial costs as:

- Time and distance costs, $C_t = \sum_{i,j \in R} C1^k_{ij}$ and $C_d = \sum_{i,j \in R} C2^k_{ij}$,
- Customer service cost, $C_{sc} = \sum_{j \in R} C3_j$,
- Fixed costs, Cf_k,

and the costs corresponding to:

- penalties, i.e., waiting time $(\sum_{i \in R} \max(0, ai - del^k_i))$ and delay $(\sum_{i \in R} \max(0, del^k_i - bi))$ at customers,
- depot delay times $(\sum_{i \in R} \max(0, del^k_0 - b0))$ and finally,
- vehicles capacity "exceeds" $(\max(0, \sum_{i \in R} d_i - Q_k))$.

We set up the following criterion for the selection of a route:

$$RN = \frac{\text{route cost}}{\text{number of customers}}$$

In order to avoid to concentrate only on routes with a large number of customers, hence with a high cost, we set this ratio which leads to calculate the average cost of a customer in a route. Hence routes with a few number or customers are the first to be studied. To select one route, we proceed as follows:

1. Sort a list L in decreasing order of routes according to their assigned RN costs.
2. **If** the first element of this list has a cost C^{first}_{RN}, $P1\%$ (P1, given parameter) greater than the second element **then** we select it. **Else**

   a - Set a list L' of routes whose costs are close to C^{first}_{RN}, i.e., within the interval $[C^{first}_{RN}(1 - \frac{P1}{100}), C^{first}_{RN}]$,

b - Sort L' in decreasing order of penalty costs,

c - Route corresponding to the first element of this list is the one to be optimized.

3. Select a move.

Selection of a type of global move Once the route is selected, we seek to determine the global move type. To specify the choice of this move, we focus on the coefficient of variation (**cov**) of the arcs of the routes. The latter will allow us to characterize the distribution of the arcs cost of the route.

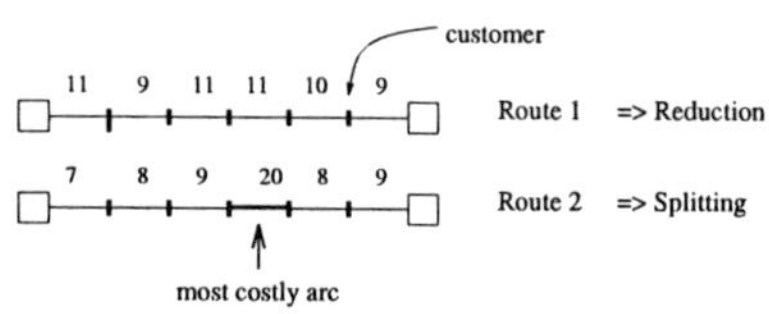

Figure 11.1 Choice between reduction and splitting

As shown in Figure 11.1, routes R1 and R2 have the same total cost. However, the high cost of R2 is due to only one arc (its fourth arc). It is better to use a splitting move on R2 at its penalized arc. Contrary to R2, arcs of R1 have a weak difference in costs between each other and the high total cost of the route is due to all the arcs of the route and not to one specified arc. Hence, we consider this route as globally not interesting. Thus we use here a reduction move. This leads to the following algorithm:

1. Compute the **standard deviation** of the arc costs on the selected route,
2. Calculate its **coefficient of variation** (cov): average standard deviation,
3. **If** cov is greater than a certain threshold P2, statistic parameter,
   then make a split on the most costly arc,
4. **else** make a reduction move.

In order to establish this cost difference between arcs, we use the variance notion

$$s^2 = \frac{\sum_{i=1}^{n}(c_i - \overline{x})^2}{n-1}$$

where

$$\overline{x} = \frac{\sum_{i=1}^{n} c_i}{n}$$

with c_i the cost of the ith arc in the route and n the number of arcs in the route. The coefficient of variation is then

$$cov = \frac{s}{\overline{x}}.$$

Here, we set

$$P2 = \frac{\sum_R cov_R}{\text{number of routes}}.$$

If the value **cov** of the selected route is greater than $P2$, this implies that the high cost of this route is due to only one arc. Otherwise, the total cost of the route is uniformly distributed over its arcs.

11.6 LOCAL MOVES AND THEIR ACTIVATING TESTS

11.6.1 Exchange

The two most penalized customers are exchanged within this move. These customers can belong either to one route or can be selected from two separate routes. In order to carry out this exchange, we extend the notion of time windows margin defined by Stan [19] to our special data set where time windows of customers are gathered in a narrow period of time.

Time Window Margin Notion. $m_{ij} = \min(b_j, b_i + t_{ij}) - \max(a_j, a_i + t_{ij})$.

When $m_{ij} \geq 0$, leaving customer i within its time window, the postman will reach customer j on time within its time window. The higher m_{ij} is, the greater the possibility to arrive within the time window of customer j (cf. Figure 11.2). When $m_{ij} < 0$, the postman arrives too early or too late at customer j.

We extend the time window margin notion of a customer i to the notion of **in going neighbours** U_i^- where $U_i^- = \{j/m_{ij} \geq 0\}$ is the set of customers from which the postman can reach "in time" customer i and **out going neighbours** U_i^+ where $U_i^+ = \{j/m_{ji} > 0\}$ is the set of customers that the postman can reach "in time" from customer i.

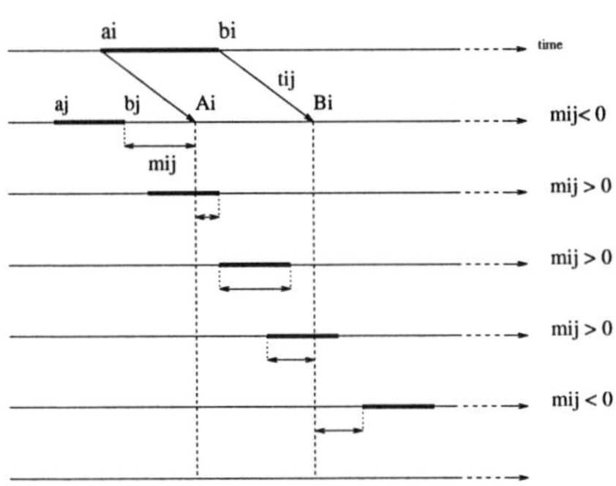

Figure 11.2 Neighbourhood determination

11.6.2 Algorithm and details

Let edge (k, m) be the most costly arc of the selected route. We evaluate the exchange of k with every out going neighbour of its predecessor and every in coming neighbour of its successor. We do the same with m and finally, we select the best solution obtained by exchanging k or m. Figure 11.3 helps to illustrate the algorithm.

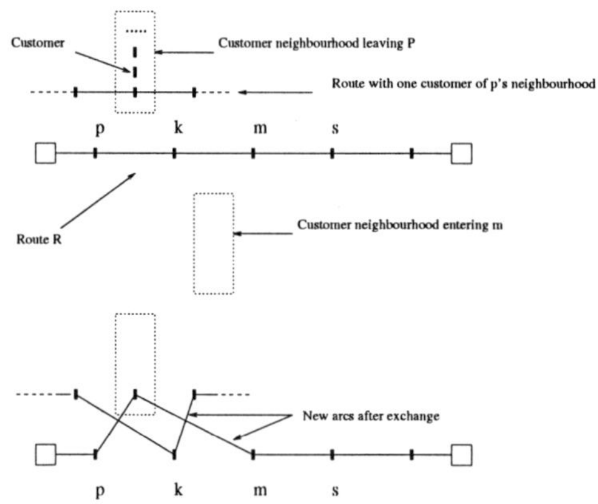

Figure 11.3 Customer exchange

Algorithm.

1. Let R be the route where an exchange has to be done and (k, m) the most costly arc of R.

2. Determine the predecessor p of k and the successor s of m.

3. Determine $E_1 = U_p^+ \cup U_m^-$.

4. Let L_1 be an empty list.

5. For each element i of E_1

   a - Evaluate the cost F_i of an exchange with k.

   b - Insert (F_i, i) in L_1.

6. Determine $E_2 = U_k^+ \cup U_s^-$.

7. Let L_2 be an empty list.

8. For each element j of E_2

   a - Evaluate the cost F_j of an exchange with m.

   b - Insert (F_j, j) in L_2.

9. Sort L_1 (resp. L_2) in increasing order of F_i (resp. F_j).

10. $min_1 = min_{(F_i,i)\in L_1}(F_i)$ and $min_2 = min_{(F_j,j)\in L_2}(F_j)$. If $min_1 < min_2$ then exchange k with the element corresponding to min_1, else exchange m with the element corresponding to min_2.

11.6.3 Reallocation

This move removes one customer from one route and inserts it in the best possible place within the set of other routes. It is applied after a series of exchange moves because it disrupts significantly the solution by altering the number of customers within two routes.

11.7 POST-OPTIMIZATION

In order to avoid passing over an important improvement of a current solution, we use the following procedures which, by their quite general character, could help us to obtain a better solution. They are used on all modified routes or newly created ones at the end of every move.

11.7.1 Or-opt procedure

This well-known procedure uses a sequence of one, two or three consecutive customers in the current solution, and moves the sequence at another location in the same route or in another route. In our case we exchange two customers. The additional cost obtained after violating one constraint could be compensated by a gain on the total time and/or distance of the route.

11.7.2 Capacity adjustment

This procedure evaluates the total cost of one route for all types of vehicles in the fleet. If a gain is obtained, then we exchange the current vehicle with the one which allows that gain. According to the relative value of penalty coefficients and vehicles fixed costs, there exist some situations where the assignment of a vehicle to one route is not the one on which we can expect the best cost. In fact, a vehicle with a larger capacity is inevitably not the faster one and respecting capacity constraint is probably not sufficient to make up for the cost of time windows constraint violation: a smaller but faster vehicle may considerably reduce the travel time cost and allows to obtain a good gain value on route cost even for the price of capacity penalty.

11.8 INTENSIFICATION AND DIVERSIFICATION: A NEW APPROACH

11.8.1 Attributes and tabu moves

The meta-heuristic tabu is characterized by the memorization of the path used during the exploration. Usually, the set of solutions that constitutes the path search is not kept in memory, but the solution attributes allowing to go from one stage to another one.

In tabu search, non-improving solutions are accepted in the goal of avoiding local minima, and tabu restrictions are used to avoid cycling. These tabu restrictions consist in setting up the opposite moves of the previous solution as forbidden (tabu) to avoid then any cycling. The whole set of the current solution is not entirely stored in a list. Only the attributes of the recently improved moves are stored. This list, called the **tabu list**, is regularly renewed.

In our application, we store the number of routes of one solution and the position of one customer into a route (solutions attributes). We have then two tabu lists : the route tabu list and the position tabu list. One containing the number of routes at each iteration and the second one, containing customer(s) that have moved to other route at each iteration, including their previous route(s).

11.8.2 Set of elite solutions

During the optimization process, we obtain solutions of different types. We decide to keep in memory the best ones: the best feasible solution and the best non-feasible solutions, solutions violating either time windows or capacity or depot due time constraints. This set will help us later to extend the tabu search exploration.

11.8.3 Intensification and diversification

Intensification strategies in tabu search are very often based on modifying the choice rules to encourage move combinations and solutions features historically found good. They may be applied to to attractive regions to search them more thoroughly. Diversification strategies, as their name suggests, are designed to drive the search into new regions. Often they are based on modifying choice rules to bring attributes that are infrequently used into the solution.

In our application, a "tool box" of moves is made up of two types of moves following the modification they bring to the solution main elements, i.e., to the number of routes or to the number of customers on a route. We can link these two types of moves to two kinds of exploration of the research space which correspond to the improvement of one or two routes, i.e., a deep exploration of the neighborhood of the current solution, and a modification of the solution structure, i.e., a large move within the solution space.

In this way, our intensification strategies are made up of the first type of moves, local moves: a series of exchanges (local exploration) followed by a series of reallocation moves. Those moves provide an intensification of the exploratory research in a fixed direction. The diversification strategies are made up of the second type of moves: reduction or splitting, according to the activating tests. However, among local moves, the reallocation move posseses a particular characteristic. On the one hand, it allows the optimization of one route and the position of a customer within the solution, but on the other hand, it offers a light regeneration of the solution by modifying one of the main parameters of a route like the number of customers. Thus this move could be used both within diversification and intensification.

11.8.4 Oscillation

"Continuous" exploration and oscillation The exploration phase may be described as a repetition of the following sequence: application of a series of intensification moves until there is no convincing results, then a diversification move. Hence, **oscillation.** This strategic oscillation provides a means to achieve an effective interplay between intensification and diversification.

The intensification phase is made up of a series of exchange and reallocation moves. The diversification phase uses either reduction or splitting according to the activating procedure. However, the use of tabu lists introduces one modification within the procedure. As reduction reduces the number of routes, if this number is set tabu we apply the splitting move instead of reduction. If a splitting move is selected and the number of routes after its application is set tabu then we apply reduction. Finally, if the two moves are tabu then we apply the more efficient one.

Remark. The oscillation method used in our algorithm alternates between intensification and semi-diversification (a series of reallocations).

Application of solutions set and dynamic intensification Once the set of elite solutions is set up, we use it to increase the size of the search space in the second stage of exploration. During this latter stage, each elite solution is a new starting solution to which we apply an intensification strategy.

After exploring the elite solutions set, if no improvement has been noticed, we modify the penalty coefficients which could contribute to the improvement of the solution quality. This is done by multiplying for example all the penalty coefficients by the same value. Then, we restart the intensification stage from the first element of the elite solutions set. If, at this level, an improvement has happened at one elite solution, we keep updating the set of elite solutions and restart the operation, intensification and penalty coefficients modification, from the begining until no further improvement is possible using any of the remaining elite solutions.

Moving from one elite solution to another is carried out when the current elite solution is not profitable. Alternation is performed until no elite solution left leads to improvement.

11.9 EXPERIMENTS AND COMMENTS

We used a data set given by the French Postal Services Company, La Poste.[1] The time windows are very tight and distributed over a same time period of a day. Figure 11.5 shows the customers' time window distribution for one of our data set. They are distributed over a period of approximately one hour and are equal to five or ten mn.

[1]The data set is available at the location http://www.prism.uvsq.fr/~pos/Benchmarks.

Dep	Veh-Typ	RN	PRN	PAMC	ACN (R)	Best Cost
1	1	11	11	41.68	9	
	3	11				2200.04
2	1	32	8	390.35	17	
	3	35				8595.84
3	1	5	4	76.51	14	
	3	7				1563.32
4	1	9	9	66.77	8	
	3	9				2991.65
5	1	2	1	27.25	13	
	3	3				587.40

Table 11.1 Dep: Depot, Veh-Typ: Vehicle Type, RN: Route Number, PRN: Predicted Route Number, PAMC: Predicted Amount of Mail to be Collected from customer, ACN: Average Customer Number per route and Best Cost: cost of the best solution.

The performance of our algorithm was evaluated on five problems corresponding to 26, 64, 75, 100 and 165 customers for a heterogeneous fleet. The program was coded in C and the computations were performed on a SUNsparc workstation.

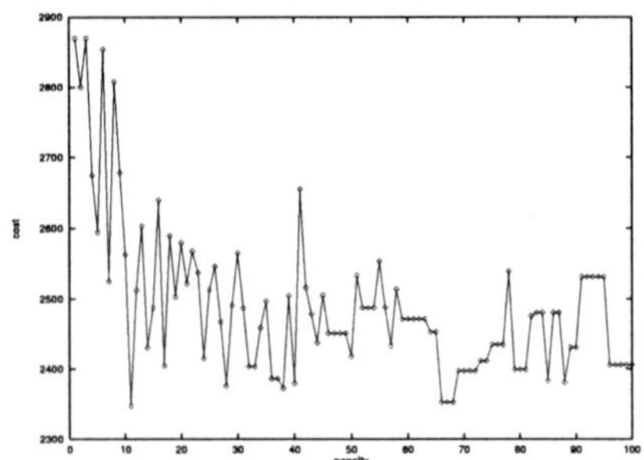

Figure 11.4 Penalty Coefficients

The intensification part of the algorithm is made up of a series of exchanges (NE) followed by a series of reallocation moves (NR). Based on experimental evidence we set the number of exchanges $NE = k1 \times NCD$ (where NCD is the number of customers in the depot and k1, $0 < k1 \leq 1$, is a parameter calculated in function of NCD and the potential of the used workstation) and $NR = k2 \times NCD$ (where $k2$ is of type $k1$, $k2 = \alpha k1$ and α is a positive constant, $0 < \alpha \leq 1/2$). The diversification phase is made up of one global move.Notice that the reduction could also be applied within the intensification phase. In fact, a reallocation move applied to a route of one customer corresponds to a reduction. Obviously, the frequent calls to reductions lead to a reduction of the total routes number. The size of the position tabu list is set to NE and the size of the route tabu list is set to 2 NE. This value intensifies the previous effect of reduction of the total number of routes. It allows us to quickly find the number of routes in a good solution. In order to avoid eleminating all routes, we decide to reduce to a predicted number of routes. The latter predicted number is estimated as follows: we compute the average time to go from one customer to another one, the average service time and the number of customers that can possibly be served between the open time of the earlier customer and the open time of the latter customer. We assume that capacity constraints are respected. The used vehicle is of average capacity and average speed among all the types of vehicles.

Table 11.1 shows the predicted number (PRN column) for our real data set. We see for depots 1, 3, 4 and 5, that this number is very close to the value (RN column) of the best solution we found during the experiments. The main difference between the value of column 3 and 4 for the depot 2 is due to the no respect of capacity constraints. The PAMC per customer is 390.35, and the estimated ACN is 17. Thus the amount of collected mail is 6635.95, which implies the violation of the capacity (4000) of the used vehicle.

To calibrate penalty values is generally not an easy task. During the experiments, we noticed that high values (from 50 to 100) generally lead to good results. But increasing these values to a very high threshold (> 100) does not improve the existing solutions. However, good solutions could sometimes be obtained for some low and specific penalty values. Experiments show that good results are obtained for penalty values close to each other. Hence, these latter coefficients have been set to the same value. Figure 11.4 shows the best feasible solutions for depot 1 after 2000 iterations for penalty values between 1 and 100.

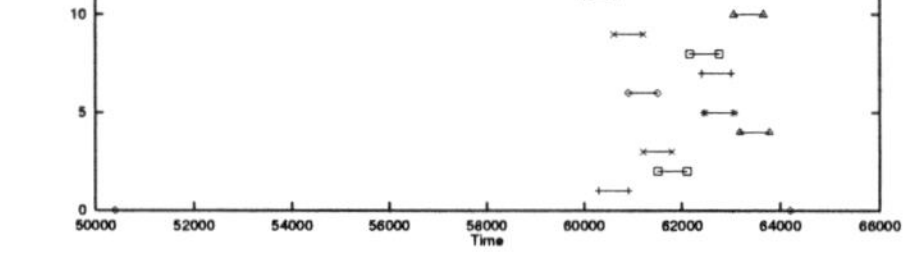

Figure 11.5 Time Windows Distribution

Figure 11.4 can be divided in two parts. The first part corresponding to the range [1,50] shows a very high instability since modifying penalty values of one unit may strongly change the value of the best feasible solution. The second part is more stable and shows good solutions. We notice that these solutions are gathered in sets of few solutions.

Finally, we conclude through these observations that good solutions are obtained when penalty values are selected within the range [50,100]. This leads to the reduction of the number of executions. Hence the global number of tests needed to find a good solution has been drastically reduced.

11.10 CONCLUSIONS

In order to solve the problem, we developed for the tabu approach a set of moves that may dramatically change the structure of the solution. Previous experience showed that the use of global moves leads quickly to the composition of a good fleet. Those moves, which increase or reduce the total number of routes, are disruptors and have to be carried out carefully. Thus, we use a selection criterion for the used global move, which is, due to its statistic foundation, precise enough to determine if the cost of the route is uniformly distributed among arcs of the route. Since they re-structure the current solution, these global moves allow often large improvements in one single iteration. Local moves are then applied in order to refine this solution.

The tests made show that the best combination is a strategy which alternates between the diversification phase (global moves) and the intensification phase (local moves).

Our method is rather general. The main ideas are, the use of two types of moves, local and global, which are chosen at each iteration of the tabu search, following some tests taking into account the history of the search (number of iterations without improvement) and the structure of the current solution (costly arcs, costly routes). If the search is trapped in a local optimum, global moves could help to escape from this local optimum and find a new solution. It acts in this case as a diversification procedure. However, local moves intensify the search around the current solution.

Once tabu search does not improve the best solution obtained until now, we apply the elite solution list which has often proved its efficiency. Varying the penalty coefficients leads to the furthest intensification and often leads to improvements.

Finally, the Variable Move Tabu Search method is efficient. We obtained good results for the French Postal Services Company application. We noticed better results with a well-proportioned data set in time and capacity (weak variance around a mean) contrary to a data set with "aberrant" values. However, to make up for this problem, we suggest the elaboration in the future of new procedures to better deal with this kind of data sets. Those procedures will for example permit the identification of the area of aberrant values of the data set and treat it separately. We also intend to extend our approach to a multi-depot case.

References

[1] L. Bodin and L. Berman. Routing and scheduling of school buses by computer. *Transportation Science*, 13:113–129, 1979.

[2] L. Bodin, B. Golden, A. Assad, and M. Ball. Routing and scheduling of vehicles on crews -the state of art. *Computers Ops Res*, 10:63–212, 1983.

[3] G. Clarke and J. Wright. Scheduling of vehicles from a central depot to a number of delivery points. *Operations Research*, 12:568–581, 1964.

[4] M. Desrochers and T. Verhoog. A matching based savings algorithm for the vehicle routing problem. Technical Report G-89-04, GERAD, Ecole des Hautes Etudes Commerciales, Montreal, 1989.

[5] M. Desrochers and T. Verhoog. A new heuristic for the fleet size and mix vehicle routing problem. Technical Report G-89-04, GERAD, Ecole des Hautes Etudes Commerciales, Ecole Polytechnique, and Erasmus University, Netherlands, Montreal and Rotterdam, 1991.

[6] J. Ferland and P. Michelon. The vehicle scheduling problem with multiple vehicle types. *Journal of the Operational Research Society*, 39:577–583, 1988.

[7] M. Fisher and R. Jaikumar. A generalized assignment heuristic for vehicle routing. *Networks*, 11:109–124, 1981.

[8] F. Gheysens, B. Golden, and A. Assad. A comparison of techniques for solving the fleet size and mix vehicle routing problem. *OR Spektrum*, 6:207–216, 1984.

[9] B. Gillett and L. Miller. A heuristic algorithm for the vehicle dispath problem. *Operations Research*, 22:340–344, 1974.

[10] B. Golden, A. Assad, L. Levy, and F. Gheysens. The fleet size and mix vehicle routing problem. *Computers Ops Res*, 11:49–65, 1984.

[11] R. Newton and W. Thomas. Bus routing in a multi-school system. *Computers Ops Res*, 1:213–222, 1974.

[12] I. Or. Traveling salesman-type combinatorial problems and their relation to the logistics of regional blood banking. PhD thesis, Department of Industrial Engineering and Management Science, Northwestern University, Evanston, IL, 1976.

[13] I. Osman and S. Salhi. *Local Search Strategies for The Vehicle Fleet Mix Problem.* In V. Rayward-Smith, I. Osman, C. Reeves, G. Smith (ed.), Modern Heuristic Search Methods, John Wiley, North-Holland, Amsterdam, 131–153, 1996.

[14] Y. Rochat and F. Semet. A tabu search approach for delivering pet food and flour in Switzerland. *Journal of the Operational Research Society*, 45:1233–1246, 1994.

[15] Y. Rochat and E. Taillard. Probabilistic diversification and intensification in local search for vehicle routing. *Journal of Heuristics*, 1:147–167, 1995.

[16] S. Salhi and G. Rand. Incorporating vehicle routing into the vehicle fleet composition problem. *European Journal of Operational Research*, 66:313–330, 1993.

[17] S. Salhi, M. Sari, D. Saidi, and N. Touati. Adaptation of some vehicle fleet mix heuristics. *OMEGA*, 20:653–660, 1992.

[18] F. Semet and E. Taillard. Solving real-life vehicle routing problems efficiently using tabu search. *Annals of Operations Research*, 41:469–488, 1993.

[19] M. Stan. *Tournées de Véhicules et satisfaisabilité logique.* PhD thesis, Université de Montréal, Montreal, 1996.

[20] H. Stern and M. Dror. Routing electric meter readers. *Computers Ops Res*, 13:209–223, 1979.

[21] E. Taillard. Parallel iterative search methods for vehicle routing problems. *Networks*, 23:661–676, 1993.

12 FLOW LINE SCHEDULING BY TABU SEARCH

Eugeniusz Nowicki and Czesław Smutnicki

Technical University of Wrocław, Institute of Engineering Cybernetics
Janiszewskiego 11/17, 50-372 Wrocław, Poland.
enowicki@ict.pwr.wroc.pl, smutnick@ict.pwr.wroc.pl

Abstract: The paper deals with a flow line scheduling problem which is an extension of two classic problems of scheduling theory, namely the flow shop and parallel shop, and can be briefly described as follows. There is a set of jobs and a set of processing centers each of which has a set of parallel identical machines. A job consists of a sequence of operations processed at successive centers, and all jobs pass through centers in the same order. At the center, a job can be processed on any machine. The transport time of jobs between centers is job dependant. We want to find a schedule that minimizes the makespan, one of the most frequently used scheduling criteria.

To show the problem, its properties, and algorithm, we consider chiefly the case with two centers. To solve the problem stated we propose a fast and easily implementable approximation algorithm based on a tabu search technique with a specific, reduced neighborhood, original search accelerator, and the technique of back jumps on the search trajectory. The neighborhood is constructed using notions of a critical path in a graph and the elimination properties of blocks of jobs. The proposed neighborhood plays an important role in proper guiding the search into the most promising regions of the solution space as well as in the reduction of the costs of the single neighborhood search. The search accelerator, designed with the help of some theoretical properties of the graph, improves additionally and significantly the speed of the algorithm by avoiding superfluous calculations. Candidate list strategy and back jumps on the search trajectory ensure proper balance between the search intensification and diversification.

Results of numerical tests (up to 300 jobs) are also provided to show exhaustive numerical characteristics of the proposed algorithm. They confirm its excellent numerical behavior. The proposed approach can be extended to cover a broad class of flexible flow line scheduling problems. Known industrial applications refer among others to chemical branches, polymer and petroleum industries, electronic industry, FMS.

12.1 INTRODUCTION

The paper deals with a flow line scheduling problem that is an extension of two classic problems of scheduling theory, namely the flow shop and parallel shop, and can briefly be described as follows. There is a set of jobs and a set of processing centers each of which has a set of parallel identical machines. A job consists of a sequence of operations processed at consecutive centers (every job is processed exactly once in either center) and all jobs pass through centers in the same order. At a center, a job can be processed on any machine. The transport time of jobs between centers is job dependent. We want to find a schedule that minimizes the makespan, one of the most frequently used scheduling criteria.

This problem has received considerable attention in the last few years due to its many possible industrial applications, e.g., in chemical industries, computer systems, telecommunication networks, FMS, etc., see the review in [7]. Some extensions of this problem are also stated, e.g., by incorporating buffers between centers [14, 16].

In spite of a simple formulation of the problem, the finding of an optimal solution is rather complicated. Strong NP-hardness of this problem determines directly the set of approaches that can be applied to solve it. In the literature, one can find exact algorithms based on branch-and-bound ($B\&B$) [12] and mixed-integer programming (MIP), as well as a variety of approximation methods. Some of these procedures have been designed for special cases, for systems that have only two centers, or systems where only one of the centers has parallel machines, etc. ([1] is one of the newest in this area.) Note the specificity of the problem locates it on the borderline between classic flow shop, job shop and parallel machine problems. Thus, the most representative tools from each of these scheduling classes are commonly used to provide solution algorithms. Computational experiments show that $B\&B$ algorithms and MIP models become practically useless for more than 10 jobs and a very small number of machines. To solve the larger instances, approximation approaches have to be used. Besides already-known *constructive* algorithms applicable to the problem stated [2, 7, 14, 16], recently more attention has been paid to approximation methods based on local search approaches. Only the latter class of methods allows us to form in a flexible way a compromise between solution time and quality of solutions obtained. In this paper we provide a local search approximation algorithm for the problem stated. To make the description of problem properties and the algorithm simpler, we consider chiefly the case with two centers. To solve the problem stated, we propose a fast and easily implementable approximation algorithm based on a tabu search technique with a specific, reduced neighborhood, an original search accelerator and the technique of back jumps on the search trajectory. The neighborhood is constructed using the idea of a critical path in a graph and the elimination properties of blocks of jobs. The proposed neighborhood plays an important role in properly guiding the search into the most promising regions of the solution space as well as in reducing the cost of the single neighborhood search. The search accelerator,

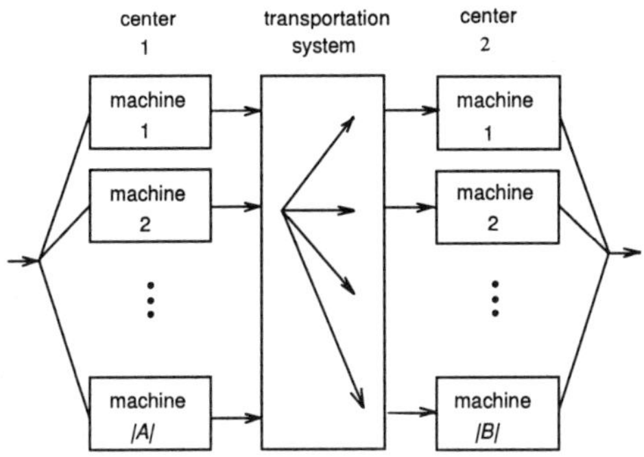

Figure 12.1 The structure of manufacturing system

designed with the help of some theoretical properties of the graph, significantly improves the speed of the algorithm by avoiding superfluous calculations. A candidate list strategy and back jumps on the search trajectory ensure proper balance between the search intensification and diversification. Results of numerical tests (up to 300 jobs) are also provided to show the complete numerical characteristics of the proposed algorithm. They confirm its excellent numerical behavior. The proposed approach can be extended to cover a broad class of flexible flow line scheduling problems.

12.2 THE PROBLEM

We consider a manufacturing system that has two processing centers, see Figure 12.1. Each center has a set of parallel identical machines. Let $\mathcal{A} = \{1, \ldots, |\mathcal{A}|\}$ and $\mathcal{B} = \{1, \ldots, |\mathcal{B}|\}$ denote the sets of these machines in the first and second center, respectively. There is a set of jobs $\mathcal{N} = \{1, 2, \ldots, n\}$ that has to be processed in this system. Each job has the same route: first, it is processed in time a_j on some machine from the set $\mathcal{A}$ at center 1, next it is transported from the center 1 to the center 2 in time t_j, and at the end it is processed in time b_j on some machine from the set $\mathcal{B}$ at center 2. We assume that machines can execute at most one job at a time and processing of a job on a machine cannot be interrupted. A feasible schedule is defined by the collections of pairs (A_j, C_j^A), (B_j, C_j^B), $j \in \mathcal{N}$, where job j is completed at center 1 on machine $A_j \in \mathcal{A}$ at time C_j^A and completed at center 2 on machine $B_j \in \mathcal{B}$ at time C_j^B, in such a way that the above constraints are satisfied. These constraints are equivalent to the following conditions:

$$A_j \in \mathcal{A}, \quad B_j \in \mathcal{B}, \quad j \in \mathcal{N}, \tag{12.1}$$

$$C_j^A \geq a_j, \quad j \in \mathcal{N}, \tag{12.2}$$

$$C_j^B \geq C_j^A + t_j + b_j, \quad j \in \mathcal{N}, \tag{12.3}$$

$$(A_i = A_j) \Rightarrow (C_j^A \geq C_i^A + a_j) \dot{\vee} (C_i^A \geq C_j^A + a_i), \quad i \neq j, \quad i,j \in \mathcal{N}, \tag{12.4}$$

$$(B_i = B_j) \Rightarrow (C_j^B \geq C_i^B + b_j) \dot{\vee} (C_i^B \geq C_j^B + b_i), \quad i \neq j, \quad i,j \in \mathcal{N}. \tag{12.5}$$

Conditions (12.1) and (12.2) are obvious. Condition (12.3) follows from the fixed technological order of job processing. Conditions (12.4)–(12.5) ensure that each machine processes at most one job at a time. The problem is to find a collections of pairs (A_j, C_j^A), (B_j, C_j^B), $j \in \mathcal{N}$ that satisfies (12.1)–(12.5) and minimizes the makespan $\max_{j \in \mathcal{N}} C_j^B$.

The formulated discrete optimization problem is rather complicated, chiefly due to the disjunctive constraints (12.4)–(12.5). Therefore we reformulate the problem stated by introducing other independent variables: sets of jobs assigned to machines and processing orders of jobs on machines (represented by permutations). This can be achieved as follows.

Let $\mathcal{N}_i$, $i \in \mathcal{A}$ be the set of jobs assigned to machine i at the center 1, and let $\mathcal{M}_i$, $i \in \mathcal{B}$ be the set of jobs assigned to machine i at center 2. These sets have to satisfy the following conditions:

$$\bigcup_{i \in \mathcal{A}} \mathcal{N}_i = \mathcal{N}, \quad \mathcal{N}_i \cap \mathcal{N}_j = \emptyset, \; i \neq j, \quad i,j \in \mathcal{A}, \tag{12.6}$$

$$\bigcup_{i \in \mathcal{B}} \mathcal{M}_i = \mathcal{N}, \quad \mathcal{M}_i \cap \mathcal{M}_j = \emptyset, \; i \neq j, \quad i,j \in \mathcal{B}. \tag{12.7}$$

Let $\alpha_i = (\alpha_i(1), \ldots, \alpha_i(|\mathcal{N}_i|))$ and $\beta_i = (\beta_i(1), \ldots, \beta_i(|\mathcal{M}_i|))$ denote a permutation on the set $\mathcal{N}_i$ and $\mathcal{M}_i$, respectively. Thus, the processing of jobs at center 1 can be determined by the sequence of $|\mathcal{A}|$ permutations $\alpha = (\alpha_1, \ldots, \alpha_{|\mathcal{A}|})$, whereas at center 2, by the sequence of $|\mathcal{B}|$ permutations $\beta = (\beta_1, \ldots, \beta_{|\mathcal{B}|})$. Next, denote by Π the collection of all such pairs (α, β); each pair provides sets of jobs assigned to machines in both centers and processing orders of jobs on all machines.

For a given $(\alpha, \beta) \in \Pi$ we can obtain a feasible schedule (A_j, C_j^A), (B_j, C_j^B), $j \in \mathcal{N}$, in the following manner. We set $A_{\alpha_i(j)} = i$, $j = 1, \ldots, |\mathcal{N}_i|$, $i \in \mathcal{A}$, and $B_{\beta_i(j)} = i$, $j = 1, \ldots, |\mathcal{M}_i|$, $i \in \mathcal{B}$. Job completion times can be found in a time $O(n)$ using the following recursive formulae:

$$C^A_{\alpha_i(j)} = C^A_{\alpha_i(j-1)} + a_{\alpha_i(j)}, \; j = 1, \ldots, |\mathcal{N}_i|, \; i \in \mathcal{A}, \tag{12.8}$$

$$C^B_{\beta_i(j)} = \max\{C^A_{\beta_i(j)} + t_{\beta_i(j)}, C^B_{\beta_i(j-1)}\} + b_{\beta_i(j)}, \; j = 1, \ldots, |\mathcal{M}_i|, \; i \in \mathcal{B}, \tag{12.9}$$

where $\alpha_i(0) = 0$, $i \in \mathcal{A}$, $\beta_i(0) = 0$, $i \in \mathcal{B}$ and $C_0^A = C_0^B = 0$. Since the scheduling criterion is a regular function, we can rephrase our problem as that of finding $(\alpha, \beta) \in \Pi$ that minimizes the makespan $C_{\max}(\alpha, \beta) = \max_{j \in \mathcal{N}} C_j^B = \max_{i \in \mathcal{B}} C^B_{\beta_i(|\mathcal{M}_i|)}$, where C_j^B, $j \in \mathcal{N}$ follows from (12.8)–(12.9). To illustrate introduced notions we provide the following example.

Example: Center 1 has two machines, $\mathcal{A} = \{1, 2\}$ and center 2 has three machines $\mathcal{B} = \{1, 2, 3\}$. We have nine jobs, $\mathcal{N} = \{1, \ldots, 9\}$, with processing times and transport times given in Table 12.1. Let us consider a solution

j	1	2	3	4	5	6	7	8	9
a_j	4	3	5	5	2	3	4	2	1
t_j	1	1	2	2	2	1	2	3	1
b_j	3	4	4	5	2	2	2	3	3

Table 12.1 Data for the example

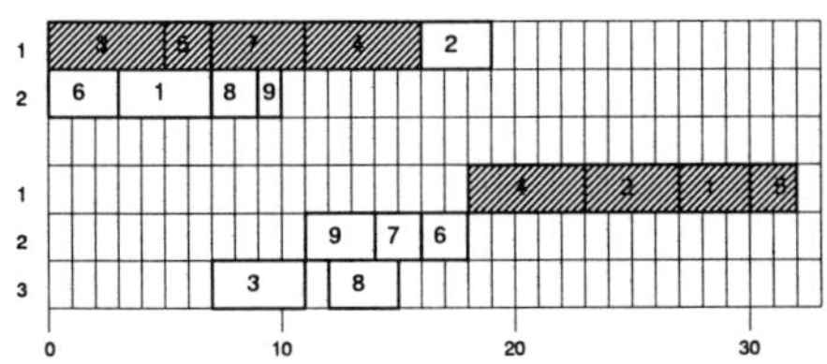

Figure 12.2 Gantt Chart for the schedule in the example

represented by $(\alpha, \beta) = ((\alpha_1, \alpha_2), (\beta_1, \beta_2, \beta_3))$, where permutations on machines have the following forms: $\alpha_1 = (3, 5, 7, 4, 2)$, $\alpha_2 = (6, 1, 8, 9)$, $\beta_1 = (4, 2, 1, 5)$, $\beta_2 = (9, 7, 6)$, $\beta_3 = (3, 8)$.

Permutations in the example determine partitions of jobs from $\mathcal{N}$ at centers as follows: (i) at center 1 on two subsets $\mathcal{N}_1 = \{2, 3, 4, 5, 7\}$, $\mathcal{N}_2 = \{1, 6, 8, 9\}$, (ii) at center 2 on three subsets $\mathcal{M}_1 = \{1, 2, 4, 5\}$, $\mathcal{M}_2 = \{6, 7, 9\}$ and $\mathcal{M}_3 = \{3, 8\}$. The feasible schedule for (α, β) from this example is given in a Gantt Chart (Figure 12.2). Job completion times follow from (12.8) and (12.9), see Table 12.1. The makespan is 32, calculated as the sum of: (i) processing times of jobs 3,5,7,4 on machine 1 at center 1, (ii) transport time of job 4 from the center 1 to 2, (iii) processing times of jobs 4,2,1,5 on machine 1 at center 2.

12.3 GRAPH REPRESENTATION

In the analysis we refer to an auxiliary graph model of type activity-to-node associated with a fixed pair $(\alpha, \beta) \in \Pi$. This graph has the form of $G(\alpha, \beta) = (\mathcal{N} \cup \mathcal{N}^*, \mathcal{H} \cup \mathcal{V}^\alpha \cup \mathcal{V}^\beta)$ where $\mathcal{N} \cup \mathcal{N}^*$ is the set of nodes and $\mathcal{H} \cup \mathcal{V}^\alpha \cup \mathcal{V}^\beta$ is the set of arcs. Each node $j \in \mathcal{N} = \{1, \ldots, n\}$ has weight a_j and represents the processing of job j at center 1. Each node $n+j \in \mathcal{N}^* = \{n+1, \ldots, n+n\}$ has weight b_j and represents the processing of job j at center 2. The set of arcs

$$\mathcal{H} = \bigcup_{j \in \mathcal{N}} \{(j, n+j)\} \tag{12.10}$$

is a result of the fixed technological processing order of jobs; each arc $(j, n+j) \in \mathcal{H}$ has weight t_j. Sets of arcs

$$\mathcal{V}^\alpha = \bigcup_{i \in \mathcal{A}} \bigcup_{j=2}^{|\mathcal{N}_i|} \{(\alpha_i(j-1), \alpha_i(j))\} \tag{12.11}$$

$$\mathcal{V}^\beta = \bigcup_{i \in \mathcal{B}} \bigcup_{j=2}^{|\mathcal{M}_i|} \{(\beta_i(j-1), \beta_i(j))\} \tag{12.12}$$

represent the processing order of jobs on all machines at center 1 and 2, respectively; each arc from these sets has weight zero. Graph $G(\alpha,\beta)$ for the data of the example is shown in Figure 12.3. Vertical arcs are zero-weighted and follow from (12.11)–(12.12), other arcs are nonzero-weighted and follow from (12.10).

It is clear that job completion times C_j^A found by (12.8) equal the length of the longest path to the node $j \in \mathcal{N}$ (including a_j) in $G(\alpha,\beta)$. Similarly, C_j^B found by (12.8)–(12.9) equals the length of such a path to node $n+j \in \mathcal{N}^*$ (including b_j). The makespan $C_{max}(\alpha,\beta)$ equals the length of the longest path (critical path) in $G(\alpha,\beta)$.

Consider a critical path L in $G(\alpha,\beta)$ interpreted as a sequence of graph nodes. This path can be naturally decomposed into two parts, $L=(L^A,L^B)$. The former part L^A is a sequence of nodes that represents the processing of jobs at center 1 on some machine, which we denote hereafter by e, $e \in \mathcal{A}$. Similarly, the latter part L^B is a sequence of nodes that represents the processing of jobs at center 2 on some machine denoted by g, $g \in \mathcal{B}$. Precisely, there exists a position f, $1 \le f \le |\mathcal{N}_e|$, in the permutation α_e and a position h, $1 \le h \le |\mathcal{M}_g|$, in the permutation β_g such that $L^A = (\alpha_e(1),\ldots,\alpha_e(f))$, $L^B = (n+\beta_g(h),\ldots,n+\beta_g(|\mathcal{M}_g|))$, and $\alpha_e(f) = \beta_g(h)$. Note that the critical path can be completely defined by values e,f,g,h. Such a representation of the critical path is advantageous for special properties introduced next and associated with so-called *blocks of jobs*. The sequence

$$\mathcal{K}^A = (\alpha_e(1),\ldots,\alpha_e(f))$$

will be called the block of jobs in center 1, whereas the sequence

$$\mathcal{K}^B = (\beta_g(h),\ldots,\beta_g(|M_g|))$$

will be called the block of jobs in center 2. (This definition of block is an extension of block notion from the paper [6].) From the definition, the last job in the block $\mathcal{K}^A$ is simultaneously the first in the block $\mathcal{K}^B$. Denoting this job by u, $u = \alpha_e(f)$, we can reformulate $C_{max}(\alpha,\beta)$ as follows

$$C_{max}(\alpha,\beta) = \sum_{j\in\mathcal{K}^A} a_j + t_u + \sum_{j\in\mathcal{K}^B} b_j. \tag{12.13}$$

The critical path in $G(\alpha,\beta)$ in Figure 12.3 has the form of $L=(L^A,L^B)$, where $L^A = (3,5,7,4)$, $L^B = (9+4, 9+2, 9+1, 9+5)$, indicated by a thick line in the figure. This path is characterised by $e=1$, $g=1$, $f=4$ and $h=1$. It provides two blocks $\mathcal{K}^A = (3,5,7,4)$ and $\mathcal{K}^B = (4,2,1,5)$.

12.4 SOLUTION METHOD

To solve the problem stated we make use of the tabu search approach TS [3, 4, 5]. Our previous research [9, 10, 11] has shown that this technique is superior to various local search methods designed for finding a near optimal solution of scheduling problems with min-max criteria. It has also been used with great success for many other combinatorial optimization problems.

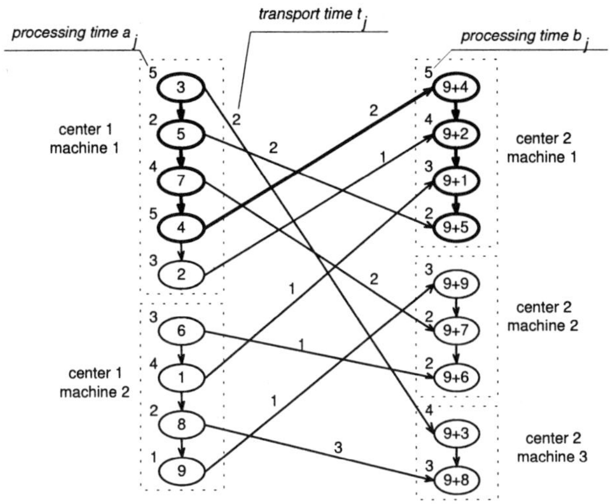

Figure 12.3 Graph for a solution from the example

The basic version of this approach refers to the following elements: *move, neighborhood, initial solution, neighbour evaluation function, searching strategy, memory of the search history, aspiration criteria, stopping rules.* Although these elements have many various forms depending on problems solved, we outline them only in the form suitable for our needs. The *move* is a small deterministic perturbation of a given solution to obtain another one, usually very close. The whole set of moves that can be performed from a given solution generates the collection of descendant solutions called the *neighborhood.* The rank of each neighbor is evaluated by the *neighbor evaluation function.*

TS starts from an initial solution. At each iteration the neighborhood of the current solution is searched in order to find the neighbor with the smallest rank value. Next, the move that leads to this neighbor is performed and the resulting solution becomes the new current one to initiate the next iteration. A memory of the search history is introduced to avoid cycling, becoming trapped in a local optimum, and to add robustness to the search. This memory records for a chosen span of time either *forbidden* moves or their *attributes*, in order to prevent future moves that would "undo" the effects of previous moves. A forbidden move *aspires* to be performed if it leads to a neighbor of rank better than the best already known. The stopping rule that ends the search traditionally consists of setting a limit on the execution time, number of iterations, or number of iterations without improving the criterion value.

12.4.1 Components of TS

Solution representation. In the problem stated, each solution is represented by the pair $(\alpha, \beta) \in \Pi$, where α is a sequence of permutations on

$\mathcal{N}_i \subseteq \mathcal{N}$, $i \in \mathcal{A}$ and β is a sequence of permutations on $\mathcal{M}_i \subseteq \mathcal{N}$, $i \in \mathcal{B}$. Each permutation is associated with a single machine.

Starting solution. This can be found by any constructive method. Here we refer to algorithms from [2, 7, 14, 16] modified appropriately to take transport times into account.

Moves. For our aim we employ the *insertion type move*, see, e.g., [9, 10, 11], which has been verified to be the most promising for scheduling problems with min-max criteria. Briefly, this insertion move operates on two chosen permutations (source and target) at a chosen center. It removes a job placed at some (source) position in the source permutation and puts it in another (target) position in the target permutation. This move is represented precisely by five elements $v = (c, k, x, l, y)$, where $c \in \{1, 2\}$ is the index of the center considered, k, l are indices of permutations of jobs assigned to machines k and l at center c, while x and y are positions in these permutations. Values k and x refer to the source, values l and y to the target permutation. Elements c, k, x, l, y are linked by obvious conditions given as follows:
(a) if $c = 1$ then the move modifies permutations α_k (source) and α_l (target), therefore $k, l \in \mathcal{A}$, $1 \leq x \leq |\mathcal{N}_k|$; $1 \leq y \leq |\mathcal{N}_l| + 1$ if $k \neq l$; $1 \leq y \leq |\mathcal{N}_l|$ and $y \notin \{x - 1, x\}$ if $k = l$,
(b) if $c = 2$ then the move modifies permutations β_k (source) and β_l (target), therefore $k, l \in \mathcal{B}$, $1 \leq x \leq |\mathcal{M}_k|$; $1 \leq y \leq |\mathcal{M}_l| + 1$ if $k \neq l$; $1 \leq y \leq |\mathcal{M}_l|$ and $y \notin \{x - 1, x\}$ if $k = l$.

Neighborhood. Denote by $(\alpha, \beta)_v$ the resulting solution obtained from (α, β) by performing the move v. All solutions $(\alpha, \beta)_v$ that can be obtained by applying moves v from a given move set $\mathcal{U}$ create the *neighborhood* $\{(\alpha, \beta)_v : v \in \mathcal{U}\}$ of this solution.

For illustration, consider the move $v = (1, 1, 2, 1, 4)$ from the solution in the example. It modifies only permutation α_1 at center 1 by changing the location of job $\alpha_1(2) = 5$. Thus after the move we obtain $(\alpha_1)_v = (3, 7, 4, 5, 2)$, $(\alpha_2)_v = \alpha_2$, $(\beta)_v = \beta$. Consider another move $v = (2, 2, 2, 1, 2)$. It modifies two permutations β_2 and β_1 at center 2 by changing the location of job $\beta_2(2) = 7$. Thus after the move we obtain $(\beta_1)_v = (4, 7, 2, 1, 5)$, $(\beta_2)_v = (9, 6)$, $(\beta_3)_v = \beta$, $(\alpha)_v = \alpha$.

Clearly, the proper selection of $\mathcal{U}$ has a significant influence on the final quality of the TS algorithm (for a more elaborate discussion of this subject, see [9]). Following the conclusions provided there, we propose a special set of moves "generated" by the critical path and blocks of jobs, as in papers [9, 10, 11]. This set has been obtained by reducing the whole (basic) set of insertion moves which can be performed from a given solution (α, β). Let $\mathcal{D}_j^A$ denote the set of all moves consisting on changing the location of job j at center 1; and $\mathcal{D}_j^B$ the appropriate move set at center 2. The basic insertion neighborhood is generated by the move set

$$\mathcal{D} = \bigcup_{j \in \mathcal{N}} \mathcal{D}_j^A \cup \bigcup_{j \in \mathcal{N}} \mathcal{D}_j^B.$$

Accordingly to the definition of the moves, sets $\mathcal{D}_j^A$ and $\mathcal{D}_j^B$ can be defined as follows

$$\mathcal{D}_{\alpha_k(x)}^A = \bigcup_{y=1;\ y \notin \{x-1,x\}}^{|\mathcal{N}_k|} (1,k,x,k,y) \cup \bigcup_{l \in \mathcal{A} \setminus \{k\}} \bigcup_{y=1}^{|\mathcal{N}_l|+1} (1,k,x,l,y)$$

for $x = 1, \ldots, |\mathcal{N}_k|$, $k \in \mathcal{A}$, and

$$\mathcal{D}_{\beta_k(x)}^B = \bigcup_{y=1;\ y \notin \{x-1,x\}}^{|\mathcal{M}_k|} (2,k,x,k,y) \cup \bigcup_{l \in \mathcal{B} \setminus \{k\}} \bigcup_{y=1}^{|\mathcal{M}_l|+1} (2,k,x,l,y)$$

for $x = 1, \ldots, |\mathcal{M}_k|$, $k \in \mathcal{B}$.

The proposed set of moves is obtained by eliminating some "non-promising" moves from set $\mathcal{D}$, such that the makespan does not decrease after the next move assigned. If the transport times are equal to zero we can apply to this aim the general property: "if in $G((\alpha,\beta)_v)$ there exists a path containing all nodes from the critical path in $G(\alpha,\beta)$ then v is non-promising" [11]. This property also holds for job-independent transport times, although it fails in the general case. In the sequel we provide some propositions that cover the general case and allow us to detect quickly the significant part of non-promising moves.

Proposition 1. *Let $(\alpha,\beta) \in \Pi$, $v = (1,k,x,l,y) \in \mathcal{D}$, and e,f,g,h be values that define the critical path in graph $G(\alpha,\beta)$. If*

(a) $\alpha_k(x) \notin \mathcal{K}^A$, *or*

(b) $\alpha_k(x) \in \mathcal{K}^A \setminus \{\alpha_e(f)\}$, $l = k$, $1 \le y < f$, *or*

(c) $\alpha_k(x) = \alpha_e(f)$, $l = k$, $f < y \le |\mathcal{N}_e|$,

then $C_{\max}((\alpha,\beta)_v) \ge C_{\max}(\alpha,\beta)$.

Proposition 2. *Let $(\alpha,\beta) \in \Pi$, $v = (2,k,x,l,y) \in \mathcal{D}$, and e,f,g,h be values that define the critical path in graph $G(\alpha,\beta)$. If*

(a) $\beta_k(x) \notin \mathcal{K}^B$, *or*

(b) $\beta_k(x) \in \mathcal{K}^B \setminus \{\beta_g(h)\}$, $l = k$, $h < y \le |\mathcal{M}_g|$, *or*

(c) $\beta_k(x) = \beta_g(h)$, $l = k$, $1 \le y < h$,

then $C_{\max}((\alpha,\beta)_v) \ge C_{\max}(\alpha,\beta)$.

Proof. Both propositions can be proved by analysing paths in graphs $G(\alpha, \beta)$ and $G((\alpha, \beta)_v)$.

Let $\mathcal{W}$ denote the set of moves obtained from $\mathcal{D}$ by elimination of moves satisfying Propositions 1 and 2. Note that $\mathcal{D}$ contains approximately $2n^2$ moves, and $\mathcal{W}$ contains approximately $|L|n$ moves, where $|L| = f + |\mathcal{M}_g| - h + 1$ is the number of nodes in the critical path. Since $|L| \ll 2n$, this reduction is significant.

For the example we have $|\mathcal{D}_j^A| = 8$ for $j \in \{1, 2, 4, 5, 7, 8, 9\}$, $|\mathcal{D}_j^A| = 9$ for $j \in \{3, 6\}$, $|\mathcal{D}_j^B| = 9$ for $j \in \{1, 2, 5, 6, 7, 8\}$, and $|\mathcal{D}_j^B| = 10$ for $j \in \{3, 4, 9\}$. Then, the set $\mathcal{D}$ contains 158 moves. Property 1 eliminates all moves from sets $\mathcal{D}_j^A$, $j \in \{2, 6, 1, 8, 9\}$, two moves from the set $\mathcal{D}_3^A$, and a single move from each set $\mathcal{D}_j^A$, $j \in \{5, 7, 4\}$. Property 2 eliminates all moves from sets $\mathcal{D}_j^B$, $j \in \{9, 7, 6, 3, 8\}$, two moves from the set $\mathcal{D}_2^B$ and a single move from each set $\mathcal{D}_j^B$, $j \in \{1, 5\}$. Then the set $\mathcal{W}$ contains 158-97=61 moves. Consequently, 97 moves from $\mathcal{D} \setminus \mathcal{W}$ (eliminated) provide solutions with the makespan value equal to or greater than $C_{\max}(\alpha, \beta)$.

Neighbor evaluation function. To evaluate each neighbor in the neighborhood, we use the makespan value. Although one might think that this method significantly increases the computational complexity of the single neighborhood search (thus algorithm running time), the more accurate evaluation of solutions allow us to lead the search trajectory quickly into a promising region of the solution space. Additionally, the calculation of the makespan values for neighbors can be performed in a very efficient manner by using the search accelerator presented below.

Search accelerator. Search accelerator is a method that offers a significant reduction of the time necessary to find makespans for all neighbors in a single neighborhood. This method will be presented for a solution (α, β) and moves from $\mathcal{D}_r^A$ for some fixed $r = \alpha_k(x)$. Moves from $\mathcal{D}_r^B$ can be analyzed in an analogous way. Since $|\mathcal{D}_r^A| \approx n$, the direct calculation of makespans requires a time of $O(n^2)$. We propose a method with a time of $O(n)$.

Besides already-known job completion times C_j^A and C_j^B (see formulae (12.8) and (12.9)), we also need some auxiliary values. Let Q_j^B denote the length of the longest path going out from the node $n + j$ (including the weight of this node) in the graph $G(\alpha, \beta)$, $j \in \mathcal{N}$. Similarly, let Q_j^A denote the length of the longest path going out from the node j, $j \in \mathcal{N}$. These values do not depend on chosen r, so calculations can be performed only once for fixed (α, β) in a time $O(n)$. To this aim we can use the following recursive formulae:

$$Q^B_{\beta_i(j)} = Q^B_{\beta_i(j+1)} + b_{\beta_i(j)}, \quad j = |\mathcal{M}_i|, \ldots, 1, \quad i \in \mathcal{B}, \tag{12.14}$$

$$Q^A_{\alpha_i(j)} = \max\{Q^B_{\alpha_i(j)} + t_{\alpha_i(j)}, Q^A_{\alpha_i(j+1)}\} + a_{\alpha_i(j)}, \quad j = |\mathcal{N}_i|, \ldots, 1, \quad i \in \mathcal{A}, \tag{12.15}$$

where $\alpha_i(|\mathcal{N}_i| + 1) = 0$, $i \in \mathcal{A}$, $\beta_i(|\mathcal{M}_i| + 1) = 0$, $i \in \mathcal{B}$ and $Q_0^A = Q_0^B = 0$.

				α_1	$\downarrow$					α_2			
	0	(3	5	7	4	2)	0	0	(6	1	8	9)	0
y	0	1	2	3	4	5	6	0	1	2	3	4	5
C^A	0	5	7	11	16	19		0	3	7	9	10	
C^B	0	11	32	16	23	27		0	18	30	15	14	
Q^A		32	27	25	21	13	0		18	15	11	9	0
Q^B		7	2	4	14	9	0		2	5	3	7	0
Q^{*A}		24	19	17									

Table 12.2 Auxiliary variables for the accelerator

Each move can be decomposed into three phases: (1) cut off the appropriate operation of job r from the graph, (2) glue the graph after cutting off, (3) make the insertion and analyse path changes. Let us consider these phases in detail. First two phases are common for all moves from $\mathcal{D}_r^A$; they change some path lengths in the graph. For the fixed r, we first calculate the length of the longest path in the graph $G(\alpha, \beta)$ in which two arcs $(\alpha_k(x-1), r)$, $(r, \alpha_k(x+1)) \in \mathcal{V}^\alpha$ are removed and a single arc $(\alpha_k(x-1), \alpha_k(x+1))$ is added. Let C denote this length of this path. Clearly C can be found in time $O(n)$. (Using the special structure of the graph, C can be found in a constant time. However, since this reduction does not change the final computational complexity of the method, this technique will not be presented here.)

After insertion (phase (3)) the longest path in $G((\alpha, \beta)_v)$ does not pass through the node r (thus has the length C) or passes through r (then its length can be found on the base of C and some paths in $G(\alpha, \beta)$). Therefore, values of makespans after the move $v = (1, k, x, l, y) \in \mathcal{D}_r^A$, $r = \alpha_k(x)$, can be found using the following formula:

(a) if $l \neq k$ then

$$C_{\max}((\alpha, \beta)_v) = \max\{C, C^A_{\alpha_l(y-1)} + a_r + \max\{Q^A_{\alpha_l(y)}, Q^B_r + t_r\}\} \tag{12.16}$$

(b) if $l = k$ and $y > x$ then

$$C_{\max}((\alpha, \beta)_v) = \max\{C, C^A_{\alpha_k(y)} + \max\{Q^A_{\alpha_k(y+1)}, Q^B_r + t_r\}\} \tag{12.17}$$

(c) if $l = k$ and $y < x$ then

$$C_{\max}((\alpha, \beta)_v) = \max\{C, C^A_{\alpha_k(y-1)} + a_r + \max\{Q^{*A}_{\alpha_k(y)}, Q^B_r + t_r\}\} \tag{12.18}$$

The value $Q^{*A}_{\alpha_k(y)}$ is defined analogously as $Q^A_{\alpha_k(y)}$, but in graph $G((\alpha, \beta)_v)$. This value can be found by applying the following recursive formula

$$Q^{*A}_{\alpha_k(j)} = \max\{Q^B_{\alpha_k(j)} + t_{\alpha_k(j)}, Q^{*A}_{\alpha_k(j+1)}\} + a_{\alpha_k(j)}, \quad j = x-1, \ldots, y \tag{12.19}$$

with the exception that for $j = x - 1$ we substitute in the right-hand side of (12.19) $Q^{*A}_{\alpha_k(j+1)}$ by $Q^A_{\alpha_k(j+2)}$ (i.e., $Q^{*A}_{\alpha_k(x)}$ by $Q^A_{\alpha_k(x+1)}$). From (12.16)–(12.17)

	$y=1$	2	3	4	5
$l=1$	29	29	–	–	35
$l=2$	24	24	28	30	31

Table 12.3 Makespans found by the accelerator

we found that for moves satisfying (a) or (b), the makespan after the move can be found in time $O(1)$. Assuming that moves (c) are analyzed in the order $y = x-1, \ldots, 1$, formulae (12.18)–(12.19) allow us to find the makespan after the move also in time $O(1)$. It means that the resulting makespan for all moves from $\mathcal{D}_r^A$ can be completed in time $O(n)$.

For illustration, let us consider the example, with the processing order shown in Figure 12.2 and moves $(1,1,4,l,y) \in \mathcal{D}_4^A$ $(\alpha_1(4) = 4)$. Values found by (12.8)–(12.9), (12.14)–(12.15) and (12.19) needed for the accelerator are given in Table 12.2 (chosen position $x = 4$ is marked by down arrow). For the reader's convenience appropriate columns are altered accordingly to orders α_1 and α_2. Next, one can verify that $C = 24$. Makespan values are given in Table 12.3. Makespans for $(1,1,4,2,y)$ are found by (12.16), for $(1,1,4,1,5)$ by (12.17) and for the remaining by (12.18)–(12.19).

The philosophy of the accelerator is relative (but the relation is very distant) to the technique of fast makespan calculation with the use of heads and tails, see [15] for the classic flow shop problem.

Short-term memory of the search history. The short-term memory of the search history is the cyclic list T of length $maxT$ (tabu list) that stores attributes of each move performed as well as visited solutions represented by one or two triples (c,i,j), $c \in \{1,2\}$, $i,j \in \mathcal{N}$. This memory is the same as was already introduced [11]. However, for the reader's convenience, a brief description is provided here.

The list is initiated by $maxT$ equal to zero elements and each new added element replaces the oldest one. Let $v = (c,k,x,l,y)$ be a move performed from (α,β). We describe the memory for $c = 1$; for $c = 2$ it acts in an analogous way. If $l \neq k$ we add to T two triples $(1, \alpha_k(x-1), \alpha_k(x))$ and $(1, \alpha_k(x), \alpha_k(x+1))$. Otherwise, we add a single triple $(1, \alpha_k(x), \alpha_k(x+1))$ if $x < y$ and $(1, \alpha_k(x-1), \alpha_k(x))$ if $x > y$.

The move $v = (1,k,x,l,y)$ from a solution (α,β) cannot be performed (has tabu status) if $l \neq k$ and at least one triple $(1, \alpha_l(j), \alpha_k(x))$, $j = 1, \ldots, y-1$ is in T or at least one triple $(1, \alpha_k(x), \alpha_l(j))$, $j = y, \ldots, |\mathcal{N}_l|$ is in T. For $l = k$ this move has tabu status if $x < y$ and at least one triple $(1, \alpha_k(j), \alpha_k(x))$, $j = x+1, \ldots, y$ is in T or if $x > y$ and at least one triple $(1, \alpha_k(x), \alpha_k(j))$, $j = y, \ldots, x-1$ is in T.

Algorithm, search strategy, aspiration function. The algorithm is designed using the general scheme from [10]. Starting from the initial solution

(α^0, β^0) we perform successive iterations, generating the sequence of solutions $(\alpha^0, \beta^0), (\alpha^1, \beta^1), \ldots$. At the i-th iteration ($i = 0, 1, \ldots$), for the given solution (α^i, β^i) we search the neighborhood created by the move set $\mathcal{W}$ defined for this solution. Denote the makespan for the solution (α^i, β^i) by $C^i = C_{\max}(\alpha^i, \beta^i)$ and the best makespan found up to iteration i by $C^{*(i)} = \min_{j \in \{0,1,\ldots,i\}} C^j$. For each move $v \in \mathcal{W}$ we calculate the makespan by using the accelerator and test its tabu status by using list T. The move v to be performed is selected as the one with the minimal makespan among unforbidden moves from $\mathcal{W}$ and those forbidden but with the makespan less than $C^{*(i)}$. Then we update T, perform this move by setting $(\alpha^{i+1}, \beta^{i+1}) = (\alpha^i, \beta^i)_v$, set $C^{i+1} = C_{\max}(\alpha^{i+1}, \beta^{i+1})$, $C^{*(i+1)} = \min\{C^{*(i)}, C^{i+1}\}$, and we go to the next iteration.

Long-term memory of the search history. In the case that C^i is strictly less than $C^{*(i-1)}$, we record, in a long-term memory represented by the cyclic list P of length $maxP$, the *profitable search region.* (We assume $C^{*(-1)} = -\infty$.) This region is characterized by the solution (α^i, β^i), by the current state of tabu list T, and by the list M containing $maxM$ best unforbidden moves from $\mathcal{W} \setminus \{v\}$. Each profitable region can be visited at most $maxM$ times, to start the search into other directions represented by the moves stored in M. We call this technique *back jump on the search trajectory,* [9]. If $maxI$ successive iterations have been performed without improving the makespan, we make the back jump to the last profitable search region stored in P. The search is restarted from the permutation and tabu list stored for this region by performing the best move taken from the list M of this region. Clearly, the list M is reduced by the move taken, and if M becomes empty this region is removed from the region list P.

12.5 COMPUTER EXPERIMENTS

Algorithm TS described in the previous section has several tuning parameters, called $maxT$, $maxM$, $maxP$, $maxI$ which have to be set experimentally. The initial tests were made assuming values $maxT = 2n/3$, $maxM = 3$, $maxP = 3$, $maxI = 1000$. Starting solutions were generated by the constructive algorithm DK1 from [2] modified to include transport times. Test instances were generated with processing times and transport times selected as random integers with uniform distribution on the intervals [1,100] and [1,10], respectively. Tests were carried out for $n = 20, 50, 100, 150, 200, 300$ and the number of parallel machines at each center is equal to 2, 4, 6 (constant) or selected randomly from 1 to 5. A sample of 10 instances was generated for each combination of jobs and machines, thus 240 instances were used.

Three parameters were observed in the experiments (see Tables 12.4 and 12.5): (1) the mean value of the relative reduction of the starting makespan $\nu^1 = 100\% \cdot (C^0 - C^{*(1000)})/C^0$, $\nu^{10} = 100\% \cdot (C^0 - C^{*(10{,}000)})/C^0$ calculated after 1000 and 10,000 iterations, respectively; (2) mean time per iteration τ (in milliseconds on a PC586/100); (3) the mean value of the relative gap to the lower bound $\eta^0 = 100\% \cdot (C^0 - LB)/LB$, $\eta^1 = 100\% \cdot (C^{*(1{,}000)} - LB)/LB$, and $\eta^{10} = 100\% \cdot (C^{*(10{,}000)} - LB)/LB$ after 0, 1000 and 10,000 iterations,

n	the number of machines in a center											
	2			4			6			random 1..5		
	ν^1	ν^{10}	τ	ν^1	ν^{10}	τ	ν^1	ν^{10}	τ	ν^1	ν^{10}	τ
20	2.8	2.9	0.6	7.6	7.8	0.5	8.9	9.8	0.4	7.1	7.5	0.6
50	0.9	1.1	2.5	3.5	3.8	1.5	5.5	5.8	1.2	2.6	2.8	2.1
100	0.3	0.4	8.2	1.4	1.5	4.1	3.1	3.2	3.0	1.4	1.5	8.3
150	0.3	0.3	19.1	1.2	1.3	8.2	2.1	2.2	5.7	0.9	1.0	18.1
200	0.2	0.3	35.3	0.8	0.9	14.1	1.3	1.4	9.7	0.7	0.8	32.1
300	0.2	0.2	73.0	0.5	0.6	29.1	0.9	1.0	19.1	0.2	0.3	69.3

Table 12.4 Relative improvements and times per iteration.

n	the number of machines in a center											
	2			4			6			random 1..5		
	η^0	η^1	η^{10}	η^0	η^1	η^{10}	η^0	η^1	η^{10}	η^0	η^1	η^{10}
20	3.1	0.2	0.1	10.1	1.7	1.5	16.5	6.1	5.1	9.1	1.4	0.9
50	1.1	0.2	0.0	4.1	0.5	0.1	6.4	0.5	0.2	3.0	0.3	0.1
100	0.4	0.1	0.0	1.6	0.1	0.0	3.3	0.1	0.0	1.5	0.1	0.0
150	0.3	0.0	0.0	1.3	0.1	0.0	2.2	0.1	0.0	1.0	0.1	0.0
200	0.3	0.1	0.0	0.9	0.1	0.0	1.4	0.1	0.0	0.8	0.1	0.0
300	0.2	0.0	0.0	0.6	0.1	0.0	1.0	0.1	0.0	0.3	0.1	0.0

Table 12.5 Relative distances to the lower bound.

respectively. LB was calculated using an appropriate single-center scheduling model with pre-emptive jobs, ready times and due dates [8].

TS improves the starting makespan up to 8.9% on average within the first 1000 iterations and up to 9.8% within the first 10,000 iterations. This improvement increases with an increasing number of parallel machines and decreases with an increasing number of jobs; the improvement for $n \geq 300$ is below 1%. Generally TS provides optimal solutions or solutions very close to optimality (relative gap to lower bound $\leq 0.1\%$), already within 1000 iterations. Only for some smaller instances ($n \leq 50$) is the gap to the lower bound greater; however, it remains three to six times less than the same gap calculated for the starting solution.

1000 iterations of TS require 0.4 to 0.6 seconds for $n = 20$ and 20 to 75 seconds for $n = 300$. This time decreases if the number of parallel machines increases, e.g., for $n = 200$ with six machines per center it is close to 10 seconds.

References

[1] B. Chen. Analysis of Classes of Heuristics for Scheduling a Two-Stage Flow Shop with Parallel Machines at One Stage. *J. Oper. Res. Soc.* 46, 234-244, 1995.

[2] F.Y. Ding and D. Kittichartphayak. Heuristics for scheduling flexible flow lines. *Computers Ind. Eng.* 26, 27-34, 1994.

[3] F. Glover. Tabu Search. Part I, *ORSA J. Comput.* 1, 190–206, 1989.

[4] F. Glover. Tabu Search. Part II. *ORSA J. Comput.* 2, 4–32, 1990.

[5] F. Glover and M. Laguna. Tabu Search. *Modern Heuristic Techniques for Combinatorial Problems*, C.Reeves, ed., Blackwell Scientific Publishing, 70–141, 1993.

[6] J. Grabowski, E. Nowicki, and S. Zdrzałka. A block approach for single-machine scheduling with release dates and due dates. *European J. Oper. Res.* 26, 278–285, 1986.

[7] J.L. Hunsucker and J.R. Shah. Comparative performance analysis of priority rules in a constrained flow shop with multiple processors environment. *European J. Oper. Res.* 72, 102–114, 1994.

[8] J. Labetoulle, E. Lawler, J.K. Lenstra and A.H.G. Rinnooy Kan. Preemptive scheduling of uniform machines subject to release dates. *Progress in Combinatorial Optimization*, Academic Press, Florida, 245–261, 1984.

[9] E. Nowicki and C. Smutnicki. A fast taboo search algorithm for the job shop problem. *Management Sci.* 42, 797–813, 1996.

[10] E. Nowicki and C. Smutnicki. A fast taboo search algorithm for the permutation flow-shop problem. *European J. Oper. Res.* 91, 160-175, 1996.

[11] E. Nowicki and C. Smutnicki. The flow shop with parallel machines. A tabu search approach. Technical Report PRE 30/95, Technical University of Wrocław, July 1995, accepted for publication in *European J. Oper. Res.*

[12] A.B. Shaukat and J.L. Hunsucker. Branch and bound algorithm for the flow shop with multiple processors. *European J. Oper. Res.* 51, 88-99, 1991.

[13] C. Sriskandarajah and S.P. Sethi. Scheduling algorithms for flexible flowshops: Worst and average case performance. *European J. Oper. Res.* 43, 43-60, 1989.

[14] T. Sawik. A scheduling algorithm for flexible flow lines with limited intermediated buffers. *Appl. Stochastic Models and Data Anal.* 9, 127-138, 1993.

[15] E. Taillard. Some efficient heuristic methods for the flow shop sequencing problem. *European J. Oper. Res.* 47, 65-74, 1990.

[16] R.J. Wittrock. An adaptable scheduling algorithm for flexible flow lines. *Oper. Res,* 36, 445-453, 1988.

13 USING LOWER BOUNDS IN MINIMUM SPAN FREQUENCY ASSIGNMENT

Stuart M. Allen[1], Steve Hurley[2], Derek H. Smith[1] and Stefan U. Thiel[2,3]

[1]Division of Mathematics and Computing, University of Glamorgan
Pontypridd, CF37 1DL, United Kingdom.
S.M.Allen@glam.ac.uk, D.H.Smith@glam.ac.uk

[2]Department of Computer Science, University of Wales
PO Box 916, Cardiff, CF2 3XF, United Kingdom.
S.Hurley@cs.cf.ac.uk

[3]Vodafone Ltd,
The Courtyard, 2-4 London Road, Newbury, RG14 1JX, United Kingdom.
Stefan.Thiel@vf.vodafone.co.uk

Abstract: The frequency assignment problem is an important NP-hard problem which involves the assignment of frequencies to a set of transmitters in such a way that certain constraints, defining the frequency separation needed between pairs of transmitters, are satisfied. In this paper we show that good assignments can be found if assignment algorithms assign a subset of the constraints initially. This partial assignment is then used as a starting assignment for the solution of the whole problem. Good results can be found if the subset chosen for the initial assignment corresponds to the subset that produces good lower bounds.

13.1 INTRODUCTION

The frequency assignment problem involves the assignment of frequencies to transmitters in a network of compatible equipment located in a geographical region. The frequencies selected for the transmitters must satisfy a set of constraints. The constraints that arise in a problem are restrictions originating from various electromagnetic compatibility considerations, examples of which are co-channel and adjacent channel response, co-site restrictions, spurious emissions and intermodulation products. Depending on the problem to be solved some or all of these restrictions need to be considered. Usually the most important restrictions generate constraints between pairs of transmitters. These are often modelled as a *constraint matrix* which gives the required frequency separation between any pair of transmitters. If the assignments are limited to discrete, not necessarily equally spaced, frequencies then this frequency separation is usually specified by giving the channel separation. Assigning frequencies in an efficient way such that the constraints are satisfied is the main objective of the frequency assignment problem.

In this paper it will be shown how lower bounds for the frequency assignment problem can be generated and how hybrids of sequential and metaheuristic algorithms can be used to find solutions which come close to these lower bounds. It will also be shown how better solutions may be obtained if certain subgraphs involved in the determination of the lower bounds are assigned first. Using this approach optimal solutions to the main variations of the well known "Philadelphia" problems have been obtained.

13.2 PROBLEM FORMULATION

The frequency assignment problem can be modelled as a quadruple

$$< X, D, C, E >$$

where

- $X = \{t_1, t_2, \dots, t_N\}$ is a finite set of N transmitters.
- $D = \{D_{t_1}, D_{t_2}, \dots, D_{t_N}\}$ is a finite set of frequency domains for the transmitters in X.
- $C = \{c_1, c_2, \dots, c_r\}$ is a finite set of constraints on a subset of the transmitters in X.
- E is the objective function which maps every assignment to a numerical value (which usually estimates some interference measure).

In this paper we consider the case when there is a single domain of frequencies available and the constraint set, C, consists of co-channel, adjacent channel and co-site constraints only (as these are the most important).

13.2.1 Constraints

When a transmitter can cause interference with a receiver, a constraint is implied on the channels assigned to the transmitter and the transmitter serving the receiver. Interference can occur when the assigned channels are the same or close together. This gives rise to constraints, of varying severity, of the following form.

- *Co-channel.* The same channel is not allowed to be assigned to certain transmitters simultaneously.
- *Adjacent channel.* Adjacent channels are not allowed to be assigned to certain pairs of transmitters simultaneously.
- *Co-site.* Any pair of channels assigned to transmitters/receivers located at the same site must be separated by a certain number.

All these constraints take the following form

$$|f_i - f_j| > \phi_{ij}$$

where f_i and f_j are the channels assigned to transmitters i and j respectively, and ϕ_{ij} is the required channel separation.

For a given problem containing N transmitters we can construct a *constraint graph* as follows:

Definition 1 *The constraint graph G has N vertices $v_1, \ldots, v_N$ representing the N transmitters. If there is a constraint between transmitters i and j then G has an edge $(v_i v_j)$ labelled ϕ_{ij}.*

13.2.2 Frequency Assignment Models

There are two main types of model used in frequency assignment. In *fixed spectrum* frequency assignment there is a fixed set of F_{fixed} frequencies. It is required to assign each transmitter a frequency from this fixed set such that all the constraints are satisfied. In large problems it will be difficult, and often impossible to assign the frequencies in such a way and therefore we determine an assignment where as many as possible of the constraints are satisfied (or some other measure is minimised).

In *minimum span* frequency assignment, the set of available channels is not fixed, and could, for example, be represented by the set of consecutive integers $D = \{1, \ldots, K+1\}$. The requirement is to satisfy all the constraints and such that K is minimal over all feasible assignments. This minimum value of K is called the *minimum span*.

13.3 LOWER BOUNDS

The frequency assignment problem is an NP-hard problem. Consequently, an efficient algorithm for its solution probably will not be found, and so heuristic methods will tend to be used for large problems. Comparisons can be made between different heuristic methods to assess their performance but these will generally not provide any information on how close to optimality these assignments are. Therefore, bounds can be generated to allow an assessment of the quality of the assignments produced by the algorithms. Several lower bounds for the frequency assignment problem have appeared in the literature [5, 15, 17]. A detailed survey of lower bounds and some new bounds can be found in [19, 1]. The following three bounds, taken from [19], will be used here.

Definition 2 *A* clique *in a constraint graph G is a maximal complete subgraph of G. If C_p is a clique in G with the label of each edge of the clique being at least p, then C_p is called a* level-p *clique.*

Proposition 1 *If G contains a level-p clique C_p then*

$$sp(G) \geq (p+1)(|V(C_p)|-1).$$

Construct a weighted complete graph G' with the vertices of G as vertices. The weights $w(v_iv_j)$ of the edges are defined as follows:

$$w(v_iv_j) = 0 \text{ if } (v_iv_j) \text{ is not an edge of } G,$$

$$w(v_iv_j) = \phi_{ij} + 1 \text{ if } (v_iv_j) \text{ is an edge of } G.$$

Let $H(G')$ denote the length of the shortest Hamiltonian path in G'.

Proposition 2

$$sp(G) \geq H(G').$$

Let $S(G')$ denote the total weight of a minimal weight spanning tree in G'. A Hamiltonian path is a spanning tree and so we have $H(G') \geq S(G')$.

Proposition 3

$$sp(G) \geq S(G').$$

Proposition 3 is normally weaker than Proposition 2, but has the advantage that $S(G')$ can be calculated by a greedy algorithm [14].

Propositions 2 and 3 should not be applied to the full constraint graph, but to a subgraph of the constraint graph. The subgraph that gives the best bound is often a level-p clique for some p, or a subgraph induced by a level-p clique with some vertices added.

13.4 STARTING ASSIGNMENTS

In Section 13.3 lower bounds were presented for the frequency assignment problem. If a suitable subgraph is chosen then the lower bounds, when calculated for the subgraph, can be tight for the whole problem. In such a case the subgraph captures the important features of the whole problem.

It seems reasonable to assume that improved assignments for the full problem might be found if the assignment algorithm first concentrated on assigning some subgraph involved in calculating the lower bound. This would allow a frequency assignment algorithm to concentrate on the hardest part of the problem.

An assignment of the subgraph is obtained using a heuristic algorithm. This assignment is referred to as a *starting assignment.* The assignment is fixed and then extended to the full problem using the same (or a different) heuristic algorithm.

We have initially investigated four possible methods of choosing the subgraph:

1. Determine the maximum level-p clique for each possible value of p. Select that clique which gives the largest bound obtained from one of Propositions 1, 2 or 3.

2. Choose some other subgraph which is known to give the best possible lower bound.

3. If two overlapping subgraphs give very close lower bounds, take the union of the subgraphs.

4. Start from the level-p clique determined by the first method. Assign the subgraph and fix the assignment. For all transmitters not in this subgraph determine the available frequencies which would not cause a constraint violation with a transmitter which has already been assigned. Add to the subgraph vertices corresponding to transmitters with the smallest available number of frequencies and take the subgraph induced by this augmented set of vertices.

Some exploration is required. Sometimes the subgraph approach allows better assignments to be found, and optimal assignments may often be attainable. On other occasions no subgraph appears to be suitable and it is better to revert to applying the algorithm to the full problem.

13.5 GENERATING MINIMUM SPAN ASSIGNMENTS

13.5.1 Sequential Algorithms

Sequential assignment methods are greedy methods which consist of three main steps or *modules* based on the work which appeared in [8]. First, the transmitters are listed in some specified order and then the first transmitter is assigned to frequency one. Second, the next transmitter to be assigned is selected (which may differ from the initial ordering). Finally, the selected transmitter is assigned to a selected frequency. Several options are available for each module.

13.5.2 Metaheuristics

In order to generate minimum span and fixed spectrum assignments we use either of two hybrid algorithms. The first is a hybrid of sequential algorithms and simulated annealing (SA), the second is a hybrid of sequential algorithms and tabu search (TS). The hybrid nature of these algorithms is based on the ideas of Costa [4]. Fuller details of the SA and TS implementations can be found in [12].

Step 1
Generate a no-violation assignment e.g. using a sequential method. Say this assignment uses frequencies $f_i \in \{1, 2, \ldots, q+1\}$ and has span q.

Step 2
Transmitters assigned with frequency $q+1$ are randomly assigned a frequency $< q+1$. This will normally introduce constraint violations.

Step 3
Perform an SA or TS algorithm (using the fixed spectrum available i.e. q frequencies) to eliminate the constraint violations introduced in Step 2 i.e. find a zero violation assignment with span $q-1$. The cost function is defined by $\mu_1 = \mu_3 = \mu_4 = 0$ and $\mu_2 = 1$.

Step 4
If a zero violation assignment is produced in Step 3, set $q = q-1$ and go to Step 2. If there are constraint violations then output q as the best span obtained and its assignment.

Figure 13.1 Structure for minimum span algorithm

The starting point for the minimum span assignment procedures is any no-violation assignment, usually generated by any sequential algorithm. The SA or TS algorithm then reassigns all transmitters which are assigned to the highest used frequency, to a randomly chosen smaller frequency. That is, it deletes the highest frequency from the frequencies available for use. This step will generally introduce a small number of violations, but preserves some features present in the original assignment. The next step then involves using SA or TS to find a valid assignment with no violations using the reduced frequency domain. Once this is achieved, the assignment found is taken as the start assignment and the process starts again. The loop terminates, when the metaheuristic is unable to reduce the number of violations to zero. Therefore, the assignment taken as the start assignment in the previous iteration contains the best zero violation assignment found. The general structure of the minimum span algorithm is given in Figure 13.1.

13.5.3 Cost Function

The cost function minimised in the algorithms is as follows:

$$E = \mu_1 e_{vio} + \mu_2 e_{sum} + \mu_3 e_{order} + \mu_4 l_{vio}$$

where e_{vio} is the number of violated constraints; e_{sum} is the sum, over all constraints violated, of the number of channels by which the constraint is violated; e_{order} is the number of distinct frequencies used; l_{vio} is the largest number of channels by which any constraint is violated. The μ_i are weights, which can include 0, that reflect the relative importance of the various factors. The results presented here are obtained with $\mu_1 = \mu_3 = \mu_4 = 0$ and $\mu_2 = 1$. The different types of constraint can also be weighted to reflect their relative importance.

13.5.4 Simulated Annealing Implementation

Representation of an Assignment Let $\{d_1, d_2, \ldots, d_{F_{fixed}}\}$ represent the set of frequencies available. A frequency assignment $f = (f_1, \ldots, f_N)$ is represented using an array of indexes $[x_1, \ldots, x_N]$ where $f_j = d_{x_j}$ for $1 \leq j \leq N$.

Generation of New Configurations The transmitters in an assignment f, which are assigned frequencies that violate one or more of the constraints, are referred to as *violating transmitters.* New configurations are generated by randomly selecting a violating transmitter i and setting $x_i = x'_i$ where $1 \leq x'_i \leq F_{fixed}$.

Starting and finishing temperatures The starting temperature is determined by first setting $t_0 = 1$. If the acceptance ratio, χ, defined as the number of accepted trial assignments (i.e. assignments that have an improved cost or pass the standard SA probability test) divided by the length of the Markov chain, is less than 0.9, double the current value of t_0. Continue this procedure until the observed acceptance ratio exceeds 0.9 (with χ reinitialised to zero prior to doubling t_0).

The algorithm terminates when the temperature, t_k, falls below t_{min} (user specified) or the number of consecutive *frozen temperatures* exceeds a user specified value, usually 10. A frozen temperature occurs when no new assignments are accepted in a Markov chain.

Annealing Schedule (reduction of t_k) The parameter t_k is calculated using

$$t_{k+1} = t_k \cdot \left(1 + \frac{\ln(1+\delta) \cdot t_k}{3\sigma(t_k)}\right)^{-1}$$

where δ is set to 0.1 and

$$\sigma(t_k) = \frac{1}{N} \cdot \sqrt{\Phi}$$

where

$$\Phi = N \cdot \sum_{i=1}^{N} \left({E_i}^k\right)^2 - \left(\sum_{i=1}^{N} {E_i}^k\right)^2 + 0.5$$

and where ${E_i}^k$ is the cost function value for the assignment obtained at iteration i, at temperature t_k, and N is the length of the Markov chain. This was found to be effective in [10, 13].

13.5.5 Tabu Search Implementation

Representation of an Assignment As in the simulated annealing algorithm a frequency assignment $f = (f_1, \dots, f_N)$ is represented using an array of indexes $[x_1, \dots, x_N]$ where $f_j = d_{x_j}$ for $1 \le j \le N$.

Neighbourhood Structure The neighbourhood of an assignment f consists of those neighbours obtained by changing the frequency assigned to a violating transmitter. A neighbour is generated by randomly selecting a violating transmitter and randomly assigning a different frequency.

Short and Long Term Memory A move to a neighbour (i, x_i') corresponding to changing the assignment of transmitter i to x_i' is said to be *tabu* if it does not satisfy the short term or long term memory condition. Suppose that at iteration j, the frequency $d_{x_i{}^j}$ assigned to transmitter i has changed. Then a move at iteration $k+1$ is tabu if

$$d_{x_i{}^{k+1}} = d_{x_i{}^t} \text{ where } k - r < t \le k, \;\; r = 1, 2, ...$$

or

$$\frac{1}{k} \sum_{j=1, d_{x_i{}^j} = d_{x_i{}^{k+1}}}^{k} 1 \; > \; \beta, \; 0 < \beta < 1.$$

In general, the non-tabu neighbour giving the lowest value of the cost function is selected.

Aspiration Criteria A tabu move is selected if the neighbour $f^{(k+1)}$ satisfies

$$E(f^{(k+1)}) \le E(f')$$

for all neighbours f' of $f^{(k)}$ and

$$E(f^{(k+1)}) < E(f^{(j)}) \; \forall \; j, \; 1 \le j \le k.$$

From computational experiments the values for the short term memory and long term memory conditions were set at $r = \frac{2*N}{5}$ and $\beta = \frac{7}{4*N}$, respectively.

13.6 MINIMUM SPAN RESULTS

The test problems used in the computational experiments are based on the so-called *Philadelphia Problem.* This problem together with its variations have appeared in the literature over many years. The original problem first appeared in Anderson [2] and later in [6, 7, 16, 18, 20], along with several variations. It is based on an area around Philadelphia in the USA and consists of 21 cells arranged as shown in Figure 13.2.

Tranceivers (transmitters and receivers) are considered to be located at the cell centres. The distance between adjacent cell centres is taken to be 1.

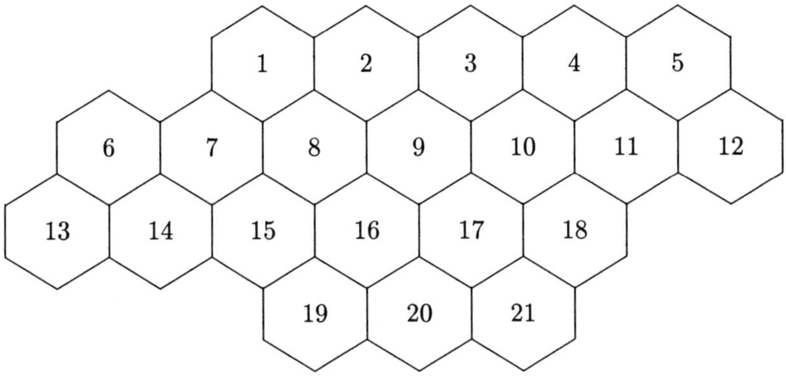

Figure 13.2 The cellular geometry of the Philadelphia problems

Constraints between tranceivers are generated by considering the distance, d, between tranceivers as follows:

d	separation
$\geq \sqrt{12}$	≥ 0
$\sqrt{3} \leq d < \sqrt{12}$	≥ 1
$1 \leq d < \sqrt{3}$	≥ 2
$0 \leq d < 1$	≥ 5

Co-sited tranceivers must be separated by at least 5 channels. The d_0 distance ($\sqrt{12}$ above) is known as the co-channel distance.

For each cell there is an associated *demand*. This indicates how many frequencies need to be assigned to the cell so that it can service its traffic requirements. Figure 13.3 shows the cell demands for the test problem P1 (with $d_0 = \sqrt{12}$). The frequency domain is a set of consecutive channels. The problem is to minimise the span.

By varying the demands within each cell and the co-channel distance, the following variations on the original problem can be constructed:

P2: $d_0^2 = 7$; demands = (8, 25, 8, 8, 8, 15, 18, 52, 77, 28, 13, 15, 31, 15, 36, 57, 28, 8, 10, 13, 8)
P3: $d_0^2 = 12$; demands = (5, 5, 5, 8, 12, 25, 30, 25, 30, 40, 40, 45, 20, 30, 25, 15, 15, 30, 20, 20, 25)
P4: $d_0^2 = 7$; demands = (5,5, 5, 8, 12, 25, 30, 25, 30, 40, 40, 45, 20, 30, 25, 15, 15, 30, 20, 20, 25)
P5: $d_0^2 = 12$; demands = (20, 20, 20, 20, 20, 20, 20, 20, 20, 20, 20, 20, 20, 20, 20, 20, 20, 20, 20, 20, 20)
P6: $d_0^2 = 7$; demands = (20, 20, 20, 20, 20, 20, 20, 20, 20, 20, 20, 20, 20, 20, 20, 20, 20, 20, 20, 20, 20)

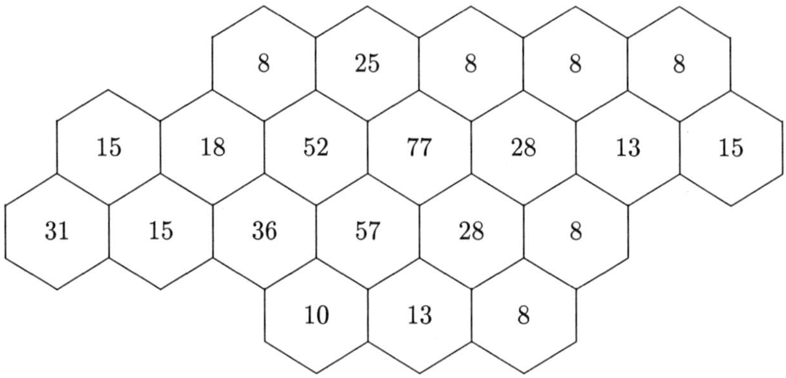

Figure 13.3 Cell demands for the Philadelphia problem P1

P7: $d_0^2 = 12$; demands $=$ (16, 50, 16, 16, 16, 30, 36, 104, 154, 56, 26, 30, 62, 30, 72, 114, 56, 16, 20, 26, 16)
P8: $d_0^2 = 12$; demands $=$ (32, 100, 32, 32, 32, 60, 72, 208, 308, 112, 52, 60, 124, 60, 144, 228, 112, 32, 40, 52, 32)

Problem	N^o*trans–* mitters	N^o constraints	Lower bound	Best seq.	TS	SA	Start
P1	481	91,835	426	447	428	428	*426* (275)
P2	481	76,979	426	475	429	438	*426* (275)
P3	470	78,635	257	284	269	260	*257* (225)
P4	470	56,940	252	268	257	259	*252* (225)
P5	420	65,590	239	250	240	*239*	– (–)
P6	420	44,790	178	230	188	200	179 (140)
P7	962	391,821	855	894	858	858	*855* (550)
P8	1924	1,568,262	1713	1800	1724	1724	*1713* (308)

Table 13.1 Comparison of results for minimum span test problems

The results are presented in Table 13.1. The cost function used here has $\mu_2 = 1$ and all other $\mu_i = 0$. *Lower bound* refers to the known optimal result or lower bound. *Best seq.* refers to the best span obtained from using sequential algorithms. Fortyeight different sequential algorithms described by Hale [8] are applied and the best result is quoted. Included in this set of fortyeight algorithms are some based on the idea of *saturation* degree, for example the DSATUR algorithm. *TS* refers to the span obtained from using the hybrid sequential/tabu search algorithm described in Figure 13.1. *SA* refers to the span obtained from using the hybrid sequential/simulated annealing algorithm described in Figure 13.1. *Start* refers to the span obtained by first assigning a subgraph of the constraint graph used in the calculation of the lower bound and extending this assignment to the complete problem. Start is not performed for P5 as the optimal solution is given in column SA. The hybrid sequential/TS or hybrid sequential/SA algorithms are used both to assign the subgraph and to extend it, to obtain the result in column *Start.* The number of vertices contained in the subgraph is given in parentheses.

We can see from Table 13.1 that when starting assignments are used then optimal solutions can be found in seven out of the eight test cases, and a solution within 1% of the lower bound for the remaining test case. The actual frequency assignments for all the test problems mentioned can be found from the examples given in [11] (P1≡E3, P2≡E4, P3≡E5, P4≡E6, P5≡E7, P6≡E11, P7≡E8, P8≡E9).

It can also be seen that neither TS nor SA outperforms the other on all problems. However, TS is probably better overall for minimum span problems.

The subgraphs used to generate the results in the *Start* column were obtained using the following methods from Section 13.4:

P1, P2, P7: Method 4 - the subgraph obtained is induced by the vertices corresponding to transmitters in cells 2, 8, 9, 15, 16, 17.

P3, P4: Method 3 - the subgraph obtained is induced by the vertices corresponding to transmitters in cells 3, 4, 5, 9, 10, 11, 12, 17, 18.

P6: Method 1 - the maximal level-0 clique induced by the vertices corresponding to transmitters in cells 2, 3, 8, 9, 10, 16, 17 is used.

P8: Method 1 - Here the maximum level-4 clique is induced by the vertices corresponding to the transmitters in cell 9. It is not, in fact, the level-p clique that gives the largest bound, but was the clique tried first.

Run times are very variable. The results for P1 and P2, for example, can easily be obtained in less than two minutes on a 133MHz Pentium PC. The other variations, particularly P3, P4 and P8, are much more demanding. However, no problem requires more than a single overnight run with the best subgraph.

In the variations P1, P2, P3, P4 of the Philadelphia problem, the optimal results only seem to be accessible if the subgraph method is used. It should be noted that the approach is not suitable for all problems, and its exploratory nature makes its application more involved than the direct approach of Figure 13.1.

13.7 FIXED SPECTRUM ASSIGNMENT

Sometimes it is necessary to produce a frequency assignment within a specified span. Earlier results for fixed spectrum algorithms were presented in [10, 13]. The ideas presented above for producing minimum span assignments (progressively reducing the span until no improvement is found) can be modified to produce good quality fixed spectrum assignments. Figure 13.4 outlines the general procedure when $q+1$ frequencies are available.

The procedure was tested on six data sets given in Table 13.2 (these data sets are available from [9]). Fifty consecutive channels were made available to

Step 1
Start with a zero violation assignment. Say this uses $p+1$ frequencies i.e. span $= p$. If $p \leq q$ then O.K. stop.

Step 2
If $p > q$, reassign each transmitter assigned a frequency $> q+1$ with a random frequency $\leq q+1$.

Step 3
Use tabu search (or simulated annealing) algorithm to minimise some interference measure, e.g. constraint violations. After termination the best solution found is output as the best assignment with span q.

Figure 13.4 Structure of fixed spectrum algorithm

allow comparison with the results presented in [3]. The range of separations imposed is $\phi_{ij} = 0, 1, 2, 10$. The cost function used here has $\mu_1 = 1$ and all other $\mu_i = 0$. *# far-site* denotes the number of far-site constraints. *# co-site* denotes the number of co-site constraints. *Search space* denotes the size of the search space. *Edge density* is the number of constraints divided by $\binom{N}{2}$.

N	# far-site	# co-site	search space	edge density
252	6658	401	10^{428}	0.223
282	10162	268	10^{479}	0.258
410	22346	411	10^{696}	0.271
450	25416	305	10^{764}	0.255
490	35535	489	10^{832}	0.301
726	74595	711	10^{1233}	0.286

Table 13.2 Fixed spectrum test problems

The results of using the procedure given in Figure 13.4 (using tabu search) are given in Table 13.3.

SCTT refers to the surrogate constraint tabu thresholding algorithm used in [3]. FASOFT refers to the system that implements the algorithms discussed here. The number in each column refers to the total number of constraint violations in the best assignment found after 10 runs. For each problem the total time taken did not exceed that cited in [3] (300, 100, 153, 176, 222, and 475 minutes for problems 252, 282, 410, 490 and 726 respectively; and using a similar machine and the gcc compiler with default settings). Note that the lower bounding techniques mentioned earlier are for minimum span problems and cannot be used in fixed spectrum problems to indicate how far from optimality a particular assignment is. However, in direct comparison with results from the SCTT technique (which uses a totally random starting assignment from

N	SCTT	FASOFT
252	20	20
282	117	114
410	365	361
450	171	130
490	615	580
726	1472	1466

Table 13.3 Comparison of fixed spectrum results

the available frequencies) the algorithm outlined in Figure 13.4 performs well. Improvements using subgraphs are sometimes possible, but were not attempted here.

13.8 CONCLUDING COMMENTS

We have presented results which show that good frequency assignments can be obtained using metaheuristic techniques. For the Philadelphia Problems improved results can be obtained if a subgraph of the constraint graph is assigned first and then used as a starting assignment for the solution of the whole problem. In fixed spectrum assignment good results were obtained when compared to previous work without assigning subgraphs first, however more work is required to fully evaluate the applicability of the starting assignment approach.

References

[1] S.M. Allen, D.H. Smith, and S. Hurley. Lower bounding techniques for frequency assignment. *Discrete Mathematics, submitted.*

[2] L.G. Anderson. A simulation study of some dynamic channel assignment algorithms in a high capacity mobile telecommunications system. *IEEE Transactions on Communications*, COM-21:1294–1301, 1973.

[3] D.J. Castelino and N.M. Stephens. A surrogate constraint tabu thresholding implementation for solving the frequency assignment problem. *Annals of Operations Research*, to appear 1998.

[4] D. Costa. On the use of some known methods for T-colourings of graphs. *Annals of Operations Research*, 41:343–358, 1993.

[5] M.B. Cozzens and D-I Wang. The general channel assignment problem. *Congressus Numerantium*, 41:115–129, 1984.

[6] N. Funabiki and Y. Takefuji. A neural network parallel algorithm for channel assignment problems in cellular radio networks. *IEEE Transactions on Vehicular Technology*, 41(4):430–437, 1992.

[7] A. Gamst. Some lower bounds for a class of frequency assignment problems. *IEEE Transactions on Vehicular Technology*, 35:8–14, 1986.

[8] W.K. Hale. New spectrum management tools. *Proc. IEEE International Symposium on Electromatic Compatibility Record*, pages 47–53, 1981.

[9] S. Hurley. Frequency assignment benchmarks. http://www.cs.cf.ac.uk/User/Steve.Hurley/f_bench.htm.

[10] S. Hurley and D.H. Smith. Fixed spectrum frequency assignment using natural algorithms. In *Proceedings of the First Conference on Genetic Algorithms in Engineering Systems*, pages 373–378, Sheffield, 1995.

[11] S. Hurley, D.H. Smith, and S.U. Thiel. Frequency assignment plans. http://www.cs.cf.ac.uk/User/Steve.Hurley/plans.htm.

[12] S. Hurley, D.H. Smith, and S.U. Thiel. FASoft: A system for discrete channel frequency assignment. *Radio Science*, 32:1921–1939, 1997.

[13] S. Hurley, S.U. Thiel, and D.H. Smith. A comparison of local search algorithms for radio link frequency assignment problems. In *ACM Symposium on Applied Computing*, pages 251–257, Philadelphia, 1996.

[14] J.B. Kruskal. On the shortest spanning subtree of a graph and the travelling salesman problem. *Proc. Amer. Math. Soc.*, 7:48–50, 1956.

[15] T.A. Lanfear. Graph theory and radio frequency assignment. Technical report, NATO EMC Analysis Project No. 5, NATO Headquarters, 1110 Brussels, 1989.

[16] R. Leese. Tiling methods for channel assignment in radio communication networks. In *3rd ICIAM Congress*, 1996.

[17] F.S. Roberts. T-colorings of graphs: recent results and open problems. *Discrete Mathematics*, 93:229–245, 1991.

[18] K.N. Sivarajan, R.J. McEliece, and J.W. Ketchum. Channel assignment in cellular radio. In *Proceedings of 39th Conference, IEEE Vehicular Technolgy Society*, pages 846–850, 1989.

[19] D.H. Smith and S. Hurley. Bounds for the frequency assignment problem. *Discrete Mathematics*, 167/168:571–582, 1997.

[20] W. Wang and C. Rushforth. An adaptive local search algorithm for the channel assignment problem (cap). *IEEE Transactions on Vehicular Technology*, 45(3):459–466, 1996.

14 A HYBRID HEURISTIC FOR MULTIOBJECTIVE KNAPSACK PROBLEMS

Foued Ben Abdelaziz, Saoussen Krichen and Jouhaina Chaouachi

LAGENDO Laboratory, Institut Supérieur de Gestion, Le Bardo 2000, Tunisia.

Abstract: Some real world situations such as cargo loading and investment choice can be modeled by the knapsack problem. When more than one objective is considered, the problem becomes a multiobjective knapsack problem. In this paper, we develop a hybrid heuristic based on the combination of a multiobjective tabu search algorithm and a multiobjective genetic algorithm, and show, through experimentations, the efficiency of the proposed method.

14.1 INTRODUCTION

Some real world situations such as cargo loading and investment choice can be modeled by the knapsack problem. When more than one objective is considered, the problem becomes a MultiObjective Knapsack Problem (MOKP), which can be formulated as:

$$\textbf{(P)}\ \text{Max } z_k(x) = \sum_{j=1}^{n} c_j^k x_j \ (k = 1, \ldots, K), x \in X$$

$$\text{with } X = \{x \mid \sum_{j=1}^{n} w_j x_j \leq w,\ x_j \in \{0,1\}, \forall j = 1, \ldots, n\}$$

where $x_j = 1$, if item j is packed in the knapsack, and 0 otherwise. Here, w_j denotes the weight (or volume) of item j and c_j^k is the utility of item j according to the criterion k.

Whereas in the uniobjective context we are looking for a single optimal solution, in the multiobjective framework we are trying to generate a set of "efficient" solutions. A solution x^* of P is called efficient if we cannot improve x^* simultaneously on all objectives, i.e., if $\forall x \in X, \forall k = 1, \ldots, K,\ z_k(x) \geq$

$z_k(x^*) \Rightarrow z_k(x) = z_k(x^*)$. Problem (P) is difficult to solve for two reasons, in particular (i) the number of efficient solutions increases with the size of the problem and mainly with the number of criteria and (ii) efficient solutions, due to the non-convexity of the feasible set, are located on the boundary and in the interior of the feasible set (supported and unsupported efficient solutions, respectively).

In the literature, special attention has been devoted to the bicriteria case. Ulungu and Teghem [9] present an exact method (the two-phase method) based on the branch and bound principle. This method is efficient only for small and medium sized problems. For more than two criteria, there is no exact procedure to generate all efficient solutions. In order to approximate the efficient set, Ulungu [10] proposes an adaptation of the simulated annealing metaheuristic to P. Gandibleux et al. [4] propose a tabu search (TS) approach for general multiobjective integer programming problems. Consequently, for specific problems such as the MOKP, one has to adapt the different principles of the general heuristic. Krichen [7] has developed a specific TS approach to MOKP where certain TS principles were adapted to the case of MOKP. Similarly, Chaouachi [2] elaborate a multiobjective adaptation of the genetic algorithm (GA) metaheuristic for problem (P). The main idea of the method is based on the observation that the natural selection is generally performed regarding more than one criterion and leads to a set of nondominated individuals rather than an optimal one. The comparison of efficient solutions generated by multiobjective TS and GA [1] shows that these two methods do not provide the same set of efficient solutions. Therefore, the combination of these heuristics seems to be an interesting way to provide a better approximation of the set of efficient solutions of P.

In this paper, we develop a hybrid heuristic based on the combination of a multiobjective TS algorithm and a multiobjective GA, and show, through experimentation, the efficiency of the proposed method. The paper is organized as follows: Sections 14.2 and 14.3 review respectively the multiobjective TS and multiobjective genetic algorithm for problem (P). Section 14.4, is devoted to the presentation of our hybrid heuristic. Computational results are illustrated and discussed in Section 14.5.

14.2 THE MULTIOBJECTIVE TABU SEARCH HEURISTIC

Tabu search has been applied successfully to a series of uniobjective combinatorial problems [5, 6]. The metaheuristic is characterized by a flexible memory which forbids returning to solutions already visited. Thus, it accepts a deterioration for a possible improvement of the solution.

In this section, we first present the adaptation of certain principles of TS for the multiobjective case, and then detail the multiobjective tabu search algorithm.

The main idea of the method is based on the fact that when we have to fill a knapsack, we naturally try to load "interesting" objects which are out of the knapsack by unloading some other objects from the knapsack.

- *Neighborhood:* To build a neighborhood for a current solution x, we proceed first by choosing an object y not currently in the knapsack with a nondominated criterion vector. Next, we unload the knapsack randomly in order to let enough space to introduce y. After introducing y, we try to load objects to fill the remaining space of the knapsack. Each configuration of unloading the knapsack, in order to introduce y, constitutes a neighbor solution of x.

- *Tabu list:* A move from a current solution x to a neighbor solution x' is defined by loaded and unloaded objects. The reversal move is stored in the tabu list in order to forbid back moves.

- *Aspiration criterion:* If a move in the tabu list provides a solution which dominates all potentially efficient solutions already generated, we eliminate the tabu status of the move.

- *Intensification:* This principle allows looking for solutions that verify some properties. In our adaptation, the intensification principle consists in looking for locally nondominated neighbor solutions in which object y is packed in the knapsack. Therefore the neighborhood principle can be considered, by itself, as an intensification process.

- *Diversification:* It consists in the exploration of other areas of the decision space. In our case, we identify the two worst criteria of the current solution and move to one of the efficient solutions according to these two criteria.

14.2.1 The Algorithm of the Multiobjective Tabu Search

To familiarize the reader with the notations used in the multiobjective tabu search procedure, we list the following:

x^0	The initial solution
x^h	The solution at iteration h
$N(x^h)$	The neighborhood of a solution x^h
T	The tabu list
$\|T\|$	The number of tabu attributes
PE	The set of potentially efficient solutions
t	The maximum number of iterations between two improvements of PE
z_i^*	The optimal value of the ith objective
d	The number of iterations between two diversifications
z_i^h	$z_i(x^h)$
z^h	The criterion vector $(z_1^h, \ldots, z_K^h)$ of the solution x^h

The multiobjective tabu search heuristic (MOTS) is described by the following steps:

Initialization:

$|T| \leftarrow \emptyset$: (initialization of the tabu list)

Generate the initial feasible solution x^0 (using the Greedy algorithm, Martello and Toth [8]). $PE \leftarrow x^0$. $h \leftarrow 1$

Iterative process:

Step 1: If h is a multiple of d then go to Step 2, else go to Step 3.

Step 2: (*Diversification*) The search is guided to other areas of the decision space. We proceed by determining efficient solutions according to the two worst criteria:

Calculate the criterion vector z^h of the current solution x^h and rank the K objectives according to decreasing values of the ratio: z_k^h/z_k^*, $k = 1, 2, \ldots, K$. Let a and b be the two worst criteria according to the previous order. Solve the following single objective knapsack problem (using the Greedy algorithm) for several values of $\lambda = (\lambda_1, \lambda_2)$ including $(.8, .2)$, $(.2, .8)$ and $(.5, .5)$:

$$\begin{aligned} max \quad & \lambda_1 z_a(x) + \lambda_2 z_b(x) \\ s.t. \quad & \sum_{j=1}^{n} w_j x_j \leq w \\ & x_j \in \{0, 1\} \end{aligned}$$

$N(x^h)$ contains the optimal solution of each problem for each λ. Go to Step 4.

Step 3: (*Neighborhood*) Generate a set of neighbors, $N(x^h)$, for the current solution x^h as described in Section 14.2.

Step 4: store in $N(x^h)$ all the elements of $N(x^h)$ such that the corresponding move does not belong to the tabu list. For each move in the tabu list, if the aspiration criteria is verified then include the provided solution in $N(x^h)$.

Step 5: $N(x^h) \leftarrow$ Non dominated solutions of $N(x^h)$.

Step 6: $PE \leftarrow PE \cup N(x^h)$.

Step 7: $x^{h+1} \leftarrow$ any $x \in N(x^h)$.

Step 8: Store the reversal move corresponding to (x^{h+1}, x^h) in the tabu list.

Step 9: If after t iterations there is no improvement of PE then $PE \leftarrow$ non dominated solutions of PE and STOP with PE as the final set of efficient solutions. Otherwise, $h \leftarrow h + 1$ and go to step 1.

14.3 THE MULTIOBJECTIVE GENETIC ALGORITHM

GAs (Davis [3]) are a class of iterative procedures which simulate the principles of natural evolution by random processes of selection, reproduction, recombination, and mutation. We propose in this section a multiobjective algorithm with an adaptation of principles of GAs to the multiobjective context. Our aim is to produce efficient solutions from a random initial population.

- *Encoding of individuals:* A binary string of length n represents a solution x to the problem: The ith item is selected for the knapsack if $x_i = 1$.

- *Evaluation function:* A vector $z(x) = (z_1(x), \ldots, z_K(x))$ of evaluation functions is computed for each admissible or non admissible individual x of the population as follows: For admissible individuals x (i.e.

$\sum_{j=1}^{n} w_j x_j \leq w$), $z_k(x) \leftarrow \sum_{j=1}^{n} c_j^k x_j$ for $k = 1, \ldots, K$. For non admissible individuals x (i.e. $\sum_{j=1}^{n} w_j x_j > w$), we penalize the criterion vector, $z_k(x) \leftarrow \sum_{j=1}^{n} c_j^k x_j - \alpha_k \max(0, \sum_{j=1}^{n} x_j w_j - w)$.

The penalty parameter α_k is to be determined by experimentations.

- *Selection and Duplication:* For each individual at any iteration, depending on its evaluation function vector, we determine its rank: the number of individuals which dominate it in the current population. Thus, well adapted individuals in an iteration are those with minimum rank. A nondominated individual will have a rank equal to 0.

  The individuals with smaller rank will be more frequently duplicated. So, the concept of dominance alone is used to perform the selection and the duplication of individuals.

- *Crossover:* The choice of the mate (partner) in the crossing process is not random but it performs the following steps: Arrange on each criterion and in an increasing order all individuals in the current population. Cross the best individuals on one criterion with the best on another criterion.

- *Mutation:* The mutation transforms one bit of an individual to its complement. In a population there is a chance of p_m (to be determined experimentally) to apply the mutation to one individual.

The algorithm of the Multiobjective Genetic Algorithm
The steps followed by the MultiObjective Genetic Algorithm (MOGA) are:
Initialization: Generate an initial population P^0 of random feasible or non feasible individuals. $PE \leftarrow P^0$
Iterative process:
Step 1: For a population P^h, at some iteration h, we proceed as follows: Compute the vector of evaluation functions $z(x)$ for each individual x of the population P^h. Determine the rank of each individual and duplicate the individuals according to their rank. Apply the crossover operator to each individual with its mate. Apply the mutation operator.
Step 2: If a fixed number of iterations is reached then $PE \leftarrow$ non dominated solutions of PE and STOP. Otherwise go to Step 3.
Step 3: $PE \leftarrow PE \cup$ non dominated solutions of P^h. $h \leftarrow h + 1$. $P^{h+1} \leftarrow$ non dominated solutions of P^h. Generate randomly other individuals for P^{h+1}. Go to Step 1.

14.4 A MULTIOBJECTIVE HYBRID HEURISTIC FOR MOKP

Experimental results show that some efficient solutions generated by MOGA are not provided by MOTS and vice versa. In fact, TS performs, first, a local search and then tries to diversify the search. However, the GA proceeds to a random search and is generally sensitive to the choice of the initial population.

The idea of mixing the two heuristics in a Hybrid Heuristic (HH) consists in keeping the structure of local search and diversification of TS and performing a

Solution Method	*Number of solutions*	*Dominated solutions*	*Same solutions*
MOTS	78	12	
MOGA	78	14	35
HH	95	0	

Table 14.1 Results of a problem instance with 100 items and three criteria

GA technique in the neighborhood of the current solution in order to intensify the search. The initial population of the GA is then used only to improve the neighborhood of the current solution. We then obtain a more efficient heuristic to solve P that avoids bad neighborhoods and is not sensitive to the choice of the initial parameters.

For a current solution x^h we generate $N(x^h)$ using the neighborhood or the diversification operator. $N(x^h)$ is then inserted as an initial population for the MOGA. The MOGA provides a set of locally efficient solutions P^h. The set of potentially efficient solutions PE is then updated by P^h. If after t iterations there is no improvement of PE the algorithm stops. Otherwise, the current solution for the next step is chosen randomly from P^h.

The algorithm of the Hybrid Heuristic
Initialization: Generate an initial feasible solution x^0. $PE \leftarrow \emptyset$.
Iterative process:
Step 1: If h is a multiple of d then apply the diversification process for the current solution x^h to obtain $N(x^h)$. Otherwise, generate a set of neighbor solutions $N(x^h)$ for the current solution x^h. $P^0 \leftarrow N(x^h)$.
Step 2: Apply the MOGA to P^0 and obtain P^h.
Step 3: $PE \leftarrow PE \cup P^h$.
$x^{h+1} \leftarrow$ a local efficient solution chosen randomly from P^h.
Step 4: If after t iterations there is no improvement of the set of potentially efficient solutions then $PE \leftarrow$ non dominated solutions of PE and STOP. Otherwise, $h \leftarrow h + 1$ and go to Step 1.

14.5 EXPERIMENTAL RESULTS

Experimentations were performed on a Pentium 166 MHZ and computing codes were written in PASCAL. We compared the HH with MOTS and MOGA. Our experimentations show that the HH is more efficient than MOTS and MOGA. As an illustration of our experimentations, we present in Table 14.1 the computational results for a problem instance with 100 items and three criteria.

35 (common) solutions were generated by the three heuristics. 12 out of 78 solutions provided by MOTS are dominated by those provided by MOGA or HH. 14 out of 78 solutions generated by MOGA are dominated by those generated by MOTS or HH. However, the HH provides a set of 95 solutions including

all efficient solutions already generated by MOTS and MOGA. The set of 95 solutions does not contain any dominated solution by MOTS or MOGA.

To test the HH in a more efficient way, we performed a library of test problems which encompassed items varying from 50 to 200 and criteria from 2 to 5. For each problem size of the library, a series of 12 problems were generated with random utilities and weights in the interval $[20, 70]$. The set of efficient solutions depends on both the number of items and a ratio r, defined in [9], by $r = \frac{w}{\sum_{i=1}^{n} w_i}$. It is shown in [9] that the size of the efficient set increases when $r \leq 0.5$ and decreases when $r \geq 0.5$. Hence if $r \simeq 0.5$, the set of efficient solutions becomes very large. The capacity of the knapsack for each problem of the library is chosen such that the ratio r is close to 0.5. Therefore, in our experimentations we provide the ratio r together with the corresponding CPU time.

Table 14.2 summarizes the data, the average number of solutions and the computational time required for each problem size. In Table 14.2, we notice that the set of efficient solutions increases with the number of items and with the number of criteria. Consequently, the CPU time required for problems with more than two criteria is relatively long.

For problems with two criteria and no more than 50 items, the hybrid algorithm provided the same solutions as those generated by an exact method: the two phases method [9]. For problems with more than two criteria, we compared our algorithm with the simulated annealing method of Ulungu [10]. The efficient solutions obtained, for the example with five criteria and 25 objects, provided in Ulungu [10], were the same.

Items	*Criteria*	*Capacity of the knapsack*	r	*Average number of solutions*	*CPU time (seconds)*
50	2	1150	0.52	53	3
50	5	1150	0.48	320	40
100	2	2100	0.48	90	100
100	3	2100	0.44	170	548
100	5	2300	0.51	398	756
150	2	3400	0.49	193	203.8
150	5	3400	0.49	985	1070.6
200	2	4500	0.3	314	300
200	5	5000	0.52	1596	1521

Table 14.2 Average measurements for 12 instances of each of the nine problem sizes

14.6 CONCLUSION

In this paper, we proposed two metaheuristics for the multiobjective knapsack problem, MOTS and MOGA, using TS and GA principles, respectively. The two metaheuristics were adapted to the multiobjective context. Based on the fact that the two previous metaheuristics do not provide the same set of efficient solutions, we built a new hybrid algorithm combining the advantages of the local search of MOTS and the random aspect of MOGA.

Our hybrid algorithm generates, for the bicriteria case, the set of all efficient solutions. For more than two and up to five criteria, we tested our algorithm on a variety of problems with number of items varying from 50 to 200. Our algorithm provided the same set of efficient solutions as those generated by the simulated annealing approach. More experimentations and comparisons with other methods are needed to decide on the performance of our hybrid algorithm.

References

[1] Ben Abdelaziz F., Chaouachi J. and Krichen S. *Métaheuristiques de résolution du problème de sac à dos multiobjectif*, In M. Itmi (ed.), Actes de la première conférence Francophone en modélisation et simulation des systèmes de production et de logistique MOSIM'97, pages 443-450, editions Hermes, 1997.

[2] Chaouachi J. "A Genetic Algorithm for Multiobjective Knapsack Problems", Mémoire de DEA, Institut Supérieur de Gestion, Tunisia, 1997.

[3] Davis L. "Handbook of Genetic Algorithms", Van Nostrand Reinhold, New York, 1991.

[4] Gandibleux X., Mezdaoui N. and Fréville A. *A tabu search procedure to solve multiobjective combinatorial optimization problems*, In R. Caballero, F. Ruiz, R.E. Steuer (eds.), Advances in Multiple Objective and Goal Programming: Proceedings of the Second International Conference on Multi-Objective Programming and Goal programming (MOPGP'96), Springer, 1997.

[5] Glover F. *Tabu Search-Part I*, ORSA J. Comput. 1, 190-206, 1989.

[6] Glover F. *Tabu Search-Part II*, ORSA J. Comput. 2, 4-32, 1990.

[7] Krichen S. "A Multiobjective Tabu Search Algorithm for Knapsack Problems with Multiple Objectives", Mémoire de DEA, Institut Supérieur de Gestion, Tunisia, 1997.

[8] Martello S. and Toth P. "Knapsack Problems: Algorithms and Computer Implementations", John Wiley & Sons, 1990.

[9] Ulungu E.L. and Teghem J. *The Two Phases method: An efficient produce to solve Bi-objective Combinatorial Optimization Problems*, Foundations of Computing and Decision Sciences 20, 149-165, 1995.

[10] Ulungu E.L. "Optimization Combinatoire Multicritère: détermination de l'ensemble des solutions efficaces et méthodes intéractives", Thèse de doctorat, Université de Mons-Hainaut, 1993.

15 HYBRID GENETIC TABU SEARCH FOR A CYCLIC SCHEDULING PROBLEM

Peter Greistorfer

Karl-Franzens-Universität, Institut für Industrie und Fertigungswirtschaft
Universitätsstraße 15/G2, 8010 Graz, Austria.
peter.greistorfer@kfunigraz.ac.at

Abstract: This paper considers a cyclic scheduling problem, which especially is applicable in the area of public transport. The problem is the so-called Irregular Polygon Scheduling Problem with the objective to maximize the distance between the nearest vertices of two different polygons. We give a mixed integer LP-formulation of this NP-hard problem, derive some bounds and present a hybrid meta-heuristic algorithm for solving the problem. The main idea is to combine a Tabu Search with a Genetic Algorithm component. Computational experience, based on comparative results obtained from CPLEX, shows the efficiency of the hybridization approach.

15.1 INTRODUCTION

We focus on a cyclic scheduling problem that has many real-world applications, especially in transport efficiency. The problem arises, e.g., in public transport timetabling and can be illustrated as follows. Consider a railway station with a number of different lines passing through it. The trains of a certain line periodically arrive at the station. Now we want to find a schedule of arrivals following an optimization criterion based on the time between consecutive arrivals of different trains. To introduce the criterion treated in this work, we give the following example. Let us assume that there are two lines to be scheduled: For line 1, there are three arrivals per hour, which are separated by 10, 14 and 36 minute intervals. For line 2, there are two arrivals per hour, which are separated by 5 and 55 minutes. The goal is to find a schedule so that the minimum time between arrivals of different lines, i.e. a safety span, is maximized. Further, assume a given schedule in which line 1 arrives at minutes 5, 15, and 29, whereas line 2 arrives at minutes 40 and 45. The resulting minimal safety span (a suboptimum) is 11 minutes, i.e. the difference between the last arrival of line 1 and the first arrival of line 2.

We designate this problem as the *max-min Irregular Polygon Scheduling Problem* (*max-min* IPSP), since it can be described by a set of irregular polygons, whose vertices lie on a common circle line. The polygons should be arranged in such a way that the minimal distance measured on the circle line between two vertices of different polygons is maximized.

Polygon Scheduling Problems (PSPs) were introduced by Cerny [4], who studied a regular PSP, i.e. a PSP in which the time between the periodic arrivals of any train on a particular line is constant. Cerny incorporated the *max-min* objective (*max-min* RPSP). In the literature polygon scheduling problems have mainly been treated from a theoretical point of view. Pioneering work was done by Guldan [16]. He presented an algorithmic approach to the *max-min* RPSP, based on the partitioning of the set of all schedules into *local regions* (sets of schedules that have the same cyclic sequence of vertices on the circle line). A polynomial algorithm based on longest paths was described that solves the problem for every region. However, the number of local regions exponentially increases with problem size. Further ideas were developed by Vince [27] who reformulated the *max-min* RPSP as a graph theory problem and developed a kind of enumeration scheme. Burkard [3] generalized the objective function also studying the *min-max* RPSP. Moreover, he gave explicit optimal solutions for the special 2-polygon case. A detailed study of the *max-min* RPSP also may be found in Greistorfer and Kellerer [15] who discussed special cases, derived bounds for the general case and presented a Tabu Search algorithm for solving the *max-min* RPSP.

The first discussion of IPSPs can be found in Brucker and Meyer [2] who proposed a polynomial algorithm for the *min-max* and the *max-min* 2-polygon IPSP. The NP-hardness of the IPSP was proved by Brucker et al. [1]. They adapted the algorithm of Guldan [16] and introduced a network flow technique to find the optimum for every local region. A detailed overview of the overall PSP literature as well as new theoretical results can be found in Hurink [17]. The main contribution in [17] is the generalization of the PSP to the so-called *weighted* PSP, a problem class that additionally considers, e.g., the number of passengers on a given line.

In Section 15.2 we shall give a precise mathematical formulation of the *max-min* IPSP and introduce an LP-formulation based on a modulo-representation of the problem. After the discussion of some bounds, the heuristic solution approaches will be presented in Section 15.3: the discrete concept of *expansion*, the detailed description of the *Tabu Search component* and the *Genetic Algorithm component*, which both lead to the hybrid meta-heuristic proposed, called the *Genetic Tabu Search.* Finally, in Sections 15.4 and 15.5 the computational results and some concluding remarks will be given.

15.2 PROBLEM FORMULATION, LP-MODEL AND BOUNDS

Mathematically the general PSP can be described as follows. Let there be $i = 1, \ldots, n$ polygons denoted P_i. It is assumed that polygon P_i has m_i vertices. All vertices lie on a common circle line $\mathcal{L}$ with length U. Choosing an arbitrary point of $\mathcal{L}$ as the origin, the PSP allows the introduction of a coordinate system, which starting from the origin, assigns the (clockwise) length of the arc of the circle to each point of $\mathcal{L}$. The coordinates of P_i, i.e. the position of its vertices, are given by $p_{i_1}, \ldots p_{i_{m_i}}$. An *arrangement* S of polygons can be given by its coordinates $\{(p_{1_1}, \ldots, p_{1_{m_1}}), \ldots, (p_{n_1}, \ldots, p_{n_{m_n}})\}$.

We start explaining the *max-min* IPSP by specifying the *distance of two points* $r, s \in \mathcal{L}$, where p_r and p_s denote their coordinates. To ensure the independence of the sequence of points, we formulate the following measure as a symmetric one:

$$d(r, s) = \min\{(p_r - p_s) \bmod U, (p_s - p_r) \bmod U\} \tag{15.1}$$

Note that in the term $(a \bmod U)$ two meanings are embedded. As commonly used, it represents the uniquely determined integer number x $(0 \leq x \leq U - 1)$ so that U is a divisor of $(a - x)$, but it also defines a so-called residue class $\{a + \lambda U \text{ with } \lambda \text{ is integer}\}$.

Each *side* of a polygon connects two consecutive vertices i_r and i_{r+1}. The integer *cost (length)* associated to a side (i_r, i_{r+1}) is abbreviated with c_{i_r} $(1 \leq r \leq m_i$ and $i_{m_i+1} = i_1)$ and is calculated according to (15.1). Since we address the *irregular* PSP, we assume that there are at least two sides with different costs. Further, the *distance between polygons* P_i *and* P_j is defined as

$$d(P_i, P_j) = \min{}_{r,s}\{d(i_r, j_s)\} \quad 1 \leq r \leq m_i,\ 1 \leq s \leq m_j \tag{15.2}$$

Then, introducing the objective, the *minimal distance of an arrangement* S is

$$d(S) = \min_{1 \leq i < j \leq n} \{d(P_i, P_j)\} \tag{15.3}$$

and for an optimal arrangement S^*, $d(S)$ is maximized, i.e. $d(S^*) = \max_S d(S)$. Note that $d(S)$ is a function of U. In that it follows from the fact that for $U_2 = \lambda U_1$, $d(S_2) = \lambda d(S_1)$ is obtained, we can identify arrangements that are of identical form and only differ regarding the circle length and the cost of sides.

The preceding definitions, especially (15.1), give rise to the following *Mixed Integer LP-formulation* (MILP). Let S be an arbitrary arrangement and $z = d(S)$. The decision variable x_i defines the amount of change for all coordinates of polygon P_i without having an effect on z. A value $x_i > 0$ $(x_i < 0)$ describes a clockwise (counterclockwise) rotation of P_i. Further, let $\delta_{i_r j_s} = p_{i_r} - p_{j_s}$, then the following must hold for all $1 \leq i < j \leq n$, $1 \leq r \leq m_i$, $1 \leq s \leq m_j$ and $0 \leq x_h < U$ $(1 \leq h \leq n)$:

$$\begin{aligned} x_i - x_j + \delta_{i_r j_s} &\geq z \bmod U \\ x_i - x_j + \delta_{i_r j_s} &\leq (U - z) \bmod U \end{aligned} \tag{15.4}$$

Introducing intervals for the L.H.S. of (15.4), allows the decomposition of the modulo-formulation. There exist $k = 1, \ldots, 4$ intervals:

$$\begin{aligned} -2U + z \leq & \text{ L.H.S. } \leq -U - z \\ -U + z \leq & \text{ L.H.S. } \leq -z \\ z \leq & \text{ L.H.S. } \leq U - z \\ U + z \leq & \text{ L.H.S. } \leq 2U - z \end{aligned} \tag{15.5}$$

The binary variables of the MILP, $y^k_{i_r j_s}$, determine whether a L.H.S. belongs to interval k or not. To facilitate reading, we set $y^k = y^k_{i_r j_s}$. Separating lower and upper interval limits yields

$$\text{L.H.S.} \leq (-U - z)y^1 - zy^2 + (U - z)y^3 + (2U - z)y^4 \tag{15.6}$$
$$\text{L.H.S.} \geq -(2U - z)y^1 + (-U + z)y^2 + zy^3 + (U + z)y^4 . \tag{15.7}$$

Together with

$$y^1 + y^2 + y^3 + y^4 = 1 \tag{15.8}$$

all conditions are established. But there is a need to reformulate (15.6) and (15.7) because both terms are non-linear. Rearranged they can be written as

$$\text{L.H.S.} + Uy^1 - Uy^3 - 2Uy^4 \leq z(-y^1 - y^2 - y^3 - y^4) \tag{15.9}$$
$$\text{L.H.S.} + 2Uy^1 + Uy^2 - Uy^4 \geq z(y^1 + y^2 + y^3 + y^4) \tag{15.10}$$

Note that because of (15.8) the R.H.S. of (15.9) and (15.10) is equal to $-z$ and z, respectively. Hence, after having substituted back the terms for L.H.S. and y^k, we can summarize the MILP model with (15.8) as follows:

$$\begin{aligned} \mathit{Max} \quad & z \\ \text{s.t.} \quad & x_j - x_i - Uy^1_{i_r j_s} + Uy^3_{i_r j_s} + 2Uy^4_{i_r j_s} - z \geq \delta_{i_r j_s} \\ & x_j - x_i - 2Uy^1_{i_r j_s} - Uy^2_{i_r j_s} + Uy^4_{i_r j_s} + z \leq \delta_{i_r j_s} \\ & y^1_{i_r j_s} + y^2_{i_r j_s} + y^3_{i_r j_s} + y^4_{i_r j_s} = 1 \\ & 0 \leq x_h < U, \; y^k_{i_r j_s} \in \{0, 1\} \\ & 1 \leq h \leq n, \; 1 \leq i < j \leq n \\ & 1 \leq r \leq m_i, \; 1 \leq s \leq m_j, \; 1 \leq k \leq 4 \end{aligned} \tag{15.11}$$

Next we discuss some lower and upper bounds. Unfortunately, one cannot employ any concept from number theory, like least common multiples or greatest common divisors, as in the case of the regular counterpart of the problem (see [15]).

Apart from the fact that the intuitive lower bound $LB = 1/n$ is trivial and weak, we are clearly more interested in deriving a good upper bound. Obviously a value U/m, with $m = \sum_{i=1}^{n} m_i$, constitutes an upper bound UB if no polygon side with cost $c_{i_r} < U/m$ exists. However, since our distance function only deals with vertices of *different* polygons, there may be more space on the circumference for finding an optimal arrangement than U/m indicates. Hence, there is a need for an adjustment that takes this fact into account. The adjustment starts as follows: Sort the cost of shorter sides following an increasing order in, say, $C_{\overline{UB}}$. Take the smallest element from $C_{\overline{UB}}$, decrease U by this value and calculate a new bound with $U'/(m-1)$. Simply imagine that this side has been shrunk to zero-length, which leads to a decrease of m by one. Then, in every iteration $C_{\overline{UB}}$ is determined while not referring to sides already examined, but according to the last U', until all sides are checked:

Adjusted *max-min* IPSP Upper Bound

Step 1: (Start) Input problem data, i.e., coordinates of polygons, m and U, compute the cost of sides as described in (15.1) and store them in, say, C.

Step 2: (Iteration) Determine $UB = U/m$ and $C_{\overline{UB}} = \{c_k \text{ with } c_k \in C,\, c_k < UB\}$. If $C_{\overline{UB}} = \emptyset$, then go to *Step 3.* If not, then take $c_{min} = \min\{c_k,\, c_k \in C_{\overline{UB}}\}$ to calculate $U' = U - c_{min}$. Set $m' = m - 1$ and $C = C \setminus c_{min}$. Go to *Step 2* with $U = U'$ and $m = m'$.

Step 3: (End) UB is an upper bound to the *max-min* IPSP.

Note that if $C_{\overline{UB}} = \emptyset$ is already true for the first iteration in *Step 2*, then there is no need to adjust the bound. For the example of the introduction, however, the adjustment actually takes place. Assuming $U = 60$ and an arbitrary arrangement to compute the cost of sides, the adjusted bound would be calculated as $UB = 15.5$, a value that is an optimal solution.

The second bound used can be seen as an algorithmic min-max bound. According to the distance measure (15.2) we calculate for all pairs of different polygons the maximum value of $d(P_i, P_j)$ while fixing P_i with $p_{i_1} = 0$ and varying all p_j from 1 to U. Since an optimal solution for the overall problem obviously is a multiple of $1/k$ (k=1, ... ,n), this method will find an optimal solution for every partial arrangement of $k = 2$ polygons if the rotations are made in 0.5-units-steps. As an upper bound to the overall problem we take the minimum optimal solution over all partial arrangements. For our instances this non-polynomial complete enumeration is convenient, but it must be emphasized that Brucker and Meyer [2] proposed an *efficient* method for the scheduling of two irregular polygons.

15.3 THE GENETIC TABU SEARCH

For the solving of the *max-min* IPSP, we combine two well-known meta-heuristic paradigms: Tabu Search (TS) and Genetic Algorithms (GAs). In the last decade TS, which was introduced by Glover [8], has turned out to be one of the most powerful tools for solving hard combinatorial problems. For a broad overview of meta-heuristics in general, we recommend the bibliography by Osman and Laporte [24]. Detailed hints regarding the pure TS or GA designs are given by Glover and Laguna [13] and Reeves [25]. New directions enforcing TS and its extensions, involving adaptive memory programming with links to Scatter Search (SCS) and Path Relinking (PR), are shown in [10, 11, 14].

Much work also has been done on the hybridization of TS with several other strategies: for example, Large Step Optimization of Martin et al. [20], which embeds Simulated Annealing (SA) and TS, and hybrid GA-TS methods, e.g., the Evolutionary TS by Costa [5]. They replace the mutation phase of a GA by a TS procedure. Similar to [5] is the Genetic Hybrid Algorithm by Fleurent and Ferland [7]. The authors use TS and classic local search to improve a randomly generated initial population of a hybrid GA or to improve crossover-constructed new individuals. Thangiah and Osman [26] introduce the idea of the so-called Genetic Sectoring Method to solve vehicle routing problems with time windows. The solutions obtained then may be improved by a TS-SA hybrid in which worsening non-tabu TS solutions are accepted or rejected according to the exponential SA acceptance criterion. In designing a hybrid GA, Kim et al. [18] use SA and its acceptance criterion to control the probability for the survival of weak offsprings. More information about GAs and links to new foundations can be found in the work of Glover et al. [12, 14] and Mühlenbein [21, 22]. An extensive overview of recent developments is provided in Osman and Kelly [23].

In our implementation the emphasis lies on the effort to design a hybrid environment, called *Genetic TS* (GTS), in which both meta-heuristic components are almost equivalently used. Another motivation for this paper comes from the interest to study the parallels between TS and GAs: Although it seems that TS works on single solutions, while iteratively scanning neighbourhoods and always choosing *one* element, the method's prime intensification strategy is to save and work with the quality solutions, which in GAs might be called an "elite" population.

Intensification is managed by operating on these solutions in one of three principal ways: (a) to "recover" them periodically during the search and to initiate searches that go from these solutions to neighbours not previously visited; (b) to keep track of the frequency that various solution attributes occur in these solutions, and then to give enhanced evaluations to moves that incorporate compatible sets of such attributes into new solutions generated; (c) to use an SCS or PR approach to move between (and beyond) these good solutions - as a way to generate solutions (offspring) that may become new starting points for improvement. All of these are population-based strategies. In addition, option (c) is also an "evolutionary" strategy, i.e., SCS and PR constitute special forms of "solution combination" methods (see, e.g., [14]). This gives a natural foundation to consider the use of hybrid procedures where a GA method is used in place of TS strategies based on (a), (b) and (c). In fact, evolutionary options in TS, though originating long ago, have only recently begun to be explored empirically (see [11]).

Other differences to be pointed out are the different coding methods, genotypical chromosome coding (GA) versus phenotypical parameter coding (TS), and the so-called problem-specific knowledge, which must be made inherent to non-GA procedures. (The issue of problem-specific knowledge is automatically handled in the evolutionary PR approach, by means of using paths in problem-specific neighborhoods to generate offspring. Here we require a modification of the genetic model to achieve a similar outcome.)

15.3.1 Expanding solutions

One useful concept to tackle the continuous *max-min* IPSP could be found within the theory of evolutionary algorithms, where variables are changed in a continuous manner following the shape of random distributions. However, from our point of view, it seems to be rather difficult to scan neighbourhoods that generally contain more equally-valued or deteriorating than improving neighbourhood solutions (which came to light during first computational experiments). The most obvious advantage of a discrete implementation is the ease with which tools can be considered to avoid reversals or cycles in the search process in order to overcome local optima.

Let us continue the example. A complete enumeration (allowing only integer changes while searching the space of $U = 60$) would not even for this small IPSP instance find the optimal solution of $d(S^*) = 15.5$. Only the worse neighbour with $d(S) = 15$ would be detected. Here we refer to the following idea: Instead of fixing the length of circumference U and searching with decimal numbers, U is allowed to increase while the coordinates of a given arrangement are *adapted.* Such a transfer from U_λ to $U_{\lambda+1} = (\lambda+1)\,U$, is called an *expansion*. The *stage* λ is a positive integer number. Doing so, i.e., turning away from continuous coordinate changes, has advantages although naturally, one cannot get rid of real numbers completely. For an expansion from S_λ to $S_{\lambda+1}$ the adapted value for a coordinate p_{i_r} is

$$p'_{i_r} = \left[p_{i_r}\frac{\lambda+1}{\lambda}\right] \tag{15.12}$$

where $[.]$ depicts the rounding-function. Assume the following arrangement for the opening example: $S = S_1 = \{(5, 15, 29), (40, 45)\}$ and $U = 60$. Then, the expanded solution is $S_2 = \{(10, 30, 58), (80, 90)\}$ and the only optimal arrangement (with P_1 unchanged) $S_2 = \{(10, 30, 58), (89, 99)\}$ with $d(S^*) = d(S_2)/2 = 31/2 = 15.5$ can be found easily. Hence,

$$d(S = S_1) = \frac{d(S_\lambda)}{\lambda} \tag{15.13}$$

Regarding this mechanism, there is one detail to be taken into account. Distances related to the shape of *one polygon*, i.e., the lengths of sides, would not be affected by the rounding, since, if there is a need for *real* rounding, the direction of rounding will be the same for all vertices of the polygon considered. This is a consequence of the fact that each of its coordinates belong to the same residue class (explained by the fact that only integer values are allowed for the length of a side). However, for distances between vertices of *different polygons*, including the minimal distance $d(S_\lambda)$, rounding actually can have an effect.

For example, let there be $U = 60$, $\lambda = 2$ and two points $1, 2 \in \mathcal{L}$, with $p_1 = 36$ and $p_2 = 39$ responsible for $d(S_2)$, then $d(S) = 3/2$ (1.50). An expansion to S_3 produces $p_1 = 54$ and $p_2 = 59$ with $d(S) = 5/3$ (1.67). Further expansions finally converge to $d(S) = 2$, which is an improvement over 1.50. The opposite happens with, e.g., $p_1 = 35$ and $p_2 = 38$. In stage 2, one starts again with $1.50 = 3/2$. But then, following the sequence ($4/3$, $5/4$, $6/5\ldots$), the minimal distance converges to the worse solution $d(S) = 1$.

For the GTS these effects are interpreted as *intensification* and *diversification*, respectively: Even though only expansions to a better solution seem to be reasonable, deteriorating expansions are also allowed as a means of reaching new regions. However, in order not to get stuck at local optima stages, a stage λ is left in any case after a given number of iterations, which is determined by the program parameter t_λ.

The next three subsections will cover the algorithmic description of the meta-heuristic approaches: the TS component, the GA component and, finally, the interplay of both components, i.e., the GTS-algorithm for the *max-min* IPSP.

15.3.2 The Tabu Search Component

Neighbourhood. Every iterative procedure needs a concept of how to construct a new solution from a given one. These necessary constructions determine the transition from a solution S to another one called S'. All solutions derived from S build the neighbourhood $N(S)$. Imagine a solution S_λ. To generate a neighbour $S'_\lambda \in N(S_\lambda)$, a polygon is either rotated clockwise or counterclockwise by 1 unit or by $(2d(S_\lambda) + 1)$ units, where $d(S_\lambda)$ is the current solution value. The cases that involve the value of the current solution enable the algorithm to improve a positive minimal distance by entering another local region.

Clearly, this is not a decomposition approach that makes use of locally optimizing procedures (see [1, 16, 17]). However, the advantage is that one is not forced to search for hopefully improving regions, which must be chosen from an exponential number of possibilities. Otherwise, the running time would increase, which compensates the gain of efficiency achieved with a decomposition approach.

Memory. Typically, tabu memory is based on a short-term recency memory that stores problem-specific key attributes [13]. Advanced techniques also include long-term aspects, e.g. the frequency count of certain attribute values. Additionally, another memory development should be pointed out: the so-called Reverse Eliminination Method, introduced in [9] and studied, e.g., in [28]. It can prevent the search process from re-visiting old solutions. For our implementation, however, there is no need to introduce attribute based tabu memory, since every solution is uniquely determined by its set of standardized coordinate values p_{i_2} with $p_{i_1} := 0$ $(i = 1, \dots, n)$, which can be stored easily. For a solution that is designated *tabu*, the set of initial coordinate values is stored in a *tabu list*. To introduce a dynamic effect the tabu list length $tSize$ is not fixed, but is a periodically varied integer, randomly chosen from the interval between $4n - [4n/3]$ and $4n + [4n/3]$. As proposed by other authors, these variations dynamically take place (see, e.g., [19]). In the GTS the list length is changed every $10\,tSize$ iterations.

Admissibility. The concept of TS is based on the use of historical search information and usually replaces $N(S)$ by another neighbourhood $\overline{N(S)}$, which embeds the search history and additionally may be interpreted as candidate list of solutions. The determination of the *admissible* solutions is embodied in the interplay between memory and tabu restrictions. Hence, in $\overline{N(S)}$, states already encountered during the search are adequately represented.

For the GTS implementation, TS admissibility is given if the neighbour solution $S'_\lambda \in N(S_\lambda)$ is not tabu, i.e. the same solution does not appear in the tabu list, or if it is *aspirated* because it outperforms the best-known solution so far. Admissible solutions are stored in $\overline{N(S_\lambda)} \subseteq N(S_\lambda)$.

15.3.3 The Genetic Algorithm Component

Encoding. A population called G consists of g_{pop} individuals. Every individual has n genes corresponding to the initial coordinates p_{i_1} $(i = 1, \dots, n)$. For the encoding we use the smallest possible gene length, γ, which is able to represent the current length of the circumference U_λ. As an encoding scheme, a standard binary representation is used. Since any decoded out-of-range values are made acceptable by using their modulo U_λ residual, operators are guaranteed to generate legal solutions.

Generation process. The control of the generation process happens due to the genetic operators reproduction, crossover and mutation. The corresponding rates are given by g_{rep} and $g_{crs} = 1 - g_{rep}$; the probability for mutation is g_{mut}.

For the *reproduction* we select $\rho = [\,g_{rep}\, g_{pop}\,]$ individuals via roulette wheel selection (RWS) to survive as members of the next generation. The remaining $(g_{pop} - \rho)$ members of the new pool are constructed by means of a 2-point *crossover*, which randomly selects two crossover points in parent chromosomes and then interchanges the bits of the inner substrings. For this, one parent is randomly chosen from the current population, the second one with an RWS. Finally, regarding *mutation*, with probability g_{mut} a bit at the locus μ, which is randomly chosen as $1 \leq \mu \leq n\gamma$ (μ integer), is inversed.

The determination of suitable operators generally is a hard task. Furthermore, any application to the *max-min* IPSP is unknown. Therefore a choice was made in favour of a direct representation underlying a genuine 2-point crossover. Basically, we restricted ourselves to the examination of 1-point and

2-point crossovers. In initial experiments it appeared that the positional bias of the 1-point crossover towards radical changes seemed to be too strong. In order not to be torn between extreme search positions, a 2-point crossover was chosen. It is also evident in literature, see [25], that increasing the number of crossover points can improve the performance of a crossover operator. The position of the crossover points, right in front of a *gene* or at its end, affects the amount of rotation of a polygon (The same holds, regarding the *chromosome*, for the choice of useful polygons to be relocated). Seen from the survival process, a "GA learns" to adapt its population members to fit into the next generation. This can be facilitated by the crossover representation used. The RWS selection ensures a moderate selective pressure that (supported by mutation) prevents the population from premature convergence. Reproduction, on the other hand, counterbalances this diversification attitude since it concentrates on the continual existence of useful information.

15.3.4 The genetic tabu search - interplay of components

We first give an overall step-by-step description of the GTS algorithm; afterwards the details will be explained:

Genetic Tabu Search for the *max-min* IPSP

Step 1: (Initialization) Input parameters and starting solution $S_{\lambda=1}$ and derive the starting population $G(t=0)$ (see 15.3.3). Set $t=0$, $\lambda=1$, $S_{best}=S_1$ and $d(S_{best})=d(S_1)$.

Step 2: (Hybrid neighbourhood) Set $t=t+1$. Perform a TS iteration (15.3.2) and a GA iteration (15.3.3), i.e., build $\overline{N(S_\lambda)}$ and $\underline{G(t)}$, respectively. Construct the hybrid neighbourhood $H=\overline{N(S_\lambda)}\cup\overline{G(t)}$, where $\overline{G(t)}\subset G(t)$ is a subset of *good* population members (decoded pool output). Select S'_λ as the best solution from H. IF S'_λ is the overall best solution so far, then set $S_{best}=S'_\lambda$ and $d(S_{best})=d(S'_\lambda)/\lambda$.

Step 3: (Termination) If *termination criterion* is true, then go to *Step 5*.

Step 4: (Update memory, pool input, expansion) Add S'_λ to the tabu list and replace a *poor* element of $G(t)$ by S'_λ (encoded pool input). If t_λ iterations are made on stage λ, then expand S'_λ and $G(t)$ to stage $\lambda:=\lambda+1$ (see 15.3.1). Set $S_\lambda=S'_\lambda$ and go to *Step 2*.

Step 5: (End) S_{best} is a GTS solution to the *max-min* IPSP with value $d(S_{best})$.

Initialization. The overall GTS starts in *Step 1* with a given solution $S_{\lambda=1}$ and initializes the starting population $G(t=0)$ by copying the encoded solution

S_1 into the pool as the first chromosome. The remaining $(g_{pop} - 1)$ members of the first generation are derived from the initial chromosome by mutation.

Hybrid neighbourhood. Now, and in every iteration t of *Step 2*, the hybrid neighbourhood H is built, based on the current TS solution S_λ and the population $G(t)$, which was constructed by a generation change just performed. In building H the (admissible) candidates of $\overline{N(S_\lambda)}$ (see 15.3.2) are supplemented by a set of decoded solutions, called $\overline{G(t)}$. The portion of GA solutions in H is given by the input variable h_{gen}. If h_{gen} stipulates a pool output of, say, g_{out} decoded binary solutions, then $\overline{G(t)}$ is constructed by copying the first- to g_{out}-best GA solution into it. In contrast to TS admissibility, GA candidates are always admissible. After all candidates of H are determined, the best one becomes the basis for the next TS iteration, a strategy well-known as the best-accept strategy. Eventually a new-best overall solution is recorded.

The motivation for freeing elements of $G(t)$ from a tabu test comes from the idea that a tabu long-term memory aspect may be embodied by the GA component. It maintains a population over a longer period and thereby also incorporates long-term information that was originally drawn from short-term tabu observations. (The types of population-based intensification strategies for TS, described at the beginning of Section 15.3, are similarly designed to operate as longer term strategies.)

Termination. The GTS terminates if an upper bound (see Section 15.2) is reached or if $t = \min\{t_{best} + t_{nip}, t_{max}\}$, where t_{best} marks the iteration in which S_{best} was found, t_{nip} is the maximum non-improvement period permitted and t_{max} denotes the maximum number of iterations allowed.

Update memory, pool input, expansion. If not terminated, before the next iteration, in *Step 4* the tabu list is updated with the initial coordinates of the (new) current TS solution S'_λ. Its binary encoded correspondence replaces an individual that was chosen with a *reverse* RWS, i.e. an RWS done with a random distribution derived from the *worse* individuals in $G(t)$. After an expansion has been possibly carried out for both components, for the TS solution and for the GA pool, a new GTS iteration starts in *Step 2*.

15.4 COMPUTATIONAL EXPERIENCE

To evaluate the procedures, we have tested two sorts of data, all based on $U = 60$ (in order to correspond with the equivalent of time). In the first group there are examples constructed by hand during the development of the GTS. Since no data sets were available to us, the second group consists of instances generated randomly from the basic inputs n, the number of polygons, and the total number of vertices, called m. The algorithms were coded in PASCAL and run, just as CPLEX [6], over a total of 26 instances on a P5/133.

The entries in the columns of Table 15.1 are to be interpreted as follows: In column (1) the name of the problem is shown. The first number before the character is n, the second one m. Randomly generated examples do not contain an h in the name. The other five columns refer to the optimality approach with the LP-formulation via the Mixed Integer Solver from CPLEX. Columns (2)

(1) *name*	(2) *rows*	(3) *vars.*	(4) $d(S^*)$	(5) *UB*	(6) *CPU*
4h13	190	253	3.500		56
4h19	403	537	2.000		837
4h33	1210	1613	1.000		9148
5h10	125	166	5.000		190
5h12	176	234	4.333		243
5h16	311	414	2.667		2797
5h18	386	514	2.500		3668
2b25	452	603	1.000		21
3b28	780	1040	1.000		425
3b35	1200	1600	0.500		2414
4a15	250	333	2.667		230
4a19	406	541	2.000		427
4a22	541	721	1.500		1276
4b25	694	925	1.500		3727
5a11	149	198	5.333		64
5a14	239	318	3.667		335
5a16	305	406	2.667		1164
5a17	350	466	2.500		1800
5a21	527	702	*1.500*	2.667	4000
6a14	246	327	4.000		1862
6a18	402	535	*2.800*	3.000	7482
6a19	453	603	*1.000*	4.000	4079
6a20	501	667	*1.000*	4.333	3743
7a14	259	344	3.667		10650
7c15	295	392	*3.500*	5.000	2451
7c16	334	444	*2.333*	4.500	8177

Table 15.1 Results of CPLEX

and (3) show the number of rows and the number of variables, respectively, followed by the optimal solution value, $d(S^*)$. To ensure reasonable computing time, the tree-memory was limited to 30 MB and the maximum number of branch-and-bound nodes was limited to 250,000. Of course, the shortcoming is that for larger instances the optimality of the solution found is not guaranteed. These cases are indicated in column (4) with italic numbers, where integer feasible solutions, obtained before break-off, are given. Then, if optimality is not ensured, the current upper bounds, additionally given by CPLEX, are presented in column (5). Column (6) shows the running times in seconds.

Table 15.2 summarizes the GTS results. The parameter settings used are the following: The maximum number of iterations was set to $t_{max} = 1500n$, which was found to be an adequate value during preliminary test runs. These runs showed that most improvements where made in the lower stages. Thus facilitating improvements at higher stages, the number of iterations given to a certain stage, should increase with increasing stages. The logarithmic function used, $t_\lambda = t_1(1 + 1.5ln(\lambda))$, and the program parameter $t_1 = 500$ ensure a non-linear increase that goes well with the computer capacities given. The maximum non-improving iterations are $t_{nip} = 750n$, i.e., no more than one half of the computing time should pass without an improvement. The size of the population $g_{pop} = 50$ is, on the one hand, limited due to storage considerations, but, on the other hand, large enough to prevent the population from premature convergence. As just mentioned, the replacement strategy is a steady-state

(1) *name*	(2) $d(S)$	(3) $d(S_\lambda)/\lambda$	(4) UB	(5) t_{best}	(6) CPU	(7) RDU
4h13	3.500	7/2	4.500	504	18	0.00
4h19	2.000	10/5	2.500	4751	26	0.00
4h33	1.000	3/3	1.000	1745	24	0.00
5h10	5.000 *	5/1	6.000	12	25	0.00
5h12	4.333	13/3	5.100	1608	36	0.00
5h16	2.500	12/5	3.867	1017	32	0.06
5h18	2.333	7/3	3.333	1529	36	0.06
2b25	1.000 *	1/1	1.000	36	4	0.00
3b28	1.000 *	1/1	1.000	41	9	0,00
3b35	0.500 *	1/2	0.500	565	10	0.00
4a15	2.667	8/3	3.000	1522	24	0,00
4a19	2.000	2/1	3.000	43	17	0,00
4a22	1.500	6/4	2.500	7396	76	0.00
4b25	1.333	4/3	1.500	1513	64	0.05
5a11	5.333	16/3	5.455	1526	36	0.05
5a14	3.667	11/3	4.818	1525	29	0.03
5a16	2.500 *	5/2	3.750	515	28	0,06
5a17	2.500	10/4	3.600	2847	46	0,00
5a21	2.000	2/1	2.667	626	35	*0.25*
6a14	3.800	19/5	4.583	7016	132	0.05
6a18	2.600	13/5	3.571	4428	68	*0,13*
6a19	2.200	11/5	3.222	7924	134	*0.32*
6a20	2.750	11/4	3.500	2845	65	*0.21*
7a14	3.500	7/2	4.286	1119	70	0.05
7c15	3.500 *	7/2	4.214	541	63	*0.17*
7c16	3.000	12/4	3.750	2938	92	*0.20*

Table 15.2 Results of GTS

incremental approach with $g_{rep} = 0.4$ and $g_{crs} = 1 - g_{rep} = 0.60$. These values were found to be a convenient choice to avoid intensive search around local optima. This aim of diversification is supported by the comparably higher mutation rate $g_{mut} = 0.03$. Finally, in order to underline the idea of equality of the TS and the GA component, the portion of population solutions in the hybrid neighbourhood was set to $h_{gen} = 0.50$.

For every problem whose solution of the first run did not reach the best-known value (optimum or upper bound), a TS parameter and a genetic parameter were modified in order to possibly achieve a better result. This was done with $t_1 = 800$ and $g_{rep} = 0.30$.

The heuristic solution value achieved with GTS in (2) is also given in (3) as the ratio of the solution value dependent on the stage to the stage itself. In (4) the (better) upper bound of Section 15.2 is shown and in column (5) t_{best}, the iteration number of the last improvement. In (6) the CPU-times, again in seconds, are given for the GTS. Column (7) estimates the quality of the overall solutions by calculating the *Relative Deviation from Upper Bound*, abbreviated with RDU. It is defined as $(UB - d(S))/UB$, where $d(S)$ is the GTS solution and UB is either the optimal value of a CPLEX-solution (4) in Table 15.1 or the best-known upper bound, i.e. values (5) in Table 15.1 or values (4) in Table 15.2.

Obviously, the first impression one might get from the study of the CPLEX entries is the fact that we deal with an NP-hard problem. Computational times are most greatly affected by the total number of vertices m (compare results of 5h10 with 5h18). This is a circumstance that also is theoretically manifested in literature (see, e.g., [17]). Since CPLEX is regarded as state-of-the-art package, the CPU times also indicate that the LP-formulation of Section 15.2 can only be seen as an initial step to provide optimal results. However, it is better to have optimal results for a number of smaller test instances rather than only dealing with heuristic solutions, even if they were obtained for large-scale problems. Further, it is indicated that the bounds given in column (4) of Table 15.2 could be improved, either by making more intensive use of theoretical results or by designing a kind of incomplete enumeration algorithm that could calculate better bounds using a subset of *important* polygons and/or sides. The idea of expansion as a basic tool appeared to be justified by the entries in (3).

From Table 15.2 it also can be deduced that the GTS performs slightly better for the hand-made instances, even though they were not made to be "easy" problems. Supposing comparable numbers m, the results also show that the neighbourhood works better for a smaller number of polygons (this is especially assumed for $\overline{N(S)}$). Introducing a higher vertex-dependence could be the right means to improve the efficiency. However, it should be pointed out that the worst results are based on bounds that may be far from optimum.

Additional testing of parameter variations revealed that the GTS is quite sensitive regarding the values of GA parameters in that population converging tendencies do not always entail useful intensification effects. This leads to situations where the search space currently considered gets trapped in non-profitable regions. So it would be of interest to try a parallel implementation yielding the possibility of a more appropriate separation of the TS component and the GA component.

To summarize, the results are satisfying since most of the problems with a known optimal solution are solved optimally or their results are within a range of 6% from optimum. Moreover, the basic idea of combining TS with a highly population-based procedure is a successful one. In most of the cases the hybrid GTS achieves better results than the pure TS method (GTS with $h_{gen} = 0$). The 6 exceptions out of 26 results are marked with a * in column (2) of Table 15.2.

15.5 SUMMARY AND FINAL REMARKS

In this paper we have considered a special polygon scheduling problem, i.e. the Irregular Polygon Scheduling Problem with the objective of maximizing the minimal distance between nearest vertices of different polygons. After the introduction, we gave a detailed mathematical formulation of the problem and discussed possibilities of dealing with the *max-min* IPSP from the theoretical point of view.

The main idea of the heuristics to be introduced, on the one hand, was the reduction of the continuous NP-hard problem to a discrete optimization problem by means of a so-called expansion mechanism. On the other hand, the GTS favours the idea of a hybrid meta-heuristic that combines elements of TS and

GA as well. In the GTS the TS neighbourhood mechanism draws information from a pool of individuals, which have been provided as encoded TS solutions generated in past iterations. In the starting phase of the GTS the population is built randomly. Then it is maintained by the operators reproduction, crossover and mutation. In addition the population is also supplemented by TS neighbourhood solutions. TS itself is based on a classic neighbourhood mechanism, which creates solutions to be checked for their admissibility using tabu memory and an aspiration criterion.

All procedures were coded in PASCAL and run over a test sample of 26 problems. The computational results achieved were compared with those obtained from the LP-formulation using the Mixed Integer Solver from CPLEX. The results exhibit a good performance of the GTS, but they also call for additional future research. In particular the derivation of better bounds should be envisaged and more attention could be paid to evaluation functions and memory structures. Furthermore, to reduce running time, efforts should also be made in the investigation of a more efficient interplay between the two optimization components and in the determination of a robust parameter constellation.

With respect to the hybridization, an interesting variant would be the examination of an evolutionary PR approach in place of a GA approach, to carry out the functions handled by the GA component in the present study. A motive for such future research is the fact that very little examination of PR has occurred in the TS literature.

Besides these procedural aspects, future work will also focus on some new problem specific elements. These could be the consideration of other objectives, or the embedding of additional constraints, e.g., the scheduling of polygons that are located on more than one circle line, i.e. the case in which the trains of more than one station have to be scheduled.

References

[1] P. Brucker, R. E. Burkard, and J. Hurink. Cyclic schedules for r irregularly occurring events. *J. Comput. Appl. Math.*, 30:173–189, 1990.

[2] P. Brucker and W. Meyer. Scheduling two irregular polygons. *Discr. Appl. Math.*, 20:91–100, 1988.

[3] R. E. Burkard. Optimal schedules for periodically recurring events. *Discr. Appl. Math.*, 15:167–180, 1986.

[4] J. Cerny. Problems for systems of regular polygons on a circumference and their application in transport. (Czech) *Matematické obzory*, pages 51–59, 1972.

[5] D. Costa. An evolutionary Tabu Search algorithm and the NHL scheduling problem. *INFOR*, 33:161–178, 1995.

[6] CPLEX Optimization. CPLEX base system with callable library, barrier and mixed integer solver. Version 4.0., 1995.

[7] C. Fleurent and J. A. Ferland. Genetic hybrids for the quadratic assignment problem. *DIMACS Series in Disc. Maths & Theor. Comp. Sci.*, 16:173–187, 1994.

[8] F. Glover. Future paths for integer programming and links to artificial intelligence. *Computers Oper. Res.*, 5:533–549, 1986.

[9] F. Glover. Tabu Search – Part II. *ORSA J. Comput.*, 2:4–32, 1990.

[10] F. Glover. Tabu Search and adaptive memory programming – Advances, applications and challenges. In R. S. Barr, R. V. Helgason and J. L. Kennington, editors, *Interfaces in Computer Science and Operations Research*, pages 1–75. Kluwer Academic Publishers, 1996.

[11] F. Glover. A template for Scatter Search and Path Relinking. In J. K. Hao, E. Lutton, E. Ronald, M. Schoenauer, and D. Snyers, editors, *Lecture Notes in Computer Science.* 1997. To appear.

[12] F. Glover, J. P. Kelly, and M. Laguna. Genetic Algorithms and Tabu Search: Hybrids for optimization. *Computers Oper. Res.*, 22:111–134, 1995.

[13] F. Glover and M. Laguna. Tabu Search. In Reeves C. R., editor, *Modern heuristic techniques for combinatorial problems*, pages 70–150. Blackwell Scientific Publications, 1993.

[14] F. Glover and M. Laguna. *Tabu Search.* Kluwer Academic Publishers, 1997.

[15] P. Greistorfer and H. Kellerer. Bounds and Tabu Search for a cyclic max-min scheduling problem. Technical report, Dept. of Industrial Management. University of Graz, 1998. Updated version.

[16] F. Guldan. Maximization of distances of regular polygons on a circle. *Aplikace Matematiky*, 25:182–195, 1980.

[17] J. Hurink. Polygon scheduling. *Discr. Appl. Math.*, 70:37–55, 1996.

[18] H. Kim, K. Nara, and M. Gen. A method for maintenance scheduling using GA combined with SA. *Computers Ind. Eng.*, 27:477–480, 1994.

[19] M. Laguna and F. Glover. Bandwith packing: A Tabu Search approach. *Management Sci.*, 39:492–500, 1993.

[20] O. Martin, S. W. Otto, and E. W. Felten. Large-step Markov chains for the TSP incorporating local search heuristics. *Oper. Res. Lett.*, 11:219–224, 1992.

[21] H. Mühlenbein, M. Georges-Schleuter, and O. Kramer. Evolution algorithms in combinatorial optimization. *Parallel Comput.*, 7:65–88, 1988.

[22] H. Mühlenbein and H.-M. Voigt. Gene pool recombination in Genetic Algorithms. In I. H. Osman and J. P. Kelly, editors, *Meta-Heuristics: Theory & Applications*, pages 53–62. Kluwer Academic Publishers, 1996.

[23] I. H. Osman and J. P. Kelly, editors. *Meta-Heuristics: Theory & Applications*. Kluwer Academic Publishers, 1996.

[24] I. H. Osman and G. Laporte. Metaheuristics: A bibliography. *Ann. Oper. Res.*, 63:513–628, 1996.

[25] C. R. Reeves. Genetic Algorithms. In Reeves C. R., editor, *Modern heuristic techniques for combinatorial problems*, pages 151–196. Blackwell Scientific Publications, 1993.

[26] S. R. Thangiah, I. H. Osman, and T. Sun. Hybrid Genetic Algorithm, Simulated Annealing and Tabu Search methods for vehicle routing problems with time windows. Technical report, Dept. of Computer Science. Slippery Rock University, 1994.

[27] A. Vince. Scheduling periodic events. *Discr. Appl. Math.*, 25:299–310, 1988.

[28] S. Voß. Observing logical interdependencies in Tabu Search — Methods and results. In V. J. Rayward-Smith, I. H. Osman, C. R. Reeves, and G. D. Smith, editors, *Modern heuristic search methods*, pages 41–59. John Wiley & Sons Ltd., 1996.

16 ADAPTIVE GENETIC ALGORITHMS: A METHODOLOGY FOR DYNAMIC AUTOCONFIGURATION OF GENETIC SEARCH ALGORITHMS

Ulrich Derigs, Martin Kabath and Markus Zils

University of Cologne, WINFORS, Pohligstr. 1, D-50969 Köln, Germany.
derigs@informatik.uni-koeln.de, kabath@informatik.uni-koeln.de, zils@informatik.uni-koeln.de

Abstract: Genetic Algorithms (GA) like other modern metaheuristics claim to be general problem solvers. Though GA have been applied successfully to a wide range of different combinatorial optimization problems, the need for careful and time-consuming tuning of GA constitutes a major drawback as a general problem solver. In this report we introduce the concept of Adaptive Genetic Algorithms (AGA) as a solution to this calibration problem, which dynamically performs an on-line autoconfiguration of the GA-parameters. To demonstrate the superior performance of AGA vs. GA in terms of solution quality, robustness and computational effort, we present computational results for three different combinatorial optimization problems. Our benchmark comprises two standard benchmark problems (Quadratic Assignment Problem and Period Vehicle Routing Problem) and one real-world problem arising in airline scheduling.

16.1 INTRODUCTION

In the last years, researchers applied successfully Genetic Algorithms (GA) to a wide range of standard combinatorial optimization problems. In spite of the obvious quality of this approach, industry seems to be reluctant to integrate GA into Decision Support Systems (DSS) for real world problems for two reasons:

- First, the development of an adequate representation scheme capturing the essential objectives and restrictions of a given, so far unstructured problem is a non-trivial task.

- Second, for any implementation of GA a large number of parameters has to be calibrated to assure fast and robust convergence to a satisfying solution quality.

In practice, the first obstacle that is an inherent step for any solution paradigm, often referred to as *art*, can be tackled by a *constructive modeling approach.* Once a coding scheme is established all different instances of a given problem can be solved. The second obstacle addresses the application of GA to a given specific problem instance. While researchers are primarily interested in fast heuristics which provide high quality solutions the practitioner's focus is in addition to that on easy implementation and robustness. To achieve the first two goals GA requires in general a very careful and time-consuming fine-tuning of parameters. Therefore GA's impact on real-world problem solving especially in non-standard, i.e., so far unstructured problem-environments is limited.

In this paper we will primarily address the practitioner's focus and contribute to the elimination of the second obstacle. First we discuss the classical GA concept and describe the so called *calibration problem.* Then as a concept to overcome this calibration problem the Adaptive Genetic Algorithms (AGA), will be introduced. Applications of this concept will be presented for three different problem classes: the Quadratic Assignment Problem (QAP), the Period Vehicle Routing Problem (PVRP) and the so-called Strategic AirCargo Problem (SACP). Here we shortly introduce the coding schemes developed and present a comparison of GA vs. AGA concerning computational effort, robustness and the quality of the generated solutions.

To show the effectiveness and the efficiency of AGA versus classical GA we concentrate on comparing rather basic approaches, i.e., we neglect the introduction of sophisticated local optimizers. We do not intend to argue that with this puristic strategy AGA will outperform other metaheuristics on any problem class. Our intention is to demonstrate that the calibration problem of the classical GA can be solved efficiently and automatically avoiding a tedious and time-consuming manual, experience-guided control by GA-modelers or practitioners. Due to this autoconfiguration in the AGA, GA becomes a valid candidate within the family of metaheuristics for tackling hard non-standard combinatorial optimization problems in real world applications.

16.2 THE CONCEPT OF GENETIC ALGORITHMS

The concept of Genetic Algorithms for solving optimization problems is based on the analogy to evolution theory in population genetics. Holland [11] adopted the idea of the *survival of the fittest* in a process of cooperation and competition among individuals to combinatorial optimization problems: The solutions of a problem are coded into *chromosomes*, a sequence of *genes.* A set of such chromosomes is called a *population.* Starting from an initial population new chromosomes are generated by standard genetic reproduction operators, e.g. crossover and mutation and are evaluated with respect to a problem specific *fitness function.* Depending on their fitness values some chromosomes survive

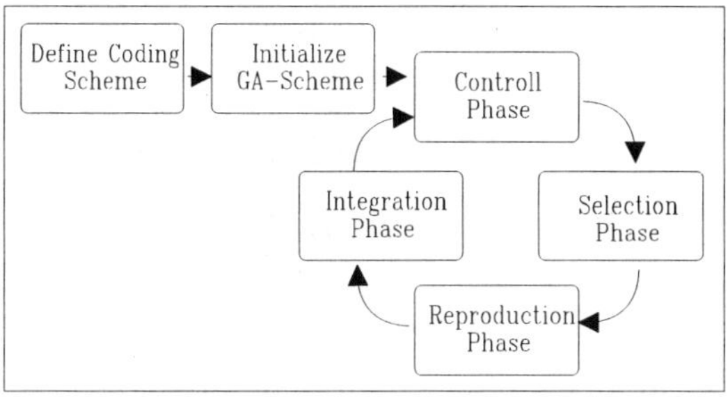

Figure 16.1 GA-Scheme

and some die out leading to a new population. Through the repetition of this reproduction process a sequence of populations is generated with the expectation to generate solutions of better quality during the course of this process. The GA-solution process can in general be structured into two stages (see Figure 16.1).

In the first static stage a coding scheme and an appropriate fitness function capturing the main objective and constraints are defined for the given problem type. Also the static parameters of the GA-scheme are initialized: e.g. population structure, size and communication scheme as well as the specification of operators and strategies to be applied in the dynamic stage.

The dynamic stage is divided into four phases which are iteratively applied until a given termination criterion is reached to produce new populations and to simulate the natural evolution process:

- **Selection phase.** In this phase a number of individuals of the current population are selected and paired for reproduction.

- **Reproduction phase.** Applying the principle genetic reproduction operators like crossover and mutation new solutions are generated by sexual reproduction.

- **Integration phase.** The new individuals are evaluated according to the defined fitness function. Then it is decided which of these offsprings will be integrated into the new population and which older individuals will be excluded from the actual population.

- **Control phase.** In this phase global metrics of the population are assessed and the communication scheme is updated. The algorithm checks if the termination criterion holds.

In the past different researchers have proposed a vast variety of strategies and operators for the different phases of the GA, leading to the problem of choice, i.e., the question which operator to apply with which application rate for a given problem instance.

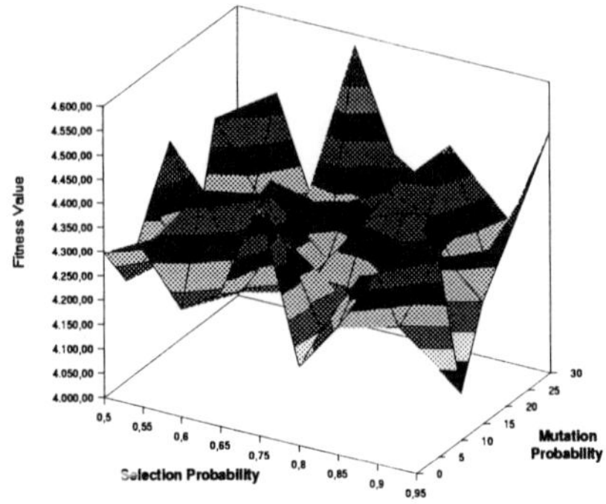

Figure 16.2 Solution Quality: SACP9

16.2.1 The calibration problem of Genetic Algorithms

Since for a single problem instance the speed, the robustness and the quality of the convergence process depends on the specific choice of strategies and combinations of operators this choice is critical for the success of this metaheuristic. To define a specific GA a number of parameters has to be fixed. This problem is known as the *calibration problem of GA*. As an example, Figure 16.2 demonstrates that different parameter sets consistently result in different solution qualities. Here the average solution quality for a specific crossover-operator (C8X) on an instance of the Strategic AirCargo problem (SACP) is plotted as a function of selection and mutation probabilities. The SACP and the C8X operator will be described in greater detail in Section 16.4.3.

Since no sound theory has been developed to guide the process of choosing strategies and operators and fine-tuning of parameters, GA-modelers must rely on their experience and/or a time-consuming trial and error calibration. Yet, there is no guarantee that this approach leads to a population containing solutions of satisfactory quality. Furthermore, there is no guarantee that the results obtained through such a calibration on one set of instances can be generalized, i.e., there is no reason to assume that the same set of parameters will be effective if applied to a different problem instance.

16.2.2 Parameter interactions

In order to guarantee that the best possible solution is obtained, theoretically, all possible parameter combinations must be analyzed. Depending on the number of potential combinations this is not a practical and feasible approach especially if a short development time is required as it is common when solving real-world problems in changing business environments. Therefore in practice, developers rely on their observations and experience to reduce the potential number of promising combinations by identifying which parameters have strong interactions [14]. Two important parameter-interactions which have been studied are:

- The interaction between **mutation probability, population size and integration strategy**: Here, the notion is that the disadvantage of smaller populations concerning the loss of genetic variability is compensated by a higher mutation probability and a smaller number of replacements in each iteration. Due to the generation of many random genes by mutation a high level of diversity is maintained [13].

- The interaction between **crossover operator, mutation probability and selection rate**: Supposedly this is the strongest interaction [14]. A high selection probability in a rank-order GA is attributed a fast convergence, while a high mutation probability is expected to maintain diversity. Different crossover-operators imply different implicit mutations. To avoid premature convergence to suboptimal extrema a specific balance between selection and mutation probability has to be identified for each individual crossover operator.

While the first interaction relates static and dynamic parameters the second only comprises parameters which are applied in the dynamic phases of the GA-scheme. If the population size remains fixed, as it is usually done in a steady-state policy, and the integration strategy is determined by the established communication scheme, the mutation probability is the only dynamic parameter that can balance the speed of convergence. In the second parameter-interaction all parameters can be changed dynamically. Even if the parameterization of only these three parameters are controlled in the experiments, the number of combinations can be still too high to probe every possible combination for identifying a good parameter setting, especially if application rates can continuously vary within an interval.

16.3 THE CONCEPT OF ADAPTIVE GENETIC ALGORITHMS

Ideally the GA-modeler should only design a coding scheme, specify a set of potential operators together with a range and discretization of their application rates and after the completion of this creative part of the development submit the tedious and tiring job of calibration to some kind of assistant, who operates using special procedural knowledge on how to produce high-quality solutions fast. Our idea of Adaptive Genetic Algorithms (AGA) is to automatically and dynamically perform an autoconfiguration of parameters. In the following we describe a first effort to device such an approach. Here we concentrate on the autoconfiguration of the strongest known interaction (see Section 16.2.2) between crossover-, mutation- and selection-operator.

Extending the classical GA-concept not only information on the solution itself is represented in the chromosomes but also information on the parameterization, the so-called environment, which was applied in the generation of this chromosome, is coded and submitted to the competition process. Thus following the GA paradigm successful parameter combinations, i.e., those which produce high quality individuals, will receive a higher probability of being inherited, so that at the end of the natural adaptation process high-quality parameter-combinations are expected to predominate resulting consistently in high-quality solutions.

The central problem of such an approach is to establish a link between the quality of parameter combinations and individual chromosomes. In the AGA concept this is done by defining the quality of the parameter setting by its ability to produce offsprings with better fitness values compared to other combinations of parameters. Based on this simple idea it is possible to define a scheme for dynamic autoconfiguration. As the classical GA the new AGA is a metaheuristic, i.e., its principle can be described independently of any special problem class. Thus, for the following description we assume an arbitrary problem class with an associated coding scheme and a given *tool-box* of operators, i.e., we assume

- C a set of applicable crossover operators
- M a set of appropriate mutation operators
- S a set of selection operators

for the specific coding scheme. To measure the effectiveness of a specific operator $j \in C \cup M \cup S$ we introduce a so-called scoreboard $B(j)$ which serves as an account, i.e., if the parameter proves to be effective it will accumulate positive votings during the adaptation process, otherwise it will be penalized by subtracting points from the scoreboard. At the beginning all operators have an identical score of 0. With every solution/individual i in a population P we associate its so-called environment $E(i) = (c(i), m(i), s(i))$ with $c(i) \in C$, $m(i) \in M$ and $s(i) \in S$. For the initial population the operators defining an environment E are independently generated at random and then randomly assigned to an individual. Due to the fact that the population will usually be much smaller in size than the number of potential operator combinations, not all combinations can be assigned to the initial population. But in principle all possible operator combinations can be generated during the course of the AGA due to the fact that the propagation of operators is determined separately for each of the three operator classes.

In the following we describe the voting-scheme and the propagation strategy of the environmental settings in detail. A pseudo-code is given in Figure 16.3. The AGA is based on a special population structure usually used in parallel implementations of GA: the so-called overlapping population structure [1, 10]. Figure 16.4 illustrates the concept of overlapping population structures with two, three and four neighborhoods.

After the scoreboard B, the population P and the parameter environment E are initialized the simulation of the natural adaptation by reproduction is performed as long as a termination criterion is not fulfilled. In general the termination criterion of an AGA-implementation should be based on qualitative criteria monitoring the process of convergence and leading to termination after stagnation in the convergence process is detected. To facilitate the analysis and the comparison with the classical GA we applied a fixed number of evaluations

```
set scoreboard B(j) := 0 for all j ∈ C ∪ M ∪ S
initialize population P
associate an environment E(i) for all i ∈ P at random
WHILE termination criterion not fulfilled DO
    select individual k ∈ P at random
    select individual l ∈ P from the neighborhood of k applying s(k)
    generate offsprings k' and l' applying c(k) and m(k)
    IF fitness(k') better than fitness(k) THEN
      B(c(k)) := B(c(k)) + 1, B(m(k)) := B(m(k)) + 1, B(s(k)) := B(s(k)) + 1
        AND P := (P \ {k}) ∪ {k'} AND E(k') := E(k)
      IF fitness(l') better than fitness(l) THEN
        B(c(l)) := B(c(l)) − 1, B(m(l)) := B(m(l)) − 1, B(s(l)) := B(s(l)) − 1
        AND P := (P \ {l}) ∪ {l'} AND E(l') := E(l)
        IF B(c(k)) − B(c(l)) > τ THEN c(l') := c(k)
        IF B(m(k)) − B(m(l)) > τ THEN m(l') := m(k)
        IF B(s(k)) − B(s(l)) > τ THEN s(l') := s(k)
ENDWHILE
```

Figure 16.3 AGA-Pseudocode

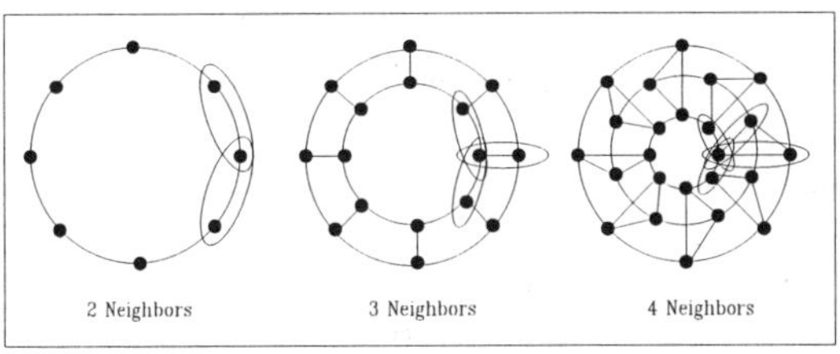

Figure 16.4 Overlapping Population Topology

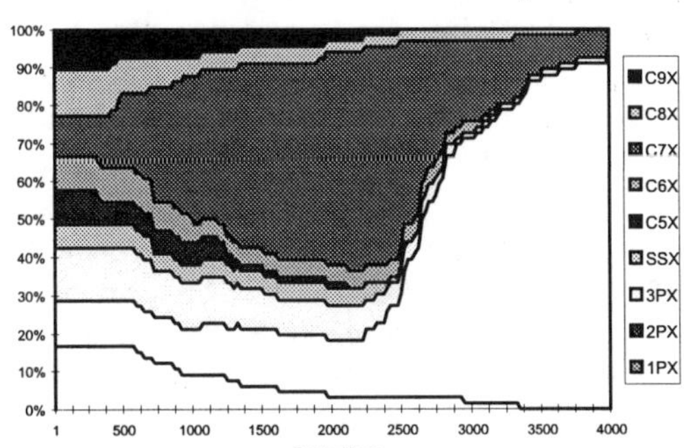

Figure 16.5 Adaptation of Crossover-Operators in the SACP9

as termination criterion in our computational tests. The concept of parallel processing in the overlapping population structure is simulated in our AGA-scheme by selecting randomly one individual at a time as a master individual for sexual reproduction with one of its neighbors. This is described in the WHILE-loop of the pseudocode (see Figure 16.3). First, the master individual k is randomly selected from the population. Based on the selection operator of its environment the slave individual l is chosen from the neighborhood of k. Two offsprings are generated applying the master's reproduction operators. One individual is assigned to be the master-offspring k' and the other the slave-offspring l'. To identify the quality of the solutions both are evaluated with respect to the fitness function.

The master k is replaced by k' if the offspring has a better fitness value. If this replacement occurs an update of the scoreboard B is performed by increasing the account for the operators of the master individual k by one. If the slave individual l is replaced by its offspring the parameters of its environment $E(l)$ are downgraded by one in the scoreboard B, because these parameters did not take part in the successful reproduction, but had a chance to prove efficiency in previous evaluations (memory function).

If a replacement of the slave individual l by its offspring l' occurs the algorithm computes the differences in the scoreboard B for all parameter pairs of k and l. If a difference exceeds a given threshold τ, the so called *acceptance rate*, the superior parameters of the master individual k are copied to the environment $E(l')$ of the new individual l'. Since the inheritance of the operators is decided separately for the different operator categories, new parameter combinations can be generated in this step.

Critical in this process is the determination of the threshold τ. Small values of τ have the effect that every time a replacement occurs all parameters would be inherited. In this randomly driven process this might also be the case for low quality parameters, which is not desired to happen. If on the other hand τ is chosen to be too large even successful parameters cannot propagate in the environments. In our AGA-implementation τ was chosen as a fixed percentage of the population size to balance the pressure of convergence with the benefits of variability. Now one could claim that this comes close to the parameterization of a meta-parameter. Essentially this is true. But the impact of τ on the solution quality is less significant than the impact of the operators, since in the worst case the AGA will only degenerate to a classical GA without dynamic autoconfiguration.

This procedure is repeated until the termination criterion applies. The idea of the *survival of the fittest* implies that at the end of this adaptation process the superior operators have propagated through the environment E of the population P while the inferior combinations have died out.

Figure 16.5 displays this typical adaptation process for one example run of a SACP-instance. The picture shows that in the beginning the 9 different crossover operators were almost equally frequent in the environment E of the population P. However during the first part of the adaptation process one crossover dominated the reproduction process while at the end another crossover operator emerged as the superior operator. This diagram illustrates the difference between the classical GA where only one operator per category (crossover, mutation, selection) is chosen to be applied during the entire process and the AGA where the operators are dynamically changed by autoconfiguration to fit the needs of the genetic search process.

Obviously, the quality of the AGA-approach depends on the quality of the given tool-box of operators. Thus for problem classes which already have been studied one should always integrate operators into the tool-box which are known to be efficient for the particular problem class. The computational studies represented in the next section will answer the question whether AGA is able to select the best combination of operators and then yield solutions which are at least comparable in quality with the best GA-implementation based on the same tool-box. Since this is the case, there is no risk but a chance to introduce "weak" operators into the tool-box and to allow nonstandard combinations, which may be successful on special "extreme" and so far not studied instances. This increased variety of operators thus might add to the robustness of the approach.

16.4 APPLICATION TO SPECIFIC PROBLEM CLASSES: COMPUTATIONAL RESULTS

The development of the AGA concept was inspired by the need of practitioners to solve instances of so far unstructured problems (see SACP in Section 16.4.3). Any proposed solution approach has to fulfill the following requirements: the guarantee to produce good solutions independent of the specific problem instance to be solved (quality and robustness) as well as the potential of rapid and efficient development (implementability and acceptance). In our empirical study of the AGA-concept we tested the following three hypotheses:

a) AGA outperforms the classical GA in terms of speed, robustness and solution quality.

b) AGA allows rapid and efficient implementation of a solution procedure generating high quality results for generic combinatorial optimization problems.

c) AGA yields satisfying results for so far non-investigated optimization models stemming from new and unstructured real world managerial problems.

Three different problem classes were used to test these hypothesis. The Quadratic Assignment Problem (QAP), a celebrated combinatorial optimization problem for benchmarking of metaheuristics, was used to test hypothesis a) and b). QAP has the property that near optimal solutions are known for a large number of instances and that the development of GA-coding schemes is a trivial task. The other two problems were selected for two reasons: Both problems have a high practical significance, yet have not received much attention by

Name	nug20	nug30	sko49	sko72	sko90
best known (b.k.)	2570	6124	23386	66256	115534
SGA best	2570	6174	23730	67530	118198
Δ best SGA vs. b.k. %	0.00%	0.82%	1.47%	1.92%	2.31%
SGA avg.	2628.22	6322.03	24272.51	69059.66	120537.41
Δ avg. SGA vs. b.k. %	2.27%	3.23%	3.79%	4.23%	4.33%
AGA best	2570	6140	23658	67232	117190
Δ best AGA vs. b.k. %	0.00%	0.26%	1.16%	1.54%	1.43%
AGA avg.	2596.00	6225.06	23872.27	67868.80	118534.84
Δ avg. AGA vs. b.k. %	0.87%	1.94%	2.08%	2.43%	2.60%
Δ best AGA vs. best SGA %	0.00%	-0.55%	-0.30%	-0.38%	-0.85%
Δ avg. AGA vs. avg. SGA %	-1.36%	-1.25%	-1.65%	-1.72%	-1.66%

Table 16.1 QAP Results

the scientific community, resulting in the lack of ready to use and established solution procedures. With the study of the Period Vehicle Routing Problem (PVRP) we tried to prove that good results can be produced rapidly (hypothesis b), i.e., in short development time. The study of the Strategic AirCargo Problem (SACP) can be regarded as the real benchmark for the AGA-concept since no mathematical decision model with an efficient solution procedure was known but management required the solution within a short period of time (hypothesis c).

To prove these hypotheses we concentrate in the following section on the demonstration of the superior effectiveness and efficiency of AGA versus classical GA. Therefore we will concentrate on comparing puristic approaches, i.e., we neglect the introduction of sophisticated local optimizers in our implementation. Such local optimizers are problem dependent and thus we will discuss the results on such refined hybridizations of AGA approaches and comparisons with classical GA and other metaheuristics seperately in subsequent problem oriented papers. The following sections comprise only overviews of the results obtained in our numerical studies. Detailed results will be presented in separate reports.

16.4.1 The Quadratic Assignment Problem

The Quadratic Assignment Problem can be formulated as follows:

$$\min_{\Phi \in P(n)} \sum_{i=1}^{n} \sum_{j=1}^{n} a_{ij} b_{\Phi(i)\Phi(j)} \tag{16.1}$$

where $A = (a_{ij})$ and $B = (b_{kl})$ are two $n \times n$ matrices and $P(n)$ is the set of all permutations of $1, .., n$. Matrix A is often referred to as a distance matrix between sites, and B as flow matrix between objects [8].

The straightforward and widely used coding scheme for the QAP is a simple permutation of length n. To guarantee that the reproduction operators do not destroy the permutation, permutation-preserving operators are implemented. In our computational study five crossover-operators widely used in GA-studies

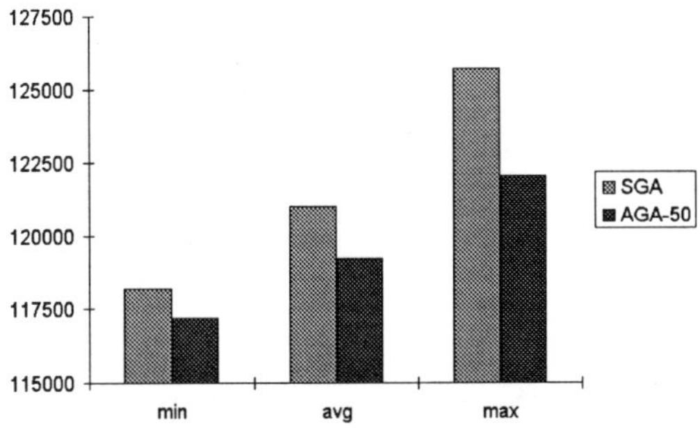

Figure 16.6 SGA vs. AGA, sko90-instance

on QAP as described in [3], namely the Partially Mapped Crossover, Cycle Crossover, Maximal Preservative Crossover, Code Based Crossover and Order Crossover, were chosen. The mutation operator consists of a simple swap of two positions of the permutation, also guaranteeing the preservation of feasibility.

In order to study hypothesis a) extensive tests were performed on 5 instances by first testing all potential parameter combinations explicitly with the classical GA and then by running the AGA on the same instances with the same parameters to be calibrated. In order to assure that an efficient GA implementation was chosen for comparison with the AGA extensive pretests were performed, covering different population structures, selection strategies and reproduction operators [15].

The most efficient GA-implementation uses a single rank-based population. As a selection operator a newly defined *diversity selection* operator was implemented. Instead of allowing multiple selection of one individual within one iteration as it is possible with rank-based selection, diversity selection aims at preserving variability in the gene-pool by using one individual only once during one iteration as a parent, hence increasing the number of different matings. In the integration phase all offsprings are integrated regardless of their quality replacing the m individuals with lowest fitness, where m is the number of offsprings. We will refer to this implementation as Sequential Genetic Algorithm (SGA).

In order to identify a good parameterization all 250 potential combinations for the five crossovers, five selection and ten mutation operators were tested for five classical QAP instances [4]. Then the identical set of operators was submitted to AGA for autoconfiguration. AGA was run using three different neighborhoods (see Figure 16.4) and five different acceptance rates. Table 16.1 summarizes absolute and average behavior for three runs with identical setup.

Our results support hypothesis a) that the AGA outperforms the SGA in terms of absolute and average solution quality (see last two rows in Table 16.1) while requiring at the most 30% of the evaluations performed by the SGA. For testing all 250 parameter combinations in three runs SGA consumed a

Solution types	Evaluation procedures
$p \times v$ daily tours	simple summation
$p \times v$ daily clusters	$p \times v$ TSP's
n delivery combinations	p VRP's

Table 16.2 Options for PVRP implementation

total of 135 million evaluations per instance compared to only 40.5 million evaluations for AGA. For the sko-90-instance 1000 evaluations consumed 3.6 seconds CPU-time on a PC-486-66MHz. In Figure 16.6 we compare the overall best solution (min), the average solution quality (avg) and the result of the best solution in the run with the lowest quality (max) for the sko90-instance. Higher robustness, the ability to produce better results on average, and the superior solution quality as displayed in Figure 16.6 is representative for all five instances tested (see Table 16.1). The absolute gap with best known solutions is at most 1.54% for the AGA, though no local optimizers were implemented (see 7th row in Table 16.1).

16.4.2 The Period Vehicle Routing Problem

The second problem which we have selected for the study of our AGA-implementation is the Period Vehicle Routing Problem (PVRP). While the standard Vehicle Routing Problem (VRP) can be described as finding the set of tours for visiting a given set of customers on a single day such that the total distance traveled is minimal PVRP considers a planning period of consecutive days, a week or a month. Here each customer can specify a set of acceptable or feasible combinations of delivery days by either explicitly giving patterns like "Monday and Thursday or Tuesday and Friday" or by an implicit description through a set of conditions to be met as for instance "twice a week with at least two days between the deliveries". In contrast to VRP, which can be seen as one of the most vital areas of combinatorial optimization, only a small number of publications discuss PVRP. Different mathematical models have been developed and heuristic solution methods proposed. To our knowledge no work on the application of metaheuristics to PVRP has been published. A survey of the work on PVRP and a discussion of the potential use of the PVRP model in DSS for various specific planing, budgeting and decision problems can be found in [12].

We assume that n customers are to be delivered on a period of p days by a fleet of v vehicles. Note that the capacities of the vehicles as well as the total period demand of the customers with its distribution over the days for every feasible delivery combination is given.

While the standard VRP can be interpreted as the combined problem of grouping the customers into so-called clusters served on one tour by the same vehicle and then determining for every cluster an optimal sequence for visiting

Problem	Days	Vehicle	Best Known	AGA best	Δ % Best
50a	2	3	524.6	535.3	2.0 %
50b	5	3	1322.9	1345.5	1.7 %
75a	5	5	840.2	847.4	0.8 %
75b	5	6	2046.2	2220.8	8.5 %
100a	5	4	831.1	843.1	1.4 %

Table 16.3 PVRP-Results

the customer, PVRP can be seen as a three phase problem with the additional first phase to select for every customer a specific, feasible delivery combination and resulting in a standard VRP for every day within the period and followed by the phases of clustering and routing for every daily subproblem.

While for the QAP-coding a permutation represents already a complete solution for the problem, PVRP requires more refined decoding strategies. Based on the three-tier view of PVRP we can distinguish three different strategies of defining an individual exploiting the semantic of a string encoding. Here an individual may represent either a complete solution consisting of clusters with routings for every single day, a set of clusters for every day or a set of feasible delivery combinations.

In the first case the fitness/objective function value can be easily computed. In the second case a traveling salesman tour (TSP) has to be determined for every cluster. This can be done optimally via dynamic programming or approximately through one of the many TSP-heuristics. In the third case we have to solve a set of VRPs. Here again besides optimal approaches a bundle of different heuristics are available. Table 16.2 summarizes the different options and their requirements. In our study we have used the same coding scheme for the three solution types: a permutation with length $m = p \times n$. Applying modular division transforms every entry in the permutation into a unique customer and day combination. Scanning the permutation from left to right a feasible delivery combination for each customer is constructed. Based on this information we can construct for every day the set of customers who receive service on this day. Constructing these sets by sequentially scanning the permutation for a second time and thereby assigning the customers consecutively to a vehicle until the weight or distance limits are reached gives the clusters of vehicles for each day as well as the individual tours if also the order of the customers in the cluster is maintained.

In our test we employed the same five crossover operators and the same mutation operator as for the QAP (see Section 16.4.1). Again we tested the AGA with several configurations for the population size and the number of neighbors. Table 16.3 summarizes the results we obtained on examples which are described in [6] by constructing delivery combinations only and solving the standard VRPs by the classical savings algorithm [7]. The results show that AGA is with the exception of problem 75b, at most 2% from best-known

Characteristic	SACP9	SACP33
# airports	9	33
# aircrafts	2 x DC8, 2 x 747	40 x 747
# Passenger Flights	8	68
# O&D-Markets	26	675
geographic scope	3 continents	global

Table 16.4 SACP characteristics

solutions [5] (see Table 16.3). Yet again, our primary goal was not to identify new best solutions but to prove that a very simple and easy to implement hybrid algorithm based on the concept of AGA is able to produce high-quality results with low computational effort.

Due to the large number of different potential hybridizations for the PVRP that we investigated, we could not enumerate all combinations and parameterizations to identify the best implementation and calibration for the classical SGA. Thus a comparison between AGA and SGA, as it was performed for the QAP, could not be generated for the PVRP.

16.4.3 The Strategic AirCargo Problem

The third problem stems from a leading internationally operating mixed cargo and passenger carrier. To fulfill a market demand given as a set of origin and destination units (O&D) a subset of flight legs from a potential network of operational feasible flight legs has to be identified. Under the constraint that the entire O&D demand can be satisfied and that the selected legs can be connected to feasible aircraft routings (so called rotations) for a given fleet of cargo aircrafts the objective function of this so called Strategic AirCargo Problem (SACP) is to minimize network costs [15]. This problem can be decomposed into two separate problems: the schedule construction and the schedule evaluation problem. The schedule construction problem consists of determining on which days in the week which aircraft should fly which route at what time. This problem is restricted by a number of operational constraints, such as minimum ground times for loading and refueling activities, payload-range constraints etc. The schedule evaluation problem is to find feasible freight routings for the cargo on the capacitated time-space-network generated by the schedule generation process and available belly-capacity on passenger flights.

The need for the solution of such problems from scratch arises in the airline's planning departments when new network structures or fleet decisions are investigated as it was the case with the SACP9-instance or the analysts have to evaluate potential benefits of strategic alliances with integration of schedules of the partner airlines (SACP33-instance). Basic descriptions of the instances are given in Table 16.4. In both cases, incrementally changing existing schedules

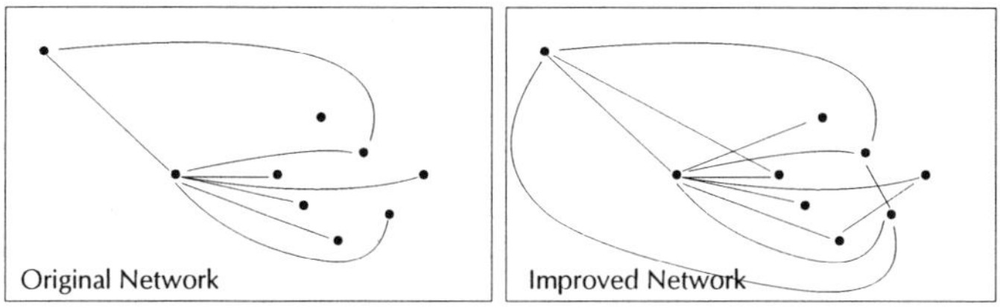

Figure 16.7 SACP9 Solution

is not a feasible approach. Schedules have to be built from scratch, which is a non-trivial task and can consume weeks of manual planning.

The AGA-concept was applied in a feasibility study to support this planning task with computer based algorithms. According to the decomposition of the problem structure the solution approach was also divided into a schedule construction phase applying AGA and a simple transport heuristic based on a greedy strategy for the schedule evaluation phase.

As a coding scheme t permutations of length n, with t being the number of days in the planning horizon ($t = 7$) and n the number of stations to be served, with two entries per position (one station key and one aircraft key) are concatenated to one chromosome. Each permutation encodes a sequence of stations as indexed in the station key. The aircraft key is a boolean variable for each individual aircraft indicating whether this station can be part of this aircraft's route. The aircraft key itself is encoded using an explicit coding table to allow multiple interpretation of an airport code in the permutation. In the decoding phase the permutations are scanned from left to right for each aircraft. The algorithm starts by identifying the first feasible station for an aircraft according to the aircraft key and then identifies the next feasible station in the permutation checking if a flight between these two stations is possible at this time. This operation is repeated until the maximum flight time of the aircraft is exceeded or the next part of the code (chromosome) represents another day. All t days of the planning period are scanned in this fashion generating a giant tour for each aircraft. At the end the algorithm checks if the first station is also the last station of this tour to assure a feasible round trip for each aircraft. If this is not the case a placement flight is performed, and if this is not possible due to the violation of a time or routing constraint the algorithm starts a backtracking search until the round trip constraint is fulfilled. Technically, this coding-scheme is implemented using the random key concept first presented in [2]. This random-key concept allows the utilization of simple crossover-operators like the n-Point-Crossover [11, 9] or Coin-Toss-Crossover [2] for always producing feasible solutions. We implemented three n-Point-Crossover operators (1PX - 3PX) and five Coin-Toss-Crossover operators with bias-probabilities of 50% to 90% (C5X - C9X). In addition a special purpose crossover operator was designed that performed a swap in each of the t segments

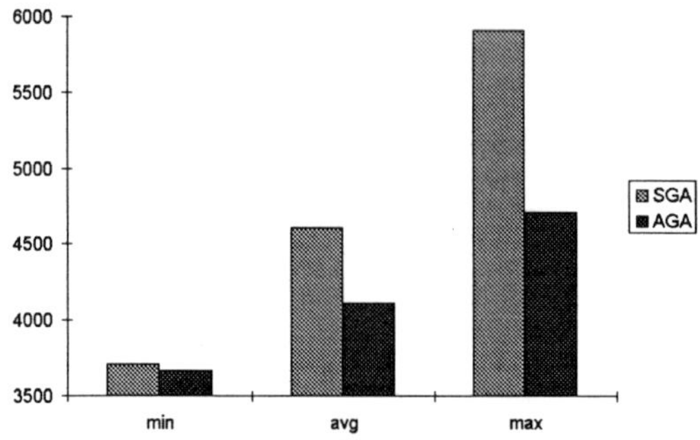

Figure 16.8 SGA vs. AGA, SACP9

of the code (SSX = Segment Swap Crossover [15]). The mutation operator performed t swaps of positions in the coding-scheme.

On the small SACP9-instance we performed extensive tests in which we compared the performance of the SGA and the AGA by probing all 630 potential parameter combinations of the nine crossover, seven mutation and ten selection operators with five repetitions of each run. The test was based on the SGA-implementation as described in Section 16.4.1. The AGA outperformed the SGA and proved to be robust against the variation of the threshold τ and the number of neighbors in the population topology. Again, Figure 16.8 demonstrates the robustness and the superior solution quality of AGA for the SACP9-instance. SACP was extensivly studied in [15]. The tests for SACP9 were performed on a PC-486-66MHz and showed a drastic difference between the necessary computational effort for AGA compared to SGA. While the AGA runs consumed only about one day of CPU-time, the complete enumeration of the equivalent sets of parameter-combinations for the SGA required approximately three months of CPU-time, which is obviously inacceptable for any industrial application.

Concerning the quantitative results of the SACP9 the AGA was able to reduce the network costs by five percent compared to the company's original network structure. In addition to that one DC8 aircraft could be saved and utilized for other purposes increasing overall productivity and the average load factor of the remaining aircrafts. As a qualitative result AGA consistently proposed a new network structure leading to a change in the analyst's Hub & Spoke planning paradigm by introducing more direct flights and collecting cargo on round trips before transferring it to the main hub as displayed in Figure 16.7. Since in the SACP33-instance a future scenario for a strategic alliance was simulated no comparison to actual networks is available. The generated schedule proposes the installation of hubs different from the hubs being operated today by the individual carriers and saved two wide-body aircrafts of type Boeing 747 worth 100 Mio. US$ each. Since this strategic problem is still under investigation no final results can be published at this point.

16.5 CONCLUSIONS

The great appeal of GA is the ease of the implementation and the wide range of problem classes that can be solved if only a coding-scheme can be developed. The goal of this study was to increase the ease and acceptance of applying GA-concepts in practice by eliminating one of the greatest drawbacks of the application of GA, the calibration problem. We introduced a method for dynamic autoconfiguration of those parameters with the greatest interaction and impact on the solution quality by presenting the concept of Adaptive Genetic Algorithms (AGA).

Originally AGA was designed to reduce the time for calibration and to achieve a higher degree of generalizability. Yet, we could empirically show that the AGA with the concept of dynamic autoconfiguration even outperforms the classical GA with respect to robustness and quality of the solutions (hypothesis a). Also our experience while developing the implementations for the three problem classes supports hypothesis b), that a rapid and efficient implementation is possible, since the same *tool-box* could be applied to different combinatorial optimization problems successfully. The results on the SACP give strong evidence that there is also no reason to reject hypothesis c), i.e., that results of high impact for unstructured real world managerial problems can be obtained with reasonable computational effort and in a short period of development time.

Our research on the application of AGA to the problems studied in this report is still far from being complete. We plan to investigate other implementations of the AGA using different coding schemes for the problem classes examined in this report. Also the extension to other problem classes will clarify whether the promising results achieved in the present study can be generalized to a broader range of combinatorial optimization problems.

References

[1] Baluja S. Structure and performance of fine-grain parallelism in genetic search. In Forrest S., editor, *Proc. 5-th Int. Conf. Genetic Algorithms*, 152–155, San Mateo, 1993. Morgan Kaufmann Publishers.

[2] Bean J.C. Genetic algorithms and random keys for sequencing and optimization. *ORSA J. Comput.*, 6:154–160, 1994.

[3] Bierwirth C. *Flowshop Scheduling mit parallelen Genetischen Algorithmen.* Dt. Univ.-Verlag, Wiesbaden, 1993.

[4] Burkard R.E., Karish S., and Rendl F. QAPLIB-A quadratic assignment problem library. *European J. Oper. Res.*, 55:115–119, 1991.

[5] Chao I., Golden B.L., and Wasil E. An improved heuristic for the period vehicle routing problem. *Networks*, 26:25–44, 1995.

[6] Christofides N. and Beasley J.E. The period routing problem. *Networks*, 14:237–256, 1984.

[7] Clarke G. and Wright J.W. Scheduling of vehicles from a central depot to a number of delivery points. *Oper. Res.*, 12:568–581, 1964.

[8] Fleurent C. and Ferland J.A. Genetic hybrids for the quadratic assignment problem. In Pardalos P.M. and Wolkowicz H., editors, Quadratic Assignment and Related Problems, DIMACS, 16:173–187, 1994.

[9] Goldberg D.E. *Genetic Algorithms in Search, Optimization and Machine Learning.* Addison Wesley, 1989.

[10] Gordon V.S. and Whitley D. Serial and parallel genetic algorithms as function optimizers. In Forrest S., editor, *Proc. 5-th Int. Conf. Genetic Algorithms*, 177–183, San Mateo, 1993. Morgan Kaufmann Publishers.

[11] Holland J.H. *Adaptation in Natural and Artificial Systems.* University of Michigan Press, Ann Arbor, 1975.

[12] Kabath M. *Das Konzept der periodischen Tourenplanung - Potentiale und Einsatzmöglichkeiten in Entscheidungsunterstützungssystemen für die Distributionslogistik.* Shaker, Aachen, 1997.

[13] Reeves C.R. Using genetic algorithms with small populations. In Forrest S., editor, *Proc. 5-th Int. Conf. Genetic Algorithms*, 92–99, San Mateo, 1993. Morgan Kaufmann Publishers.

[14] Schaffer D.J., Caruana R.A., Eshelman L.J., and Das R. A study of control parameters affecting online performance of genetic algorithms for function optimization. In Schaffer D.J., editor, *Proc. 3-rd Int. Conf. Genetic Algorithms*, San Mateo, 1989. Morgan Kaufmann Publishers.

[15] Zils M. Genetische Algorithmen zur strategischen Flotteneinsatzplanung in der Luftfrachtindustrie. Master's thesis, WINFORS, University of Cologne, 1996.

17 THE LAVISH ORDERING GENETIC ALGORITHM

Emanuel Falkenauer

Department of Applied Mechanics, Brussels University (ULB)
50, avenue F. Roosevelt, B-1050 Brussels, Belgium.
efalkena@ulb.ac.be

Abstract: It has often been assumed that ordering problems form a special class of problems amenable to a largely uniform treatment. Since successful *ordering* Genetic Algorithms (GAs) have been defined for optimization of some ordering problems, it has become customary to assume that ordering problems are well solved by ordering GAs fitted with "standard" operators.

The success of the standard ordering GAs on some problems led to an indirect approach of solving numerous combinatorial problems as follows: define a *decoder* of a permutation of items that builds a solution to the original problem, and use the ordering GA to find a permutation that decodes into a good solution of the original problem. Such lavish applications of the standard ordering GA have become pervasive to many types of problems.

In this paper, we construct a very simple way of transforming a function defined over bitstrings to a function defined over permutations, and vice versa. As a result, there is a simple way of transforming any function to an "ordering problem".

The implication of this is that should there indeed be a standard way of efficiently solving ordering problems, then the standard ordering GA would be capable of efficiently optimizing all functions. That, however, would contradict the No Free Lunch Theorem for search. Consequently, many ordering problems are not amenable to solution by standard ordering GAs, i.e., the lavish use of the standard ordering GA is unwarranted.

We support this claim by experiments in which we attempt to optimize the easy 32-bit ONE-MAX function with an ordering GA, using the simple transformation between permutations and bitstrings. As expected, standard ordering crossover operators perform extremely poorly on that simple function. In order to illustrate how much an ordering crossover for ONE-MAX would have to differ from the standard ones, we then construct an "exotic" tailored crossover that performs well.

17.1 INTRODUCTION

It has often been assumed that ordering problems, i.e., those in which the aim is to find a good *permutation* of a set of items, form a special class of problems amenable to a largely uniform treatment. Given the fact that reasonably successful *ordering* Genetic Algorithms (GAs) have been defined in the past for optimization of some ordering problems, such as the Traveling Salesman Problem (TSP), it has become customary to assume that ordering problems are well solved by the ordering GAs fitted with "standard" operators.

The success of the ordering GAs on some problems led to an indirect approach of solving numerous combinatorial problems the following way: define a *decoder* of a permutation of items that builds a solution to the original problem, and use the ordering GA to find a permutation that leads, after decoding, to a good solution of the original problem.

The arguments of validity of this indirect approach have usually been based on Goldberg's theory of *o-schemata*, the ordering equivalents of Holland's schemata [4]. That theory uses a reformulation of the Schema Theorem, showing that the theorem holds for o-schemata as well, at least for the Partial Matching Crossover (PMX) of Goldberg. Based on that theory, it has been assumed that the standard ordering GA was *efficiently* exploring a search space of permutations, and the indirect approach followed.

In spite the fact that *tailored* ordering crossover operators, such as the Edge-based Crossover of Whitley [7], have been proposed for some ordering problems and shown to yield significantly better results than "standard" ordering crossovers, lavish applications of the standard ordering GA to all sorts of problems have become pervasive.

We have previously shown that the indirect approach cannot work on the special class of grouping (or partitioning) problems [1]. Indeed, the ordering GA, fitted with a decoder to yield solutions to a grouping problem, is unable to capture the regularities of these problems. The experimental results confirmed our skepticism. Despite this, grouping problems were perhaps the favorite target of the indirect approach.

In this paper, we go further still. In Section 17.3, we construct a very simple way of transforming a function defined over bitstrings to a function defined over permutations, and vice versa. As a result, there is a simple way of transforming any function to an "ordering problem".

The implication of this is that should there indeed be a standard way of efficiently solving ordering problems, then the standard ordering GA would in fact be capable of efficiently optimizing all functions. That, however, would contradict the No Free Lunch theorem for search of Wolpert and Macready [8]. Consequently, it must be the case that many ordering problems are *not* amenable to solution by standard ordering GAs, i.e., the lavish use of the ordering GAs is unwarranted.

We support this claim by experiments in which we attempt to optimize the GA-easy 32-bit ONE-MAX function with an ordering GA, using the simple transformation between bitstrings and permutations. In Section 17.4, we use two standard crossovers, with extremely disappointing results. In Section 17.5, we then design an "exotic" crossover based on the structure of the target problem. Although it is unclear whether o-schemata are transmitted at any reasonable rate by that crossover, the performance is excellent. We draw conclusions from the experiments in Section 17.6.

17.2 ORDERING GAS WITH DECODER

There are many hard optimization problems that are naturally cast as "ordering problems". Perhaps the best known example is the Traveling Salesman Problem (TSP) [2], where the aim is to find an *order* of visiting each of a set of cities exactly once, such that the total cost of the travel is minimal. In short, the aim in TSP is to find the best *permutation* of the cities.

An ordering GA searches for a good permutation of items. It usually uses the permutations as representation of solutions, but not necessarily [5, 7]. Since the representation is usually quite different from simple bitstrings, several crossover and mutation operators have been proposed for ordering GAs (as opposed to the approach proposed by Holland). While most of the ordering GAs have been designed for the TSP and closely related problems (e.g., flowshop scheduling), they have quickly been picked up as a means of addressing various *other* problems. The approach is based on the observation that for many combinatorial problems, heuristics exist that construct different solutions to the problem depending on the order in which the entities involved are handled. In this context, the heuristic can be seen as a *decoder* that transforms a permutation of items into a solution to the problem being optimized.

In this approach, the permutation is seen as the genotype that the decoder transforms into a solution to the original problem (the phenotype), just as the DNA in plants and animals must be decoded into the living creature by the developmental process during gestation.

Perhaps the oldest use of that approach can be found in Smith [6]. In that paper, a variation of the two-dimensional Bin Packing problem is considered: a set of rectangular items of various sizes must be packed into a given rectangular area. Smith proposed to tackle the problem by taking the items one by one and putting each item as close as possible to the items already in the area, following a packing heuristic. The solution generated thus depends on the order in which the items are packed, and Smith used an ordering GA to find the order (permutation) of items that leads to the best packing.

Since [6], this indirect approach has been used to optimize other difficult problems, including one-dimensional Bin Packing, Equal Piles (or Load Balancing), Graph Coloring, Quadratic Assignment, Scheduling with Resource Allocation, and many others.

While the nature of the problems being tackled in this indirect manner varies widely, most of the ordering GAs used for finding a good genotype permutation rely on quite a standard set of genetic operators. The rationale usually offered is that those operators have previously been found to be efficient in searching the space of permutations (often for TSP), which means that they are good in solving "ordering problems". Consequently, since the problem at hand is cast as an "ordering problem", those operators should equally efficiently find a permutation that decodes into a good solution to the problem being solved. In order to show that such an approach has in fact *little* chance of working correctly in general, we construct one more decoder below.

17.3 FROM ORDER TO BITS AND BACK

There are numerous ways of transforming a permutation to a bitstring and vice versa. One computationally fast method is the following transformation between permutations of $N+1$ items and bitstrings of N bits.

17.3.1 Permutation to bitstring

Given a permutation of $N+1$ items, the corresponding bitstring of N bits is constructed as follows. Starting with the empty bitstring, the first N items in the permutation are scanned from the left to the right. Each time the next item in the permutation is smaller than the current item, a 1 is appended to the bitstring, otherwise, a 0 is appended. For instance, the permutation 4 5 6 7 8 3 2 1 0 transforms into the bitstring 0 0 0 0 1 1 1 1. Calling the transformation *p2b* (for Permutation-to-Bitstring), we write

$$p2b(4\,5\,6\,7\,8\,3\,2\,1\,0) = 0\,0\,0\,0\,1\,1\,1\,1.$$

17.3.2 Bitstring to permutation

Given a bitstring of length N, the corresponding permutation of $N+1$ items (numbered 0 through N) is constructed as follows. Set L and C to the number of 1s in the bitstring. Set the first item in the permutation to L, and then scan the bitstring from left to the right. For each 1 encountered, decrement C and append the resulting number to the permutation. For each 0 encountered, increment L and append the resulting number to the permutation. For instance, the above bitstring, 0 0 0 0 1 1 1 1, is transformed as follows:

Step one:	L=C=4	Permutation: 4 ...
Step two:	L=5	Permutation: 4 5 ...
Step five:	L=8	Permutation: 4 5 6 7 8 ...
Step six:	C=3	Permutation: 4 5 6 7 8 3 ...
Step nine:	C=0	Permutation: 4 5 6 7 8 3 2 1 0.

Calling the transformation *b2p* (for Bitstring-to-Permutation), we write

$$b2p(0\,0\,0\,0\,1\,1\,1\,1) = 4\,5\,6\,7\,8\,3\,2\,1\,0.$$

Of course, since there are $(N+1)!$ different permutations of $N+1$ items and only 2^N bitstrings of length N, the transformation is not (and cannot be) bijective. For instance, the permutation 0 1 2 3 8 7 6 5 4 also transforms to the bitstring 0 0 0 0 1 1 1 1. Thus in this transformation, several permutations transform to one bitstring. In other words, *b2p* selects for each bitstring B one of the permutations P for which $p2b(P) = B$.

17.4 SOLVING ONE-MAX WITH A STANDARD ORDERING GA

The important point of the *p2b* transformation is that each permutation transforms into one bitstring. On the other hand, each bitstring transforms to one permutation, i.e., for every bitstring there is a corresponding permutation. Consequently, given a function defined over bitstrings, the transformation represents a way of optimizing the function with an *ordering* GA as follows.

Given a function $\mathcal{F}$ defined over N bits, set up an ordering GA that searches for permutations of $N+1$ items. Each time a permutation P is evaluated, the value $\mathcal{F}(p2b(P))$ is used for the value of the permutation, i.e., the permutation is *decoded* to yield the bitstring that represents a solution to the original problem. Needless to say, an optimal permutation P_{opt} found by the ordering GA is one that satisfies $\mathcal{F}(p2b(P)) = \text{optimum}(\mathcal{F})$, and the corresponding optimal bitstring is $p2b(P_{opt})$.

We applied this strategy to the 32-bit ONE-MAX problem, where the objective function to be maximized (or minimized) is simply the number of 1s in the bitstring. Note that ONE-MAX is a notoriously *trivial* function with no epistasis or deception.

In the experiments, we used two standard ordering crossover operators, the attribute "standard" being due to the rate at which they seem to be used by GA practitioners. The mutation operator was the standard permutation of two items selected at random.

The first crossover was the PMX (Partially Mapped Crossover) of [3]. It proceeds as follows. Given two parents, a crossing section is selected at random, e.g.,

$$0\,1\,2\,3\,|\,4\,5\,6\,7\,|\,8\,9$$

and

$$9\,1\,7\,3\,|\,5\,4\,8\,2\,|\,6\,0.$$

The corresponding genes in the crossing section define a mapping, in this case $4 \leftrightarrow 5$, $5 \leftrightarrow 4$, $6 \leftrightarrow 8$ and $7 \leftrightarrow 2$. The first child is constructed by copying the crossing section from the first parent and all the genes from the second parent that do not appear in the crossing section of the first, yielding

$$9\,1\ \ \,3\,|\,4\,5\,6\,7\,|\ \ \,0.$$

Finally, the remaining places are filled by the mapping's images of the genes that previously occupied those positions in the second parent. In this case, this means inserting 2 for the 7 that disappeared from the third position, and inserting 8 for the 6 that disappeared from the ninth position, yielding

$$9\,1\,2\,3\,|\,4\,5\,6\,7\,|\,8\,0.$$

The second child is obtained by permuting the roles of the parents.

The second crossover operator we used simply imposes the order of the items in the crossing section in one parent, to the corresponding items of the second parent. Thus for instance, the parents

$$0\,1\,2\,3\,|\,4\,5\,6\,7\,|\,8\,9$$

and

$$9\,1\,7\,3\,5\,4\,8\,2\,6\,0,$$

would yield the child

$$9\,1\,4\,3\,5\,6\,8\,2\,7\,0.$$

Despite the purported efficiency with which the standard ordering crossovers explore the space of permutations, their performance on the easy ONE-MAX function was truly disastrous: even after *20 million* function evaluations, neither crossover was able to set all of the 32 bits to one. Note that the figure is conservative, as we simply stopped the GA after that number of evaluations, suspecting that finding the optimum would be impractical.

Note that the crossover operators used *do* transmit o-schemata, the ordering equivalents of Holland's schemata (PMX transmits well the "type a" o-schemata, while the other crossover transmits well the "type r", see [3]). Thus according to the theory, this approach *should* work satisfactorily.

17.5 THE BIT-BASED ORDERING CROSSOVER

We then designed the following "exotic" ordering crossover, which we called BIT (for BIT-based). First compute the bitstrings corresponding to the parental permutations via the *p2b* transformation described in Section 17.3.1. Then apply the classic two-point crossover to those bitstrings, yielding two child bitstrings. Finally, transform the child bitstrings using the transformation *b2p* described in Section 17.3.2 to obtain the child permutations.

It is unclear to what extent the *o-schemata* are transmitted by BIT. Nevertheless, switching to BIT from the classic crossover operators while keeping the rest of the GA unchanged led to a stunning performance: over a sample of 20 trials, the GA with BIT was able to set all of the 32 bits to 1 in just 1816 function evaluations. That was under *2 seconds* per optimization on a modest 166 MHz Pentium PC. Note that no local optimization was used, which means that the performance was only due to the recombining power of the BIT crossover.

It is important to note that the non-bijectivity of the above transformation does not invalidate the example, since evaluating the cost function via the transformation to bitstrings is as valid as any other method of assigning integers to permutations. Even more importantly, most (if not all) GAs using the ordering-GA-plus-decoder approach suffer the same problem, which is indeed one of the drawbacks of the approach.

17.6 CONCLUSIONS

We have questioned the widely held belief that ordering GAs, using "standard" crossover operators, efficiently search the space of permutations *whatever* the actual ordering problem being solved. Based on that belief, the ordering GAs have been lavishly applied to all sorts of problems by use of a decoder mapping each permutation to a solution of the original problem.

We experimented with that approach by designing a decoder that transforms a permutation to a bitstring. We then applied the resulting "*ordering* GA for bitstrings" to the trivial 32-bit ONE-MAX function. The performance of the standard ordering GA on ONE-MAX is simple to summarize: that approach does not work.

We then designed an "exotic" ordering crossover operator, BIT, which takes into account the actual problem being solved. Although it is unclear to what extent o-schemata are being transmitted by BIT, the ONE-MAX became trivial for the ordering GA.

Of course, the BIT operator does not and is not intended to have any practical value: a simple standard bit-based GA with two-point crossover would achieve exactly the same performance on ONE-MAX. We constructed the bit operator solely for the sake of illustration of how **NON**standard an ordering operator would have to be in order to perform correctly on the trivial ONE-MAX problem.

In these experiments, we took a *simple* optimization problem, transformed it into an "ordering problem", and applied standard ordering GAs. Despite the popular belief of validity of such an approach, the results were disastrous.

The conclusion, as suspected, is that there is no such thing as an "ordering problem", not in the "solvable by a standard ordering GA" sense. Lavish applications of the standard ordering GA to various problems using a decoder of permutations, based solely on the assumption that the GA efficiently solves the resulting "ordering problem", are unwarranted.

The failure of standard ordering crossover operators and the success of the BIT operator on the ONE-MAX function confirms once again that in a successful GA, the operators *must* be designed in such a way that they exploit whatever structure there may be in the problem being optimized, be it an "ordering problem" or not. Indeed, the indisputable validity of the Schema Theorem (and its ordering equivalent) does *not* guarantee a successful GA as long as the schemata it manipulates do not represent meaningful regularities of the search space.

References

[1] E. Falkenauer. A new representation and operators for GAs applied to grouping problems. *Evolutionary Computation*, 2:123–144, 1994.

[2] M.R. Garey and D.S. Johnson. *Computers and Intractability – A Guide to the Theory of NP-completeness.* W. H. Freeman, San Francisco, CA., 1979.

[3] D. Goldberg. *Genetic Algorithms in Search, Optimization and Machine Learning.* Addison–Wesley Publishing Company Inc., 1989.

[4] J.H. Holland. *Adaptation in Natural and Artificial Systems.* University of Michigan Press, Ann Arbor, MI., 1975.

[5] N.J. Radcliffe and P.D. Surrey. Fitness variance of formae and performance prediction. In D. Whitley and M. Vose, editors, *Foundation of Genetic Algorithms 3 (FOGA–94)*, pages 51–72, San Francisco, CA., 1994.

[6] D. Smith. Bin packing with adaptive search. In J.J. Grefenstette, editor, *Proceedings of an International Conference on Genetic Algorithms and their Applications*, pages 165–170, Hillsdale, NJ., 1985. Carnegie-Mellon University.

[7] D. Whitley, T. Starkweather, and D'A. Fuquay. Scheduling problems and traveling salesmen: The genetic edge recombination operator. In D.J. Schaffer, editor, *Proceedings of the Third International Conference on Genetic Algorithms*, pages 133–140, San Mateo, CA., 1994.

[8] D.H. Wolpert and W.G. Macready. No free lunch theorems for search. Technical Report SFI-TR-95-02-010, The Santa Fe Institute, 1995.

18 FITNESS LANDSCAPES AND PERFORMANCE OF META-HEURISTICS

Cyril Fonlupt[1], Denis Robilliard[1],
Philippe Preux[1] and El-Ghazali Talbi[2]

[1]LIL, Université du Littoral-Côte d'Opale
BP 719, 62228 Calais Cedex, France.
fonlupt@lil.univ-littoral.fr

[2]LIFL, Université de Lille 1, Bâtiment M3
59655 Villeneuve d'Ascq Cedex, France.
talbi@lifl.fr

Abstract: We perform a statistical analysis of the structure of the search space of some planar, euclidian instances of the traveling salesman problem. We want to depict this structure from the point of view of iterated local search algorithms. The objective is two-fold: understanding the experimentally known good performance of metaheuristics on the TSP and other combinatorial optimization problems; designing new techniques to search the space more efficiently. This work actually led us to design a hybrid genetic algorithm that competes rather well with other local search heuristics for the TSP, notably Jünger et al.'s version of ILK. This work also opens promising horizons to the study of other combinatorial optimization problems such as the quadratic assignment problem.

18.1 INTRODUCTION

Metaheuristics have shown their ability to reach good approximations of optimal solutions to numerous $\mathcal{NP}$-hard problems. Some authors have tried to explain these performances by a study of the topology of the research space associated to the problem (see e.g. [12, 6, 31]). We wish to go on further on this kind of studies and use them to design better search methods.

Metaheuristics come in different brands, such as simulated annealing [18], tabu search [8, 9], evolutionary algorithms (including all variants such as genetic algorithms [10], evolution programming [26]), ant colony systems [5] being the most famous. (See [28, 1] for recent reviews of metaheuristics.) Basically, metaheuristics are composed of a loop. The body of the loop is based on the following elements:

- a current state which is a point, or a collection of points of the research space,
- a neighborhood defined by some operator Ω that is applied on the current state to get the set of potential next states. To cope with the case of algorithms involving more than one operator such as evolutionary algorithms, we consider in this case that Ω is the composition of the different operators. This operator can be quite complicated. It might even be a metaheuristic.
- a transition rule (also called a pivoting rule, or a selection strategy) which chooses the actual next state among the set of potential next states,
- a halting criterion to decide when to stop the algorithm.

The initial state is either taken at random in the research space, or obtained from a construction heuristic.

The theoretical background of local search is not yet shedding much light on their actual behavior [35, 36]. Given a starting point, we can neither predict the number of iterations that will be done before reaching a local optimum, nor predict the quality of this local optimum. Assuming $\mathcal{P} \neq \mathcal{NP}$, the number of iterations might grow exponentially with the size of the problem, and there are local optima of arbitrary bad quality. We have no formal result about the proportion of local optima, nor the number of global optima, nor their distribution in the search space.

However, experimental studies have shown that the situation is not so hopeless in real applications. Indeed, in the case of the Traveling Salesman Problem (TSP), metaheuristics are providing excellent local optima: for instances of up to 10^5 cities, solutions which length are not more than 1% in excess to that of the shortest tour length are reached within minutes on typical workstations [14]. Metaheuristics do also perform very well on problems such as vehicle routing problem, quadratic assignment problem, graph coloring problem, set partitioning problem, VLSI layout design, ...

Facing these experimental evidences, we have felt that it would be useful to understand the performance of metaheuristics. To meet this goal, we study the dynamics of a great number of walks of a very simple metaheuristic as it is solving different instances of a problem. This metaheuristic is an ascent walk (or descent whether maximizing or minimizing the objective function): in the previous algorithmic schema, the transition rule either deterministically chooses the best next state (SDW: Steepest Descent Walk), or one better neighbor (RDW: Random Descent Walk). In both cases, the walk stops when no more improvement is possible from the current state. The current state is then a local optimum for Ω. Though very simple, this algorithm is at the basis of metaheuristics, so that our study is relevant with regards to the goal. Both tabu search and simulated annealing have much in common with these algorithms, though the pivoting rule is not so crude in their cases. The metaheuristic is probing the search space and we collect information along the walk in order to figure out what the landscape looks like. In this paper, we study the TSP. We observe that all walks reach a certain, very small, area of the search space, where lots of local optima are concentrated around one, or more, shortest tour(s). In itself, this result is not new since the discovery of a "deep valley" by Kirkpatrick and Toulouse in 1985 [19]. However, the way we obtain this result is very different, relying on statistics. One may feel that this approach is very weak and does not prove anything. Though perfectly aware of this weakness, we think that the fact that we obtain the same conclusions as earlier works is re-assuring. We also get some confidence from the fact that when we use this knowledge to derive a new search algorithm to take advantage of it, we obtain an algorithm which performance can be compared with some of the best heuristics to date, such as ILK. Furthermore, we have since then used the same approach to study the structure of an other combinatorial problem (the Quadratic Assignment Problem). Again, we are able to use the knowledge of the structure of the problem to improve search algorithms. So we feel confident that this approach to study the structure of search space can be used to tackle many combinatorial optimization problems, even if one has to be very careful with the conclusions he/she can draw from it.

Weinberger [32] has initiated studies on random walks (in a random walk, the pivoting rule takes a neighbor at random) in various search spaces. Though this algorithm is completely useless for the purpose of optimization, the analysis of these walks is able to bring some interesting information about the structure of the space, the distribution of local optima and their quality. We consider this first work, its followers (such as [33, 30, 13]), and some other works dealing with the study of the correlation of the cost of neighborhing points [23, 11] as complementary to ours.

In the following, we first present our study of the structure of the planar, euclidian TSP. From it, we derive a search algorithm and provide experimental results and comparisons with other metaheuristics in Section 18.3. We conclude by a discussion and perspectives of this work.

18.2 THE STRUCTURE OF THE PLANAR EUCLIDIAN TSP

18.2.1 The Traveling Salesman Problem

The Traveling Salesman Problem has a very long history. Though, as it is now suspected, the TSP is rather atypical in its structure, it is often used due to its intuitive appeal. We define it very briefly, pointing the reader to Lawler et al.'s TSP book [20] for (much) more precisions, and also the book from Jünger et al. [16].

Given a set of n cities $c_{i,0\leq i<n}$ in a d dimensional space, one has to find the shortest tour that goes through each city. From any city, all other cities may be reached. The problem may be defined either by the coordinates of the cities, or by the distance matrix that provides the distance between any two cities. n is the size of the instance.

Obviously, the shortest tour is one that visits each town once and only once. Thus, it is a hamiltonian tour.

In this paper, we restrict ourselves to the study of planar, symmetric TSP where the distance between cities is euclidian. Most studies of the TSP are limited in this manner as most applications of the TSP can be seen as two, or three, dimensional in our physical world. Even in this case, TSP is $\mathcal{NP}$-hard [7]. We name 2ETSP this class of instances of the TSP.

A tour is coded by the set of n edges that it is composed of. There are $N = (n-1)!/2$ different tours when taking account of symmetries and redundancies in the coding of solutions. This is also the number of points in the search space for the algorithms that are considered in this paper.

The basic move of local search is the 2-opt [4] move which consists in removing two edges from a tour and reconstruct a new different tour (there is always one and only one way to reconstruct a new valid tour that is different from the original). The 3-opt proposed by Bock in 1958 exchanges up to 3 edges [21]. We can go on and define the k-opt which exchanges k edges. In practice, values of $k > 4$ are seldomly used. The LK-move, defined by Lin and Kernighan in 1973 [22], is based on the exchange of edges, the number of edges to exchange being evaluated to get the best possible improvement. Since 1973, it is the best move known for the TSP and is at the core of the most successful algorithms to date for the TSP. We will not go further in the description of the LK move and refer the interested reader to [14, 15].

We define two distances to measure the difference between two tours. First, we define an intuitive distance:

Definition 1 *Let $t1$ and $t2$ be two tours. We denote by $\delta(t1, t2)$ the number of edges shared by both $t1$ and $t2$.*

[19] already used such a distance, though normalized by dividing it by n, the size of the instance.

Obviously, we have: $0 \leq \delta(t1, t2) \leq n$ for any two valid tours for an instance of size n.

We would like the distance between two tours not to be merely intuitive, but related to the operator used in the search algorithm. This distance encompasses some sort of "minimal algorithmic difficulty" to obtain a point out of a given point. The definition is stated having 2-opt in mind. It can be straightforwardly modified to deal with any other operator.

Definition 2 *Let $t1$ and $t2$ be two tours. We denote by $\Delta_{2opt}(t1, t2)$ the minimum number of applications of 2-opt to obtain $t2$ out of $t1$.*

Obviously, Δ_{2opt} is symmetric, that is $\Delta_{2opt}(t1, t2) = \Delta_{2opt}(t2, t1)$.

This distance is very attractive but we know of no polynomial way to compute it. So, we use δ as an approximation of Δ as did [19, 2]. [2] prooves the following inequalities: $\delta(t1, t2)/2 \leq \Delta_{2opt}(t1, t2) \leq \delta(t1, t2)$.

This all implies that when in the sequel we say that two tours are close to each other (that is, their distance δ is much smaller than n), then we can go from one to the other in a very small amount of iterations (again, small with regards to the size of the instance).

18.2.2 The landscape of the 2ETSP

The landscape is involved by the operator Ω, that is, the definition of neighborhood used in the algorithm. The walk itself is also due to the pivoting rule.

Let us first define the notion of landscape (also called fitness landscape, energy landscape, or cost surface). Let $\mathcal{G} = (\mathcal{E}, \mathcal{V})$ be the graph which vertices $\mathcal{V}$ are the N points of the search space. The edges of the set $\mathcal{E}$ connect all couple of points (x, y) such that $x \in \Omega(y)$, or $y \in \Omega(x)$, that is two points are connected if one can be obtained by the application of the operator on the other. Then, considering this graph as the ground floor, we elevate each point x to an altitude $z(x)$ equal to its cost, the length of x in our case. We obtain a surface, or landscape, made of peaks, valleys, plateaus, cliffs, ... The search algorithm aims at finding the lowest point(s). Local optima are points such that all their neighbors are higher than them. The problem lies in the difficulty to have a realistic view of this landscape: it is a geometrical object of N dimensions, which turns to be many thousands even for very moderate values of n. Thus, one has to devise a way to investigate the structure of this object to make something of it. The approach we propose is to study the trajectories of a great number of simple steepest or random descent walks and, if statistical significance holds, figure out some properties of the landscape.

We run 10^4 SDW and RDW starting from points sampled at random and we examine the length of the walks, the cost of the local optima that are reached, the distance between each others. We use instances from the TSPLIB [29] for which we know the shortest tour, att48, kroA100, tsp225. Though these instances are very few and of small size, the results obtained here are in accordance with what we observed on many other instances of larger size while running the algorithm described below (see Section 18.3). The number of runs (10^4) was set in order to get some confidence in their statistical significance. We obtain the following conclusions:

1. starting from a tour t_{init}, the length of the walk is nearly equal to $\delta(t_{\text{init}}, t_{opt})$, and the length of random descent walk is nearly four times longer (t_{opt} is the shortest tour as provided in the TSPLIB). According to Johnson [14], the length of SDW scales with n. However, for the small instances we use, this does not contradict our observations.

2. all local optima are gathered in a somewhat restricted region of the search space. For all local optima that are reached t_{lo}, $\delta(t_{\mathrm{lo}}, t_{opt}) < n/3$ (see Fig. 18.1). Furthermore, the distance between the local optima themselves is small. As the number of tours that are at a certain distance from t_{opt} scales like $n^{\delta(t,t_{opt})}$, it is clear that the region of the space where $\delta(t, t_{opt}) < n/3$ only holds a minority of points of the search space, resulting in a high concentration of local optimum in this region. Such a concentration was dubbed as a "deep valley" by Kirkpatrick and Toulouse [19], or "massif central" by Kauffman [17]. This experimental result is is accord with [19, 3], as well as with Stadler and Schnabl conjecture that there exists a massif central in the 2ETSP [31].

3. the distribution of the quality of the local optima that are reached by either the SDW and the RDW are identical. The distribution for local optima obtained by multiple runs of the RDW starting from the same initial random tour is also the same.

4. basins of attraction are highly intertwined: walks (either SDW, or RDW) starting from two neighboring points typically ends on two different local optima. From a visual point of view, basins of attraction should not be viewed as crater-like structures. Rather, they are more like very long and very thin canyons.

5. there is no correlation between the distance of the local optimum to the shortest tour and the length of the walk that finds it.

6. there is no correlation between the quality of the local optimum and the length of the walk that finds it.

Statistical/empirical studies such as these have severe drawbacks at first. Relying on intuition that can be faulty, it does not have the certainty of a mathematical proof. One may ask whether the big valley is a mirage or an actual fact. The fact that other authors have all obtained the same conclusion about the valley is on the positive side. Furthermore, to assess our conclusions, we had two distinct approaches. First, we designed instances having several well distinct, optimal tours. This distance between these shortest tours can be tuned from very small (the shortest tours are twin peaks in the landscape) to rather large (two shortest tours share at most 4 common edges), with all intermediate possibilities. Our intuition is that a big valley surrounds a shortest tour. As we expected, the local optima obtained by multiple runs of SDW either gathered in one region around the twin peaks, or they clearly gathered in different areas when the optimal tours are very distant from each others, or, in the intermediate case, the big valley gets wider and wider (the distribution of distances of couples of local optima shows that local optima are at increasing distances from each others, see Fig. 18.2.) Clearly, this all does not make impossible the case where an optimal tour would be left unfound. However, in the instances we used, the multiple runs did not find it. But they did find a "good" region.

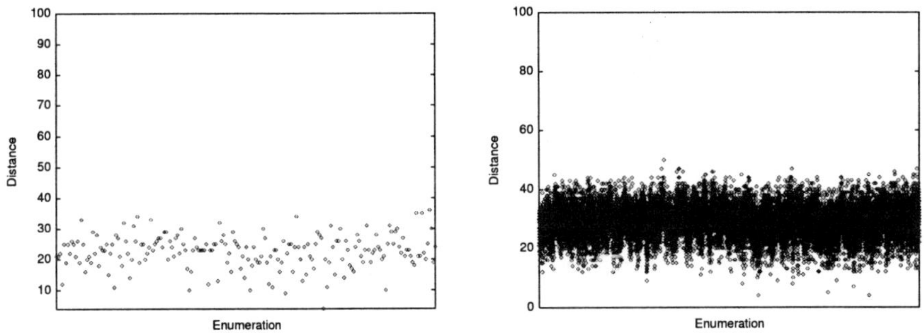

(a) Plot of the distribution of the distances to the optimal tour given in the TSPLIB for the local optima that are found after SDW and RDW. In terms of distance, these local optima are very close to the best tour. The distance between any local optimum and the best tour is less than 1/3 of the size of the instance

(b) Plot of the distribution of the distance between any couple of local optima that are found after the 10^4 RDW. All local optima are concentrated in a small area. The distance between any two local optima is less than 2/3 of the size of the instance.

Figure 18.1 Massif Central of Local Optima. These figures were measured on kroA100. Note that the same results were obtained for every instance we measured.

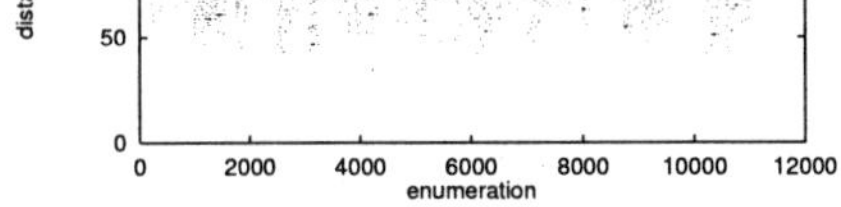

Figure 18.2 Plot of the distribution of the distance between any couple of local optima that are found after 200 RDW. The darker stripe, centered at y-coordinate 100, stands for RDW performed on instance lin318. The lighter cloud, spreading from y-coordinate 50 to 200, stands for RDW performed on a 300 cities instance with 2 global optima sharing only 4 edges.

A second approach relates to our original goal, the design of better heuristics. If using this knowledge about the structure of the search space we are able to design an algorithm that takes advantage of it to find local optima efficiently, this would also be a sign that this structure might be realistic.

Basically, we know that a short tour is to be found within a valley. Thus, the idea is to first find a valley with mere steepst descent, and then work inside the valley by recombining the local optima that have been found. At this stage, we can iterate the whole process to progress further inside the valley. We end-up with some kind of hybrid genetic algorithm. However, we have the advantage over standard approaches (see [27] for a review) to begin recombination in an area of the search space where it is really useful, instead of recombining tours of very low quality. This speeds-up our hybrid GA a lot as will be seen below.

Our approach has much to be compared with Martin et al.'s Chained Local Optimization (CLO) [25, 24] and Johnson's Iterated LK [14]. In the next section, we first present our hybrid GA, briefly review Johnson's version of ILK and Jünger et al.'s ILK, and provide some experimental results comparing the 3 algorithms. Note has to be taken that Johnson's ILK algorithm, together with Martin's CLO, are the best local search algorithms for the TSP to date, according to Johnson's review [14].

18.3 A HYBRID GA THAT SEARCHES THE TSP VALLEY

We have made our benchmarks with small populations of 10 individuals. Before all, the initial generation is optimized with 3-opt before attempting recombination. New individuals are then created by cross-over using a slightly refined version of the well-known enhanced-edge recombination (EER). We don't recall here the details of EER, and refer the reader to [34]. Our modification to EER relies on the fact that edges not belonging to any parents may have to be introduced in order to satisfy the tour constraint. It has been noted that this fact tend to decrease much the fitness of children. To try to limit the effect of this problem, we begin the construction of the tour with the shortest edge common to both parents (so we are never forced to replace it) and, when faced with the constraint, we use a ranked selection to choose the new edge with a probability inversely proportional to its length. Eventually the children are optimized with 3-opt. We use a "malthusian" replacement for the population : firstly we delete doublons, to keep diversity as the population size is small, then parents and children are sorted by fitness and the best individuals form the new population. We have chosen to limit the number of generations in such a way that the number of 3-opt calls is equal to the instance size, in order to compare our results with Johnson's ILK performances given in [14]. Results are shown in the GA column of Tab. 18.3 : average percent excess is given relatively to the optimum values given in the TSPLIB, and is the average computed on at least 10 runs. Results for 3-opt and Jünger et al.'s ILK comes from [16].

Running time are ranging from 40 s. to 10 mn. on Pentium 200 processor. These running times are longer than those of Johnson, but shorter than those of Jünger et al. Clearly, there is still room for improvement when compared

instance	Average percent excess			
	3-opt	Jünger et al.'s ILK	**GA**	Johnson's ILK
lin318	2.67	0.53	**0.41**	0.03
pcb442	3.01	0.90	**0.58**	0.03
u724	4.20	0.67	**0.49**	n.a.
pr1002	3.80	1.51	**0.68**	0.12

to the top of the art heuristic from Johnson. Nonetheless, notice that the performances of our algorithm are rather less sensitive to the increase of instance size than the others. Jünger et al.'s ILK consists in repeating successive independant runs of ILK, then selecting the best result. Johnson's implementation can be roughly described by saying that the result of a run is altered by a 4-opt random move and becomes the new starting solution for the next ILK local search. We think the difference in performance is explained by staying in the big valley (Johnson's and our algorithms), rather than starting again from bad solutions and being trapped by local optima at the border of the valley (Jünger et al.'s ILK) : this gives experimental evidence that a simpler 3-opt, well-used, can beat ILK which is much harder to implement.

18.4 CONCLUSION AND PERSPECTIVES

We have investigated the structure of the search space of instances of planar euclidian traveling salesman problem using 2-opt and 3-opt operators. This study confirms Kirkpatrick and Toulouse study with regards to the existence of a "deep valley" where a large number of local optima are gathered around (a) shortest tour(s). From that point, we do not imply that all local optima are concentrated in the valley. We merely mean that this structure is meaningful for steepest descent algorithms, thus metaheuristics such as tabu search and simulated annealing. Though very minute with regards to the overall search space, the valley still contains a large number of points. This valley is very attractive for steepest descent walks. There might be more than one valley when there are global optima very disimilar from each others. In this case, steepest-descent walks are attracted by one or the others. Basins of attraction seems to be highly intertwined, giving a canyon-like structure to the landscape, rather than a crater-like structure.

A lot of work has still be be done to have a good description of the landscape. It would be very interesting to be able to have indicators to predict the structure of the landscape of a given instance.

Current work on the structure of the QAP seems to imply that the 2ETSP is an extreme case of that of the QAP.

We have used this intuitive knowledge about the landscape of the 2ETSP to design a new hybrid genetic algorithm. We think that recombination is misused in standard GA's. Instead of recombining solutions taken at random, which are typically very bad solutions in the case of 2ETSP, it is much more efficient to first obtain good solutions, and only then, recombine them. These good solutions are obtained via steepest-descent walks started with random points, based on 2, or 3-opt. The local optima that are obtained are then recombined. The key idea is that we first come into the local optima valley and then, the recombination works inside the valley. The new points are optimized again via steepest descent walks giving new, better, local optima that can be recombined again, and so on. After a series of iterations, no improvement is made any longer the algorithm keeps on generating the same local optima at each iteration. This algorithm is very efficient and fast. Instances with many hundreds cities are solved to optimality within a few seconds on standard workstations. Its performance are comparable to that of Martin et al.'s chained local optimization and Johnson's ILK, the two top of the art heuristics for the TSP. A lot of work has still to be done to enhance our algorithm. For example, ILK uses LK while we use 2-opt, or 3-opt; the recombination operator is not very sophisticated: recombining more than two solutions seems attractive; more work on the implementation is needed to tackle instances of many tens of thousands cities. Despite the current limitations due to implementation, we think that our algorithm is promising for the future. Furthermore, based on an analysis of the structure of the problem search space, we think we have used an approach that can be brought to other problems. Some results on the QAP have already been obtained along this way.

References

[1] E. Aarts, J.K. Lenstra (eds), *Local Search in Combinatorial Optimization*, Wiley, 1997

[2] K.D. Boese, *Models For Iterative Global Optimization*, PhD Dissertation, UCLA, USA, 1996

[3] K.D. Boese, A.B. Kahng, S. Muddu, A New Adaptive Multi-Start Technique For Combinatoiral Global Optimizations, *Operations Research Letters*, **16**, 101–113, 1994

[4] G.A. Croes, A Method For Solving The Traveling Salesman Problems, *Operations Research*, **6**, 791–812, 1958

[5] M. Dorigo, L. Gambardella, Ant Colony System: A Cooperative learning approach to the traveling salesman problem, *Evolutionary Computation*, **1**(1), 1997

[6] J. Frank, P. Cheeseman, J. Stutz, When Gravity Fails: Local Search Topology, *Journal Of Artificial Intelligence Research*, **7**, 249–281, 1997

[7] M.R. Garey, D.S. Johnson, *Computers And Intractability: A Guide To The Theory Of NP-Completeness*, Freeman, 1979

[8] F. Glover, Tabu Search: part I, *ORSA Journal on Computing*, **1**, 190–206, 1989

[9] F. Glover, Tabu Search: part II, *ORSA Journal on Computing*, **2**, 4–32, 1990

[10] D. Goldberg, *Genetic algorithms in Search, Optimization, and Machine Learning*, Addison-Wesley, 1989

[11] L.K. Grover, Local Search And The Local Structure Of NP-Complete Problems, *Operations Research Letters*, **12**, 235–243, 1992

[12] A. Hertz, B. Jaumard, M. Poggi de Aragão, Local optima topology for the k-coloring problem, *Discrete Applied Mathematics*, **49**, 257–280, 1994

[13] W. Hordijk, A Measure Of Landscapes, *Evolutionary Computation*, **4**(4), 1996

[14] D.S. Johnson, The Traveling Salesman Problem: A Case Study, in [1], 215–310, 1997

[15] D.S. Johnson, J.L. Bentley, L.A. McGeoch, E.E. Rothberg, *Near-Optimal Solutions To Very Large Traveling Salesman Problems*, to appear

[16] M. Jünger, G. Reinelt, G. Rinaldi, The Traveling Salesman Problem, *Network Models*, Handbooks in Operations Research and Management Science, Volume 7, North-Holland Amsterdam, 225–330, 1995

[17] S. Kauffman, Adaptation on Rugged Fitness Landscapes, and Principles of Adaptation in Complex Systems, in D.L. Stein, *Lectures In The Sciences Of Complexity* volume I, Addison-Wesley, 1989

[18] S. Kirkpatrick, C.D. Gelatt Jr., M.P. Vecchi, Optimization by Simulated Annealing, *Science*, **220**(4598), 671–680, 1983

[19] S. Kirkpatrick, G. Toulouse, Configuration Space Analysis Of The Traveling Salesman Problem, *Journal de Physique*, **46**, 1277–1292, 1985

[20] E.L. Lawler, J.K. Lenstra, A.H.G. Rinnooy Kan, D.B. Shmoys (eds), *The Traveling Salesman Problem, A Guided Tour of Combinatorial Optimization*, Wiley, 1985

[21] S. Lin, Computer Solutions Of The Traveling Salesman Problem, *Bell System Technical Journal*, **44**, 2245–2269, 1965

[22] S. Lin, B. Kernighan, An Effective Heuristic Algorithm For The Traveling Salesman Problem, *Operations Research*, **21**, 498–516, 1973

[23] B. Manderick, M.D. Weger, P. Spiessens, The Genetic Algorithm and the Structure of the Fitness Landscape, in R. Belew (ed), *Proc. International Conf. on Genetic Algorithms*, Morgan-Kaufman, 1991

[24] O. Martin, S.W. Otto, Combining Simulated Annealing With Local Search Heuristics, Annals of Operations Research, **63**, 57–75, 1996

[25] O. Martin, S.W. Otto, E.W. Felten, Large-step Markov Chains for the TSP Incorporating Local Search, *Operation Research Letters*, **1**, 219–224, 1992

[26] Z. Michalewicz, *Genetic Algorithm + Data Structure = Evolution Program*, Springer-Verlag, 1995

[27] H. Mühlenbein, Genetic Algorithms, in [1], 137–171, 1997

[28] C. Reeves (ed), *Modern Heuristic Techniques for Combinatorial Optimization*, Blackwell, 1993

[29] G. Reinelt, TSPLIB: A Traveling Salesman Problem Library, *ORSA Journal on Computing*, **3**, 376–384, 1991. The TSPLIB is available on the web.

[30] P. Stadler, Towards A Theory Of Landscapes, in R. Lopès-Peña, R. Capovilla, R. Garcia-Pelayo, H. Waelbroeck, F. Zertuche (eds), *Complex Systems and Binary Networks*, Springer-Verlag, 1995

[31] P. Stadler, W. Schnabl, The Lansdscape Of The Traveling Salesman Problem, *Physics Letter A*, **161**, 337–344, 1992

[32] E.D. Weinberger, Correlated and Uncorrelated Fitness Landscapes And How To Tell The Difference, *Biological Cybernetics*, **63**, 325–336, 1995

[33] E.D. Weinberger, Local Properties of Kauffman's N-k model: A Tunably Rugged Energy Landscape, *Physical Review A*, **44**(10), 6399–6413, 1995

[34] D. Whitley, T. Starkweather, D. Shaner, Scheduling Problems and Traveling Salesman: The Genetic Edge Recombination Operator, *Proc. 3rd ICGA*, Morgan Kaufmann, 133–140, 1989 Chapter 22, 350–372, 1991

[35] M. Yannakakis, The Analysis Of Local Search Problems And Their Heuristics, in *Proc. STACS'90*, Lecture Notes In Computer Science, **415**, 298–311, 1990

[36] M. Yannakakis, Computational Complexity, in [1], 19–55, 1997

19 A NETWORK-BASED ADAPTIVE EVOLUTIONARY ALGORITHM FOR CONSTRAINT SATISFACTION PROBLEMS

María-Cristina Riff

INRIA-CERMICS, BP. 93. 06902 Sophia-Antipolis, France.
mcriff@sophia.inria.fr

Abstract: We are interested on defining a general evolutionary algorithm that repairs to solve Constraint Satisfaction Problems and which takes into account both advantages of the systematic and traditional methods and of a characteristics of the CSP. We use the knowledge about properties of the constraint network to define a fitness function, and three operators *arc-mutation*, *arc-crossover* and *constraint dynamic adaptive crossover*. The number of constraint checks has also taken into consideration for designing the operators. The algorithm has been tested by running experiments on randomly generated 3-coloring graphs. The results suggest that the technique may be successfully applied to solve CSP.

19.1 INTRODUCTION

In this paper, we focus on two research areas: Constraint Satisfaction Problems (CSP) with finite domains, and Evolutionary Methods. CSPs are NP-hard problems. A (finite domain) CSP consists of a set of variables, their related domains, and a set of constraints between them. The goal is to find a value for each variable, from their respective domains, such that the constraints are satisfied [7, 9, 5]. We are interested in the stochastic approaches to solve CSP. The most popular method in this category is *min-conflicts hill-climbing* [13]. This method creates a complete, but inconsistent assignment and then repairs constraint violations until a consistent assignment is achieved. This method is guided by a simple ordering heuristic for repairing constraint violations. Minton et al. [13] empirically demonstrated that *min-conflicts* is considerably more efficient than traditional constructive backtracking methods. In order to improve the search performance researchers have proposed many different heuristic algorithms [11].

The stochastic methods are proposed, in general, to solve large-scale contraint satisfaction problems. Using an appropriate heuristic algorithm can make a tremendous difference in the time required to solve a CSP, [11].

Evolutionary methods are based on the evolution theory and they are in the category of stochastic search methods for optimization problems [10]. It has been applied to solving Constraint Satisfaction Optimization Problems [20], and CSP [2, 3, 14, 15, 16]. These approaches have concentrated on studying the genetic representation and the reproduction mechanisms. Most of these techniques use fixed representations, operators and control parameters. Since evolutionary algorithms are based on the idea of evolution, it is more than natural to expect some self-adapting and adaptability characteristics of these techniques. Here we incorporate an adapting mechanism into designing a new crossover operator, called *Constraint Dynamic Adaptive Crossover.* On the other hand, the effort of the constraints research community has traditionally attemped to develop techniques for improving the algorithm performance, using the knowledge about the constraints [4, 7, 5, 1], for example, by pruning the search space. In order to use the knowledge on constraints in the design of an evolutionary algorithm we concentrate on the constraint network, which we represent by a constraint matrix.

The remainder of this paper is organized as follows. After defining some notions on CSP and on Evolutionary Algorithm in Section 19.2, an approach to compute the evaluation function for CSP is presented in Section 19.3. The adaptive crossover operator *Constraint Dynamic Adapting Crossover* is presented in Section 19.4. We then address the graph 3-coloring problem subject to the restriction that adjacent nodes in the graph must be colored differently in Section 19.5. In the same section we define a model to estimate the number of constraint checks in order to have an idea of the complexity of our algorithm. A discussion and further issues are given in Section 19.6. We complete the paper with directions for future research and conclusions in Section 19.7.

19.2 PRELIMINARIES

Here we will give the basic notions on CSP, constraint matrix, instantiation and partial instantiation. Then we will present the structure of an evolutionary algorithm and its components, that will be useful in the following sections of the paper. We restrict our attention here to binary CSPs, in which the constraints involve two variables. It is possible to convert CSP with n-ary constraints to another equivalent binary CSP [12].

19.2.1 Concepts on CSP

A *Constraint Satisfaction Problem* (CSP) is composed of a set of *variables* $V = \{X_1, \ldots, X_n\}$, their related *domains* $D_1, \ldots, D_n$ and a set θ containing η *constraints* on these variables. The domain of a variable is a set of values to which the variable may be instantiated. The domain sizes are $m_1, \ldots, m_n$, respectively, with $\mathbf{m}$ being the maximum of the m_i. Each variable X_j is *relevant* (in the sequel we denote "*being relevant for*" by $\triangleright$), to a subset of constraints $C_{j_1}, \ldots, C_{j_k}$ where $\{j_1, \ldots, j_k\}$ is some subsequence of $\{1, 2, \ldots, \eta\}$. A constraint which has exactly one relevant variable is called a *unary constraint.* Similarly, a *binary constraint* has exactly two relevant variables. A binary CSP is associated with a constraint graph, where nodes represent variables and arcs represent constraints. If two values assigned to variables that share a constraint are not among the acceptable value-pairs of that constraint, an *inconsistency* or constraint violation has occured. A solution to the CSP consists of an instantiation of all the variables which do not violate any constraint, i.e., a consistent labeling of each variable with a value from its domain. The simplest algorithm is the brute force algorithm (generate and test), which simply tries every possible combination of values.

Definition 19.2.1 *(Constraint Matrix)*
A Constraint Matrix $\mathbf{R}$ *is a* $\eta \times n$ *rectangular array, such that:*

$$\mathbf{R}_{\alpha j} = \mathbf{R}[\alpha, j] = \begin{cases} 1 & \text{if variable } X_j \triangleright C_\alpha \\ 0 & \text{otherwise} \end{cases}$$

Remark 19.2.1 *With binary constraints, there are only two non-zero entries in each row of* $\mathbf{R}$.

Definition 19.2.2 *(Instantiation)*
An Instantiation $\mathbf{I}$ *is a mapping from a n-tuple of variables* $(X_1, \ldots, X_n) \rightarrow D_1 \times \ldots \times D_n$*, such that it assigns a value from its domain to each variable in* V*, e.g.* $(X_1, \ldots, X_n) \rightarrow (14, 18, \ldots, p, q)$

Remark 19.2.2 *An instantiation* $\mathbf{I}$ *corresponds to a chromosome (individual) in our evolutionary algorithm.*

Definition 19.2.3 *(Constraint Arity)*
We define the Constraint Arity *for a constraint* C_α*,* a_α*, as the number of relevant variables for* C_α.

Definition 19.2.4 *(Partial Instantiation)*
Given $V_p \subseteq V$*, a* Partial Instantiation $\mathbf{I_p}$ *is a mapping from a j-tuple of variables*$(X_{p_1}, \ldots, X_{p_j}) \rightarrow D_{p_1} \times \ldots \times D_{p_j}$*, such that it assigns a value from its domain to each variable in* V_p*, e.g.* $(X_{p_1}, \ldots, X_{p_j}) \rightarrow (20, \ldots, s)$

Note: $\mathbf{I_p}$ *satisfaction of* C_α *is given if all of their relevant variables are instantiated.*

19.2.2 Notions on Evolutionary Algorithm

The structure of an evolutionary algorithm is shown in Figure 19.1. Our research effort has been mainly concentrated on the evaluation function (or fitness function), on the selection algorithm, and on the genetic operators.

```
Begin /* procedure Evolutionary Algorithm */
 t = 0
 initialize population P(t)                    (1)
 evaluate fitness of indivuals in P(t)         (4)
 while (not termination-condition) do
       begin
       t=t+1
       Parents = select-parents-from P(t-1)    (5)
       Children = alter Parents                (3)
       P(t) = Children
       evaluate P(t)                           (4)
       end
 endwhile
End /* procedure Evolutionary Algorithm */
```

Figure 19.1 Structure of an EA and its components

In our algorithm the initial population is generated randomly. The variable values are selected from their domains with a uniform probability distribution. We have selected a non-binary genetic representation because it is a more natural representation for this kind of problems. The selection algorithm is biased to the better evaluated individuals.

19.3 FITNESS FUNCTION

Evolutionary methods have generally been applied to optimization problems. Although in CSP there is no objective to the optimized, a fitness value for a given solution may be derived from the number of satisfied constraints [6]. In [17] we propose a fitness function specifically defined for CSP, which we describe briefly in the next definitions.

Definition 19.3.1 *(Variables involved)*
For a CSP and an instantiation $\mathbf{I}$ *of its variables, we define a set of variables involved* $\mathbf{E}_{\alpha(\mathbf{I})} \subseteq V$ *for each constraint* C_α $(\alpha = 1, \dots, \eta)$*, as follows:*

- $\mathbf{E}_{\alpha(\mathbf{I})} = \phi$ *iff* C_α *is satisfied*
- $X_i \in \mathbf{E}_{\alpha(\mathbf{I})}$ *if* $X_i \rhd C_\alpha$ *and* C_α *is not satisfied under* $\mathbf{I}$
- $X_j \in \mathbf{E}_{\alpha(\mathbf{I})}$ *if* $X_i \rhd C_\alpha$ *and* $\exists C_\beta \neq C_\alpha$ *such that both* $X_i and X_j \rhd C_\beta$*, and* C_α *is not satisfied under* $\mathbf{I}$

This definition shows that there are variables which are relevant for more than one constraint. When C_α is violated the variables involved in this violation are the variables directly connected by C_α and the other variables connected to the relevant variables of C_α. More precisely, for a certain instantiation $\mathbf{I}$, the effect of changing the value of one variable could be reflected in many constraints. It is this effect that we have incorporated in the evaluation function. Before defining our evaluation function, we need the following definition:

Definition 19.3.2 *(Error-evaluation)*
For a CSP with a constraint matrix $\mathbf{R}$ *and an instantiation* $\mathbf{I}$*, we define the Error-evaluation* $\mathbf{e}(C_\alpha, I)$ *for a constraint* C_α *by:*

$$\mathbf{e}(C_\alpha, \mathbf{I}) = \textit{Number of variables in } \mathbf{E}_{\alpha(\mathbf{I})}$$

$$\mathbf{e}(C_\alpha, \mathbf{I}) = \begin{cases} \sum_{w=1}^{n} \mathbf{R}[\alpha, w] \left(\sum_{\beta=1}^{\eta} \mathbf{R}[\beta, w] \right) & \textit{if } C_\alpha \textit{ is violated} \\ 0 & \textit{otherwise} \end{cases} \tag{19.1}$$

Remark 19.3.1 *If a binary non-satisfied constraint* C_α *has* X_k *and* X_l *as relevant variables, (it has exactly two), then:*

$\mathbf{e}(C_\alpha, \mathbf{I})$ = *(Number of relevant variables for* C_α*) + (Propagation Effect* X_k *and* X_l*)*
where Propagation Effect X_k *and* X_l *in a binary constraint network, is defined as the number of constraints* C_β, $\beta = 1, \ldots, \eta, \beta \neq \alpha$ *that have either* X_k *or* X_l *as relevant variables.*

Definition 19.3.3 *(Contribution to Fitness)*
Given a CSP, we define the contribution of C_α to the fitness function $\mathbf{c}(C_\alpha)$ *by :*

$$\mathbf{c}(C_\alpha) = \mathbf{e}(C_\alpha, \mathbf{I_j}), \textit{ when } C_\alpha \textit{ is violated under } \mathbf{I_j}.$$

Remark 19.3.2 *The value* $\mathbf{c}(C_\alpha)$ *does not depend on the instantiation values, but the fact that the instantiation violates the constraint.*

Finally, our fitness function is defined as the sum of the *error-evaluations* (equation 19.1) of all constraints in the CSP, that is:

Definition 19.3.4 *(Fitness Function)*
For a CSP with constraint matrix $\mathbf{R}$*, an instantiation* $\mathbf{I}$*, and* Error-evaluation $\mathbf{e}(C_\alpha, \mathbf{I})$ *for each constraint* $C_\alpha, (\alpha = 1, \ldots, \eta)$*, we define a* Fitness Function $\mathbf{Z}(\mathbf{I})$ *as:*

$$\mathbf{Z}(\mathbf{I}) = \sum_{\alpha=1}^{\eta} \mathbf{e}(C_\alpha, \mathbf{I}) \tag{19.2}$$

The objective is to minimize the evaluation function $\mathbf{Z}(\mathbf{I})$, which equals zero when all constraints are satisfied. The Evaluation Function reflects that it is more important to satisfy a constraint depending on variables which also affect other constraints.

19.4 OPERATOR: CONSTRAINT DYNAMIC ADAPTING CROSSOVER

19.4.1 Idea

In [19] we introduced a sexual operator based on crossover for binary CSP, called *arc-crossover*, for the exploitation (i.e., it constructs new individuals selecting quasi-solutions from the current population). The name *arc*-crossover is due to the fact that each constraint on the constraint network is represented by an arc. It performs a crossover between two randomly selected individuals, the parents, to create a new individual (their child). The child inherits its variable values using a greedy procedure, which analyzes each constraint (arc) according to a predefined order. The first arc to be analyzed is the hardest one to be satisfied, that is the constraint with the largest *error-evaluation function* value. Because the *error-evaluation function* depends only on the network structure, arc-crossover will analyze the constraints following the same priority order each time.

Constraint Dynamic Adapting Crossover is based on the idea that there is no fixed crossover point. It constructs an offspring using the same arc-crossover greedy procedure, but analyzes each constraint according to a dynamic priority. The constraint dynamic priority not only takes into account the network structure, but also the current values of the parents. Thus, the new operator also reflects the adaptivity concept. The operator needs current information about all arcs in order to compute the priorities of the arc of the child. For that we define two *partial fitness functions.* The first one is the *partial crossover fitness function* which allows us to guide the selection of a combination of variable values by constraint. The second one is the *partial mutation fitness function* for choosing a new variable value. Before describing the new operator the following definitions are needed:

Definition 19.4.1 *(S_α)*
Given a CSP as before with a partial instantiation $\mathbf{I_p}$. For a completely instantiated constraint C_α (i.e. both/all relevants variables to C_α are instantiated) we define a set $\mathbf{S}_\alpha \subseteq \theta$ by $C_\beta \in \mathbf{S}_\alpha$ iff

- *$\exists X_i : X_i \rhd C_\alpha$ and $X_i \rhd C_\beta$ (C_α, C_β have a common variable)*
- *$\forall X_j$ relevant to C_β, X_j is instantiated*

This definition shows that the fact of changing the value of one variable would be reflected in other constraints. It is used by the algorithm when the variables are instantiated. Note:obviously $C_\alpha \in \mathbf{S}_\alpha$

Definition 19.4.2 *(Partial crossover fitness function)*
Given a CSP as before with a partial instantiation $\mathbf{I_p}$, and the sets $\mathbf{S}_\alpha$, and the functions $\mathbf{e}(C_\alpha, \mathbf{I_p})$. For all completely instantiated constraints C_α, we define the partial crossover fitness function, *$\mathbf{cff}(C_\alpha)$ for C_α as:*

$$\mathbf{cff}(C_\alpha, \mathbf{I_p}) = \sum_{C_\gamma \in \mathbf{S}_\alpha} \mathbf{e}(C_\gamma, \mathbf{I_p}) \tag{19.3}$$

Remark 19.4.1 *We extend the domain of the functions defined in Section 19.3 from* $\mathbf{I}$ *only to* $\mathbf{I}$ *and* $\mathbf{I_p}$ *by calculating these functions only considering the constraints involved (related to* C_α*) and whose variables are instantiated in* $\mathbf{I_p}$

Definition 19.4.3 *(*$\mathbf{M_j}$*)*
Given a CSP as before with a partial instantiation $\mathbf{I_p}$*. For an instantiated variable* X_j *we define a set of constraints* $\mathbf{M_j} \subseteq \theta$ *by* $C_\alpha \in \mathbf{M_j}$ *iff:*

- $X_j \rhd C_\alpha$
- C_α *is completely instantiated under* $\mathbf{I_p}$ *(i.e.* $\forall X_k \rhd C_\alpha : X_k$ *is instantiated)*

Definition 19.4.4 *(Partial Mutation Fitness Function)*
Given CSP as before with a partial instantiation $\mathbf{I_p}$*, the sets* $\mathbf{M_j}$ *for all instantiated variables under* $\mathbf{I_p}$*, and the functions* $\mathbf{e}(C_\alpha, \mathbf{I_p})$*. For all completely instantiated constraints* C_α*, we define the* partial mutation fitness function, **mff** *for* X_j *as:*

$$\mathbf{mff}(X_j, \mathbf{I_p}) = \sum_{C_\gamma \in \mathbf{M_j}} \mathbf{e}(C_\gamma, \mathbf{I_p}) \tag{19.4}$$

Remark 19.4.2 *This function is calculated only considering the constraints involved (related to* X_j*) and whose variables are instantiated in* $\mathbf{I_p}$

19.4.2 Constraint Dynamic Adaptive Operator

Until now the best results found for the CSP resolution using evolutionary algorithms work with asexual operators [15, 3], rather oriented in the exploration than in the exploitation, owing in part to the epistatic characteristic of CSPs. Furthermore, using a classical crossover operator could produce a performance degradation of the evolutionary algorithm for CSP. Our aim here is to add to the exploitation concept an adapting mechanism for improving the search performance. In order to incorporate the adaptivity idea we need the following definitions:

Definition 19.4.5 *(Number of Violations)*
Given a CSP and its Error-evaluation $\mathbf{e}(C_\alpha, \mathbf{I_1})$ *for each constraint* C_α*, under an instantiation* $\mathbf{I_1}$*, and* $\mathbf{e}(C_\alpha, \mathbf{I_2})$ *for each constraint* C_α*, under an instantiation* $\mathbf{I_2}$*. We define for each constraint* C_α *the* number of violations $\mathbf{nv}(C_\alpha, \mathbf{I_1}, \mathbf{I_2})$ *as:*

$$\mathbf{nv}(C_\alpha, \mathbf{I_1}, \mathbf{I_2}) = \begin{cases} 0 & \mathbf{e}(C_\alpha, \mathbf{I_1}) = \mathbf{e}(C_\alpha, \mathbf{I_2}) = 0 \\ 1 & \textit{either } \mathbf{e}(C_\alpha, \mathbf{I_1}) \neq 0 \textit{ or } \mathbf{e}(C_\alpha, \mathbf{I_2}) \neq 0\textit{, but not both} \\ 2 & \mathbf{e}(C_\alpha, \mathbf{I_1}) \neq 0\textit{, and } \mathbf{e}(C_\alpha, \mathbf{I_2}) \neq 0 \end{cases}$$

Definition 19.4.6 *(Constraint Adapting Priority)*
Given a CSP, its Contribution to the fitness function $\mathbf{c}(C_\alpha)$ *and its number*

of violations $\mathbf{nv}(C_\alpha, I_1, I_2)$ *for each constraint* C_α*, under some instantiations* $\mathbf{I_1}$,$\mathbf{I_2}$. *We define a Constraints Adapting Priority,* $\mathbf{P_{ca}}(\mathbf{I_1}, \mathbf{I_2})$ *as an ordered sequence of the constraints (in what follows we denote "prior to" by "*$\preceq$*"), as:*
$\mathbf{P_{ca}}(\mathbf{I_1}, \mathbf{I_2}) = \langle C_{k_1}, \ldots, C_{k_\eta} \rangle$ *with* $C_{k_i} \preceq C_{k_{i+1}}, \forall i = 1, \ldots, \eta - 1,$
such that $C_{k_i} \preceq C_{k_j}$ *if either*

- $\mathbf{nv}(C_{k_i}, \mathbf{I_1}, \mathbf{I_2}) = \mathbf{nv}(C_{k_j}, \mathbf{I_1}, \mathbf{I_2})$ *and* $\mathbf{c}(C_{k_i}) \geq \mathbf{c}(C_{k_j})$
- *or* $\mathbf{nv}(C_{k_i}, \mathbf{I_1}, \mathbf{I_2}) > \mathbf{nv}(C_{k_j}, \mathbf{I_1}, \mathbf{I_2})$

$\mathbf{P_{ca}}(\mathbf{I_1}, \mathbf{I_2})$ *is an ordered* η*-tuple of constraints. We are roughly ordering the constraints according to their number of violations and the contribution to the fitness function of their child* $\mathbf{Z}(\mathbf{I_3})$.

It means that the first constraint to be considered is the constraint violated for both parents and which has the largest $\mathbf{c}(C_\alpha)$. After considering all violated constraints for both parents, *Constraint Dynamic Adaptive Crossover* continues with the constraints violated for only one parent ordered by their *error-evaluation function.* The last constraints to be considered will be the satisfied constraints for both parents. Contrary to the arc-crossover operator that works with $\mathbf{P}$, a static priority, the new operator constructs its priority dynamically ($\mathbf{P_{ca}}$) using information from the current parents.
At the beginning the child does not have any instantiated variables. After analyzing the first constraint C_α, it has a_α (Definition 19.2.3), instantiated variables, therefore, we have the first $\mathbf{I_p}$, and so forth. Once most of the constraints have been analyzed, in the current $\mathbf{I_p}$ the next constraint to be considered could already have instantiated some of their relevant variables. In this case the child variable value will be the parent variable value with the smaller **mff**. The whole process is shown in Figure 19.2.

19.5 COMPARING GENETIC OPERATORS

The aim of the experiments was to investigate the effect of incorporating our adaptive operator in an evolutionary algorithm, and to compare it with the others confirmed algorithms presented in the literature. The algorithm has been tested by running experiments on randomly generated 3-coloring graphs, subject to the constraint that adjacent nodes must be colored differently. We used the same procedure for generating test problems as [13]. We generate *solvable* problems with n nodes and m arcs as follows:
- Create three groups of nodes, each with $\frac{n}{3}$ nodes
- Randomly create m arcs between nodes in different groups
- Accept the graph if it has no unconnected components.

19.5.1 Number of constraint checks

In this section we estimate the number of constraint checks for our algorithm which uses *arc-mutation* and *arc-crossover* and for an algorithm which uses

```
Procedure Constraint Dynamic Adaptive Crossover (Parent1, Parent2)
Begin
For each Cα in the order given by Pca
Analyze Cα (xi, xj) from Parent1 and Parent2
if ((Xi ⊣ Ip)^a and (Xj ⊣ Ip)) then
   if (nv(Cα, Parent1, Parent2) = 0) then
      if (Z(Parent2) > Z(Parent1)) then
         Ip(Xi, Xj) = (xi1, xj1)
      else
         if (Z(Parent1) > Z(Parent2)) then
            Ip(Xi, Xj) = (xi2, xj2)
         else
            Ip(Xi, Xj) = random((xi1, xj1), (xi2, xj2))
   else
      if (nv(Cα, Parent1, Parent2) = 1) then
         if (Cα ⊨ Parent1)^b then
          Ip(Xi, Xj) = (xi1, xj1)
         else Ip(Xi, Xj) = (xi2, xj2)
      else
        Ip(Xi, Xj) = argmin_{s1∈S1}(cff(Cα, (Ip ∪ (xi1, xj2))),
                                   cff(Cα, (Ip ∪ (xi2, xj1))))^c
else
   if ((Xi ⊣ Ip) or (Xj ⊣ Ip)) then
      if (Xi ⊣ Ip) then k=i else k=j
    Ip(Xk) = argmin_{s2∈S2}(mff(Xk, (Ip ∪ xk1)), mff(Xk, (Ip ∪ xk2)))^d
End
```

[a] X_i not-instantiated in $\mathbf{I_p}$
[b] $\models$: satisfied by
[c] $argmin_{s \in S}\{a_s\}$ give s^* such as $a_{s^*} \leq a_s, \forall s \in S$.
$S_1 = \{(x_{i_1}, x_{j_2}), (x_{i_2}, x_{j_1})\}$
[d] $S_2 = \{x_{k_1}, x_{k_2}\}$

Figure 19.2 Structure of procedure Constraint Dynamic Adaptive Crossover

$(\#, r, b)$ and mutation. Given an individual *arc-mutation* selects randomly the variable to be changed. The variable value is chosen evaluating for each value from its domain the partial mutation fitness function as given in Definition 19.4.4. The value selected will have the smallest **mff**. *Arc-crossover* uses the same greedy procedure as constraint dynamic adaptive crossover but with a static priority. $(\#, r, b)$ is a specialized asexual operator defined by Eiben in [15] as: An asexual operator selects a number of positions in the parent, then selects new values for these positions. The number of modified values, the criteria for identifying the positions of the values to be modified and the criteria for defining the new values for the child are the defining parameters of the asexual operator. Therefore an asexual operator is denoted by the triple (n,p,g) where n indicates the number of modified values, and the values for p and g are chosen from the set {r,b}, where r indicates uniform random selection and b indicates some heuristic-based biased selection. For this kind of problems the best parameters for this operator were (n,p,g) $= (\#, r, b)$ where # means that the number of values to be modified is chosen randomly but is at most $\frac{1}{4}$ of all positions.

The model The parameters in the model are:

- p_c crossover probability
- p_m mutation probability
- t_p population size
- n number of variables
- m domain size
- p_1 probability of connectivity, i.e., the probability that a constraint exists between two variables.

Consequences of the model:

- Average number of constraints $= \frac{n(n-1)p_1}{2}$
- average connectivity for each variable $= p_1(n-1)$

arc-crossover. We analyse the worst case for *arc-crossover*, i.e., when the constraint C_α is violated for both parents and the two variables involved in C_α are not instantiated. In this case *arc-crossover* must choose between the two combination of values from their parents, with two **cff** evaluations. Each constraint is tested only once for combination of values. So, we have the inequality:

- $2\eta \leq$ number of constraints checks by arc-crossover$\leq 4\eta$

So, the minumum number of constraints checks for arc-crossover is 2η when at least one of the parents satisfies all the constraints in the network.

Remark 19.5.1 *The estimation of the number of constraints checks for constraint-dynamic, adaptive crossover is the same as for arc-crossover.*

arc-mutation. tests with **mff** all the others values of the variable domain therefore,

- Number of constraints checks by arc-mutation $= 2\eta(m-1)p_m$

Thus, the number of constraints checks for the entire algorithm (N_{valgo}) is:

- $N_{valgo} \leq \left(4\eta\frac{p_c}{2-p_c} + 2\eta(m-1)p_m\right) t_p$

Constraints	N'_{valgo}	CPU'time	N_{valgo}	CPUtime
30	11141	67	7609	18
45	185671	1144	44967	110
60	843709	4891	422426	1053
90	1401437	6372	465359	917
120	1311153	5908	250505	632
150	1060255	5938	219715	452
180	1070010	5961	219847	580
210	1022748	5693	201619	419
240	936505	5296	220700	419
270	1153107	5087	252820	673

Table 19.1 CPU time $[sec]$ for solving 100 graphs and the average number of constraints checks.

$(\#, \mathbf{r}, \mathbf{b})$. is the asexuel operator. Therefore, the number of constraints checks for $(\#, r, b) = \frac{\eta m}{2}$ For the algorithm with $(\#, r, b)$ and mutation the number of constraints checks is equal to

- $N'valgo = \frac{t_p p_c \eta m}{2}$

because the mutation operator does not test any constraint. Obviously, the number of constraints checks using our operators is larger than using asexuel operators, but the performance of an algorithm is related to:

- Percentage of cases that the algorithm finds a solution
- The number of generations to find a solution

We have tested the two algorithms for the three coloring problems where problem instances are characterized by a connectivity percentage in the range of $[10..90]$. Table 19.1 shows the CPU time $[sec]$ for solving 100 graphs and the average number of constraints checks.

The performance of our operators is better than an algorithm using $(\#, r, b)$ and mutation, because our algorithm converges more quickly to a solution, and it has been able to solve more problems.

19.5.2 3-Coloring Random Solvable CSPs with Connectivity between [4,6]

We have generated random CSPs with different topologies with a degree of connectivity between $[4, 6]$ for 30 variables. All of them have at least one solution. The maximum number of iterations is fixed at 500 and the population size is 20. We have compared three algorithms which differ in their operators. We have compared four algorithms which differ in their operators. All of them work with our fitness function and the selection algorithm, [18].

The first one *with-UAX* works with *uniform adaptive crossover* defined by Warwick and Tsang in [21] as: UAX has an extended string representation. The extra binary string is used as a template to control the creation of the offspring string during the crossover process. Successful strings will have the opportunity as parents to pass on their crossover points to their offspring. The first stage of the crossover operator randomly selects two parents from the matepool. Parent strings are cut at points where the binary values match and alternating sections (starting from a randomly select parent) are used to create an offspring.

The second one *with-Arc_operators* works with the arc-crossover and arc-mutation operators. The third one *with-*$(\#, r, b)$ works with $(\#, r, b)$. The last one works with Constraint Dynamic Adaptive Crossover (*CAC*).
For each connectivity we have generated 100 different graphs. The maximum number of iterations is fixed at 500. The graph in Figure 19.3 shows the average generations for the 100 graphs in each connectivity. We can see that using the *CAC* operator the number of generations needed to find a solution is strongly reduced, and the new algorithm has a more uniform behaviour. Furthermore, for the *with-*$(\#, r, b)$ the problems with a connectivity equal to 4.7 are very hard. Here, the new operators lead to a general improvement. Figure 19.3 also shows the percentage of solutions found by the three algorithms for each connectivity. *CAC* found 97% of the solutions in the worst case, in contrast to *with-*$(\#, r, b)$, which found 70%. In addition, *with-Arc_operators* found 83%, and *with-UAX* only 15%. All of them work with pmutation equal to 0.5. The real improvement detected using the new operator for values between [4.5,5.3] concering the connectivity. It is known that this is the region where harder 3-coloring problem instances may be expected, [8].

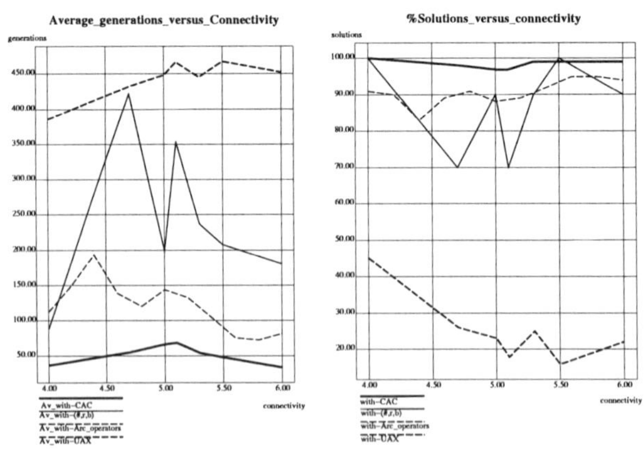

Figure 19.3 Comparison: Average number of generations and Percentage of solutions found for *CAC*, *Arc-operators*, $(\#, r, b)$, *UAX* for different connectivities

19.6 DISCUSSION AND FURTHER ISSUES

In our approaches we have incorporated some topics typically addressed in the constraint research community. First of all, the concept of *structure* has been included in the fitness function and in the partial fitness functions. An evolutionary algorithm repairs chromosomes searching for a solution, that means that a chromosome which is violates the stronger connected constraints will likely be more difficult to repair. Another important concept is *decomposition* or subproblems, the basic idea is to partition a constraint graph into subgraphs to improve the search performance [4]. We used this idea for designing *Constraint Dynamic Adaptive Crossover.* Minimizing the number of constraint checks is the goal for any algorithm that solves CSP in a systematic way. We have used this concept to compute the partial fitness functions in such a way that they only take into account possible local effects of the change. That is, if we change a variable value, the *partial mutation fitness function* will be computed only considering the variables joined to this variable by a constraint. In the same way, when we modify the values of a constraint, the *partial crossover fitness function* will be calculated only analyzing the constraints which share some variable with it.

19.7 CONCLUSION

A new evolutionary algorithm has been presented which takes into account the structure of constraint network, and incorporates the idea of adaptivity in order to define a new operator for CSP. Our research allows us to conclude that the structure of a constraint network is very important to guide the search. The positions of the variables in the chromosome for the problems with high degree of interaction (epistasis), such as the CSPs, can lead to performance degradation when the algorithm works with a one-point or two-point crossover. However, for *Constraint Dynamic Adaptive Crossover* the order is not a relevant factor for its performance. The result suggests that our algorithm is a good option for solving CSPs.

There is a variety of ways in which the techniques presented here can be extended. The principal advantage of our method is that it is general, i.e., the approach is not related to a particular problem. Now our research is directed towards incorporating the constraints difficulty concept into the evolutionary algorithm.

Acknowledgements

I wish to gratefully acknowledge the discussions with Bertrand Neveu.

References

[1] Affane M.S. and Bennaceur H. A labelling arc consistency method for functional constraints. In Eugene Freuder, editor, *Proceedings of Constraint Processing CP96*, pages 16–30, 1996.

[2] Bowen J. and Dozier G. Solving constraint satisfaction problems using a genetic/systematic search hybrid that realizes when to quit. In *Proceedings*

of the Sixth International Conference on Genetic Algorithms, pages 122–129, 1995.

[3] Bowen J., Dozier G. and Bahler D. Solving small and large scale constraint satisfaction problems using a heuristic-based microgenetic algorithm. In *Proceedings of the First IEEE Conf on Evolutionary Computation*, pages 306–311, 1994.

[4] Dechter R. Enhancement schemes for constraint processing: backjumping, learning, and cutset decomposition. In *Artificial Intelligence*, pages 273–312, 1990.

[5] Freuder E. A sufficient condition of backtrack-free search. In *Journal of the ACM*, pages 24–32, 1982.

[6] Freuder E. The many paths to satisfaction. In M. Meyer, editor, *Constraint Processing*, pages 103–119, 1995.

[7] Haralick R.M. and Elliott G.L. Increasing tree search efficiency for constraint satisfaction problems. In *Artificial Intelligence*, pages 263–313, 1980.

[8] Kanefsky B., Cheeseman P. and Taylor W. Where the really hard problems are. In *Proceedings of IJCAI-91*, pages 163–169, 1991.

[9] Kumar V. Algorithms for constraint satisfaction problems:a survey. In *AI Magazine*, pages 32–44, 1992.

[10] Michalewicz Z. *Genetic Algorithms + Data Structures = Evolution Programs.* Springer-Verlag, 1994.

[11] Minton S. Automatically configuring constraint satisfaction programs: A case study. In *Constraints*, 1(1), pages 7–43, 1996.

[12] Petrie C., Rossi F. and Dhar V. On the equivalence of constraint satisfaction problem. Act-ai-222-89, MCC Corporation, Austin, Texas, 1989.

[13] Philips A., Minton S., Johnston M. and Laird P. Minimizing conflicts: a heuristic repair method for constraint satisfaction and scheduling problems. In *Artificial Intelligence*, pages 161–205, 1992.

[14] Raué P-E., Eiben A.E. and Ruttkay Zs. Solving constraint satisfaction problems using genetic algorithms. In *Proceedings of the First IEEE Conf on Evolutionary Computation*, pages 542–547, 1994.

[15] Raué P-E., Eiben A.E. and Ruttkay Zs. Ga-easy and ga-hard constraint satisfaction problems. In M. Meyer, editor, *Constraint Processing*, pages 267–283, 1995.

[16] Raué P-E., Eiben A.E. and Ruttkay Zs. Self-adaptivity for constraint satisfaction: Learning penalty functions. In B. Porto, editor, *Proceedings*

of the Third IEEE Conf on Evolutionary Computation, pages 258–261, 1996.

[17] Riff M.-C. From quasi-solutions to solution: An evolutionary algorithm to solve csp. In E. Freuder, editor, *Proceedings of Constraint Processing CP96*, pages 367–381, 1996.

[18] Riff M.-C. Using the knowledge of the constraints network to design an evolutionary algorithm that solves csp. In B. Porto, editor, *Proceedings of the Third IEEE Conf on Evolutionary Computation*, pages 279–284, 1996.

[19] Riff M.-C. Evolutionary search guided by the constraint network to solve csp. In *Proceedings of the Fourth IEEE Conf on Evolutionary Computation*, pages 337–342, 1997.

[20] Tsang E. Applying genetic algorithms to constraint satisfaction optimization problems. In *Proceedings of ECAI-90*, pages 649–654, 1990.

[21] Warwick T. and Tsang E. Using a genetic algorithm to tackle the processors configuration problem. In *Proceedings of ACM Symposium on Applied Computing (SAC)*, pages 217–221, 1994.

20 APPLYING THE ANT SYSTEM TO THE VEHICLE ROUTING PROBLEM

Bernd Bullnheimer, Richard F. Hartl and Christine Strauss

Department of Management Science, University of Vienna
Bruenner Str. 72, A-1210 Vienna, Austria.
bullnhei@pom.bwl.univie.ac.at, hartl@pom.bwl.univie.ac.at, strauss@pom.bwl.univie.ac.at

Abstract: In this paper we use a recently proposed metaheuristic, the Ant System, to solve the Vehicle Routing Problem in its basic form, i.e., with capacity and distance restrictions, one central depot and identical vehicles. A "hybrid" Ant System algorithm is first presented and then improved using problem-specific information (savings, capacity utilization). Experiments on various aspects of the algorithm and computational results for fourteen benchmark problems are reported and compared to those of other metaheuristic approaches such as Tabu Search, Simulated Annealing and Neural Networks.

20.1 INTRODUCTION

The Ant System, introduced by Colorni, Dorigo and Maniezzo [6, 10, 12] is a new distributed metaheuristic for hard combinatorial optimization problems and was first applied to the well known Traveling Salesman Problem (TSP). It has further been applied to the Job Shop Scheduling Problem [7], to the Graph Colouring Problem [8] and to the Quadratic Assignment Problem [18].

Observations on real ants searching for food were the inspiration to imitate the behaviour of ant colonies for solving combinatorial optimization problems. Real ants are able to communicate information concerning food sources via an aromatic essence, called pheromone. They mark the path they walk on by laying down pheromone in a quantity that depends on the length of the path and the quality of the discovered food source. Other ants can observe the pheromone trail and are attracted to follow it. Thus, the path will be marked again and will therefore attract more ants. The pheromone trail on paths leading to rich food sources close to the nest will be more frequented and will therefore grow faster.

The described behaviour of real ant colonies can be used to solve combinatorial optimization problems by simulation: artificial ants searching the solution space simulate real ants searching their environment, the objective values correspond to the quality of the food sources, and an adaptive memory corresponds to the pheromone trails. In addition, the artificial ants are equiped with a local heuristic function to guide their search through the set of feasible solutions.

In this paper we present the application of the ant system to the Vehicle Routing Problem (VRP) with one central depot and identical vehicles. The remainder of the paper is organized as follows: In Section 20.2 we present the VRP and the ant system algorithm to tackle it. A "hybrid" ant system algorithm, using the 2-opt heuristic and problem specific information, is developed in Section 20.3 and Section 20.4, respectively. Experiments on various aspects of the algorithm and computational results for fourteen benchmark problems are presented in Section 20.5. We conclude with a discussion of the results in Section 20.6.

20.2 ANT SYSTEM FOR VRPS

The VRP can be represented by a complete weighted directed graph $G = (V, A, d)$ where $V = \{v_0, v_1, v_2, \ldots, v_n\}$ is a set of vertices and $A = \{(v_i, v_j) : i \neq j\}$ is a set of arcs. The vertex v_0 denotes the *depot*, the other vertices of V represent *cities* or *customers*, and the nonnegative weights d_{ij}, which are associated with each arc (v_i, v_j), represent the distance (or the travel time or the travel cost) between v_i and v_j. For each customer v_i, a nonnegative demand q_i and a nonnegative service time δ_i is given ($q_0 = 0, \delta_0 = 0$). The aim is to find minimum cost vehicle routes where

- every customer is visited exactly once by exactly one vehicle
- all vehicle routes begin and end at the depot
- for every vehicle route the total demand does not exceed the vehicle capacity Q
- for every vehicle route the total route length (including service times) does not exceed a given bound L.

The VRP is a very complicated combinatorial optimization problem that has been studied since the late fifties because of its central meaning in distribution management. Problem specific methods (e.g. [5, 15]) as well as metaheuristics like tabu search (e.g. [13]), simulated annealing (e.g. [19]), genetic algorithms (e.g. [17]) and neural networks (e.g. [14]) have been proposed to solve it.

The VRP and the TSP are closely related. As soon as the customers of the VRP are assigned to vehicles, the VRP is reduced to several TSPs. For that reason, our approach is highly influenced by the TSP ant system algorithm by Dorigo et al. [12].

To solve the VRP, the artificial ants construct vehicle routes by successively choosing cities to visit, until each city has been visited. Whenever the choice of another city would lead to an infeasible solution for reasons of vehicle capacity or total route length, the depot is chosen and a new tour is started. For the selection of a (not yet visited) city, two aspects are taken into account: how good *was* the choice of that city, an information that is stored in the pheromone trails τ_{ij} associated with each arc (v_i, v_j), and how promising *is* the choice of that city. This latter measure of desirability, called visibility and denoted by η_{ij}, is the local heuristic function mentioned above. In the case of the VRP (or the TSP) it is defined as the reciprocal of the distance, i.e., $\eta_{ij} = 1/d_{ij}$.

With $\Omega = \{v_j \in V : v_j \text{ is feasible to be visited}\} \cup \{v_0\}$, city v_j is selected to be vistited after city v_i according to a *random-proportional rule* [11] that can be stated as follows:

$$p_{ij} = \begin{cases} \frac{[\tau_{ij}]^\alpha [\eta_{ij}]^\beta}{\sum_{h \in \Omega} [\tau_{ih}]^\alpha [\eta_{ih}]^\beta} & \text{if } v_j \in \Omega \\ 0 & \text{otherwise} \end{cases} \tag{20.1}$$

This probability distribution is biased by the parameters α and β that determine the relative influence of the trails and the visibility, respectively.

After an artificial ant k has constructed a feasible solution, the pheromone trails are laid depending on the objective value L_k. For each arc (v_i, v_j) that was used by ant k, the pheromone trail is increased by $\Delta\tau_{ij}^k = 1/L_k$. In addition to that, all arcs belonging to the so far best solution (objective value L^*) are emphasized as if σ ants, so-called *elitist ants*, had used them. One elitist ant increases the trail intensity by an amount $\Delta\tau_{ij}^*$ that is equal to $1/L^*$ if arc (v_i, v_j) belongs to the so far best solution, and zero otherwise. Furthermore, part of the existing pheromone trails evaporates (ρ is the trail persistence).[1] Thus, the trail intensities are updated according to the following Formula (20.2), where m is the number of artificial ants:

$$\tau_{ij}^{new} = \rho\tau_{ij}^{old} + \sum_{k=1}^{m} \Delta\tau_{ij}^k + \sigma\Delta\tau_{ij}^* \tag{20.2}$$

Concerning the initial placement of the artificial ants, it was found that the number of ants should be equal to the number of cities in the TSP, and that each ant should start its tour from another city.[2] The implication for the VRP is that as many ants are used as there are customers in the VRP (i.e., $m = n$), and that one ant is placed at each customer at the beginning of an iteration. After initializing the basic ant system algorithm, the two steps, *construction of vehicle routes* and *trail update*, are repeated for a given number of iterations.

[1]Elitist ants improved the results obtained for the TSP and were therefore also used for the VRP. Trail evaporation is used to avoid early convergence. For a more detailed description the reader is referred to [12].

[2]These are results of experiments by Dorigo et al. [12] as well as our own experiments.

20.3 "HYBRID" ANT SYSTEM ALGORITHM

Hybridization in general means combining ideas of two different methods in one approach. Such proceeding is common practice for hard combinatorial optimization problems and has been successfully applied to other problems such as timetabling [3] or production scheduling [20].

The *2-opt-heuristic* for the TSP [9] is an exchange procedure that generates a so-called *2-optimal* tour. A tour is called 2-optimal if there is no possibility to shorten the tour by exchanging two arcs. In vehicle routing, 2-opt is used in the *Sweep-Algorithm* [15], where customers are first clustered and then 2-optimal vehicle routes for each cluster are generated. The same can be done with solutions constructed by artificial ants: at the end of an ant system iteration, each vehicle route generated is checked for 2-optimality and is improved if possible. Only then the total objective value is calculated and the trails are updated. The "hybrid" ant system[3] for the VRP can be described by the schematic algorithm given in Figure 20.1.

I	Initialize
II	For I^{max} iterations do:
	(a) For each ant $k = 1, \ldots, m$ generate a new solution using (20.1)
	(b) Improve all vehicle routes using the 2-opt-heuristic
	(c) Update the pheromone trails using (20.2)

Figure 20.1 "Hybrid" Ant System Algorithm

The resulting quantitative improvements achieved by the "hybrid" ant system are shown in Section 20.5. In the next section, the "hybrid" ant system is further improved by including problem-specific information in step II (a) of the algorithm, the construction of vehicle routes.

20.4 PROBLEM SPECIFIC IMPROVEMENTS

The close relation between the VRP and the TSP, and thus the corresponding ant system approaches, has been mentioned above. One major difference, namely the existence of one distinct city in the VRP, the depot, has been taken into account. But the VRP has some further characteristics that can be included in an ant system algorithm for the purpose of improving the quality of the solutions.

In the VRP, not only the relative location of two cities is important[4], but also their relative location to the depot v_0 is essential for the tour length. The

[3]It is questionable whether the addition of the 2-opt approach deserves the name hybrid method or whether it is only a post-optimization. We use the term *hybrid* following Goldberg's schematic of a "hybrid using a batch scheme" [16], p.203.

[4]This information is included in the visibility.

so-called *savings*[5] measure the favourability of combining two cities v_i and v_j in a tour and can be quantified by: $\mu_{ij} = d_{i0} + d_{0j} - d_{ij}$. High savings μ_{ij} indicate that visiting customer v_j after v_i is a good choice. This can be used to improve the quality of the ant system algorithm if high savings lead to a high probability of selection, i.e., if $p_{ij} \sim \mu_{ij}^{\gamma}$ where the parameter γ regulates the relative influence of the savings.

Furthermore, for a capacity restricted problem as the VRP, it seems reasonable to assure a high degree of capacity utilization of the vehicles. Let Q_i be the total capacity used including the capacity requirement of customer v_i, then high[6] values $\kappa_{ij} = (Q_i + q_j)/Q$ indicate high capacity utilization through the visit of customer v_j after visiting v_i. This again can be used for the ant system by giving those customers a high probability of being selected: $p_{ij} \sim \kappa_{ij}^{\lambda}$. The parameter λ determines the relative influence of κ_{ij}. The probability distribution for selecting customer v_j to be visited next after customer v_i can thus be extended to:

$$p_{ij} = \begin{cases} \dfrac{[\tau_{ij}]^{\alpha}[\eta_{ij}]^{\beta}[\mu_{ij}]^{\gamma}[\kappa_{ij}]^{\lambda}}{\sum_{h\in\Omega} [\tau_{ih}]^{\alpha}[\eta_{ih}]^{\beta}[\mu_{ih}]^{\gamma}[\kappa_{ih}]^{\lambda}} & \text{if } j \in \Omega \\ 0 & \text{otherwise} \end{cases} \qquad (20.1')$$

20.5 COMPUTATIONAL RESULTS

The ant system for VRPs was tested on fourteen benchmark problems described in [4]. These problems contain between 50 and 199 customers in addition to the depot. The customers in problems C1-C10 are randomly distributed in the plane, while they are clustered in problems C11-C14. Problems C1-C5 and C6-C10 are identical, except that for the latter the total route length is bounded, whereas for the former it is not. The same is true for the clustered problems: problems C13-C14 are the counterparts of problems C11-C12 with additional route length constraint. For the problems with bounded route length, all customers require the same service time $\delta = \delta_1 = \cdots = \delta_n$.

Before the results for these test problems are presented at the end of this section, we illustrate some of our experiments[7] as well as the stepwise improvement of the ant system in problem C1, which contains 50 randomly distributed customers. As a summary of the results we present the deviation from the best known solution[8] for the best and the average solution out of 30 runs in Table 20.1.

---

[5]See the *Savings-Algorithm* in [5]. That approach starts with depot-customer-depot tours. Then, according to decreasing savings, tours are combined as long as no restrictions are violated.

[6]$\kappa_{ij} \leq 1$ for a feasible solution.

[7]For each experiment we simulated 30 independent ant system runs of 50 iterations each. As we used one ant per customer, the number of solutions generated per iteration was equal to the number of customers, thus a total of 2500 solutions was generated per run.

[8]These solutions are not always the optimal but the "best published" solutions as only for some of them optimality has been proven. In the following there is no distinction made regarding this aspect.

method	Ø	dev.	best	dev.
NN	646.22	23.18%	599.66	14.31%
AS	617.47	17.70%	590.74	12.61%
HAS	592.32	12.91%	564.44	7.59%
HAS-sav	554.36	5.67%	542.61	3.43%
HAS-cap	563.52	7.42%	542.85	3.48%
HAS-1	546.11	4.10%	532.88	1.58%
HAS-5	540.42	3.01%	524.61	0.00%

Table 20.1 Comparison of Results

The basic ant system algorithm (denoted by AS in Table 20.1) solved problem C1 just satisfactorily in the first experiment.[9] The best solution the ants found by selecting the customers according to the probability distribution given in Formula (20.1) was 12%, the average over 30 runs was 17% above the optimum. To see whether the pheromone trails contribute at all to the results, we tested $\alpha = 0$, a setting that could be described as a *stochastic nearest neighbour heuristic* (denoted by NN). The results showed clearly that using the trail information does contribute to the quality of the solution: without it the average objective value was 23% above the optimum. The "hybrid" ant system (HAS) on the other hand, generated much better solutions (dev. 7%) than the basic ant system.

Through the problem-specific features described in Section 20.4, i.e., through the use of Formula (20.1′) for the selection probabilities, a further reduction of route lengths was achieved. In three tests we studied the sole influence of respectively, savings ($\gamma = 5, \lambda = 0$, denoted by HAS-sav), capacity utilization ($\gamma = 0, \lambda = 5$, HAS-cap), as well as their combined influence ($\gamma = \lambda = 5$, HAS-1).[10] Both features improved the performance of the ant system algorithm, with the savings yielding better results (avg. dev. 5.6% as compared to 7.4%), and worked best when applied simultaneously (avg. dev. 4%). As a consequence of the reduced influence of the pheromone trails compared to visibility, savings and capacity utilization, the adaptive effect almost vanished. Therefore, all terms were weighted equally and the parameter setting $\alpha = \beta = \gamma = \lambda = 5$ was chosen, which lead to the best results where the ant system (HAS-5) found the optimal solution (total length 524.61).

In order to analyze the ant-specific contribution to the quality of the results, we further compared the "hybrid" ant system (HAS-5) with a stochastic local search procedure[11]. The latter uses visibility, savings and capacity utilization for tour construction ($\alpha = 0, \beta = \gamma = \lambda = 5$, i.e., no pheromone trails are used), and the 2-opt heuristic for tour improvement.

[9]The parameter setting $\alpha = 1$, $\beta = 5$ and $\rho = 0.75$ lead to good results for the TSP (cf. Footnote 2) as well as the VRP and was chosen, if not indicated otherwise.
[10]The other two parameters were kept at $\alpha = 1$ and $\beta = 5$.
[11]Recall the similar comparison between the basic ant system (AS) and NN.

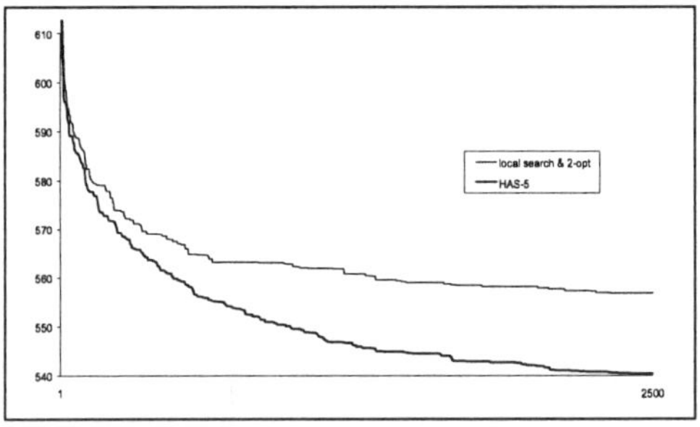

Figure 20.2 Ant System vs. Local Search

Figure 20.2 depicts the continuous reduction of objective values (50 iterations $\hat{=}$ 2500 solutions, averaged over 30 runs) for both methods. The graph shows clearly that the *local search & 2-opt procedure* is outperformed by the "hybrid" ant system. In the early phase of the search, the two methods look almost identical. The trail intensities are still close to their initial value τ_0 and have therefore hardly any effect on the selection probabilities. Thus, the artificial ants select the customers in this stage primarily according to visibility, savings and capacity utilization, which is also done in the local search procedure. Later, when trail intensities for some arcs increase because of frequent use, and decrease for others because of evaporation, the ants use this accumulated information. Thus, the solution space is reduced and better solutions are generated, whereas the local search is still based on initial data only.

σ	Ø	dev.	best	dev.
0	559.74	6.70%	552.04	5.23%
10	550.12	4.86%	528.20	0.68%
30	544.17	3.73%	525.13	0.10%
50	540.42	3.01%	524.61	0.00%
70	545.40	3.96%	530.26	1.08%
90	548.94	4.64%	531.84	1.38%

Table 20.2 Influence of Elitist Ants

In further tests we studied the influence of the elitist ants. In [12] the ant system performance for a TSP with 30 cities first increased with the number of elitist ants (up to an optimal range around 8) and then decreased again. For the VRP we found a similar phenomenon: introducing elitist ants and increasing their number brought better results, but only up to a range around 50, i.e., the number of "regular" ants / customers. The use of more elitist ants lead to poorer performance, caused by massive exploration of suboptimal tours early in the search. The results for various numbers of elitist ants are illustrated in Table 20.2.

Furthermore, we looked at the initial placement of the artificial ants. As the VRP has one distinct city, namely the depot, starting the search from

initial placement	Ø	dev.	best	dev.
depot	550.84	5.00%	527.98	0.64%
customer	540.42	3.01%	524.61	0.00%
random	545.76	4.03%	531.90	1.39%

Table 20.3 Influence of Initial Placement

there is another possibility because the depot is per definition included in every vehicle route. Alternatively, choosing the starting points for the artificial ants randomly is also possible. The comparison of these options, which is illustrated in Table 20.3, confirms our assumption that placing one ant at each customer is best.

ρ	Ø	dev.	best	dev.
0.99	545.68	4.02%	531.66	1.34%
0.95	544.33	3.76%	525.13	0.10%
0.75	540.42	3.01%	524.61	0.00%
0.50	544.41	3.77%	524.63	0.00%
0.25	548.42	4.54%	524.93	0.06%

Table 20.4 Influence of Trail Persistence

Finally, the influence of the trail persistence was subject of further tests (cf. Table 20.4). The results underline our early findings that $\rho = 0.75$ is a good setting. Higher values prevent efficient exploration of the search space as the trail intensities on arcs belonging to suboptimal vehicle routes are kept too high for too long. For lower values the learning effect diminishes and even though the finding of very good solutions is possible, the average quality of the algorithm decreases.

Table 20.5 compares the computational results for the fourteen test problems. For each problem the columns give the problem size n, the vehicle capacity Q, the maximal route length L, the service time δ and the objective value of the optimal solution. In the last three columns, the best solutions obtained with the ant system, the deviation from the optimum and the number of vehicles used are shown. According to our findings, we set $\rho = 0.75$ and used $m = n$ ants, initially placed at the customers $v_1, \ldots, v_n$. For all problems $I^{max} = 100$ iterations were simulated with $\sigma = n$ elitist ants. The random problems were solved using HAS-5 ($\alpha = \beta = \gamma = \lambda = 5$). For the problems C11-C14, where the customers are clustered, we found that the savings do not really contribute to an improvement. The reason is that cities belonging to different clusters, which are located behind each other, might be combined to a tour because of high savings (which result from being located in line with the depot). Thus, we used HAS-cap ($\alpha = 1, \beta = 5, \gamma = 0$ and $\lambda = 5$) for the clustered problems.

The computational results show that reasonably good solutions can be obtained by the ant system. Especially the results on the clustered problem instances C11-C14 seem to be better. There the deviation from the optimum

Random problems								
Prob.	n	Q	L	δ	optimal solution	Ant System	dev.	vehicles used
C1	50	160	∞	0	524.61[a]	524.61	0.00%	5
C2	75	140	∞	0	835.26[a]	870.58	4.23%	10
C3	100	200	∞	0	826.14[a]	879.43	6.45%	8
C4	150	200	∞	0	1028.42[a]	1147.41	11.57%	12
C5	199	200	∞	0	1291.45[b]	1473.40	14.09%	16
C6	50	160	200	10	555.43[a]	562.93	1.35%	6
C7	75	140	160	10	909.68[a]	948.16	4.23%	12
C8	100	200	230	10	865.94[a]	886.17	2.34%	9
C9	150	200	200	10	1162.55[a]	1202.01	3.39%	14
C10	199	200	200	10	1395.85[b]	1504.79	7.80%	19
Clustered problems								
Prob.	n	Q	L	δ	optimal solution	Ant System	dev.	vehicles used
C11	120	200	∞	0	1042.11[a]	1072.45	2.91%	9
C12	100	200	∞	0	819.56[a]	819.96	0.05%	10
C13	120	200	720	50	1541.14[a]	1590.52	3.20%	12
C14	100	200	1040	90	866.37[a]	869.86	0.40%	11

[a] Taillard [24]
[b] Rochat and Taillard [22]

Table 20.5 Ant System Results

ranged from 0.05% to 3.20%. Most random problems were solved within a 5% range, only for problems C4 and C5 the ant system showed higher deviations.

As run times are another criterion for the quality of an algorithm the proposed method is compared to other metaheuristic approaches for which run times were reported in Table 20.6. Tabu search (the sequential algorithm from

Random problems								
Prob.	Tabu Search [21]		Simulated Annealing [19]		Neural Networks [14]		Ant System	
C1	0.00%	0.9	0.65%	0.1	2.78%	0.9	0.00%	0.6
C2	0.27%	16.8	0.40%	59.4	—	—	4.23%	2.4
C3	0.17%	33.9	0.37%	102.9	8.14%	6.5	6.45%	11.3
C4	2.52%	27.2	2.88%	71.6	5.47%	13.2	11.57%	28.5
C5	3.64%	16.3	6.55%	22.9	8.51%	23.2	14.09%	82.2
C6	0.00%	3.2	0.00%	11.6	1.06%	4.3	1.35%	0.2
C7	0.00%	23.1	0.00%	5.2	—	—	4.23%	3.5
C8	0.27%	8.6	0.09%	6.1	3.28%	18.4	2.34%	7.3
C9	1.40%	15.6	0.14%	983.6	8.73%	27.2	3.39%	26.6
C10	1.79%	52.0	1.58%	40.3	13.22%	52.4	7.80%	57.3
Clustered problems								
Prob.	Tabu Search [21]		Simulated Annealing [19]		Neural Networks [14]		Ant System	
C11	0.14%	6.3	12.85%	4.4	5.79%	4.2	2.91%	16.2
C12	0.00%	1.2	0.79%	0.8	0.68%	1.7	0.05%	10.1
C13	0.59%	2.0	0.31%	76.2	4.37%	31.3	3.20%	4.3
C14	0.02%	9.4	2.73%	5.0	1.55%	8.5	0.40%	3.1
Ø	0.77% Sun Sparc 4		2.09% VAX 8600		5.30% VAX 8600		4.43% Pentium 100	

Table 20.6 Deviation and Run Times for several Metaheuristic Approaches

[21][12]) outperforms all other metaheuristics with an average deviation of 0.77%. The ant system (4.43%) performs not as good as Osman's simulated annealing

[12] A comparison on basis of run times on different machines is not perfectly meaningful. To ensure maximum comparability we did not include their parallel implementation.

approach [19] with 2.09% but better than Ghaziri's neural networks approach [14], where the average deviation was 5.30% with only 12 out of 14 problems tested. Run times (given in CPU minutes in Table 20.6) for all algorithms are more or less similar and vary with the problem size in a range from approximately one minute for the smallest to approximately one and a half hours for the largest problem.

20.6 DISCUSSION AND CONCLUSION

The presented contribution shows the application and the improvement of an ant system algorithm to the VRP. The computational results confirm the positive experiences made with the ant system by applying it to the TSP [1, 11, 23]. Although some very good solutions for the VRP instances were obtained, the best-known solutions for the fourteen test problems could not be improved. For practical purposes deviations up to 5% are more than acceptable as uncertainty about travel costs, demands, service times etc. makes perfect planning impossible. As the ant system can compete with other vehicle routing metaheuristics in terms of run times, the presented approach is an alternative to tackle VRPs.

Tabu Search performs much better, but nevertheless the results for the ant system also indicate that there still is much potential for improvement. The superiority of tabu search for VRPs can be explained by two facts: tabu search is an excellent method that has been studied and improved a lot since its introduction, and, much more VRP-related research has been done on tabu search (cf. [13, 19, 21, 22, 24]) than on any other method. Therefore, we are certain that future work on the ant system approach will help to further improve its quality for vehicle routing, even though our current version can not yet compete with the best tabu search algorithms.

Primarily, a more detailed analysis of parameter values is necessary. A metaheuristic could be used to guide the search through the parameter space. Also an automatic adjustment of the parameters done in Evolution Strategies might be of use for the ant system. In addition to that, more elaborated local search procedures exchanging customers not only within but also among tours should be considered. Another very interesting aspect is the use of candidate lists. In the current version of the ant system all feasible customers have the chance to be selected. For many of them the probability of being selected is very low because of large distances, low trail levels or both. Concentrating only on the more promising candidates should yield better results. Moreover, the algorithm seems to be well suited for parallel implementation [2].

A more radical change of the existing algorithm would be to use the ants only to cluster the customers and subsequently, to apply a local search to find good tours among them. A similar idea using a genetic algorithm as a cluster builder has been proposed in [17].

Besides these methodological considerations, additional modifications of the algorithm to extensions of the VRP, e.g., multiple depots or problems with time windows are of interest.

Acknowledgements

The authors would like to thank Marco Dorigo, Vittorio Maniezzo and Gerhard Waescher as well as two anonymous referees for their comments that helped to improve the quality of this paper.

References

[1] B. Bullnheimer, R.F. Hartl, and C. Strauss. A new rank based version of the ant system: a computational study. *Working Paper No.1, SFB Adaptive Information Systems and Modelling in Economics and Management Science*, Vienna, 1997.

[2] B. Bullnheimer, G. Kotsis, and C. Strauss. Parallelization Strategies for the Ant System. Paper presented at *Conference on High Performance Software for Nonlinear Optimization: Status and Perspectives (HPSNO'97)*, Ischia (Italy), 4-6 June 1997.

[3] E.K. Burke, D.G. Elliman, and R.F. Weare. A Hybrid Genetic Algorithm for Highly Constrained Timetabling Problems. In *Proc. 6-th Int. Conf. Genetic Algorithms (ICGA'95)*, pages 605-610, Morgan Kaufmann, 1995.

[4] N. Christofides, A. Mingozzi, and P. Toth. The Vehicle Routing Problem. In N. Christofides, A. Mingozzi, P. Toth, and C. Sandi, editors, *Combinatorial Optimization*, pages 315-338, Wiley, 1979.

[5] G. Clarke, and J.W. Wright. Scheduling of Vehicles from a Central Depot to a Number of Delivery Points. *Oper. Res.* 12 (1964), pages 568-581.

[6] A. Colorni, M. Dorigo, and V. Maniezzo. Distributed Optimization by Ant Colonies. In F. Varela, and P. Bourgine, editors, *Proc. Europ. Conf. Artificial Life (ECAL'91)*, pages 134-142, Elsevier Publishing, 1991.

[7] A. Colorni, M. Dorigo, V. Maniezzo, and M. Trubian. Ant system for Job-Shop Scheduling. *JORBEL - Belgian Journal of Operations Research, Statistics and Computer Science* 34 (1994) 1, pages 39-53.

[8] D. Costa, and A. Hertz. Ants can colour graphs. *J. Oper. Res. Soc.* 48 (1997), pages 295-305.

[9] G.A. Croes. A Method for solving Traveling-Salesman Problems. *Oper. Res.* 6 (1958), pages 791-812.

[10] M. Dorigo. *Optimization, Learning and Natural Algorithms.* Doctoral Dissertation, Politecnico di Milano, Italy (in Italian), 1992.

[11] M. Dorigo, and L.M. Gambardella. Ant Colony System: A Cooperative Learning Approach to the Traveling Salesman Problem. *IEEE Trans. Evol. Comput.* 1 (1997) 1, pages 53-66.

[12] M. Dorigo, V. Maniezzo, and A. Colorni. Ant System: Optimization by a Colony of Cooperating Agents. *IEEE Trans. Sys., Man, Cybernetics* 26 (1996) 1, pages 29-41.

[13] M. Gendreau, A. Hertz, and G. Laporte. A Tabu Search Heuristic for the Vehicle Routing Problem. *Management Sci.* 40 (1994), pages 1276-1290.

[14] H. Ghaziri. Supervision in the Self-Organizing Feature Map: Application to the Vehicle Routing Problem. In I. Osman, and J. Kelly, editors, *Meta-Heuristics: Theory & Applications*, pages 651-660, Kluwer Academic Publishers, 1996.

[15] B.E. Gillett, and L.R. Miller. A Heuristic Algorithm for the Vehicle Dispatch Problem. *Oper. Res.* 22 (1974) pages 340-347.

[16] D. Goldberg. *Genetic Algorithms in Search, Optimization, and Machine Learning.* Addison-Wesley, 1989.

[17] H. Kopfer, G. Pankratz, and E. Erkens. Entwicklung eines hybriden Genetischen Algorithmus zur Tourenplanug. *Oper. Res. Spekt.* 16 (1994), pages 21-31.

[18] V. Maniezzo, A. Colorni, and M. Dorigo. The Ant System applied to the Quadratic Assignment Problem. *Technical Report IRIDIA / 94-28*, Universite Libre de Bruxelles, Belgium, 1994.

[19] I. Osman. Metastrategy simulated annealing and tabu search algorithms for the vehicle routing problem. *Ann. Oper. Res.* 41 (1993), pages 421-451.

[20] E. Pesch. *Learning in Automated Manufacturing.* Physica, 1994.

[21] C. Rego, and C. Roucairol. A Parallel Tabu Search Algorithm Using Ejection Chains for the Vehicle Routing Problem. In I. Osman, and J. Kelly, editors, *Meta-Heuristics: Theory & Applications*, pages 661-675, Kluwer Academic Publishers, 1996.

[22] Y. Rochat, and E. Taillard. Probabilistic Diversification and Intensification in Local Search for Vehicle Routing. *J. Heuristics* 1 (1995), pages 147-167.

[23] T. Stuetzle, and H. Hoos. The MAX-MIN Ant System and Local Search for the Traveling Salesman Problem. *Proc. ICEC'97 - 1997 IEEE 4-th Int. Conf. Evolutionary Computation*, IEEE Press, pages 308-313.

[24] E. Taillard. Parallel Iterative Search Methods for Vehicle Routing Problems. *Networks* 23 (1993), pages 661-673.

21 COOPERATIVE INTELLIGENT SEARCH USING ADAPTIVE MEMORY TECHNIQUES

Lutz Sondergeld and Stefan Voß

Technische Universität Braunschweig, Institut für Wirtschaftswissenschaften
Abteilung Allgemeine Betriebswirtschaftslehre
Wirtschaftsinformatik und Informationsmanagement
Abt-Jerusalem-Straße 7, D - 38106 Braunschweig, Germany.
l.sondergeld@tu-bs.de, stefan.voss@tu-bs.de

Abstract: Ant colony methods have recently attracted attention for their application to several types of optimization problems, especially those with a "graph related formulation". Like other heuristics the ant system was also inspired by the adaptation of biological processes. However, first results have not been very promising for further research on that specific branch of a much broader field of science, that we will draw attention to in this paper, the intelligent agent systems. Besides the experience with ant systems intelligent agent systems may provide a useful paradigm for search processes designed to solve complex problems. These systems are particularly relevant for parallel processing applications and also offer useful strategies for sequential heuristic search. Respective methods can be interpreted as a set of specific formulas (to monitor "ant traces") that embody components of strategic principles being fundamental to adaptive memory programming (AMP) processes, as notably represented by tabu search.

From a conceptual view we show that the more general framework of intelligent agents, which does not restrict its operation to the limited perspectives embodied in ant colony methods, may provide improved efficiency. Specifically, we examine the use of agents that are more heterogeneous characterized by mechanisms of communication between the agents which are more variable and dynamic. Furthermore, these intelligent agents make fully use of adaptive memory ideas from AMP. The conceptual idea of our AMP system model is exemplified on a classical combinatorial optimization problem, the traveling salesman problem.

21.1 INTRODUCTION

Common to most combinatorial optimization problems is their difficulty, i.e., their computational complexity when they are to be solved exactly. Notably, when NP-hardness comes into play [9], the use of heuristics becomes more and more significant. Within the class of heuristics special attention is given to the class of so-called meta-heuristics, i.e. heuristic frames that can be used to deduce hybrids of heuristics adapted to different kinds of combinatorial optimization problems. In this respect approaches such as genetic algorithms, neural networks, tabu search, and adaptive memory programming are being found increasingly relevant [6, 8, 13]. All of these concepts of "intelligent search" have one thing in common: the use of some kind of more or less intelligent techniques to deduce dynamic adaptations and adjustments of optimization strategies (like, e.g., move selection formulas, search restrictions, etc.) from information of the search process itself.

The purpose of this paper is to show how to develop a heuristic framework for cooperative search that embeds such adaptive memory programming elements at a high level. That is, we intend to motivate some means of improvements for ant colony methods. The point of departure for our conceptual approach is the so-called ant system, a dynamic optimization process reflecting the natural interaction between ants searching for food [2, 3, 4], which is one of the recently proposed methods in the category of intelligent search. Within this concept the ants' operations are influenced by two different kinds of search criteria. The first one is the local visibility of food, i.e. the attractiveness of food in each ant's neighborhood. Additionally, each ant's way through its food space is affected by the other ants' trails as indicators for possibly promising directions. The intensity of trails itself is time-dependent: As time goes by, parts of the trails "are gone with the wind", meanwhile the intensity may be increased by new and fresh trails. With the quantities of these trails changing dynamically an "autocatalytic" optimization process is started with the hope of forcing the ants' search into most promising regions. This process may be conceived as a form of interactive learning, which can easily be modeled for most kinds of optimization problems by using simultaneously and interactively processed search trajectories.

This paper is organized as follows. Section 21.2 describes the key aspects of the ant system for a simple application on the well-known *traveling salesman problem* (TSP). Section 21.3 identifies some essential structural deficiencies of the basic ant system. Outgoing from this discussion Section 21.4 sketches the general concept of a multiple agent approach that is dedicated to overcome most of these deficiencies followed by different FIND strategies in applications for the TSP (Section 21.5). Some results and conclusions are given in Sections 21.6 and 21.7.

21.2 APPLICATIONS OF ANT SYSTEMS FOR THE TSP

One of the first applications of the basic ant system approach can be found for the traveling salesman problem [4]. Given a graph $G = (V, E)$ with node set V consisting of n nodes and costs c_{ij} indicating the distance between two nodes i and j, a minimum cost n-node tour including each node has to be found:

$$\text{Minimize } Z(x) = \sum_{i=1}^{n} \sum_{j=1}^{n} c_{ij} x_{\pi(i)\pi(j)}$$

where π is a permutation of $V = \{1, 2, ..., n\}$.

Whereas Dorigo et al. [4] investigate a construction path orientated ant system we also propose a so-called search path orientated ant system.

21.2.1 Construction path orientated ant system

The implementation of a basic variant of the *construction path orientated* ant system is simple. At the beginning at each of the n nodes a certain number m of ants is placed (e.g. either randomly or on purpose). Each of these ants successively chooses $n - 1$ nodes - laying down its corresponding trail - until a complete traveling salesman tour is reached. For each iteration, we consider respective trail values for all edges, i.e., $\tau_{ij}(t)$ is the trail value along edge (i, j) in iteration t. Given these values, respective values for the next iteration $t + 1$ are determined as follows. Let $\Delta\tau_{ij}^{k}(t)$ indicate the (change in the) trail value laid down by the k-th ant in iteration t. Then the change in the trail from iteration t to iteration $t + 1$ for an edge (i, j) in the graph is computed as follows:

$$\tau_{ij}(t+1) = \rho\tau_{ij}(t) + \Delta\tau_{ij}(t), \quad \Delta\tau_{ij}(t) = \sum_{k=1}^{m} \Delta\tau_{ij}^{k}(t)$$

where $(1 - \rho)$ serves as a trail *evaporation* parameter to describe the time-dependency of the trails. The search criteria visibility and trails are combined in a priority function or a move selection rule (MSR) for selecting node j if an ant is actually placed on node i:

$$MSR = \max_{j}((\tau_{ij})^{\alpha}(\frac{1}{c_{ij}})^{\beta})$$

The parameters α and β are appropriate weight factors for both criteria, i.e., they describe an appropriate way of combining trails and visibility. With trails τ_{ij} recomputed after each run, the ants are placed again at their starting node and the whole search process is repeated (when clear from the context we may skip the reference to a specific iteration t for the trail values). It is obvious that for this construction path orientated implementation of the ant system no further specific restrictions are necessary, when combined, e.g., with an insertion heuristic, where for each possible inclusion only those nodes j are taken into account that are not yet visited.

21.2.2 Search path orientated ant system

Alternatively, as a new idea a *search path orientated* variant of an ant system might be used, too. That means, the ants start from a feasible traveling salesman configuration performing edge exchange operations as neighborhood transition steps with according trail computations. In addition, for this case appropriate search restrictions with respect to these exchange operations must be used in order to prevent the ants from cycling. One well-known opportunity to operate with such restrictions is provided by tabu search [6, 7, 12].

21.3 FUNDAMENTAL CRITICISM ON THE ANT SYSTEM

It is important to keep in mind that the ant system approach is an instance of an intensification procedure. Here, the primary theme is to reinforce good past choices. However the way that reinforcement occurs is limited by a model that considers only a restricted type of neighborhood and choice rule process. In addition, ant systems have several significant deficiencies.

Failing forcible mechanism for introducing elements from good solutions into currently generated ones. The trace mechanism of the ant system does not assure that any particular portion of the links from attractive solutions will be introduced into solutions currently generated. The formulas for encouraging links from past solutions to be included in present solutions are arbitrary in this regard, and their effect is somewhat unpredictable.

Failing perception and distinction between attractive and less attractive traces. The ant system applies weights that encourage the inclusion of links from less attractive as well as those from more attractive solutions. (To qualify as an intensification strategy, of course, the method must give greater encouragement to links from more attractive solutions.) This "democratic" weighting that introduces bias to favor inferior past choices evidently dilutes the bias to favor superior past choices. In addition, it is doubly ill-advised, because it also competes against the part of the current evaluation that is independent of the past.

To see the effect of this bias more clearly, suppose the total evaluation of a move (which is a trace move in the ant system approach) is divided into three parts: (a) influence of present conditions, (b) influence of good past choices, (c) influence of poor (and mediocre) past choices. Whenever (a) and (b) are included in the choice rule, the question is why (c) should also be included where the effect of (c) is to duplicate less effective choices made in the past. A more strategic form of diversification would seek choices that were not made in the past, and to emphasize such choices that yield higher evaluations. (Such evaluations can be provided both by present and past conditions, as by favoring the choice of elements that often looked attractive in the past, but that did not quite manage to get chosen. This is the type of approach we often use in our own thinking to "draw upon" the past, and can also guide the type of intensification that recovers complete solutions from the past.)

Failing distinction between different solutions produced. This is perhaps the most significant deficiency of all, from a practical standpoint. Ant methods fail, in an important sense, to distinguish between different solutions produced. The result is that they can lead to "hovering about" a fairly decent solution that is somewhat less attractive than the best solutions found – especially if this less attractive solution is relatively easy to find, and hence its links may readily be retraced. Then, each time such a solution is visited, and each time other reasonably good solutions are visited that share its accessible links, the weighting for the trace on these links will be increased. As older traces decay, the links of more exceptional solutions, which are harder to find and less often visited, will receive less weight. Consequently, the method devotes most of its bias to retracing the links of "popular" (but unexceptional) solutions. This effect is magnified as a result of the previously discussed deficiency.

No distinction between information generated at different times. Ant systems make no distinction with respect to the point of time when collecting and transmitting information. Since information collected might differ with respect to the level of consolidation this might be crucial as well.

No handling of possible conflicts between diverging trails. There are possible conflicts between the information held in the adaptive memory. From the ant system it is obvious that diverging trails can lead in different directions resulting in "confusing" the ants and reduced success of the optimization process.

No real division of labor. Ant systems provide no element of dividing labor among the ants. Strictly speaking, the ants operate concurrently but also isolatedly. That means, one of the simplest and nevertheless most important arguments for introducing parallel processes - reduced effort by hand-in-hand work - cannot be applied for the ant system. Consider the case of the TSP where m ants concurrently build a feasible traveling salesman tour but each of them solely with single-chain traces instead of possible multi-chain ones. Furthermore, the ant has a lack of flexibility by the restricted neighborhood or a smaller freedom of choice in each single construction step.

Critical efficiency through manifold work. This kind of deficiency closely relates to the previous one. It should be noted that for all heuristic approaches with such parallel search techniques the immanent possible loss of efficiency should be observed. This is due to the fact that multiple agents do the same thing just from different points of view or different departure points. Since it seems probable that "four eyes see more than two" (but there is no guarantee for it), we might have additional effort without appropriate gain in heuristic performance. Defining efficiency as the computational effort to realize a certain level of performance (usually: solution quality) the efficiency might decrease by the factor m (number of ants) in comparison to a sequential approach in the worst case. This is the case if the use of $m - 1$ additional agents leads to no significant improvement in solution quality or earlier realization of the best solution found.

21.4 CONCEPT OF A COOPERATIVE AGENT SYSTEM

21.4.1 Basic requirements

Cooperative multi-agent systems follow the principle of "four eyes see more than two". The key idea is to apply simultaneously multiple agents (say *nb_agents*, corresponding to the parameter m above) for the optimization of one and the same problem with the hope that the parallel work provides information that may be embedded in the search process leading to a gain in efficiency, i.e., either to find better solutions or at least good solutions faster (or to be more specific: with less effort). A further advantage strived for results from the dynamic interplay between intensification and diversification. Placing (initializing) the agents at different places of the solution space leads to a diversification of the search process, i.e., a broad scattering of the search over the solution space. With time proceeding, the agents' information is consolidating and the search focuses on most promising regions of the solution space – the search gets an intensifying character. Thereby, such relevant information should take effect both on the agents appearence and their future behavior. Information may find its way in the ongoing optimization process, e.g., within (re-)initialization phases, or by restricting and prioritizing specific operations. It is noteworthy, that the whole process has to be run on (say a number of *nb_levels*) different levels or stages with corresponding adaptations of relevant strategic control parameters.

The basic scheme of an agent system is shown below. We do not want to prescribe further details referring to the type of agent (homogeneous/heterogeneous), the type of clustering (implicit, explicit, link based, etc.) or even the number of agents within each level, since this is a part of the user- and problem-dependent strategy.

for (*nb_levels*) **do**
    initialize the agents (information based)
    while not (stopping rule) **do**
        let the agents communicate, cooperate (information based),
        and be coordinated (information based)
        while the information (base) is updated continuously
    adapt relevant control parameters (strategy dependent)

21.4.2 Communication, coordination and cooperation within agent systems

For intelligent agent systems the three "C"s communication, coordination and cooperation play an important part. In this respect, **cooperation** is the generic term for any kind of "working-together hand-in-hand" of agents that is necessary or at least helpful for generating good solutions. **Communication** in the meaning of mutual information exchange itself is a form of cooperation that is used to yield a raise in algorithmic performance. **Cooperation** in a narrower sense includes all operations of agents that cannot be performed without a partner agent like, e.g., connection of agents or inheritance of agent characteristics. Coordination means a strategic element of a heuristic that lets the agents behavior depend on the behavior of the other agents. Coordination is principally based on communication and cooperation. However, it may have a broader meaning. Well-known coordination principles for heuristic agent search are, e.g.:

- to scatter agents over the solution space (diversification)
- to concentrate the agents' search (intensification)
- to set up clusters for agents (diversification)

Usually, coordination assumes the existence of a superior coordination instance taking all agents into account simultaneously, in contrast to sometimes much simpler bi- or multi-lateral communication and cooperation processes.

21.4.3 Information-based coordination and cooperation within agent systems

If one defines agents as representatives of (not necessarily feasible) solutions and heuristic search as the iterative transition of different states of these solutions performed by corresponding operations, information about the optimization process is either related to these solution states or the operations themselves. To simplify matters, we will thereby take over the denotation "trails" from the basic ant system approach and designate all kinds of process-oriented databases as trails. Most heuristic approaches using solution-related information go back on specific good solutions found during the search process (so-called elite solutions). Useful information (in particular for clustering techniques) is in many cases the identification of "strongly determined" and/or "consistent" parts of solutions (e.g., edges, chains, subsets of variables). Since parts of solutions are always representable by corresponding variables, we will use the notion of strongly determined and consistent variables, in general. Furthermore, we consider those variables in context with the elite solution concept. The notion of strongly determined variables refers to those variables that cannot change their values in the elite solutions except by significantly damaging feasibility or solution quality, while consistent variables are those that "often" receive the same values (i.e., within a certain percentage of all elite solutions) – or values that lie within the same restricted range – in the elite solutions. With the information about strongly determined and consistent variables, different clusters may be built by correspondingly constraining different combinations of those variables, and finally reapplying the heuristic on these elite solutions.

Implicit clustering. For the standard case, a kind of implicit clustering is the most flexible and therefore recommendable form of clustering. With this type of clustering, clusters are not fixed statically but implicitly by the underlying evaluation functions of the heuristic processes. That is, the algorithm is forced to build clusters, however, without systematic or strategic control over the structure of these clusters, which makes this approach more flexible than in the case of explicit clustering. A fraction of all consistent variables detected gives the number of consistent variables that have to stay in a solution during the intensification phases. Additional consistent variables in a solution are allowed as well as the exchange of one consistent variable against another (for the case that swap moves are used), but the fraction remains the lower bound for the number of consistent variables in any solution. The strategic advantage of such an optimization frame is its high level application of AMP, both by the use of the elite solutions as information provider for the clustering and the high flexibility of the control parameters. More generally, techniques within the area of AMP could be, e.g., the flexible investigation for long term information to optimize the search process or use of long term information-based dynamic move selection formulas. Besides these points, the dynamic handling of operational restrictions and the dynamic variation of different sets of neighborhoods (various neighborhoods) might be considered as techniques of AMP.

Cluster-based information embedding. Even if it is most recommendable to use implicit clustering, the explicit variant is also conceivable. Especially, if besides the information provided by the elite solution further information exists – like, e.g., trails – that additionally should be embedded, another variant with a conditioned-based connection of information is proposed. That is, clusters are built in the same way as for the original form but now not fixing variables themselves. Instead only those trails are enabled to take effect on further ants search that correspond to the clustered variables. This approach profits from the fact that the search process of the agents is not disturbed by conflicting trails anymore. Simultaneously, the clusters provide more room for dynamic components and are less restrictive than for the approach with implicit clustering.

21.5 A COOPERATIVE AGENT SYSTEM FOR THE TRAVELING SALESMAN PROBLEM

While it seems preferable not to restrict attention to "trace neighborhoods and choice rules" of the type used by simple agent systems, a more general form of tracing is more appropriate. The idea is quite straightforward, and consists simply of generating a "multi-chain" trace. Since the ant system approach was first designed by reference to the TSP, we can consider the corresponding form of the trace for this problem.

21.5.1 Cooperation, communication and coordination for the traveling salesman problem

Agents that represent partial solutions may cooperate to build new (partial) solutions; the degree of flexibility for this cooperation depends on the definition of appropriate operation rules. A multi-chain trace for the TSP may be easily modeled by starting with a collection of chains, where an isolated node is a special instance of a chain in which both endpoints are the same. We assume that members of the collection are pairwise node-disjoint. We maintain that condition by repeatedly introducing any edge that joins an endpoint of one chain to an endpoint of a different chain, thus producing a chain that is the union of the joined chains and the added edge. When all nodes belong to a single chain, a cycle is created by joining the chain's endpoints and the method stops. It seems that this simple process is much more appropriate than the kind of tracing used by the ant system approach. Specifically, if we use a constructive process that seeks to include certain collections of preferred edges, we obtain a greater degree of freedom by a multi-chain trace than by a single chain trace. In general, this same freedom of choice is valuable for applying various biases with the decision rules.

Forced cooperation. We first want to propose an implementation of a cooperative agent system for the TSP that follows the multi-chain concept in a simple form. Our agent system works with agents that initially are only single edges growing stepwise by edge-connection to chains of edges and from chains by chain-merging to larger chains until we obtain feasible solutions. Thereby, we operate with different groups of agents; each group tries to construct a feasible solution. However, these groups do not work isolatedly, as their agents are allowed to have access on chains of other groups. Note that we have an implicit diversification effect by different initialization of the number of agents for each population (which at the beginning might be performed randomly). After constructing feasible solutions, a set of elite solutions is built with the definition of consistent edges or chains. The whole process is then restarted with these consistent variables such that the optimization process is not restricted too much (i.e., not too many variables are pre-defined) and with different populations simultaneously the clustering concept could be realized.

Possible cooperation. The second approach derived from the basic concept is designed for a greater flexibility of the agents. Instead of groups (populations) of agents that build chains that must be connected and, therefore, restrict the search process more than might be intended, we resolve the concept of disjoint populations for the benefit of a global pool of concurring agents. Following this approach, agents may and will become "incompatible" (i.e., a later simple connection is not possible anymore since agents already have common nodes that are not end nodes). Nevertheless, we allow this for two reasons: (a) along with such incompatibility an implicit diversification element is involved and (b) in any case, each of the agents is able to find a feasible solution. That means, in later phases of the search when some agents are initialized with consistent chains, these agents are no obligatory contractor for other agents. When an agent has completed its work (i.e., a feasible tour was built) this agent is taken out of the race. Furthermore, we also want to embed information from (yet feasible) elite solutions. Concluding, we initialize agents as consistent edges or chains of edges taken from these elite solutions. However, these chains are not obligatory since no other agent is forced to connect to these chains. The prime argument for subsequently using this extended variant instead of the original one is that clusters with not conflicting chains would imply to successively connect all chains to one feasible tour which means that each of the members of one population must be inserted in the solution. This would be a limitation beyond our intention.

21.5.2 Information-based cooperation and coordination

As basic information that is obtained systematically we use the classical edge-related trails from the ant system concept. Furthermore, we use good solutions (elite solutions) found so far both as a basis for re-initialization and also as a move selection rule factor.

Information-based cooperation and coordination by edge trails. The influence of information on the operational process of heuristic search in most cases is defined in specific rules that say what to do next and what not. Let us consider first the basic MSR that we use for the TSP with trail information as a prioritizing information factor. The definition of a basic MSR is primarily orientated to add the next cheapest edge to any of the agents:

$$MSR = \min_{nb\_agents} \min_{i,j}(c_{ij})$$

with i endnode of an existing chain, j a node, that is either endnode of a yet existing chain or any other node. It is noteworthy, that there exists a variety of more complex selection rules – e.g., by considering the difference between the best and the second best edge to add for each chain – but the MSR mentioned above is sufficient for our purpose of conceptional testing of different agent system strategies.

Ant trails (ant information). For a rudimentary implementation of an ant system we keep track of all construction moves (edge insertion moves) in the sense that trails are laid corresponding to the edges that are added to a partial solution. In doing so we can modify the MSR by simply multiplying the trail factor.

$$MSR = \min_{nb\_agents} \min_{i,j}\left((c_{ij})^{\alpha} \circ (\tau_{ij})^{\beta}\right)$$

with τ_{ij} representing the trail of edge (i, j) between nodes i and j. The trail computation follows the scheme described in Section 21.2. The functional combination $\circ$ of both kinds of information may follow the basic ant system concept initially proposed by [4]. In our case, we simply subtract the trails from the edge values. However, the main difference is that we use multiple agents that might cooperate (by being connected to one) even if they must not.

Condition-based embedding of trail information. One of the basic deficiencies of the ant system was its failing ability to distinct between promising and less-promising trails. For this purpose we determine only those trails to take effect that seem to lead to good solutions using the elite solution concept. We could even overwrite trails that do not correspond with the properties of elite solutions in this way. However, in doing so we would loose some part of the long-term memory information. With other words, information that might be helpful would be neglected without necessity.

Information-based coordination by solution initialization. A very simple form of using information deduced from the elite solutions is to split these solutions in chains and to initialize these chains as new agents. Since the number of possible partial solutions is usually by far larger than the number of agents, it is recommended to use either prioritizing rules (e.g., by the length of the chains) or simply probabilistic techniques. In our case we use the latter opportunity.

21.5.3 Further implementation issues

Post-optimization. On each feasible solution we let a post-optimization procedure run performing steepest descent/mildest ascent 2-opt edge-exchanging steps using a simple static tabu search approach [7]. By using tabu search with a comparably small and also fixed number of tabu search operations (here: 100 steps) we just want to ensure that any better solution nearby the current solution found by the construction orientated agent system is found with at least a high probability. Spending more effort in techniques for local optimization would embed the risk of diluting results for the conceptual testing of the agent system approach.

Elite solutions. Implemented is a variant of the elite solution concept that accepts the best five feasible solutions found up to a specific point.

Starting point for information to take effect. Generating a broad basis of independent information is important. Re-initialization with (parts of) good solutions too early or using non-consolidated trail information is not advisable. So we initialize each agent randomly as a 1-element chain on the first ten levels independent from the re-initialization strategy in later stages of the search, even the trail information starts to take effect at this point for the first time.

Reduced neighborhood. Since for larger instances the effort for completely evaluating a solution's neighborhood would be extremely costly, we reduce the number of moves taken into consideration appropriately. That is, for constructing a feasible solution nearest-neighbor-nodes are tried to be embedded first; even for the post-optimization 90 from 100 operations are restricted with respect to the set of edges to be possibly exchanged. Only the last ten steps are performed with a full exploration of the neighborhood.

21.6 NUMERICAL RESULTS

In preliminary computational studies we have applied our ideas to some well-known combinatorial optimization problems: the TSP, the Steiner tree problem in graphs, the quadratic assignment problem, and the p-median problem. The ideas and methods have not yet been fully customized and fully tuned for any of the problems and it has to be admitted that more fine-tuning is required to make the approach truly efficient. However, to underline our purpose that improvements are possible by putting more "intelligence" into the basic ant system, we report some results of our different (combinations of) agent system strategies for several instances of the TSP with size up to 439 nodes (Pr76, Lin105, Pr152, D198, Pr299, Pr439). For all these instances the number in the name refers to the number of nodes in a symmetric TSP with Euclidean distances [10].

Thereby, two classes of parameters may be distinguished (see Table 21.1):

1. Parameters with fixed values: parameters that are fixed and cannot change their values or are initialized with control values that were not changed during our test process due to logical reason or practical limitation.

2. Control parameters: parameters that are set by the (external) program initialization and, therefore, have to be tested for their influence on algorithmic performance behavior.

The different combinations of *nb_levels* and *nb_agents* are chosen to guarantee a comparable time base for all variants (the more agents the less levels) in order to be able to investigate whether more agents are worth being used. Dependent from the kind of information used, six different test runs (TR) are performed (see Table 21.2). Note that with the different combinations of *nb_agents* and *nb_levels* we obtain $3 \cdot 6 = 18$ results for each problem instance. Since we are interested in relative performance of the approaches tested here, we measure performance simply by comparing which approach produced a better solution over all parameter combinations. In this respect solution stands for the best solution value provided by at least one agent out of the same population. In doing so, we can summarize the results in two tables given below. The matrix of Table 21.3 shows the relative performance by counting the number of times in which one approach produced a better solution in direct comparison with another one. For example, the value 4 for 'TR5(5) > TR6(10)' means that in four of six cases TR5 with five agents produced a solution that outperformed the corresponding result for TR6 with ten agents.

fixed parameters		**value**
deviation		0.1
ρ	"evaporation" parameter for the ant system; preliminary numerical experiments show that fixing ρ to 0.5 provides quite stable algorithmic properties with respect to average solution quality and variance.	0.5
pre-specified control parameters		
α, β	weight factors of the classical ant system	1
nb_agents	number of agents	5,10,15
nb_levels	number of levels (depending on *nb_agents*)	$nb\_levels(5) = 60$ $nb\_levels(10) = 45$ $nb\_levels(15) = 30$
t	value (in %) for determ. of consistent chains	0.1, 0.2, 0.4
split_length	length of partial chains	randomly in [2, n]

Table 21.1 Parameters

run	description
TR1	Simple agent system without extensions (in this version we let the basic approach run, however, without any further technique for more specific use of information provided by all agents.
TR2	Agent system using ant trails
TR3	Agent system with conditioned use of ant trails
TR4	Agent system with re-initialization strategy
TR5	Agent system with re-initialization strategy using ant trails
TR6	Agent system with re-initialization strategy using conditioned ant trails

Table 21.2 Test runs

nb_agents		5						10						15					
	>	TR1	TR2	TR3	TR4	TR5	TR6	TR1	TR2	TR3	TR4	TR5	TR6	TR1	TR2	TR3	TR4	TR5	TR6
5	*TR1*	-	0	3	1	1	1	2	2	3	1	1	1	2	2	2	2	2	2
	TR2	0	-	3	1	1	1	2	2	3	1	1	1	2	2	2	2	2	2
	TR3	1	1	-	0	0	1	2	2	3	1	1	1	1	1	1	2	2	2
	TR4	5	5	6	-	0	3	6	6	6	3	3	4	6	6	6	3	3	3
	TR5	5	5	6	0	-	3	6	6	6	3	3	4	6	6	6	3	3	3
	TR6	5	5	6	1	0	-	5	5	5	1	1	3	5	5	5	2	2	2
10	*TR1*	1	1	2	0	0	1	-	0	1	0	0	0	1	1	1	1	1	1
	TR2	1	1	2	0	0	1	0	-	1	0	0	0	1	1	1	2	2	2
	TR3	1	1	1	0	0	1	0	0	-	0	0	0	1	1	1	2	2	2
	TR4	5	5	5	3	3	5	6	6	6	-	0	3	5	5	4	4	4	4
	TR5	5	5	5	3	3	5	6	6	6	0	-	3	5	5	5	4	4	4
	TR6	5	5	5	2	2	3	6	6	6	1	1	-	5	5	5	3	3	4
15	*TR1*	0	0	1	0	0	1	1	1	2	1	1	1	-	0	0	2	2	2
	TR2	0	0	1	0	0	1	1	1	2	1	1	1	0	-	0	2	2	2
	TR3	0	0	1	0	0	1	1	1	2	1	1	1	0	0	-	2	2	2
	TR4	4	4	4	3	3	4	5	4	4	2	2	3	4	4	4	-	0	3
	TR5	4	4	4	3	3	4	5	4	4	2	2	3	4	4	4	0	-	3
	TR6	4	4	4	3	3	4	5	4	4	2	2	3	4	4	4	1	1	-

Table 21.3 Comparative performance

nb_agents		*nb_levels* 5	10	15	20	25	30	35	40	45	50	60
5	TR1	0.091	0.086	0.086	0.084	0.054	0.050	0.050	0.050	0.050	0.048	0.047
	TR2	0.091	0.086	0.086	0.084	0.054	0.050	0.050	0.050	0.050	0.048	0.047
	TR3	0.091	0.086	0.086	0.084	0.050	0.050	0.050	0.050	0.050	0.050	0.050
	TR4	0.089	0.087	0.065	0.058	0.039	0.039	0.039	0.039	0.039	0.035	0.035
	TR5	0.089	0.087	0.065	0.058	0.039	0.039	0.039	0.039	0.039	0.035	0.035
	TR6	0.089	0.087	0.065	0.058	0.042	0.042	0.042	0.042	0.042	0.042	0.042
10	TR1	0.092	0.086	0.084	0.084	0.048	0.048	0.048	0.048	0.048	-	-
	TR2	0.092	0.086	0.084	0.084	0.048	0.048	0.048	0.048	0.048	-	-
	TR3	0.092	0.086	0.084	0.084	0.053	0.053	0.053	0.053	0.053	-	-
	TR4	0.086	0.085	0.056	0.045	0.040	0.037	0.037	0.036	0.036	-	-
	TR5	0.086	0.085	0.056	0.045	0.040	0.037	0.037	0.036	0.036	-	-
	TR6	0.086	0.085	0.056	0.045	0.041	0.041	0.039	0.039	0.039	-	-
15	TR1	0.084	0.082	0.082	0.082	0.049	0.049	-	-	-	-	-
	TR2	0.084	0.082	0.082	0.082	0.049	0.049	-	-	-	-	-
	TR3	0.084	0.082	0.082	0.082	0.049	0.049	-	-	-	-	-
	TR4	0.086	0.084	0.053	0.046	0.041	0.041	-	-	-	-	-
	TR5	0.086	0.084	0.053	0.046	0.041	0.041	-	-	-	-	-
	TR5	0.086	0.084	0.063	0.057	0.038	0.038	-	-	-	-	-

Table 21.4 Time-related performance (average deviation from optimality in %)

A further analysis refers to the algorithms' performance for different points in time. By way of explanation we give the average deviation values over all problem instances for different levels (see Table 21.4).

From the numerical results several conclusions may be drawn: The use of elite solutions as starting information is more effective than the other approaches as can be seen from the superiority of TR4, TR5 and, partially, TR6. The simple ant system does not work well here. This might be due to insufficient parameter tuning or a "wrong" form of embedding the information (trail) part. However, in conditioned combination with partial solutions the trail information might lead to an improvement in performance. The best number of agents strongly depends on the number of levels chosen; both parameters should be set while keeping the size of the problem under consideration in mind. Otherwise, a too small number of agents provide not sufficient information or the algorithm stops too early. To summarize, we may conclude that "information helps".

21.7 CONCLUSIONS

In this paper, we have put the well-known ant system into the light of more general cooperative agent systems. Several shortcomings have been pointed out and ideas for overcoming them have been proposed. The discussion of the deficiencies of ant colony systems reveals that these methods, in their orthodox form, fail to provide very good results. In fact, this is just one more occurrence of a basic reality: Nature may provide interesting ideas for the design of heuristics, but one should not stick too closely with analogies, especially if very difficult problems are to be solved; it is much more valuable to address heuristic design issues by thinking in terms of basic algorithmic principles. This is exactly what happens with the multi-agent search framework described in this paper.

We can see that cooperative search with multiple agents investigating, providing and, finally, using long-term based information might be helpful for realizing better algorithmic performance than conventional (single-pass) heuristics do. However, the critical point is the agent system's strategy for processing the richness of information in an appropriate way before being re-used for further search progress. Thereby, the simple and easy to implement concept of the so-called elite solutions as a central information pool seems to be the key to a leap in performance, even if it should be noted that combinations with other types of information – as, e.g., search trails – might be likewise successful. Our approach was implemented for some combinatorial optimization problems to show its specific multi-chain property. The heuristic frame proposed here is a general one such that applications to most types of other problems are conceivable. So current work concentrates on other variants, e.g., for the quadratic assignment problem or the vehicle routing problem (and a comparison with competitive intelligent search approaches, e.g., [11]).

Acknowledgements

This paper would not have been possible without the inspiring discussions with Prof. Fred Glover. Furthermore, the comments of the referees as well as the assistance of Kai Gutenschwager for some final polishing of the paper are appreciated.

References

[1] J.E. Beasley, OR-library: distributing test problems by electronic mail. J. Oper. Res. Soc. 44 (1990) 1069–1072.

[2] A. Colorni, M. Dorigo and V. Maniezzo, Distributed optimization by ant colonies. Proceedings of the first European Conference on Artificial Life, Paris, (1991) 134–142.

[3] A. Colorni, M. Dorigo and V. Maniezzo, An investigation of some properties of an ant algorithm. Proceedings of the Second Conference on Parallel Problem Solving from Nature, Brussels, (1992) 509–520.

[4] M. Dorigo, V. Maniezzo and A. Colorni, Ant System: optimization by a colony of cooperating agents. IEEE Transactions on Systems, Man, and Cybernetics, Vol. B-26 (1996), 29–41.

[5] F. Glover, Heuristics for integer programming using surrogate constraints. Decision Sciences 8 (1977) 156–166.

[6] F. Glover, Tabu search and adaptive memory programming - advances, applications and challenges. In: Interfaces in Computer Science and Operations Research, eds. R.S. Barr, R.V. Helgason and J.L. Kennington (Kluwer, Boston, 1996) 1–75.

[7] F. Glover and M. Laguna, Tabu search. In: Modern Heuristic Techniques for Combinatorial Problems, ed. C.R. Reeves (Blackwell, Oxford, 1993) 70–150.

[8] I.H. Osman and J.P. Kelly, Meta-Heuristics: an overview. In: Meta-Heuristics: Theory & Applications, eds. I.H. Osman and J.P. Kelly (Kluwer, Boston, 1996) 1–21.

[9] C.H. Papadimitriou and K. Steiglitz, Combinatorial optimization: algorithms and complexity, Prentice Hall, New York (1982).

[10] G. Reinelt, TSPLIB – A traveling salesman problem library. ORSA Journal on Computing 3 (1991) 376–384.

[11] Y. Rochat and E.D. Taillard, Probabilistic diversification and intensification in local search for vehicle routing. Journal of Heuristics 1 (1995) 147–167.

[12] S. Voß, Tabu search: applications and prospects. In: Network Optimization Problems, eds. D.-Z. Du and P.M. Pardalos (World Scientific, Singapore, 1993) 333–353.

[13] S. Voß, Intelligent Search, Manuscript, TH Darmstadt (1993).

22 THE MAX-MIN ANT SYSTEM AND LOCAL SEARCH FOR COMBINATORIAL OPTIMIZATION PROBLEMS

Thomas Stützle and Holger Hoos

TU Darmstadt, FB Informatik, FG Intellektik
Alexanderstr. 10, 64283 Darmstadt, Germany.
stuetzle@informatik.tu-darmstadt.de, hoos@informatik.tu-darmstadt.de

Abstract: In this paper we present an extension of the $\mathcal{MAX}$–$\mathcal{MIN}$ Ant System and apply it to Traveling Salesman Problems and Quadratic Assignment Problems. The extension involves the use of a modified choice rule and a hybrid scheme allowing ants to improve their solution by local search. The computational results show that this algorithm can be used to efficiently find near-optimal solutions to hard combinatorial optimization problems and that it is one of the best methods for the solution of structured quadratic assignment problems.

22.1 INTRODUCTION

Ant Colony Optimization (ACO) is a population-based, cooperative search metaphor inspired by the foraging behavior of real ants. One of the basic ideas of ACO is to use the equivalent of the *pheromone trail* used by real ants as a medium for cooperation and communication among a colony of artificial ants. The seminal work on ACO is the Ant System [8, 10] that was first proposed for solving the Traveling Salesman Problem (TSP). In the Ant System, the ants are simple agents that are used to construct solutions, guided by the pheromone trail and heuristic information based on intercity distances. Since the work on the Ant System, several extensions of the basic algorithm have been proposed, among those are the Ant-Q [14], the Ant Colony System (ACS) [15], the $\mathcal{MAX}$–$\mathcal{MIN}$ Ant System [32, 34] and the rank-based version of the

Ant System [5]. All these extensions are in some sense "greedier" than the Ant System, but differ significantly in main aspects of the search control.

The $\mathcal{MAX}$–$\mathcal{MIN}$ Ant System [34, 32] yields a significantly better solution quality than the Ant System for TSPs if only tour constructions are allowed. Additionally, the performance of ACO approaches can be considerably enhanced by adding a local search phase [32, 33, 9, 27], in which some ants are allowed to improve their solutions by a local search procedure. Thus, we obtain a hybrid algorithm consisting of a construction mechanism and a subsequent local search phase.

The Ant System, and also the $\mathcal{MAX}$–$\mathcal{MIN}$ Ant System ($\mathcal{MM}$AS), can be seen as adaptive sampling algorithms — adaptive in the sense that they consider the experience gathered in earlier iterations of the algorithm. In case local search procedures are used, the information of earlier iterations is collected indirectly by the pheromone trails associated with *features* of local optimal solutions. For TSPs these features are arcs, which are part of local optimal solutions. In case of Quadratic Assignment Problems (QAP) we use the location of specific items as solution features. These features are combined in subsequent iterations by a global probabilistic construction mechanism biased by the pheromone trails, directing the sampling of new solutions towards promising regions of the search space. Important issues associated with adaptive sampling algorithms are which solutions (here ants representing the solutions) should be allowed to perform a local search (see [33] for a first discussion of this issue for the application of the $\mathcal{MM}$AS to the TSP) and how powerful the local search method should be [19, 37].

In this paper we present an extension of the $\mathcal{MM}$AS using a modified selection rule for the construction of solutions, the *pseudo-random-proportional rule.* This selection rule was first proposed for Ant-Q [14] and is also used in the Ant Colony System [15], the successor of Ant-Q. The main effect of this selection rule is to yield a faster detection of high quality solutions by the algorithm. We apply the modified $\mathcal{MM}$AS to the TSP [24, 31], and to the QAP [11, 23], which models many practical problems arising in Location Science. $\mathcal{MM}$AS has been applied to TSP before in [34, 33], but the results presented here are significantly better, and the application of $\mathcal{MM}$AS to QAP is new. Both problems are $\mathcal{NP}$-hard [17], but whereas very large Euclidean TSPs can be solved to optimality, general instances of QAP with $n > 25$ are considered intractable. Apart from the practical interest of these problems, they also serve as a standard testbed for combinatorial optimization methods that try to find near-optimal solutions to $\mathcal{NP}$-hard problems. $\mathcal{MM}$AS is a *general purpose algorithm* that can be applied with minor modifications to combinatorial optimization problems yielding high quality solutions competitive with the best algorithms for both TSPs and QAPs. The computational results presented on symmetric and asymmetric TSP instances from TSPLIB [30] *and* QAP instances of QAPLIB [6] support this claim.

The paper is organized as follows: To make the paper self-contained, we briefly introduce $\mathcal{MM}$AS and its application to TSPs in Section 22.2. Subsequently, in Section 22.3, we describe how $\mathcal{MM}$AS is applied in combination with local search procedures to TSPs. In Section 22.4, we briefly introduce QAP and present the modifications for $\mathcal{MM}$AS that have to be made for its application to QAP. The computational results along with a discussion of specific issues concerning the selection of ants for local search are presented in Section 22.5, and we conclude after having discussed recent related work in Section 22.6.

22.2 THE $\mathcal{MAX}$–$\mathcal{MIN}$ Ant System FOR TSPS

22.2.1 The traveling salesman problem

In the TSP a given set of n cities has to be traversed so that every city is visited exactly once and the tour ends in the initial city. The optimization goal is to find a shortest possible tour. More formally, a TSP can be represented by a complete weighted directed graph $G = (\mathcal{V}, \mathcal{A}, d)$ with n nodes, $\mathcal{V}$ being the set of nodes, $\mathcal{A}$ being the set of arcs and $d : \mathcal{A} \mapsto \mathbb{N}$ a weight function associating a positive integer weight d_{ij} with every arc (i, j) that corresponds to the distance between cities i and j. The problem then is to find a shortest Hamiltonian cycle in the graph. For symmetric TSPs, the distances between nodes are independent of the direction of traversing the cities, i.e., $d_{ij} = d_{ji}$ for every pair of nodes. In the more general asymmetric TSP (ATSP), for at least one pair of nodes we have $d_{ij} \neq d_{ji}$. The TSP is a $\mathcal{NP}$-hard optimization problem [17] and is extensively studied in the literature [24, 31].

22.2.2 The $\mathcal{MAX}$–$\mathcal{MIN}$ Ant System

The $\mathcal{MM}$AS shares some important features such as the tour construction mechanism with the Ant System, see [32, 34] for a more detailed discussion. The Ant System is a population-based approach using a number of m ants. To solve TSPs in the Ant System and $\mathcal{MM}$AS a positive real value τ_{ij} is associated with each arc (i, j), imitating the pheromone trails. Additionally, the ants may also use heuristic information that, in the case of the TSP, is some function of the distance between cities. Initially, each ant is set on a randomly chosen city. Then in each step an ant selects from its current city i one of the cities it still has not visited. In the Ant System the ants make this selection according to the following probability distribution:

$$p_{ij} = \begin{cases} \dfrac{\tau_{ij}^{\alpha} \cdot \eta_{ij}^{\beta}}{\sum_{k \text{ not visited}} \tau_{ik}^{\alpha} \cdot \eta_{ik}^{\beta}} & \text{if city } j \text{ is not yet visited} \\ 0 & \text{otherwise} \end{cases} \tag{22.1}$$

where η_{ij} is a local heuristic function which in the Ant System and $\mathcal{MM}$AS is defined for TSPs as $\eta_{ij} = 1/d_{ij}$. η_{ij} is relatively high if the distance between nodes i and j is small, thus shorter arcs are preferred by the heuristic function. The probability distribution for the selection of the next city is biased by parameters α and β that determine the relative influence of the trail strength and the heuristic information. To keep track of the cities already visited, every ant maintains a list that is used to store its partial tour. After all ants have constructed a complete tour and have calculated the tour length L_k, the trails are updated. In $\mathcal{MM}$AS only the ant corresponding to the best tour found in an iteration is allowed to update the trails, whereas in Ant System all ants lay down some pheromone. The chosen ant lays down a constant quantity $Q = 1$ of pheromone. Thus, the trail intensities are updated according to the formula

$$\tau_{ij}^{new} = \rho \cdot \tau_{ij}^{old} + \Delta\tau_{ij} \tag{22.2}$$

where ρ with $0 < \rho < 1$ is the persistence of the trail, thus $1 - \rho$ models the evaporation. The lower ρ is, the faster information gathered in previous iterations is forgotten. The amount $\Delta\tau_{ij}$ is equal to $1/L_{\text{best}}$, L_{best} being the tour length of the trail updating ant if arc (i, j) is used by the updating ant on its tour, otherwise it is zero. Thus, frequently used arcs contained in short tours receive a higher amount of pheromone and will be selected more often in future cycles of the algorithm. This can be interpreted as a learning of *good* arcs. The two basic steps *tour construction* according to (22.1) and *trail update* according to (22.2) are then simply repeated for a given number of iterations. We refer to the complete cycle of tour construction and trail update as one iteration, similar to the notion of generation in genetic algorithms. The most important part in the Ant System is the treatment of the trail intensities. If some trails have very high intensities, the ants are very likely to choose the next arc among those. In practice, the long term effect of the trail intensities is to successively reduce the size of the effective search space by concentrating the search on a relatively small number of arcs.

The main differences between the $\mathcal{MM}$AS and the Ant System are that in $\mathcal{MM}$AS only one ant[1] is allowed to provide a feedback mechanism by updating the trails and that the trails are limited to an interval between some maximum and minimum possible values, $\tau_{\max}$ and $\tau_{\min}$. A minor difference is that in the $\mathcal{MM}$AS the trails are initialized to their maximum value $\tau_{\max}$. The introduction of minimum and maximum trail limits is motivated by the following observation: Especially for larger TSP instances with $n \geq 50$, the Ant System gives a rather poor solution quality, which can be enhanced by introducing some kind of elitism such as allowing only the best ant to update the trails. Yet, then it may occur that the differences between the trail intensities on arcs get so high that the same tours are constructed again and again, i.e., premature stagnation occurs. This may even lead to the fact that many, possibly better tours cannot be constructed anymore. To alleviate this problem, in $\mathcal{MM}$AS explicit maximum and minimum trail strengths $\tau_{\min}$ and $\tau_{\max}$ on the arcs are used to counteract stagnation of the search. Using limits on the trails, the differences among the trails on the arcs are limited and exploration of new, possibly better tours is increased. In [32, 34] we showed that this mechanism

[1]Choosing only one ant for trail update was proposed also for Ant-Q and ACS. Other possibilities like choosing several of the best ants as proposed in [5] may also be used.

is effective and allows the $\mathcal{MM}$AS to find significantly better tours than the Ant System and an improved tour quality for longer runs compared to ACS.

22.2.3 Modifications to the $\mathcal{MAX}$-$\mathcal{MIN}$ Ant System

Compared to $\mathcal{MM}$AS as presented in Section 22.2.2, we considered the modification of the selection rule for the next city and adopted the *pseudo-random-proportional rule* used in the ACS [15, 9]. Here with some probability p_0, the best possible choice is made, i.e., the ant chooses a city for which $\tau_{ij}^{\alpha} \cdot \eta_{ij}^{\beta}$ is maximal. Otherwise with probability $1 - p_0$, the next city is selected according to (22.1). The parameter p_0 determines the degree of exploration an ant performs while constructing a tour. If $p_0 = 1$, the ant will always choose the best possible action and not do any exploration, for $p_0 = 0$, this selection rule corresponds to the selection rule used in Ant System. The main advantage of using the pseudo-random-proportional rule is that good solutions can be found considerably faster.

When using the pseudo-random-proportional rule in $\mathcal{MM}$AS, for higher values of p_0, tighter maximum and minimum trail strengths have to be chosen. An intuitive explanation for this may be that for high values of p_0, the best possible choice is almost always made; in the cases where the choice is based on (22.1) and the intervals for the trails are rather wide, it may happen that also in the probabilistic selection according to (22.1), a city with highest $\tau_{ij}^{\alpha} \cdot \eta_{ij}^{\beta}$ will nearly always be chosen. Hence, nearly no exploration takes place. To limit the influence of the trails the trail limits have to be tighter, providing a better balance between exploitation and exploration. The probabilistic selection according to (22.1) still is biased by the trails, giving a preference for arcs with higher trails. To determine the tightness of the trail limits, we introduce a parameter *max-factor* such that $\tau_{\max} = \tau_{\min} \cdot$ *max-factor*.

22.2.4 Adding local search to the $\mathcal{MAX}$-$\mathcal{MIN}$ Ant System

Constructive algorithms usually result in a poor solution quality compared to local search algorithms, see [31] for an extensive investigation of this issue for TSPs. Yet, it has been noted that repeating local searches from randomly generated initial solutions still results in a considerable gap to the optimal solution for most problems. Thus, to find solutions of very high quality, some hybrid method is needed. An often used idea is to combine a probabilistic, adaptive construction heuristic with local search heuristics [4, 12], where the construction heuristic provides good starting points for the local search. The $\mathcal{MM}$AS is such an adaptive construction heuristic. It can be used to identify promising regions of the search space using the pheromone trails as an adaptive memory of previously found solutions and to construct new promising starting points for the local search. Thus, we have such a hybrid algorithm, in which ants are used to construct initial solutions for a subsequent local search, see Figure 22.1 for an algorithmic frame. Another advantage of using $\mathcal{MM}$AS as a construction heuristic is that by generating good initial solutions, the subsequent local search needs much fewer iterations to reach a local optimum. Thus, for a given time limit more local searches can be done than by starting from randomly generated solutions.

```
procedure MAX--MIN Ant System with local search
(1) Initialize the pheromone trails and the parameters
(2) while ( termination condition not met) do
  for k := 1 to m do
    construct a solution for ant k
  end for;
  Select ants to improve their solution by local search
  Apply local search to selected ants
  Update the pheromone trail, ∀τij τmax ≥ τij ≥ τmin
(3) return best solution found
end MAX--MIN Ant System;
```

Figure 22.1 General algorithmic frame for the $\mathcal{MAX}$–$\mathcal{MIN}$ Ant System with local search procedures.

22.3 THE $\mathcal{MAX}$–$\mathcal{MIN}$ Ant System APPLIED TO TSPS

While the principles of applying $\mathcal{MM}$AS to TSPs have been described before, in this section we present some important implementation aspects. To speed up the tour construction, we use candidate sets [31] of size 20 that contain for a given city i a list of the 20 nearest neighbors. If possible, an ant chooses the next city among the members of the candidate set, otherwise it chooses the closest city. For a simple local search procedure for symmetric TSPs, we implemented the so-called `2-opt` algorithm and a version in which moves corresponding to node insertions are also made, denoted as `2-h-opt` [2, 31]. For ATSP the `2-opt` heuristic is not directly applicable, as in a 2-opt move one part of the tour is inverted. So we use a particular 3-opt move that allows the re-insertion of arcs without reversing the direction of any partial tour, called `reduced 3-opt`. As we are not solving very large ATSP instances, we use a straightforward implementation that is much slower than more sophisticated implementations. We use `reduced 3-opt` as a first-improvement heuristic, whereas `2-opt` and `2-h-opt` are run as best-improvement heuristics, i.e., they select the best possible move in each iteration.

For larger TSPs a straightforward implementation of the local search procedures is computationally too expensive. As $\mathcal{MM}$AS, especially for symmetric TSPs, is run on larger problems we restricted the number of 2-opt moves to the 40 nearest neighbors [21] and search for an improving move via a fixed radius near neighbor search, as described in [2]. Another improvement concerns the update of the trail matrix. In the Ant System and $\mathcal{MM}$AS the trail matrix has n^2 entries, thus, the update according to (22.2) would require time $O(n^2)$.

To quicken the trail update, we only modify the trails leading from a node to members of its candidate set.

22.4 THE $\mathcal{MAX}$–$\mathcal{MIN}$ Ant System FOR THE QAP

In the QAP, n units have to be assigned to n locations. Knowing that a flow of value a_{ij} has to go from unit i to unit j and that b_{rs} is the distance between locations r and s, the aim is to find an assignment that minimizes the sum S of the products flow×distance:

$$S = \sum_{i=1}^{n} \sum_{j=1}^{n} a_{ij} b_{\phi(i)\phi(j)}$$

where $A = (a_{ij})$ and $B = (b_{rs})$ are two $n \times n$ matrices, $\Phi(n)$ is the set of all permutations (assignments) of $\{1, \ldots, n\}$, and $\phi(i)$ gives the location of unit i in the current solution $\phi \in \Phi(n)$.

When applying $\mathcal{MM}$AS to QAPs, the trails refer to specific locations for units, i.e., τ_{ij} gives the desirability of assigning unit i to location j. For the construction of a solution, essentially two steps have to be considered. First, we have to choose a unit that has to be assigned in the second step to one of the free[2] locations. In $\mathcal{MM}$AS the two steps are performed as follows: we choose randomly some free unit that is placed on some free location according to the pseudo-random-proportional rule. If a location is chosen probabilistically, this is done according to (22.1) without using any heuristic information.

For a simple local search heuristic for QAP, we implemented a greedy 2-opt heuristic whose neighborhood consists of all solutions that can be obtained by exchanging two units. In every step it explores the complete neighborhood $N(\phi)$ and chooses the move that gives the highest decrease of the objective function value, hence in every iteration $O(n^2)$ moves are explored. Note, that for QAP the evaluation of the moves can use the information available from the preceding iteration. The first iteration of the local search has a computational cost of $O(n^3)$, whereas the following ones have a computational cost of $O(n^2)$, see [35] for details. As another possibility we consider applying short runs of Robust Tabu Search [19].[3] The reason for doing so is that for most problem instances contained in QAPLIB, the best-known solutions have been found by Tabu Search approaches. Thus, by adding Tabu Search a higher solution quality than by adding a simple greedy 2-opt procedure should be expected.

A first application of the Ant System to QAP is presented in [27]. Independently from our approach, another new Ant Colony Optimization approach has been proposed for QAPs [16]. In Section 22.5.3 we will compare their results with ours, showing that both approaches result in a comparable solution quality. Yet, the approach of [16] differs strongly from the initial version of the Ant

[2] A unit is free if it is not yet assigned to a location and a location is free if no unit is assigned to it in the partial solution.

[3] A hybridization of genetic algorithms with short Tabu Searches was earlier proposed in [13].

Colony System, and is fine tuned for the application to QAPs. On the other side, our approach is a straightforward application of $\mathcal{MM}$AS as also used for TSPs.

22.5 EXPERIMENTS AND COMPUTATIONAL RESULTS

22.5.1 Parameter settings for the $\mathcal{MAX}$–$\mathcal{MIN}$ Ant System

For $\mathcal{MM}$AS various parameters have to be set, but good parameter settings can be found that give excellent results for both the TSP and QAP. The parameter value for α is set to $\alpha = 1.0$ throughout all experiments. The parameter values for ρ are set to $\rho = 0.95$ in case of TSPs and $\rho = 0.8$ in case of the application to QAPs. Note, that the higher ρ is, the slower the learning of the solution features is. The number of ants is set to ten for the TSPs and five for the QAP. In the case of the application to TSPs we discuss the role of the heuristic information and show that with additional local search, the heuristic information may be neglected, i.e., we may set $\beta = 0.0$ without significant solution quality degradation for most problem instances.

After each iteration the trails are modified according to (22.2). Hence, also for the trail interval, reasonable values around the expected amount of trail laid down by the best ant has to be found. To find a suitable trail interval as determined by $\tau_{\min}$ and $\tau_{\max}$ we calculate an estimate of the optimal solution *estimate-opt* by constructing a random solution and a subsequent run of a local search procedure and then chose $\tau_{\min} = 2/\textit{estimate-opt}$.[4] The upper trail limit is determined by the parameter *max-factor*, see Section 22.2.3. Throughout all experiments we set *max-factor* $= 5$. As we set a fixed interval for the trail limits, we modified the parameter value for p_0 according to the problem instance. For both TSP and QAP we chose p_0 in such a way that roughly 15 cities or units are chosen according to (22.1), i.e., $p_0 = (n - 15)/n$. In some sense this is similar to the modifications made in iterated Lin-Kernighan [20] to the current solution: In iterated Lin-Kernighan the current solution is modified by a random 4-opt move that cannot be reversed directly by the Lin Kernighan local search procedure. In $\mathcal{MM}$AS most of the time an ant makes the best possible choice and only sometimes a randomized choice leads to a slightly modified solution with respect to the one that would be obtained by setting $p_0 = 1$.

For the selection of the ant to update the trails, we depart from the initial approach to always choose the best ant of the iteration. It has been shown that, in combination with local search, for $\mathcal{MM}$AS it is advantageous to sometimes choose the best found solution for updating the trails [32]. On the other hand, if only the best solution is chosen, the algorithm may get stuck early in suboptimal solutions. Thus, we decided for alternatingly choosing the best ant of the iteration and the overall best ant for the trail update.

---

[4]Another possibility would be to use the best found solution to determine $\tau_{\min}$ and $\tau_{\max}$ dynamically during the run of the algorithm.

22.5.2 The $\mathcal{MAX}$-$\mathcal{MIN}$ Ant System for traveling salesman problems

22.5.2.1 Adapting the number of ants that perform local search. When combining $\mathcal{MM}$AS with local search, two specific issues are which and how many ants should be allowed to perform local search. In [33] it was shown that definite answers do not exist to these questions when using only the choice rule of (22.1). It was conjectured that the number of ants that is allowed to perform a local search should be increased during the run of the algorithm. Here we pursue such an approach. We start with a fixed number of ants, here 10, and first allow only one ant, the ant that has constructed the best tour, to perform a local search. The number of ants that are allowed to improve their tours by local search is then successively increased by one after every `incr-ls` iterations. In Table 22.1, we compare different settings of `incr-ls` for a rather short run time on benchmark problems. If `incr-ls` = 0, all ants are allowed to perform local search. As can be seen from Table 22.1, adapting the number of ants in the way outlined above actually improves the solution quality obtainable within a fixed time limit. The best results are obtained for values for `incr-ls` from 25 to 100. This result is a consequence of the fact that in early iterations the local search starts from rather bad initial solutions and takes a longer time to find the local minimum. Additionally the number of iterations performed to learn good trail values is too limited. On the other hand, the best ant may often have an initial solution that is very close to the previous best one and is more likely to again yield the previously found best solution. Thus, it is reasonable to increase the number of ants that perform a local search during the run of the algorithm. For the experimental results presented in Section 22.5.2.2, we chose `incr-ls` = 25 as this value seemed to yield reasonable performance for short runs.

`incr-ls`	`d198` 10 sec	`lin318` 20 sec	`pcb442` 30 sec	`att532` 60 sec	`rat783` 180 sec
0	0.48	1.99	2.24	2.13	2.11
1	0.43	1.73	1.86	1.89	2.06
3	0.43	1.46	1.71	1.67	1.85
5	0.45	1.24	1.54	1.77	1.51
10	**0.31**	1.02	1.36	1.26	1.60
25	0.34	**0.64**	1.22	**1.12**	1.51
50	0.40	0.78	**1.06**	1.16	1.40
100	0.45	0.76	1.21	**1.12**	**1.27**
∞	0.42	0.99	1.16	1.28	1.68

Table 22.1 $\mathcal{MM}$AS for short run times, dynamic scheme of choosing ants for a local search, see text for details. Averages over 10 runs of $\mathcal{MM}$AS. We give the instance name, the number that appears is the problem size, and the time limit in seconds allowed on a Sun Sparc 5 Workstation. The results are indicated as percent deviation from the known optimal solution, best results are indicated in boldface.

22.5.2.2 Results for Longer Runs with Local Search. In this section we present results obtained with $\mathcal{MM}$AS and `2-h-opt` as a local search procedure in case of symmetric TSPs and `reduced 3-opt` in case of ATSPs. The solution quality of `2-h-opt` usually is slightly better than the one obtained with `2-opt`. We compare the results of $\mathcal{MM}$AS with those obtained by our own implementation of the ACS using the same local search procedure. For ACS we used the parameter settings given in [9].[5] The experiments were run using the measurement protocol proposed for the First Competition for Evolutionary Computation.[6] In this protocol a fixed number of steps, each with a computational complexity of $O(n^2)$, is allowed, where n is the problem size. Every tour construction and every improving step found by the local search procedure is counted as one step, in case of `reduced 3-opt` one complete neighborhood scan is counted as n steps. For all algorithms we allowed $n \cdot 1000$ steps on symmetric TSPs and $2 \cdot n \cdot 2500$ steps on ATSPs. Thus, the comparison among the algorithms is fair, especially, as for all the same local search procedure was used. Additionally we give the computation times measured on a single processor of a Sun UltraSparc II with 167Mhz. To investigate whether the heuristic information is still necessary if local search procedures are used, we run $\mathcal{MM}$AS once with ($\beta = 1.0$) and once without heuristic information. Both approaches are compared to an *iterated local search* approach. We apply the random 4-opt move used in the iterated Lin-Kernighan algorithm [20, 21, 26] to modify the local optimal solution and use `2-h-opt` as the local search procedure. For the iterated local search approach we use the same local search as for $\mathcal{MM}$AS and allowed the same computation time as needed by $\mathcal{MM}$AS.

The computational results show that both $\mathcal{MM}$AS and ACS are competitive with the iterated local search approach, yielding for nearly all instances a higher solution quality. Compared to ACS, $\mathcal{MM}$AS yields better average results for most problems. Furthermore, the computation times for $\mathcal{MM}$AS are, especially for the larger instances, lower than for ACS. The reason for this difference is that ACS generates more solutions that are often already locally optimal and cannot be improved by the local search. Hence, ACS performs more iterations and constructs more tours than $\mathcal{MM}$AS, leading to higher run times. On the other side, solutions constructed by $\mathcal{MM}$AS are more diverse and offer more diversity to find better local minima by the local search.

The results show that for symmetric TSP instances the use of heuristic information gives, apart from instance `att532` slightly better results. On the contrary, for ATSPs the heuristic information slightly degrades the solution quality. It seems that the heuristic information in case of symmetric TSPs is of more use, whereas for ATSPs it may sometimes be misleading. We also verified that for ACS, neglecting the heuristic information results only in a small

[5]In [9] a `3-opt` local search algorithm is used, giving better results than `reduced 2-h-opt`. In the meanwhile we also implemented a fast 3-opt procedure using don't look bits [2]. First results show a considerable quality improvement. For latest results, visit our WWW-page at http://www.intellektik.informatik.tu-darmstadt.de/~tom/tsp.html.

[6]See also http://homepages.ulb.ac.be/~gseront/ICEO/2ndICEO.html and [3].

Problem instance	iterated local search	ACS	$\mathcal{MMAS}$ no heuristic	$\mathcal{MMAS}$ + heuristic
`d198.tsp` 15780	(0.0690%) 169	(0.0196%) 170	(0.0139%) 169	**(0.00697%)** 162
`lin318.tsp` 42029	(0.211%) 443	(0.198%) 506	(0.124%) 443	**(0.107%)** 429
`pcb442.tsp` 50778	(0.438%) 601	(0.354%) 896	(0.389%) 601	**(0.353%)** 589
`att532.tsp` 27686	(0.316%) 1266	(0.249%) 1477	**(0.214%)** 1250	(0.240%) 1266
`rat783.tsp` 8806	(0.660%) 2203	(0.498%) 3088	(0.386%) 2203	**(0.280%)** 2165
`ry48p.atsp` 14422	(0.326%) 20	(0.235%) 20	**(0.0%)** 20	**(0.0%)** 20
`ft70.atsp` 38673	(0.050%) 61	**(0.0114%)** 60	(0.0137%) 61	(0.0212%) 61
`kro124.atsp` 36230	(0.822%) 190	(0.756%) 187	**(0.0%)** 190	(0.0127%) 190
`ftv170.atsp` 2755	(0.396%) 966	(0.428%) 982	**(0.119%)** 966	(0.189%) 966

Table 22.2 Computational results for benchmark instances from TSPLIB comprising symmetric TSPs (upper half) and ATSPs (lower half). The known optimal solution is indicated below the instance name, and the number in the identifier is the problem size, except `kro124.atsp` has 100 cities. We give the average percentage deviation over 25 runs from the optimal solution and the maximal computation times in seconds, best average results are indicated in boldface.

decline in solution quality. This is a noticeable result, as it was shown that for the Ant System and ACS without local search, the heuristic information is very important [10, 9]. In case local search is allowed, the heuristic information is replaced by the information regarding which arcs are contained in the local optimal solutions, making the heuristic information of less use.

22.5.3 Quadratic Assignment Problems

In this section we present results obtained on some problem instances from QAPLIB [6]. As noted in [35], the problem instances may be classified in basically two types: Purely randomly generated problems without structure, for many of these instances the entries of the matrices A and B are uniformly distributed, and more structured real world problems. Additionally, in [35] randomly generated problem instances are proposed that exhibit a structure similar to the one observed in real world problems. It was also shown that the performance of heuristic methods depends strongly on the problem type. For real world problems in [35] it was shown that a Genetic Hybrid method (GH) [13] gave best results, whereas for randomly generated problems with uniform matrix entries robust Tabu Search [36] and Reactive Tabu Search [1] give the highest solution quality for the same computation times.

For the application of $\mathcal{MMAS}$ to QAP, we especially investigate the issue which local search procedure should be added to the $\mathcal{MMAS}$. We consider two

Problem instance	Robust Tabu Search	GH + 250 TS	HAS-QAP	$\mathcal{MMAS}$ + 250 TS	$\mathcal{MMAS}$ + 2-opt
sko72	0.146	0.143	0.277	**0.103**	0.301
sko81	0.135	0.136	0.144	**0.077**	0.153
sko90	**0.128**	0.196	0.231	**0.128**	0.366
tai50a	1.125	**1.049**	2.800	1.147	2.213
tai60a	1.203	1.159	3.070	**0.980**	2.211
tai80a	0.899	0.796	2.689	**0.741**	1.808
tai100a	0.894	—	—	0.684	1.66
bur26a-h	0.002	0.004	**0.0**	**0.0**	**0.0**
kra30a	0.268	**0.134**	0.629	0.135	0.821
kra30b	0.023	0.054	0.0711	**0.015**	0.121
tai35b	0.065	0.107	0.026	**0.024**	0.047
tai40b	0.531	0.2109	**0.0**	0.201	**0.0**
tai50b	0.345	0.2142	0.1916	0.125	**0.001**
tai60b	0.471	0.291	0.048	0.043	**0.014**
tai80b	1.031	0.827	**0.667**	0.725	0.741
tai100b	0.512	—	—	0.336	**0.334**

Table 22.3 $\mathcal{MMAS}$ compared to Genetic Hybrids, HAS-QAP and our implementation of Robust Tabu Search for unstructured (upper half) and structured (lower half) QAP instances. The number in the problem identifier indicates the dimension of the problem. Results for genetic hybrids and HAS-QAP are taken from [16]. The average solution quality over 10 runs for each algorithm is presented as percentage deviation from the best known solution values for the instances, see text for more details.

possibilities: One is the application of a simple 2-opt greedy local search, while the second consists of running short Tabu Searches of length $4n$ as in GH [13]. We allow for a total of 250 applications of Tabu Search. For the greedy 2-opt local search a total of 1000 local search applications were allowed, resulting in computation times comparable (or less) to the 250 applications of Tabu Search. Here all ants perform a local search in every iteration. We compare our results to those obtained in [16] with HAS-QAP and their re-implementation of GH. Additionally, we present results of our own implementation of Robust Tabu Search, run for $1000 \cdot n$ iterations. For GH, 250 Tabu Searches of length $4n$ were also allowed and in HAS-QAP, 1000 applications of a first-improvement local search with a limit of two complete scans of the neighborhood were done. Note, that in case of QAP the computational burden of the local search is so high, that the overhead caused by $\mathcal{MMAS}$ is negligible. Hence, as [16] report that GH and HAS-QAP have roughly comparable run times, their results are comparable to ours also with respect to computation times.

The computational results presented were obtained with two slight modifications for $\mathcal{MMAS}$. We use the *smoothing of trails* approach [34]. This approach increases the trails proportionally to the difference between τ_{ij} and $\tau_{\max}$ if stagnation of the algorithm is detected. We noted that the best results were obtained with a re-initialization of the trails to $\tau_{\max}$. To achieve a stronger search diversification, after a new initialization of the trails we choose only the best ant of the iteration for trail update for a given number of iterations, here four iterations. After that we continue with alternatingly choosing best ant of the iteration and the global best solution for the trail update.

The computational results show that both ant based approaches, $\mathcal{MM}$AS + 2-opt and HAS-QAP, have a comparable behavior.[7] $\mathcal{MM}$AS appears to be better on larger instances of `taixxa`, this may be due to the local search procedure chosen. Comparing the average solution quality of $\mathcal{MM}$AS + 2-opt and $\mathcal{MM}$AS + 250 Tabu Searches, we observe that using Tabu Search as a local search operator pays off in the case of unstructured instances (see upper part of Table 22.3), whereas for most of the structured QAP instances the simple 2-opt algorithm gives better results than the more sophisticated Tabu Search. Hence, for these instances it seems to be better to apply a rather simple local search procedure more often in order to identify promising regions of the search space. Comparing $\mathcal{MM}$AS + Tabu Search with GH, the combination of Tabu Search with $\mathcal{MM}$AS gives better solution quality on most instances. Also note that on nearly all instances, the combination of $\mathcal{MM}$AS with TS gives better average results than Robust Tabu Search, indicating the usefulness of $\mathcal{MM}$AS as a diversification mechanism for Tabu Search. As a general result, $\mathcal{MM}$AS seems to be one of the best approaches for the solution of structured real world QAPs.

22.6 RELATED WORK

Many algorithms have been proposed and applied to both the TSP and QAP. We cannot discuss all of these approaches but wish to indicate some rather recent work on these two problems. For the TSP the most efficient approaches are the iterated Lin-Kernighan algorithm [20, 21, 26] and the Genetic Local Search approach of [29]. Yet, both algorithms use the powerful Lin-Kernighan heuristic — it is regarded as the uncontested champion for TSPs — as the local search operator. It has been noted [37] that for the same time limit, a much better solution quality can be obtained using the Lin-Kernighan local search heuristic than by applying, for example, 2-opt. We are convinced that these results might also occur in $\mathcal{MM}$AS and that the solution quality can still be significantly improved. Ant System has been introduced using TSP as an example application. With ACS [15, 9] very good results are obtained for symmetric TSP and especially for ATSPs with an extension of ACS by a 3-opt procedure.[8] For TSP another promising approach is the extension to ASPARAGOS presented in [18]. Like our approach, she uses a very simple local search procedure for a spatially structured genetic algorithm. For ATSPs impressive results are reported solving all problems used here to optimality in all runs with computation times of only few seconds.

[7]Note, that in [16] it was shown that Reactive Tabu Search [1] performs best for instances `taixxa`.

[8]More interesting work an Ant Colony Optimization can be found at the Ant Colony Optimization page http://iridia.ulb.ac.be/dorigo/ACO/ACO.html.

For QAP apart from HAS-QAP [16], two other very promising approaches have been proposed. In [29] a genetic local search algorithm for QAPs is presented, extending earlier work of the authors on TSPs. Their approach is in the same spirit as our application of $\mathcal{MM}$AS to QAPs as they adapt their successful genetic algorithm for the TSP to the QAP using only minor modifications. A new feature they present is a very fast local search procedures inspired by *don't look bits* [2]. Especially for larger problems they obtain very good results with comparably low computation times. Another new approach is based on Scatter Search and uses Tabu Searches as the local search operator [7]. They obtain very high quality solutions but apparently at the cost of rather high run times.

22.7 CONCLUSION

In this paper we have presented an enhancement of the $\mathcal{MM}$AS using the pseudo random proportional rule first proposed for Ant-Q and ACS. We discussed a simple approach of how to adapt the number of ants that are allowed to perform a local search dynamically to increase the solution quality obtained for short runs of the algorithm. Furthermore, we showed that the heuristic information may be neglected when additional local search is performed giving only a minor degradation of solution quality. Compared to ACS, better solution quality could be obtained using $\mathcal{MM}$AS, while both approaches behave better than the iterated local search algorithm used for comparison. To show that $\mathcal{MM}$AS is a general purpose algorithm that can be easily extended to other problems, we presented its application to QAP. For QAP, $\mathcal{MM}$AS is among the best approaches for the solution of structured QAP instances. For the largest QAPLIB instance `tai256c`, we could even find a new best solution with an objective function value of 44759294 in some longer runs of our algorithm [22]. For the other structured problems very good results were obtained, finding the best known solutions in nearly all runs for problem instances up to $n = 60$. Especially for QAP, the local search method that is combined with $\mathcal{MM}$AS in a hybrid algorithm has a strong impact on the performance. For unstructured problem instances the use of short Tabu Search runs, instead of simple 2-opt, has shown to be particularly effective to obtain competitive results with other procedures, whereas for structured instances with 2-opt a better performance could be achieved. On the other side, for TSPs general knowledge seems to be that it is always best to choose Lin-Kernighan for the local search, if available [21]. Yet, if one cannot use this sophisticated local search procedure, which is much more difficult to implement than, e.g., `2-opt` or `3-opt`, $\mathcal{MM}$AS in its presented form is a viable alternative for the solution of TSPs. Generally, the $\mathcal{MM}$AS seems to be a very promising tool providing a general adaptive framework for the solution of hard combinatorial optimization problems.

In future we will add a more powerful local search procedure like the Lin-Kernighan algorithm to $\mathcal{MM}$AS to improve the solution quality on TSPs. Also for attacking larger QAPs efficiently, choosing faster local search procedures is an important issue. Another important issue is to identify the problem classes which can be efficiently solved using ACO approaches. For many problems it has been shown that their search space provides a structure in the way that locally optimal solutions are clustered in a small part of the search space [4]. It is on problems exhibiting such a behavior that current Ant Colony Optimization approaches work specifically well. Thus, we conjecture that current ACO approaches will prove to be particularly competitive on structured problem domains.

References

[1] R. Battiti and G. Tecchiolli. The Reactive Tabu Search. *ORSA J. Comput.*, 6(2):126–140, 1994.

[2] J.J. Bentley. Fast Algorithms for Geometric Traveling Salesman Problems. *ORSA J. Comput.*, 4(4):387–411, 1992.

[3] H. Bersini, M. Dorigo, S. Langerman, G. Seront, and L. Gambardella. Results of the First International Contest on Evolutionary Optimisation. Technical Report IRIDIA/96-18, IRIDIA, Université Libre de Bruxelles, 1996.

[4] K.D. Boese, A.B. Kahng, and S. Muddu. A New Adaptive Multi-Start Technique for Combinatorial Global Optimization. *Oper. Res. Letters*, 16:101–113, 1994.

[5] B. Bullnheimer, R. Hartl, and C. Strauss. A New Rank Based Version of the Ant System: A Computational Study. POM Working Paper 3/97, University of Viena, 1997.

[6] R.E. Burkard, S. Karisch, and F. Rendl. QAPLIB - A Quadratic Assignment Problem Library. *European J. Oper. Res.*, 55:115–119, 1991. Updated Version: http://fmatbhp1.tu-graz.ac.at/

[7] V.-D. Cung, T. Mautor, P. Michelon, and A. Tavares. A Scatter Search based Approach for the Quadratic Assignment Problem. In *Proc. IEEE Int. Conf. Evol. Comp. (ICEC'97)*, pages 165–170, 1997.

[8] M. Dorigo. *Optimization, Learning, and Natural Algorithms.* PhD thesis, Politecnico di Milano, 1992.

[9] M. Dorigo and L.M. Gambardella. Ant Colony System: A Cooperative Learning Approach to the Traveling Salesman Problem. *IEEE Trans. Evolut. Comp.*, 1:53–66, 1997.

[10] M. Dorigo, V. Maniezzo, and A. Colorni. The Ant System: Optimization by a Colony of Cooperating Agents. *IEEE Trans. Systems, Man, and Cybernetics - Part B*, 26(1):29–41, 1996.

[11] A.N. Elshafei. Hospital Layout as a Quadratic Assignment Problem. *Operations Research Quarterly*, 28:167–179, 1977.

[12] T.A. Feo and M.G.C. Resende. Greedy Randomized Adaptive Search Procedures. *J. Global Optim.*, 6:109–133, 1995.

[13] C. Fleurent and J.A. Ferland. Genetic Hybrids for the Quadratic Assignment Problem. In P.M. Pardalos and H. Wolkowicz, editors, *Quadratic Assignment and Related Problems*, pages 173–189. American Mathematical Society, 1994.

[14] L.M. Gambardella and M. Dorigo. Ant-Q: A Reinforcement Learning Approach to the Traveling Salesman Problem. In *Proc. 12th Int. Conf. Mach. Learning*, pages 252–260, 1995.

[15] L.M. Gambardella and M. Dorigo. Solving Symmetric and Asymmetric TSPs by Ant Colonies. In *Proc. IEEE Int. Conf. Evolut. Comp. (ICEC'96).* IEEE Press, 1996.

[16] L. Gambardella, E.D. Taillard and M. Dorigo. Ant Colonies for the QAP. Technical Report IDSIA-4-97, 1997.

[17] M.R. Garey and D.S. Johnson. *Computers and Intractability: A Guide to the Theory of NP-Completeness.* Freeman, San Francisco, CA, 1979.

[18] M. Gorges-Schleuter. Asparagos96 and the Travelling Salesman Problem. In *Proc. IEEE Int. Conf. Evolut. Comp. (ICEC'97)*, pages 171–174, 1997.

[19] W.E. Hart. *Adaptive Global Optimization with Local Search.* PhD thesis, University of California, San Diego, 1994.

[20] D.S. Johnson. Local Optimization and the Travelling Salesman Problem. In *Proc 17th Coll. Automata, Lang. and Prog.*, volume 443 of *LNCS*, pages 446–461. Springer, 1990.

[21] D.S. Johnson and L.A. McGeoch. The Travelling Salesman Problem: A Case Study in Local Optimization. In E.H.L. Aarts and J.K. Lenstra, editors, *Local Search in Combinatorial Optimization*, pages 215–310. John-Wiley and Sons, Ltd, 1997.

[22] S. Karisch, personal communication.

[23] T.C. Koopmans and M.J. Beckman. Assignment Problems and the Location of Economic Activities. *Econometrica*, 25:53–76, 1957.

[24] E.L. Lawler, J.K. Lenstra, A.H.G. Rinnooy Kan, and D.B. Shmoys. *The Travelling Salesman Problem.* John Wiley & Sons, 1985.

[25] Y. Li, P.M. Pardalos, and M.G.C. Resende. A Greedy Randomized Adaptive Search Procedure for the Quadratic Assignment Problem. In P.M. Pardalos and H. Wolkowicz, editors, *Quadratic assignment and related problems*, pages 237–261. American Mathematical Society, 1994.

[26] S. Lin and B.W. Kernighan. An Effective Heuristic Algorithm for the Travelling Salesman Problem. *Oper. Res.*, 21:498–516, 1973.

[27] V. Maniezzo, M. Dorigo, and A. Colorni. The Ant System Applied to the Quadratic Assignment Problem. Technical Report IRIDIA/94-28, Université Libre de Bruxelles, Belgium, 1994.

[28] P. Merz and B. Freisleben. A Genetic Local Search Approach to the Quadratic Assignment Problem. *Proc. Seventh Int. Conf. Genetic Algorithms (ICGA'97)*, pages 465–472, 1997.

[29] P. Merz and B. Freisleben. Genetic Local Search for the TSP: New Results. In *Proc. IEEE Int. Conf. Evolut. Comp. (ICEC'97)*, pages 159–164, 1997.

[30] G. Reinelt. TSPLIB — A Traveling Salesman Problem Library. *ORSA J. Comp.*, 3:376–384, 1991.

[31] G. Reinelt. *The Traveling Salesman: Computational Solutions for TSP Applications*, Lecture Notes in Computer Science 840. Springer, 1994.

[32] T. Stützle and H. Hoos. Improving the Ant-System: A Detailed Report on the $\mathcal{MAX}$–$\mathcal{MIN}$ Ant System. Technical Report AIDA–96–12, FG Intellektik, TH Darmstadt, August 1996.

[33] T. Stützle and H. Hoos. The $\mathcal{MAX}$–$\mathcal{MIN}$ Ant System and Local Search for the Traveling Salesman Problem. In *Proc. IEEE Int. Conf. Evolut. Comp. (ICEC'97)*, pages 309–314, 1997.

[34] T. Stützle and H. Hoos. Improvements on the Ant System: Introducing the $\mathcal{MAX}$–$\mathcal{MIN}$ Ant System. In G.D. Smith, N.C. Steele, R.F. Albrecht, editors, *Artificial Neural Networks and Genetic Algorithms*, pages 245–249, Springer Verlag, Wien, 1998.

[35] É.D. Taillard. Comparison of Iterative Searches for the Quadratic Assignment Problem. *Location Science*, 3:87–105, 1995.

[36] É.D. Taillard. Robust Taboo Search for the Quadratic Assignment Problem. *Parallel Computing*, 17:443–455, 1991.

[37] N.L.J. Ulder, E.H.L. Aarts, H-J. Bandelt, P.J.M. van Laarhoven, and E. Pesch. Genetic Local Search Algorithms for the Travelling Salesman Problem. In *Proc. 1st Int. Workshop Parallel Problem Solving from Nature (PPSN)*, Lecture Notes in Computer Science 496, pages 109–116. Springer, 1991.

23 TOWARDS AN EVOLUTIONARY METHOD - COOPERATING MULTI-THREAD PARALLEL TABU SEARCH HYBRID

Teodor Gabriel Crainic[1,3] and Michel Gendreau[2,3]

[1]Département des sciences administratives, Université du Québec à Montréal.
theo@ctr.umontreal.ca

[2]Département d'informatique et recherche opérationnelle Université de Montréal.
C.P. 6128, succursale Centre-Ville, Montréal, Québec, Canada, H3C 3J7.
michelg@crt.umontreal.ca

[3]Centre de recherche sur les transports, Université de Montréal.

Abstract: We present a first version of a hybrid metaheuristic that combines a cooperative multi-thread parallel tabu search procedure and a genetic search engine. The two algorithms evolve independently while systematically and asynchronously exchanging solutions. The method appears to be the first to propose a cooperation scheme where the initial population of the genetic algorithm is an elite set of solutions obtained by the parallel metaheuristic, while the best individuals generated during the genetic search enrich the pool of solutions available to all tabu search threads. Experimentation with instances of a multicommodity, capacitated, fixed cost network design formulation leads to an initial assessment of the performances of the method.

23.1 INTRODUCTION

Parallel computing architectures offer the possibility to design procedures that exploit more efficiently the solution space. This may be achieved through the acceleration of some tedious computational phase of the algorithm, the decomposition of the problem domain or search space, or the design of a multi-thread parallel search with various degrees of synchronization and cooperation. See, for example, Gendron and Crainic [8], Grama and Kumar [12], Crainic et al. [6], Crainic and Toulouse [5], and the references cited in these papers.

We are particularly interested in cooperative multi-thread tabu search (*TS*) methods and their application to multicommodity, capacitated, fixed cost network design problems. Recently, we developed such a parallel approach based on several tabu search threads that cooperate by an asynchronous exchange of information about the best solutions identified so far. The individual searches use the sequential tabu search method of Crainic, Gendreau and Farvolden [4], but differ in the search parameter settings and initial solutions.

Cooperation among the tabu search threads gradually builds up a pool of good solutions, an *elite* population. It is then natural to ask the question whether this population may be improved while the parallel tabu search (*PTS*) proceeds. The class of metaheuristics collectively known as *genetic algorithms* - *GA*, see, e.g., [13, 11, 22, 7, 18], offers an interesting approach to create a procedure to answer this question.

Parallel hybrid metaheuristics are increasingly becoming the vehicle of choice for developing efficient procedures for complex problems. Several authors have explored the idea of combining genetic principles and some other metaheuristic procedure in a parallel setting. Thus, Mahfoud and Goldberg [17] and Brown et al. [2] used simulated annealing principles either to improve population individuals or to control the application of GA operators. Others, such as Lin et al. [15] and Varanelli and Cohoon [21], started from parallel multiple-path simulated annealing methods and used genetic operators either to enrich the current solution or to control the exchanges among the annealing threads. As for hybrids that include tabu search ideas, Liepins and Baluja [14] present what is probably the first implementation of the "temporal memory" and "intensification" mechanisms in genetic search and introduce them into a cooperative multi-thread parallel GA procedure to reinforce the selection of desired characteristics within each population. Moscato [19] presents another cooperative multi-thread procedure where genetic operators are used to exchange information among the cooperating "agents" and a memoryless tabu search works towards increasing population diversity.

In this paper we present a first version of a Parallel Tabu Search – Genetic Algorithm hybrid that departs significantly from the paradigms already proposed in the literature. In our approach, a cooperative multi-thread parallel tabu search procedure and a genetic search engine proceed independently but cooperate by systematically and asynchronously exchanging solutions. Moreover, the initial GA population is a set of good solutions obtained by the parallel tabu search metaheuristic, while the best individuals generated during the genetic search become members of a pool of solutions available to all TS threads.

To our knowledge, the method presented in this paper is the first to propose such a cooperation scheme as a basis for building parallel tabu – genetic search hybrids. Our objective is to describe a first implementation of this strategy. We also make an initial assessment of its performances, and identify promising directions for further research and developments.

The paper is organized as follows. Section 23.2 recalls the formulations of the problem and the basic ideas of the parallel tabu search method, and presents the basic GA-PTS cooperation scheme. The description of the genetic method is detailed in Section 23.3. Experimental results are reported and analyzed in Section 23.4. We conclude with a number of general remarks and further research perspectives.

23.2 FORMULATIONS

We first recall the network design formulation and the sequential tabu search method. The multi-thread parallel implementation is then briefly described. The section concludes with the presentation of the cooperation scheme between the parallel tabu search and the genetic algorithm.

23.2.1 Model and sequential tabu search

Network design is a well–known problem, of both practical and theoretical significance [1]. A particular class of design models, the fixed charge, capacitated, multicommodity network design formulations with linear costs (CMND), appears prominently when addressing issues in infrastructure construction or improvement, transportation service network design, telecommunication network design and dimensioning, power system design, etc. Here, several "commodities" (goods, data packets, people, etc.) have to be moved, from their respective origins to their respective destinations, over the links of a network with limited capacities. Furthermore, other than the usual "transportation" costs related to the volume of each commodity flowing through a given link, a "fixed" (construction or utilization) cost is payed as soon as a link is used or capacity is added. The trade-off between the variable and fixed costs inherent in the selection of any given solution, as well as the interplay between the limited capacity (and the resulting competition among the various commodities) and the fixed costs of using the links of the network, makes for a formulation that presents both a broad modelling appeal and considerable algorithmic challenges when one attempts to solve realistically sized instances [16].

The general form of the arc–based formulation of the CMND that we address makes use of two sets of decision variables: x_a^p, the amount of flow of commodity $p \in P$ on arc $a \in A$, and y_a, the network design variables. It may be written as follows.

$$\text{Minimize } z(x,y) = \sum_{a \in D} f_a y_a + \sum_{p \in P} \sum_{a \in A} c_a^p x_a^p$$

$$\begin{aligned} \text{Subject to } \sum_{j \in N} x_{ji}^p - \sum_{j \in N} x_{ij}^p &= d_i^p \qquad \forall i \in N, p \in P \\ \sum_{p \in P} x_a^p &\leq u_a y_a \qquad \forall a \in A \end{aligned}$$

$$\begin{aligned} x_a^p &\geq 0 \qquad \forall a \in A, p \in P \\ y_a &= \{0,1\} \quad \forall a \in A, \end{aligned}$$

where:

$$d_i^p = \begin{cases} -w^p & \text{if } i = r^p, \\ w^p & \text{if } i = s^p, \\ 0 & \text{otherwise,} \end{cases}$$

and $G = (N, A)$ is the graph describing the network, N the set of nodes, $A = \{a = (i,j) \,|\, i \in N, j \in N\}$ the set of (design) arcs, and P the set of commodities to be distributed. For the sake of simplicity, we assume all arcs to be design arcs and define a commodity $p \in P$ as the flow between an origin–destination pair (r^p, s^p) with a transportation demand of w^p. We also have c_a^p, the unit cost of moving flow of commodity $p \in P$ on arc $a \in A$, f_a, the fixed cost of introducing capacity on arc $a \in A$, and u_a, the total capacity of arc $a \in A$.

The corresponding path-based formulation (CMND-P) is

$$\text{Minimize } z(h,y) = \sum_{a \in D} f_a y_a + \sum_{p \in P} \sum_{l \in L^p} k_l^p h_l^p$$

$$\begin{aligned} \text{Subject to } \sum_{l \in L^p} h_l^p &= w^p \qquad \forall p \in P && (23.1) \\ \sum_{p \in P} \sum_{l \in L^p} h_l^p \delta_{al}^p &\leq u_a y_a \quad \forall a \in A && (23.2) \\ h_l^p &\geq 0 \qquad \forall p \in P, l \in L^p, && (23.3) \\ y_a &= \{0,1\} \quad \forall a \in A. && (23.4) \end{aligned}$$

where

L^p = the set of paths for commodity $p \in P$,
h_l^p = the flow of commodity $p \in P$ on path $l \in L^p$,
δ_{al}^p = 1 if arc a is in the lth path for commodity p, and 0 otherwise,
k_l^p = $\sum_{a \in A} c_a^p \delta_{al}^p$, the cost unit of path $l \in L^p$ for commodity $p \in P$.

The sequential tabu search starts with the path-based CMND-P formulation and is based on the exploration of the space of extreme points $\tilde{H}$ of the polyhedron H defined by equations (23.1), (23.2), and (23.3). The adjacency relationships in $\tilde{H}$ define a natural neighborhood structure, while the pivoting rules of the simplex method provide an efficient way to determine the neighbors of any given solution. Furthermore, for any given path flow pattern $\tilde{h} \in \tilde{H}$, a minimum cost network design relative to the path flows, denoted $y(\tilde{h})$, can be obtained by setting

$$y_a(\tilde{h}) = \begin{cases} 0 & \text{if } \sum_{p \in P} \sum_{l \in L_a^p} \tilde{h}_a^p = 0 \,, \\ 1 & \text{otherwise,} \end{cases}$$

where L_a^p stands for the set of paths using design arc $a \in A$, $L_a^p \subset L^p$.

The method follows a rather straightforward pattern: following an initialization phase, the search proceeds with a sequence of local searches and diversification phases until a predetermined stopping criterion is met. Each local search exhibits the main structure of the primal simplex method with column generation, and is made up of several alternating series of pivot moves (the impact on the "true" objective function - the actual cost of the design solution - is used to select the best move) and path generation phases.

A neighborhood exploration sequence of pivot moves followed by a column generation phase is called a *column generation cycle.* Local search is terminated when no improvement to the current local best solution is obtained after a number of consecutive column generation cycles. The method then proceeds to diversify the search. The diversification strategy is based on a long term *frequency* memory structure that records the number of iterations that an arc has been in the basis (i.e., it belonged to at least one basic path). To diversify, one then selects a small number of often used arcs and closes them. During the tabu tenure of these arcs, all paths that contain them are not allowed to enter the basis (unless, of course, the aspiration criterion overrides the tabu status). Furthermore, the closed arcs are not available during the column generation phases. The length of the tabu tenure is defined in terms of a number of column generation cycles, following which the normal termination criterion of the local search procedure is activated.

Complete descriptions of the model and method, as well as calibration and performance results are presented and analyzed in Crainic et al. [4].

23.2.2 Parallel tabu search

In our implementation of cooperative multi-thread parallelism, each process independently executes the tabu search procedure we just described. The particular setting of a number of important search parameters (e.g., length of the short term tabu tenure, number of consecutive moves without improvement before the column generation cycle is activated, number of different paths generated at each iteration of the column generation cycle, etc.) differentiates each tabu search thread from the others. Moreover, different initial solutions are generally used for each thread. Thus, each thread follows a different *search strategy.*

The cooperation aspect of this parallelization scheme is achieved through asynchronous exchanges of information among the tabu search threads. The cooperation is characterized by what information to share, when it is communicated, between which threads it is exchanged, as well as by the utilization each search thread makes of the received information [20].

Threads exchange information about their respective best solutions identified so far. When a search thread improves its "local" best overall solution Z_{best}, it sends out the value of the solution, $z(h, y_{best})$, as well as its *context* defined as the vector of design variables y_{best}. Information exchanges are carried out through a *central memory* or *pool of solutions.* In this scheme, whenever a thread desires to send out information, it sends it to the pool. Similarly, when a thread desires to receive information, it reaches out and takes it from the pool. Communications are initiated exclusively by the individual threads irrespective of their role as senders or receivers of information. No broadcasting is taking place and there is no need for complex mechanisms to select the threads that will receive or send information.

The pool is thus an efficient implementation device that allows for a strict asynchronous mode of exchange, with no pre-determined connection pattern, where no process is interrupted by another for communication purposes, but where any thread may access at all times the data previously sent out by any other search thread. The pool also constitutes a set of good, "elite", solutions, continuously improved by the individual tabu search threads.

To complete the description of the cooperative process, one has to define when individual threads request information and how the external information is used locally. In the current implementation, a solution is requested from the pool before a diversification operation. Then, if the imported solution is better than the current local best, a substitution is performed and the diversification proceeds from the new solution. Otherwise, the imported solution is discarded and the procedure proceeds as usual. Note that local memories are not touched when the external solution is substituted. Thus, the search proceeds according to the information gathered locally.

Complete description of the procedure and some of its variants, as well as full implementation details are given in Crainic and Gendreau [3].

23.2.3 The structure of the PTS-GA hybrid

The cooperative multi-thread approach briefly recalled in the previous subsection belongs to the p-C MPDS strategy in the taxonomy proposed by Crainic et al. [6]: collegial control distributed among the p processes, asynchronous communications, multiple initial points, diverse search strategies. The PTS-GA hybrid is built according to the same parallelization strategy.

The genetic search engine is not actually linked to the various tabu searches. Instead, it communicates directly with the solution pool. The GA is initialized with the contents of the pool, once a sufficient number of solutions exist in the pool. Later during the search, asynchronous communications link the pool and the GA. The level of transfers from the PTS to the GA is controlled such that the GA population is not overloaded with foreign individuals, while ensuring a reasonable level of GA population diversity.

Migration thus occurs at intervals (a fixed number of generations in the current implementation) controlled by the GA and on request from the genetic engine. It is a two-way migration. In one direction, the solution transmitted from the PTS pool to the GA is extracted by the same mechanism used to return solutions to the individual tabu search threads: solutions are ranked according to their objective function value, and are randomly selected according to rank-biased probabilities. In the opposite direction, the GA sends its best individual not yet sent to the PTS so that the pool is enriched.

Thus, the PTS-GA hybrid may be viewed as a 2-process cooperative multi-thread parallelization strategy. From the PTS point of view, the GA is just another cooperating search thread that contributes to the enrichment of the solution pool. In genetic terms, the hybrid is a coarse-grained parallel method that implements an island model where two populations evolve independently, according to different strategies, and communicate their best individuals through an asynchronous migration policy.

23.3 THE EVOLUTIONARY ENGINE

Each individual considered by the genetic engine corresponds to a solution of the multicommodity network design problem. It is represented as a binary vector of the status of the design arcs: the *context vector* $\bar{y}$. It is assumed, of course, that an optimal assignment of flows to the network is associated to each particular design $\bar{y}$. Consequently, the fitness $\mathcal{F}$ of a feasible individual may be measured by the corresponding value $z(x(\bar{y}), \bar{y})$ of the objective function of the network design CMND formulation. This value may be obtained by solving the associated capacitated multicommodity minimum cost network flow problem and adding to the resulting cost of moving the flows the sum of the fixed costs of the open and used arcs.

The application of genetic operators may yield infeasible individuals. Unfeasibility may result either from too many arcs being closed, which disconnects the network, or from insufficient capacity to move all demands. In both cases, the amount of flow on artificial arcs measures the degree of unfeasibility. In the current implementation, infeasible individuals are rejected.

To generate a new individual, one selects two solutions in the current population and combines them. The selection criterion does not work with the actual fitness values; only the ranks according to these values are considered. We thus have a criterion that is independent of the specific problem context and should help avoid the overwhelming selection of a "super-individual" (which could lead to premature convergence).

The two selected parents are combined to yield two new individuals. We identify the two parents as P_1 and P_2, with associated evaluations $\mathcal{F}_1$ and $\mathcal{F}_2$, respectively, and $\mathcal{F}_1 \geq \mathcal{F}_2$, and the two offspring O_1 and O_2.

The first, *uniform* crossover operator generates the two new individuals by randomly exchanging genes between the two parents, according to a uniform distribution:

$$O_1(i) = P_1(i) \text{ and } O_2(i) = P_2(i), \ \forall i: \ m(i) = 1$$

$$O_1(i) = P_2(i) \text{ and } O_2(i) = P_1(i), \ \forall i: \ m(i) = 0$$

where the mask vector $m_i = \{0, 1\}$ is defined such that $Prob(m(i) = 1) = 1/2, \ i \in A$.

With the second operator, we aim to incorporate within the evolution mechanism some of the knowledge regarding the characteristics of good solutions gathered from the previous generations of the population. To achieve this goal, we construct a *frequency* vector f that indicates how often arcs were open in the best solutions found during the history of population evolution. Frequencies (appropriately scaled) may then be used to define a new mask $m_i = \{0, 1\}$ such that $Prob(m(i) = 1)$ is proportional to $f(i),\ i \in A$. The *frequency* crossover then aims to transmit to at least one offspring (the first one) the characteristics, in terms of open arcs, of the best parent (P_1), according to the evolution history as captured by the frequency count:

1. $\forall i$ such that $m(i) = 1$ and $P_1(i) = 1$, $O_1(i) = P_1(i)$ and $O_2(i) = P_2(i)$
2. $\forall i$ such that $m(i) = 1$ and $P_1(i) = 0$, $O_1(i) = P_2(i)$ and $O_2(i) = P_1(i)$
3. $\forall i$ such that $m(i) = 0$, $O_1(i) = P_2(i)$ and $O_2(i) = P_1(i)$

The evolutionary engine may then be described as follows:

- Get the initial population.
- At each generation;
  - evaluate the fitness of the population,
  - select two individuals,
  - select (eventually) and apply a crossover operator, and
  - add new feasible individuals to the population.
- When the population reaches a specified threshold, a number of least fit individuals are eliminated.
- At specific moments, import a good solution from the parallel tabu search pool and send in return the fittest individual not yet sent.

The procedure is stopped when the PTS procedures stops, that is, when all the tabu search threads have reached their stopping criterion (a given number of diversification phases).

23.4 COMPUTATIONAL EXPERIMENTATION

Computational experiments were run to determine if the addition of a genetic thread improves the performance of the parallel tabu search, and to assess the sensitivity of the hybrid algorithm to the parameters of the genetic search engine. These experiments were performed on a network of Sun UltraSparc 1 workstations, by using the PVM environment. The sequential tabu search procedure is in Fortran, while the PTS pool management and the GA are in C++.

Ten benchmark problems covering a wide range of network sizes (number of nodes, links, and commodities) were extracted from a larger set that is fully described in Crainic et al. [4]. The problems are presented in Table 23.1. All links are design arcs, and each commodity corresponds to an origin-destination pair. All problems have tight capacities and dominant fixed costs, the experimentations with the sequential tabu search procedure having shown that these characteristics make problems more difficult to solve. These experimentations, as well as those for the parallel tabu search procedure, have also shown that the problems of Table 23.1 are representative of the behaviour of the larger set. Our experiments with the hybrid procedure support this argument.

	1	2	3	4	5	6	7	8	9	10
Nodes	25	25	100	20	20	20	30	30	30	30
Links	100	100	400	230	300	300	520	520	700	700
Commodities	10	30	10	40	40	200	100	400	100	400

Table 23.1 Test Problem Dimensions

For all runs, four processors were used for the tabu threads and one for the genetic algorithm. Computational results are reported in Table 23.2 for two population sizes ($N = 100$ and 200) and three combinations of the crossover operators (frequency-based and uniform). For all runs, the GA was initiated with the first 50 solutions sent to the pool by the tabu search threads. If feasible, both offspring are added to the population. When the population reaches its maximum size, it is reduced by half by eliminating the individuals with the worst fitness values. All reported times are total wallclock times in seconds.

It is interesting to note that a large population does not seem to be necessary for good performances. Experiments with populations of maximum size = 500 further strengthened this observation. This could be explained in part by the particular origin of the population. Here, individuals represent the context vectors – the open arcs – of feasible solutions to network design problems. But, since problem instances of interest are tightly capacitated, the number of feasible individuals is relatively smaller than it is usual in GA applications. A larger population is not therefore a guarantee for increased population diversity. In fact, it is this issue of diversity among the individuals of a given population that prompted our choice of evolution policy: keep the parents and the feasible offspring.

With respect to the crossover operators, no combination dominates the others. In fact, the positive conclusion here is that using a frequency-based operator yields solutions at least as good as the more classical uniform crossover. This is an interesting indication that the integration of memory features to genetic search may be useful to address difficult problem instances (problems with many commodities, in our case). This result recalls the conclusions of Liepins and Baluja [14], and points to interesting research avenues. It is also noteworthy that the effect of using the frequency-based operator alone yields shorter solution times.

Problem	Crossover	Hybrid N=100		Hybrid N=200		Parallel tabu	
	Freq - Unif	Value	Time	Value	Time	Value	Time
1	0.0 - 1.0	51394	89.0	51284	117.8		
1	0.5 - 0.5	51416	90.7	51239	88.7	51234	237.9
1	1.0 - 0.0	51406	90.5	51394	88.7		
2	0.0 - 1.0	86594	227.5	86594	256.5		
2	0.5 - 0.5	86594	247.9	86594	224.9	86594	217.8
2	1.0 - 0.0	86594	220.4	86594	227.7		
3	0.0 - 1.0	70428	985.6	72701	947.0		
3	0.5 - 0.5	70659	980.7	70428	947.0	71040	1285.6
3	1.0 - 0.0	70458	950.3	71726	733.9		
4	0.0 - 1.0	643362	455.5	643362	523.8		
4	0.5 - 0.5	644032	429.8	644032	393.8	644032	928.9
4	1.0 - 0.0	644032	416.0	644032	374.6		
5	0.0 - 1.0	605394	423.2	605394	715.9		
5	0.5 - 0.5	605394	401.6	605394	409.6	605394	746.5
5	1.0 - 0.0	605394	351.3	605394	367.1		
6	0.0 - 1.0	126568	7989.2	124802	6423.6		
6	0.5 - 0.5	127102	6241.9	126992	5841.9	125466	10805.7
6	1.0 - 0.0	126780	6089.2	126992	5014.8		
7	0.0 - 1.0	104514	2565.5	104514	2817.2		
7	0.5 - 0.5	104514	2479.2	104514	2561.4	104514	3515.9
7	1.0 - 0.0	104514	2372.3	104514	2317.9		
8	0.0 - 1.0	167280	31267.2	169637	36098.3		
8	0.5 - 0.5	171338	27790.1	172144	28054.5	169185	66580.7
8	1.0 - 0.0	168234	24763.1	170220	27850.0		
9	0.0 - 1.0	57644	7230.2	57644	7921.8		
9	0.5 - 0.5	58117	6844.3	57718	6822.2	58117	3515.9
9	1.0 - 0.0	57644	6487.0	57644	6721.8		
10	0.0 - 1.0	147655	48455.2	147976	47231.2		
10	0.5 - 0.5	148716	47318.7	148716	43131.2	150104	43620.0
10	1.0 - 0.0	146231	43817.0	148716	41928.2		

Table 23.2 Computational Results with Four TS and One GA Search Threads

Turning to the relative performance of the hybrid compared to the parallel tabu procedure, one first observes that computation times are often significantly shorter for the hybrid algorithm. This follows directly from the cooperation between the PTS and the GA. Recall that the hybrid stops according to the same criterion as the PTS: once all tabu search threads have accomplished a given number (three, in our case) of diversification phases. A tabu search thread diversifies when it cannot improve its best solution for a number of consecutive column generation cycles. Shorter computation times thus imply that the time required to realize that a solution cannot be improved and that the search has to diversify has been reduced. But nothing has been changed to the PTS procedure; it is only the composition of the PTS pool that has been modified by the addition of good GA-generated solutions. Thus, the cooperation between the PTS and the GA changes the sequence of solutions examined by the tabu search threads and alters the behaviour of the parallel tabu search procedure. The results hints that the GA component may help in finding very good solutions earlier in the search.

To further clarify the influence of cooperation, we examined the relative behaviour of the three procedures – the PTS, the GA and the PTS-GA hybrid – on the larger problems (it appears rather clearly that the addition of a genetic process is not beneficial for small-sized problem instances). The three procedures have been run for the same wallclock time. First the GA has been run alone, without migration, starting from a random population and from the initial elite population built by the PTS. In all experiments, starting from the elite population has yielded better solutions (by a margin of 4% to 7%) than starting from purely random ones. Moreover, when starting from the elite populations, the good solutions are found very early in the search (of the order of 10% of the total time allowed), which corroborates the earlier conclusion. We then compared the GA alone (without migration) but starting from the elite population and the hybrid. The hybrid identified better solutions (by a margin of the order of 3% to 5%).

It is noteworthy that the best solutions identified by the PTS-GA hybrid were found in the parallel pool. This indicates that the exchange mechanism has imported solutions from the GA to the PTS pool, which have then been improved by one or more tabu search processes. It is a further indication of the value of cooperation between the two metaheuristic procedures.

As the results displayed in Table 23.2 indicate, the addition of a genetic process to the cooperative multi-thread tabu search is not always beneficial. Actually, it is certainly not interesting for small-sized problem instances with few products. On the other hand, it helps achieve better solutions, especially in the case of the more difficult problems, with a larger number of commodities.

23.5 CONCLUSIONS AND PERSPECTIVES

We have presented a first version of a multi-thread, cooperative parallel tabu search – genetic algorithm hybrid and applied it to address a difficult class of formulations: the fixed cost, capacitated, multicommodity network design problem.

The hybrid method appears very promising. Experimental results indicate that better quality solutions, in some cases significantly better, may be obtained. Moreover, since improvements seem to appear early in the evolution process, execution times can also be reduced.

These encouraging results prompt us to continue the development. Among the issues that we plan to address are:

- The adjustment of individuals and evaluation following generation. Infeasible individuals could be repaired, for example, by using greedy procedures. However, the sequential tabu search offers the means not only to perform this task, but also to improve the newly generated individuals.
- Explore more complex GA mechanisms, such as population diversity measures, dynamic population and migration control, new operators for mutation, selection, etc.
- Explore the impact of other quality measures to use in crossover operators: stability (number of generations arcs have been continuously open in the last best solutions), frequency vectors from "champion" individual tabu search threads, global PTS memories, etc.
- Continue the analysis of the cooperation between the parallel tabu search and the genetic engine, especially when more processors are dedicated to the PTS procedure.

- Parallelize the evolutionary engine.

More broadly, a fascinating area of research deals with other methodological approaches to population manipulation and evolution, such as scatter search (Glover [9]) and adaptive memory (Glover [10]) techniques in particular. We plan to explore and compare these techniques in the future.

Acknowledgements

Funding for this project has been provided by the Natural Sciences and Engineering Research Council of Canada, through its Research and Strategic Grant programs, and by the Fonds F.C.A.R. of the Province of Québec. We would like to take this opportunity and express our appreciation of the dedicated work of Mr. Anh-Tuan Nguyen and Mr. Jean-Phillipe Metz who undertook the implementation and experimentation activities.

References

[1] R.K. Ahuja, T.L. Magnanti, and J.B. Orlin. *Network Flows – Theory, Algorithms, and Applications.* Prentice-Hall, Englewood Cliffs, NJ., 1993.

[2] D.E. Brown, C.L. Huntley, and A.R. Spillane. A Parallel Genetic Heuristic for the Quadratic Assignment Problem. In J.D. Schaffer, editor, *Proceedings of the Third International Conference on Genetic Algorithms*, pages 406–415. Morgan Kaufmann Publishers, San Mateo, CA, 1989.

[3] T.G. Crainic and M. Gendreau. A Cooperative Parallel Tabu Search for Capacitated Network Design. Centre de recherche sur les transports, Université de Montréal, Montréal, QC, Canada, 1998.

[4] T.G. Crainic, M. Gendreau, and J.M. Farvolden. Simplex-based Tabu Search for the Multicommodity Capacitated Fixed Charge Network Design Problem. Publication CRT-96-07, Centre de recherche sur les transports, Université de Montréal, Montréal, QC, Canada, 1996.

[5] T.G. Crainic and M. Toulouse. Parallel Metaheuristics. In T.G. Crainic and G. Laporte, editors, *Fleet Management and Logistics*, pages 205–251. Kluwer, Norwell, MA, 1998.

[6] T.G. Crainic, M. Toulouse, and M. Gendreau. Towards a Taxonomy of Parallel Tabu Search Algorithms. *INFORMS Journal on Computing*, 9:61–72, 1997.

[7] D.B. Fogel. Evolutionary Programming: An Introduction and Some Current Directions. *Statistics and Computing*, 4:113–130, 1994.

[8] B. Gendron and T.G. Crainic. Parallel Branch-and-Bound Algorithms: Survey and Synthesis. *Operations Research*, 42:1042–1066, 1994.

[9] F. Glover. Genetic Algorithms and Scatter Search: Unsuspected Potentials. *Statistics and Computing*, 4:131–140, 1994.

[10] F. Glover. Tabu Search and Adaptive Memory Programming – Advances, Applications and Challenges. In R.S. Barr, R.V. Helgason, and J. Kennington, editors, *Interfaces in Computer Science and Operations Research.* Kluwer, Norwell, MA: 1–75, 1997.

[11] D.E. Goldberg. *Genetic Algorithms in Search, Optimization and Machine Learning.* Addison-Wesley, Reading, MA, 1989.

[12] A. Grama and V. Kumar. Parallel Search Algorithms for Discrete Optimization Problems. *ORSA Journal on Computing*, 7:365–385, 1995.

[13] J. Holland. *Adaptation in Natural and Artificial Systems.* University of Michigan Press, Ann Arbor, MI, 1975.

[14] G.E. Liepins and S. Baluja. apGA: An Adaptive Parallel Genetic Algorithm. In O. Balci, R. Sharda, and S.A. Zenios, editors, *Computer Science and Operations Research : New Developments in Their Interfaces*, pages 399–409. Pergamon Press, New York, NY, 1992.

[15] F.T. Lin, C.Y. Kao, and C.C. Hsu. Incorporating Genetic Algorithms into Simulated Annealing. In *Proceedings of the Fourth Int. Symp. on AI*, pages 290–297, 1991.

[16] T.L. Magnanti. Modeling and Solving Network Design Problems. 1993. Presented at NETFLOW93, San Miniato, Italy, October 3-7 (see Technical Report TR-21/93, Dipartimento di Informatica, Università degli Studi di Pisa, 155–159).

[17] S.W. Mahfoud and D.E. Goldberg. Parallel Recombinative Simulated Annealing: A Genetic Algorithm. *Parallel Computing*, 21:1–28, 1995.

[18] Z. Michalewicz. *Genetic Algorithms + Data Structures = Evolution Programs.* Springer-Verlag, Berlin, 1992.

[19] P. Moscato. An Introduction to Population Approaches for Optimization and Hierarchical Objective Functions: A Discussion on the Role of Tabu Search. *Annals of Operations Research*, 41:85–121, 1993.

[20] M. Toulouse, T.G. Crainic, and M. Gendreau. Communication Issues in Designing Cooperative Multi Thread Parallel Searches. In I.H. Osman and J.P. Kelly, editors, *Meta-Heuristics: Theory & Applications*, pages 501–522. Kluwer, Norwell, MA, 1996.

[21] J.M. Varanelli and J.P. Cohoon. Population-Oriented Simulated Annealing: A Genetic/Thermodynamic Hybrid Approach to Optimization. In D. Eshelmann, editor, *Proceedings of the Sixth International Conference*

on Genetic Algorithms, pages 174–181. Morgan Kaufmann Publishers, San Mateo, CA, 1995.

[22] L.D. Whitley. A Genetic Algorithm Tutorial. *Statistics and Computing*, 4:65–85, 1994.

24 PARALLEL TABU SEARCH FOR LARGE OPTIMIZATION PROBLEMS

El-Ghazali Talbi, Zouhir Hafidi and Jean-Marc Geib

LIFL URA-369 CNRS / Université de Lille 1
Bâtiment M3, 59655 Villeneuve d'Ascq Cedex, FRANCE.
talbi@lifl.fr, hafidi@lifl.fr, geib@lifl.fr

Abstract: This paper presents a parallel implementation of a tabu search algorithm based on adaptive parallelism. Adaptive parallelism demonstrates that massively parallel computing using a hundred of heterogeneous machines is feasible to solve large optimization problems. The parallel tabu search algorithm includes different tabu list sizes and new intensification/diversification mechanisms. Encouraging results have been obtained in solving the quadratic assignment problem. We have improved the best known solutions for some large real-world problems.

24.1 MOTIVATION AND GOALS

Many interesting combinatorial optimization problems are NP-hard, and they cannot be solved exactly within a reasonable amount of time. Consequently, heuristics are applied to solve such instances, usually derived from real-world problems. Tabu search (TS) is a general purpose heuristic (meta-heuristic) that has been proposed by Glover [7]. TS has achieved widespread successes in solving practical optimization problems in different domains (resource management, process design, logistics, telecommunications, etc.). Promising results of applying TS to a variety of academic optimization problems (traveling salesman, quadratic assignment, time-tabling, job-shop scheduling, etc.) are reported in the literature [8]. Solving large problems motivates the development of a parallel implementation of TS.

The proliferation of powerful workstations and fast communication networks with a constantly decreasing cost/performance ratio have led to the emergence of heterogeneous workstation networks and homogeneous clusters of processors (DEC Alpha farms, IBM SP/2, Cray T3D, CM-5, etc.) as platforms for high performance computing. These parallel platforms are generally composed of an important park of machines shared by many users.

Many parallel implementation of TS algorithms have been proposed in the literature. In general, they do not use advanced programming tools (load balancing, adaptive parallelism, checkpointing, etc.) to efficiently use the machines. Most of them are developed for dedicated parallel homogeneous machines. Load analysis of networks of workstations during long periods of time showed that only a small percentage of the available power was used [13, 23]. There was a substantial amount of idle time. In addition, a workstation may belong to an owner who will not tolerate external applications degrading the performance of the machine. Therefore, dynamic adaptive scheduling of parallel applications is essential.

Our aim is to develop a parallel implementation of TS, which can benefit greatly from a platform having combined computing resources of massively parallel machines (MPPs) and networks of workstations (NOWs). For this purpose, we use a dynamic scheduling system (MARS[1]), which harnesses idle cycles of time (keeping in mind the ownership of workstations), and supports adaptive processor allocation to dynamically reconfigure the set of processors hosting the parallel TS.

The testbed optimization problem we used is the quadratic assignment problem (QAP), one of the hardest among the NP-hard combinatorial optimization problems. The parallel TS algorithm includes different tabu list sizes and intensification/diversification mechanisms based on frequency, long-term memory and restricted neighborhood.

The remainder of the paper is organized as follows. In Section 24.2, we describe the sequential version of the heuristic algorithm. The parallel implementation of the algorithm will be detailed in Section 24.3. Finally, Section 24.4 will present results of experiments for several standard instances from the QAP-library.

24.2 A SEQUENTIAL HYBRID HEURISTIC FOR THE QAP

24.2.1 Sequential tabu search

A combinatorial optimization problem is defined by the specification of a pair (X, f), where the solution space X is a discrete set of all (feasible) solutions, and the objective function f is a mapping $f : X \longrightarrow R$. A neighborhood N is a mapping $N : X \longrightarrow P(X)$, which specifies for each $S \in X$ a subset $N(S)$ of X of neighbors of S.

The most famous local search optimization method is the *descent method.* A descent method starts from an initial solution and then continually explores the neighborhood of the current solution for a better solution. If such a solution is found, it replaces the current solution. The algorithm terminates as soon

[1]Multi-user Adaptive Resource Scheduler

as the current solution has no neighboring solution of better quality. Such a method generally stops at a local but often not global minimum.

Unlike a descent method, which might make a downhill move at each iteration of the search, TS selects the best neighbour solution even if this results in a worst solution than the current one. However, this strategy may result in cycling. So a *tabu list* is introduced to keep information about recently examined moves (short-term memory). A tabu move applied to a current solution may appear attractive because it gives, for example, a solution better than the best found so far. We would like to accept the move in spite of its status by defining *aspiration conditions.* Other advanced techniques may be used such as *intensification* to encourage the exploitation of a promising region in the search space, and *diversification* to encourage the exploration of new regions (long-term memory). To use TS for solving an optimization problem, we must define in the **input** the following items:

- S_0 : an initial solution
- $|T_i|$: size of the tabu list T_i (static or dynamic)
- $nbmax$: maximum number of iterations between two improvements of f

The **output** of the algorithm represents the best solution found during the search process. The following is a description of a sequential basic TS algorithm (without intensification/diversification mechanisms) [10]:

Step 1 : Initialization
- Choose an arbitrary initial solution $S_0 \in X$ $(S = S_0)$
- $nbiter = 0$ /* current iteration */
- $bestiter = 0$ /* iteration where the best solution has been found */
- $bestsol = S_0$ /* best solution found */
- $T_i = \emptyset$ /* The tabu list is empty */

Step 2 : Iteration
While $(nbiter - bestiter < nbmax)$ **Do**
- $nbiter = nbiter + 1$;
- Generate a set $V^* \subset N(S)$ containing solutions $t \in N(S)$, which are either not tabu or the aspiration condition holds
- Choose a solution S^* minimizing f over V^*
- Update the tabu list T_i ;

If $f(S^*) < f(bestsol)$ **Then** $bestsol = S^*$; $bestiter = nbiter$;
- $S = S^*$;

24.2.2 A tabu search based heuristic for the QAP

The heuristic algorithm has been used to solve the quadratic assignment problem (QAP). The QAP represents an important class of combinatorial optimization problems with many applications in different domains (facility location, data analysis, task scheduling, image synthesis, etc.). The first formulation was given by Koopmans and Beckmann in 1957 [11]. The QAP can be defined as follows. Given a set of objects $O = \{O_1, O_2, ..., O_n\}$, a set of locations $L = \{L_1, L_2, ..., L_n\}$, a flow matrix C, where each element c_{ij} denotes a flow cost between the objects O_i and O_j, and a distance matrix D, where each element d_{kl} denotes a distance between location L_k and L_l, find an object-location bijective mapping $M : O \longrightarrow L$, which minimizes the objective function f:

$$f = \sum_{i=1}^{n} \sum_{j=1}^{n} c_{ij}.d_{M(i)M(j)}$$

The QAP is NP-hard [5]. Finding an ϵ-approximate solution is also NP-complete [18]. This fact has restricted exact algorithms (such as branch and bound) to small instances ($n < 22$) [2]. An extensive survey and recent developments can be found in [14].

To apply the heuristic algorithm to the QAP, we must define for the TS sequential task the neighborhood structure of the problem and its evaluation, the short-term memory to avoid cycling and the long-term memory for the intensification/diversification phase.

- **Neighborhood structure and evaluation**: To represent a solution of the QAP, we use a representation that is based on a permutation of n integers $s = (l_1, l_2, ..., l_n)$ where l_i denotes the location of the object O_i. We use a pair exchange move in which two objects of a permutation are swapped.

  We use the formulae reported in [22] to efficiently compute the variation in the objective function due to a swap of two objects. The evaluation of the neighborhood can be done in $O(n^2)$ operations.

- **Short-term memory**: The tabu list contains pairs (i, j) of objects that cannot be exchanged (recency-based restriction).

  The aspiration function allows a tabu move if it generates a solution better than the best found solution. The total number of iterations is limited to $1000n$.

- **Long-term memory**: We use a frequency-based memory as a long-term memory that complements the information provided by recency-based memory. A matrix $F = (f_{i,k})$ represents the long-term memory. Let S denote the sequence of all solutions generated. The value $f_{i,k}$ represents the number of solutions in S for which $s(i) = k$. This quantity identifies the number of times the object i is mapped on the location k. The different values are normalized by dividing them by the average value, which is equal to $\frac{1+nb\_iterations}{n}$.

  If no better solution is found in $100n$ iterations, the *intensification* phase is started. The intensification phase starts from the best solution found in the current region and an empty tabu list. The use of the frequency-based memory will penalize non-improving moves by assigning a larger penalty to swaps with greater frequency values.

  A simulated annealing based process is used in the intensification phase. The relevance of our approach compared to pure simulated annealing is that it exploits memory of TS for selecting moves. For each move $m = (i, j)$, a penalty value P_m is introduced in the objective function which is equal to $Max(f_{i,s(i)}, f_{j,s(j)})$.

```
/* Simulated annealing with a penalty in the objective function */
Step 1 : Initialization
  Choose the best found solution S0 (S = S0)
  nbiter = 0 /* current iteration */
  T := Tmax /* Initial temperature Tmax = 10 */
Step 2 : Iteration
  Repeat
    nbiter := nbiter + 1 ;
    For j:=1 To NB_MAX Do /* NB_MAX = n */
      Generate a random move m which transforms S to S'
      ΔS = f(S') - f(S)
      If (ΔS < 0) or (random(0, 1) < exp(-(ΔS*Pm)/T)) Then
        S := S';
    End For
    T := a.T ; /* annealing schedule with a = 0.9 */
  Until T < Tmin /* Tmin = 0.5 */
```

The *diversification* phase is started after $100n$ iterations are performed without any improvements of the restarted TS algorithm. Diversification is performed in $10n$ iterations. The search will be forced to explore new regions by introducing a tabu restriction to the least frequently used entries. The incentive value associated to a move $m = (i, j)$ is $I_m = Min(f_{i,s(i)}, f_{j,s(j)})$. The tabu restriction is defined as $I_m < 1$. When the diversification phase terminates, the tabu status based on long-term memory is dropped.

24.2.3 *Restricted neighborhood*

In this section we present a version of the proposed heuristic for a particular real-world problem class: charts of grey densities. To generate a grey frame of density $\frac{m}{n}$ $(m \leq n)$, the problem consists of mapping m black cases and $n - m$ white cases on a grid which contains $n = n_1 * n_2$ square cases (numbered from 1 to n). The black (or white) cases must be spread uniformly on the grid to have nice frames. The problem was presented as a QAP in [22]. The ith element $(i \leq m)$ of a solution π gives the location where a black case has to be placed in the grid. The distance and the flow matrices are defined as follows:

$$d_{ij} = \begin{cases} 1 \; if \; i \leq m \; and \; j \leq m \\ 0 \; otherwise \end{cases}$$

$$c_{ij} = c_{n_2*(r-1)+s \;\; n_1*(t-1)+u} =$$

$$\max_{v,w \in \{-1,0,1\}} \frac{1}{(r - t + n_1 * v)^2 + (s - u + n_2 * w)^2}$$

$$(i, j) \in [1, n]$$

$$(r, t) \in [1, n1], (s, u) \in [1, n2]$$

where the cases i and j are located at the coordinates (r, s) and (t, u).

Exchanging two black or two white cases does not change the value of the objective function, and then many neighboring solutions have the same fitness value. The search reaches a plateau of the fitness landscape and the tabu search algorithm is then unable to grasp any information to guide itself towards the next solution to visit. Therefore, the properties of these problems must be considered in order to skip local optima. The exploration of the neighbourhood is restricted by swapping only black cases with white ones. By doing so, the neighbourhood size decreases from $O(n^2)$ to $O(m(n-m))$.

24.3 PARALLEL IMPLEMENTATION

24.3.1 Parallel tabu search

An extensive review of parallel TS may be found in [24, 4]. Many classifications of parallel TS algorithms have been proposed. They are based on many criteria: number of initial solutions, identical or different parameter settings, control and communication strategies. We have identified two main categories (Figure 24.1):

- **Domain decomposition**: Parallelism in this class of algorithms relies exclusively on:
  The decomposition of the search space (high-level decomposition): the main problem is decomposed into a number of smaller subproblems, each subproblem being solved by a different TS algorithm [20].
  The decomposition of the neighborhood (low-level decomposition): the search for the best neighbour at each iteration is performed in parallel, and each task evaluates a different subset of the partitioned neighborhood [19, 3]. A high degree of synchronisation is required to implement this class of algorithms.

- **Multiple tabu search tasks**: This class of algorithms consists of executing multiple TS algorithms in parallel. The different TS tasks start with the same or different parameter values (initial solution, tabu list size, etc.). Tabu tasks may be independent (without communication) [21] or cooperative [12, 17].

Parallelizing the decomposition of the search space or the neighborhood is problem-dependent. This assumption is strong and is met only for a few problems. The second class of algorithms is less restrictive and therefore more general. A parallel algorithm that combines the two approaches (two-level parallel organization) has been proposed in [1].

Implementations of parallel TS algorithms fall into three categories, depending on whether the number and/or the location of work (tasks, data) depends on the load state of the parallel machine (Table 24.1) :

- **non-adaptive processor allocation**: This category represents parallel implementations in which both the number of tasks of the application

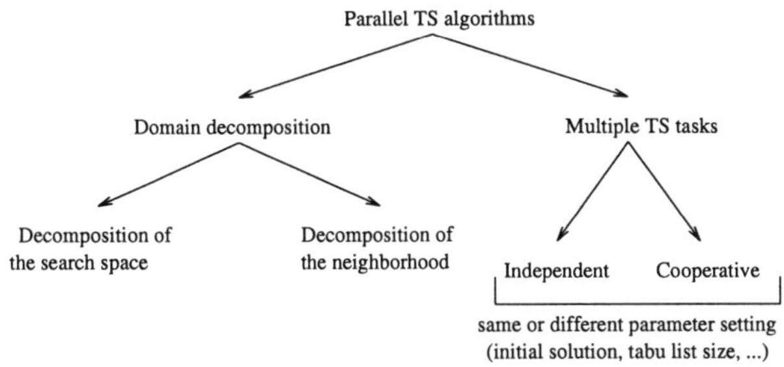

Figure 24.1 Hierarchical classification of parallel TS strategies.

and the location of work (tasks or data) are generated at compilation time (static scheduling). The allocation of processors to tasks (or data) remains unchanged during the execution of the application regardless of the current state of the parallel machine. Most of the proposed algorithms belong to this class.

An example of such an approach is presented in [15]. The neighborhood is divided into equal size partitions depending on the number of workers, which is equal to the number of processors of the parallel machine. In [3], the number of tasks generated depends on the size of the problem and is equal to n^2, where n is the problem size.

When there are noticeable load or power differences between processors, the search time of the non-adaptive approach presented is derived by the maximum execution time over all processors (highly loaded processor or the least powerful processor). A significant number of tasks are often idle waiting for other tasks to complete their work.

- **semi-adaptive processor allocation**: To improve the performance of the parallel non-adaptive processor allocation mechanisms, dynamic load balancing must be introduced [15, 1]. This class represents applications for which the number of tasks is fixed at compilation time, but the locations of work (tasks, data) are determined and/or changed at run-time. Load balancing requirements are met in [15] by a dynamic redistribution of work between processors. During the search, each time a task finishes its work, it proceeds to a work-demand. Dynamic load balancing through partition of the neighborhood is done by migrating data.

  However, the degree of parallelism in this class of algorithms is not related to load variation in the parallel system: when the number of tasks exceeds the number of idle nodes, multiple tasks are assigned to the same node. Moreover, when there are more idle nodes than tasks, some of them will not be used.

- **adaptive processor allocation**: A *parallel adaptive processor allocation* refers to a parallel execution with a dynamically changing set of tasks. Tasks may be created or killed as a function of the load state of the parallel machine. A task is created automatically when a processor becomes idle. When a node becomes busy, the task is killed. As far as we know, no work has been done on adaptive processor allocation for parallel TS.

	tasks or data	
processor allocation	**number**	**location**
non-adaptive	static	static
semi-adaptive	static	dynamic
adaptive	dynamic	dynamic

Table 24.1 Implementation mechanisms for parallel TS algorithms.

24.3.2 A parallel adaptive implementation

In this paper, a straightforward parallel TS approach has been used to introduce adaptive processor allocation. It consists of multiple independent TS algorithms running in parallel. This requires no communication between the sequential tasks. The algorithms are initialized with different solutions. Different parameter settings are also used (such as size of the tabu list). The size of the tabu list varies between $\frac{n}{2}$ and $\frac{3n}{2}$. The maximum number of TS tasks is set to n and each TS task is initialized with a different tabu list size from $\frac{n}{2}$ to $\frac{3n}{2}$ with an increment of 1.

The programming style used is the master/slave paradigm (SPMD model). The master task generates work to be processed by the slaves. Each slave task receives work from the master, computes a result and sends it back to the master. The master/slave paradigm works well in adaptive dynamic environments because:

- when a new node becomes available, a slave task can be started there,
- when a node becomes busy, the master task gets back the pending work, which was being computed on this node, to be computed on the next available node.

The master implements a central memory through which pass all communications, and that captures the global knowledge acquired during the search. The number of slaves created initially by the master is equal to the number of idle nodes in the parallel platform. Each slave implements a sequential TS task. The initial solution is generated randomly and the tabu list is empty. The parallel TS algorithm reacts to two events (Figure 24.2):

- **Transition of the load state of a node from idle to busy**: If a node hosting a slave becomes loaded, the master *folds up* the application by withdrawing the slave. The concerned slave puts back all pending work to the master and dies. The pending work is composed of the current solution, the best local solution found, the short-term memory, the long-term memory and the number of iterations done without improving the best solution. The master updates the best global solution if it is worse than the best local solution received.

- **Transition of the load state of a node from busy to idle**: When a node becomes available, the master *unfolds* the application by starting a new slave on it. Before starting a sequential TS, the slave task receives the values of the different parameters from the master: the best global solution and an initial solution which may be an intermediate solution found by a folded TS task, which constitute a "good" initial solution. In this case, the slave receives also the state of the short-term memory, the long-term memory and the number of iterations done without improving the best solution.

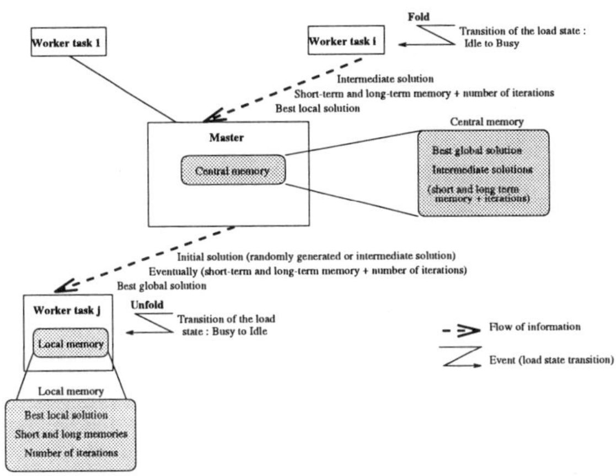

Figure 24.2 Architecture of parallel TS.

The parallel run-time system to be used must support dynamic adaptive scheduling of tasks, where the programmer is kept from the complex task of managing the availability of nodes and the dynamics of the target machine. Piranha (under *Linda*) [6], CARMI/Wodi (under *PVM*) [16], and MARS (under PVM/PM^2) [9] are representative of such scheduling systems. We have used the MARS dynamic scheduling system.

24.4 COMPUTATIONAL RESULTS

For our experimental study, we collected results from a platform combining a network of heterogeneous workstations and a massively parallel homogeneous machine. The network is composed of 126 workstations (PC/Linux, Sun4/Sunos, Alpha/OSF, Sun/Solaris, etc.) owned by researchers and students of our University. The parallel machine is composed of 16-Alpha processors connected by a crossbar interconnection network. The parallel adaptive TS competes with other applications (sequential and parallel) and owners of the workstations.

To evaluate the performance of the parallel TS in terms of solution quality and search time, we have used standard QAP problems of different types (QAP library):

- random and uniform distances and flows : Tai35a, Tai100a
- random flows on grids : Nug30, Sko100a-f, Tho150, Wil100
- real-life or real-life like problems : Bur26d, Ste36b, Esc128, Tai150b, Tai256c

The parallel TS algorithm was run 10 times to obtain an average performance estimate. Table 24.2 shows the best known, best found, worst, average value and the standard deviation of the solutions obtained for the chosen small instances ($n < 50$). The search cost was estimated by the time in seconds to find the best solution.

We always succeed in finding the best known solutions for small problems. This result shows the efficiency and the robustness of the parallel TS.

Instance	Best known	Best found	Worst	Average	Standard Deviation	Average search time (sec)
Tai35a	2 422 002	2 422 002	2 422 002	2 422 002	0	566
Nug30	6124	6124	6124	6124	0	337
Bur26d	3 821 225	3 821 225	3 821 225	3 821 225	0	623
Ste36b	15 852	15 852	15 852	15 852	0	763

Table 24.2 Results for small problems ($n < 50$) of different types.

Table 24.3 shows the best results obtained for five runs of the parallel TS for large problems. More runs are needed to have an average performance study. For random-grid problems, we found the best known solutions (for *Sko100c*) or solutions very close to best known solutions. The most difficult instance for our algorithm is the random-uniform *Tai100a*, in which we obtain a gap of 0.32% above the best known solution. For this class of instances, it is difficult to find the best known solution but simple to find "good" solutions.

For the third class of instances (real-life or real-life like), the best known solutions for *Tai150b* and *Tai256c* (generation of grey patterns) have been improved.

According to the results, we observe that the algorithm is well fitted to a large number of instances but its performance decreases for large uniform-random instances (Tai100a). Work is still to be done to improve the efficiency of the

Instance	Best known	Best found	Gap	Search time (mn)
Tai100a	21 125 314	21 193 246	0.32%	117
Sko100a	152 002	152 036	0.022%	142
Sko100b	153 890	153 914	0.015%	155
Sko100c	147 862	147 862	0%	132
Sko100d	149 576	149 610	0.022%	152
Sko100e	149 150	149 170	0.013%	124
Sko100f	149 036	149 046	0.006%	125
Wil100	273 038	273 074	0.013%	389
Tho150	8 134 030	8 140 368	0.078%	287
Esc128	64	64	0%	230
Tai150b	499 348 972	**499 342 577**	-0.0013%	415
Tai256c	44 894 480	**44 810 866**	-0.19%	593

Table 24.3 Results for large problems ($n \geq 50$) of different types.

algorithm by introducing intensification mechanisms based on *path relinking*, where S represents a small subset of elite solutions.

Performance results are very encouraging for large real-world problems, such as *Tainnc*. Table 24.4 shows the results obtained by the parallel TS with restricted neighborhood. The instances are denoted $Greyn_1\_n_2\_m$. We have improved the best known solutions for three instances and found the best known solutions for the others. For example, the *Tai256c* instance has been improved by 0.053% relative to our best solution, against 0.24% relative to the best known solution, and the search time is considerably reduced.

Instance	Problem size	Density	Best known	Best found	Gap	Search time (mn)
Grey8_8_13	64	13/64	1 855 928	1 855 928	0%	< 1
Grey16_16_64	256	64/256	19 065 968	**19 050 432**	-0.081%	213
Grey16_16_86	256	86/256	38 702 396	**38 381 310**	-0.83%	250
Grey16_16_92	256	92/256	44 894 480	**44 787 190**	-0.24%	392
Grey16_16_98	256	98/256	51 486 642	51 486 642	0%	17
Grey16_16_128	256	128/256	90 565 248	90 565 248	0%	15

Table 24.4 Results for grey densities problems. *Grey8_8_13* and *Grey16_16_92* correspond respectively to *Tai64c* and *Tai256c* instances of the QAP-library

24.5 CONCLUSION AND FUTURE WORK

A parallel TS algorithm which can benefit greatly from a platform having combined computing resources of MPPs and NOWs has been proposed, with special emphasis in solving efficiently large-scale optimization problems.

An experimental study has been carried out in solving the QAP. The parallel TS algorithm includes different tabu list sizes and intensification/diversification mechanisms (e.g., frequency based long-term memory or restricted neighborhood). The performance results obtained for several standard instances from the QAP-library are very encouraging in terms of robustness. The parallel algorithm have always succeeded in finding the best known solutions for small problems ($n < 50$).

The best known solutions of the large real-life problems "charts of grey densities" (*Tai256c, ...*) and the real-life problem instances *Tai150b* have been improved. The parallel algorithm often produces best known or close to best known solutions for large random problems. Other sophisticated intensification/diversification mechanisms to improve the results obtained for large random-uniform problems (*Taixxa*) are under evaluation.

The parallel adaptive TS is used to solve other optimization problems: set covering, and independent set, multiconstraint knapsack. We plan to improve our framework to provide adaptive parallelism to other meta-heuristics (such as simulated annealing, genetic algorithms, and ant colonies).

References

[1] P. Badeau, M. Gendreau, F. Guertin, J-Y. Potvin, and E. Taillard. A parallel tabu search heuristic for the vehicle routing problem with time windows. *RR CRT-95-84, Centre de Recherche sur les Transports, Universit de Montral*, Dec 1995.

[2] A. Brungger, A. Marzetta, J. Clausen, and M. Perregaard. Joining forces in solving large-scale quadratic assignment problems in parallel. In A. Gottlieb, editor, *11th Int. Parallel Processing Symposium*, Geneva, Switzerland, Apr 1997. Morgan Kaufmann.

[3] J. Chakrapani and J. Skorin-Kapov. Massively parallel tabu search for the quadratic assignment problem. *Annals of Operations Research*, 41:327–341, 1993.

[4] T.D. Crainic, M. Toulouse, and M. Gendreau. Towards a taxonomy of parallel tabu search algorithms. *INFORMS Journal on Computing*, 9:61–72, 1997.

[5] M.R. Garey and D.S. Johnson. *Computers and intractability: A guide to the theory of NP-completeness.* W.H. Freeman and Co., San Francisco, 1979.

[6] D. Gelernter and D.L. Kaminsky. Supercomputing out of recycled garbage: Preliminary experience with Piranha. In *6th ACM International Conference on Supercomputing*, Jul 1991.

[7] F. Glover. Tabu search - part I. *ORSA Journal on Computing*, 1:190–206, 1989.

[8] F. Glover and M. Laguna. Tabu search. In C. Reeves, editor, *Modern Heuristic Techniques for Combinatorial Problems*, pages 70–150. Blackwell Scientific Publications, 1992.

[9] Z. Hafidi, E.G. Talbi, and J-M. Geib. MARS: Adaptive scheduling of parallel applications in a multi-user heterogeneous environment. In *European School of Computer Science ESPPE'96: Parallel Programming Environments for High Performance Computing*, pages 119–122, Alpe d'Huez, France, 1996.

[10] A. Hertz and D. de Werra. The tabu search metaheuristic: How we use it? *Annals of Mathematics and Artificial Intelligence*, pages 111–121, 1989.

[11] T.C. Koopmans and M.J. Beckmann. Assignment problems and the location of economic activities. *Econometrica*, 25:53–76, 1957.

[12] M. Malek, M. Guruswamy, M. Pandya, and H. Owens. Serial and parallel simulated annealing and tabu search algorithms for the traveling salesman problem. *Annals of Operations Research*, 21:59–84, 1989.

[13] D.A. Nichols. Using idle workstations in a shared computing environment. *ACM Operating System Review*, 21(5):5–12, 1987.

[14] P. M. Pardalos, F. Rendl, and H. Wolkowicz. The quadratic assignment problem: A survey and recent developments. *DIMACS Series in Discrete Mathematics and Theoretical Computer Science*, 16:1–42, 1994.

[15] C.S. Porto and C. Ribeiro. Parallel tabu search message-passing synchronous strategies for task scheduling under precedence constraints. *Journal of Heuristics*, 1:207–223, 1996.

[16] J. Pruyne and M. Livny. Parallel processing on dynamic resources with CARMI. In *Proc. of the Workshop on Job Scheduling for Parallel Processing IPPS'95*, 1995.

[17] C. Rego and C. Roucairol. A parallel tabu search algorithm using ejection chains for the vehicle routing problem. In I.H. Osman and J.P. Kelly (editors), *Meta-Heuristics: Theory & Applications*, pages 661–675. Kluwer, 1995.

[18] S. Sahni and T. Gonzales. P-complete approximation problems. *Journal of the ACM*, 23:556–565, 1976.

[19] E. Taillard. Robust taboo search for the quadratic assignment problem. *Parallel Computing*, 17:443–455, 1991.

[20] E. Taillard. Parallel iterative search methods for vehicle routing problems. *Networks*, 23:661–673, 1993.

[21] E. Taillard. Parallel taboo search techniques for the job shop scheduling problem. *ORSA Journal on Computing*, 6:108–117, 1994.

[22] E. Taillard. Comparison of iterative searches for the quadratic assignment problem. *Location Science*, 3:87–103, 1995.

[23] M.M. Theimer and K.A. Lantz. Finding idle machines in a workstation-based distributed system. *IEEE Transactions on Software Engineering*, 15:1444–1458, 1989.

[24] S. Voß. Tabu search: Applications and prospects. In D.-Z. Du and P. M. Pardalos (editors), *Network Optimization Problems*, pages 333–353. World Scientific, 1993.

25 SEQUENTIAL AND PARALLEL LOCAL SEARCH ALGORITHMS FOR JOB SHOP SCHEDULING

Huub M.M. ten Eikelder[1], Bas J.M. Aarts[1,2],
Marco G.A. Verhoeven[2] and Emile H.L. Aarts[1,2]

[1]Eindhoven University of Technology
Department of Mathematics and Computing Science
P.O. Box 513, 5600 MB Eindhoven, The Netherlands.
wsinhte@win.tue.nl

[2]Philips Research Laboratories
P.O. Box 8000, 5600 JA Eindhoven, The Netherlands.
aartsb@natlab.research.philips.com, aarts@natlab.research.philips.com,
mverhoev@natlab.research.philips.com

Abstract: We describe a fast sequential tabu search algorithm for the job shop scheduling problem. The algorithm uses an efficient way to recompute the path lengths in the graph representation after a step to a neighbouring solution. For a parallel algorithm, that consists of independent runs of the sequential algorithm, we give a probabilistic analysis of the possible speed-up.

25.1 INTRODUCTION

In this paper we are concerned with job shop scheduling, which is an important model in scheduling theory [6], that serves as a proving ground for new algorithmic ideas and provides a starting point for the more complicated, practically relevant scheduling models. In this paper we address some novel issues in the design and analysis of sequential and parallel local search algorithms. More specifically in Section 25.2 of this paper, we present a fast tabu search algorithm for job shop scheduling. The algorithm is based on a neighbourhood structure that has been proposed by Nowicki and Smutnicki [9] and on the method of

Taillard [11] to use a fast computable lower bound for the cost of neighbouring solutions. Taillard's method requires that the length of the longest path up to and from each node (in the graph representation) is known. After a step to a next solution, which corresponds to a new graph, these lengths have to be recomputed. However, the old and new graph only differ in the direction of a few arcs. This observation leads to so-called Bow-tie algorithm, which allows a faster recomputation of the various path lengths. In this way we obtain a computation time decrease of 35 to 40 percent. Finally in Section 25.3 we describe a probabilistic analysis that allows one to compute the speed-up of a parallel algorithm consisting of independent runs.

An instance of the job shop scheduling problem consists of three finite sets, J jobs, M machines and O operations. For each operation $a \in O$ there is a job $j(a) \in J$ to which it belongs, a machine $m(a) \in M$ on which it must be processed and a processing time $d(a) \in \mathbb{N}$. Furthermore, for each operation $a \in O$ its successor in the job is denoted by $sj_a \in O$, except for the last operation in a job, which has no successor. The problem is to find a start time $s(a) \geq 0$ for each $a \in O$ such that the cost function

$$\max_{a \in O} \, (s(a) + d(a))$$

is minimal, subject to

(1) $$\forall_{a, sj_a \in O} \; s(sj_a) \geq s(a) + d(a)$$

and

(2) $$\forall_{a,b \in O | a \neq b \wedge m(a)=m(b)} \; s(b) \geq s(a) + d(a) \vee s(a) \geq s(b) + d(b).$$

25.2 A TABU SEARCH ALGORITHM

To apply local search to the job shop scheduling problem we use the *disjunctive graph* representation of Roy and Sussmann [10]. There an instance of the job shop scheduling problem is represented by a disjunctive graph. A feasible solution is obtained by orienting the edges in the graph such that the resulting digraph is acyclic.

Neighbourhood structures

Most neighbourhoods for the job shop scheduling problem are based on reversing machine capacity arcs in the digraph, i.e., reversing the order in which operations are processed on a machine. Van Laarhoven, Aarts and Lenstra [13] gave the following neighbourhood structure. Given a solution s, represented by a digraph D, a neighbouring solution is obtained by choosing two adjacent operations a and b, with $m(a) = m(b)$, and reversing arc (a, b) in D. In [13] it is shown that only arcs on a longest path in D should be reversed. Matsuo, Suh and Sullivan [8] and Nowicki and Smutnicki [9] have given smaller versions of this neighbourhood structure. They showed that it is not useful to reverse the so-called *internal arcs*, defined as follows.

Definition 25.2.1 *A block is a maximal sequence of adjacent operations that are processed on the same machine and belong to a longest path. An operation* *of a block is called* internal *if it is neither the first nor the last operation of that block. An* internal arc *is an arc between two internal operations of a block, an arc between the first operation of a first block and an internal operation of that block, or an arc between an internal operation of a last block and the last operation of that block.*

We also use the neighbourhood introduced by Nowicki and Smutnicki [9]. The digraph obtained by reversing the arc (a, b) in D will be denoted by $rev(D, a, b)$. The set of all neighbours of a solution D is denoted by $\mathcal{N}_1(D)$. It has been shown in [9] that, if for a solution D holds that $\mathcal{N}_1(D) = \emptyset$, then D is a global optimum.

Neighbourhood search strategy

The time needed to search a neighbourhood depends on the size of the neighbourhood and on the time complexity of the cost computation of a neighbour. Since the size of $\mathcal{N}_1(D)$ for a digraph D is already rather small, the time complexity can only be improved further by fast cost computations. For the job shop scheduling problem the computation of the cost $f(D)$ of a digraph D has time complexity $\mathcal{O}(N)$, when D has N vertices. Taillard [11] proved that for a digraph D it is possible to compute a *lower bound* for the cost of a neighbour of D in constant time, resulting in a significantly faster search for large instances.

Before presenting this lower bound, we define for an acyclic digraph $D = (V, E)$ and a vertex $a \in V$ the function $st(a, D)$ as the length (i.e., sum of node-weights $d(c)$ for nodes c on the path) of the longest path up to and without a and the function rt as the length of the longest path starting from and including a. Hence in the solution of the job shop scheduling problem represented by a digraph D, the start time $s(a)$ of an operation a is equal to $st(a, D)$. Moreover let $mx(D, a, b)$ be the length of a longest path through a or b in D and $\overline{mx}(D, a, b)$ be the length of a longest path not through a or b in D. Finally let $f_2(D)$ denote the length of a second longest path in D —which can be equal to $f(D)$—. The following theorem generalizes an observation made by Taillard [11].

Theorem 1 *Given are a digraph D, operations a and b with $m(a) = m(b)$ and (a, b) on a longest path in D. Let $D' = rev(D, a, b)$. Then,*
$mx(D', a, b) \geq f(D) \;\Rightarrow\; f(D') = mx(D', a, b)$,
$f_2(D) < mx(D', a, b) < f(D) \;\Rightarrow\; f(D') = mx(D', a, b)$ *and*
$f_2(D) \geq mx(D', a, b) \;\Rightarrow\; f(D') = \max\{mx(D', a, b), \overline{mx}(D, a, b)\}$.

Proof. See Taillard [11] or Verhoeven and Aarts [14].

Theorem 2 *The lower bound $mx(D', a, b)$ given in Theorem 1 can be computed in constant time if $st(c, D)$ and $rt(c, D)$ are available for all vertices c.*

Proof. See Verhoeven and Aarts [14].

We also use that lower bound instead of the exact cost for selecting the next solution from the neighbourhood. The advantage of this search strategy is its speed, since for every neighbour a lower bound for the cost can be computed in constant time. The disadvantage, however, is the loss of accuracy, since a neighbour with the smallest lower bound is not necessarily a neighbour with the lowest cost. The neighbourhood search is performed by a procedure *Propose* $(D, rt, st, tl, globmin, m)$ that, given a digraph D, corresponding functions st and rt, and a tabu list tl, returns a non-internal arc m on a longest path in D that is not on tl, or as close to the end of tl as possible. (In the last case the tabu list is shortened such that the found arc is no longer on it.) If several arcs satisfy this condition, then an arc $m = (m_0, m_1)$ with the smallest value of the lower bound $mx(D', m_0, m_1)$ (with $D' = rev(D, m_0, m_1)$) is selected. If such an arc cannot be found (as indicated by the value of $globmin$), then $\mathcal{N}_1(D) = \emptyset$, and a global minimum has been found, see [9].

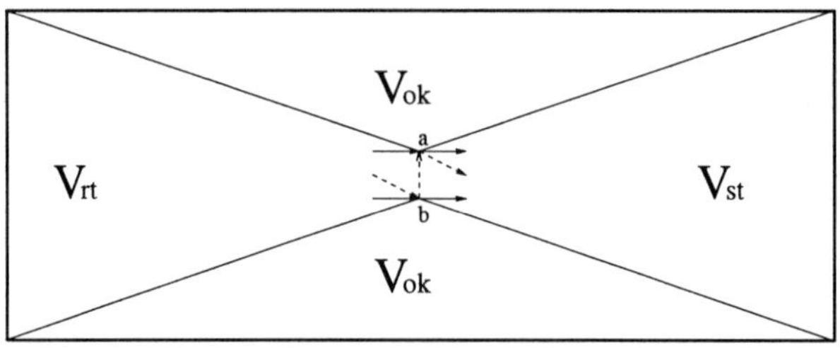

Figure 25.1 A digraph D' obtained from digraph D by the reversal of arc (a, b). The values $st(c, D')$ and $rt(c, D')$ for $c \in V \setminus V_{st}$ and $c \in V \setminus V_{rt}$ are equal to $st(c, D)$ and $rt(c, D)$, respectively.

The Bow-tie algorithm

Computation of the lower bound of Taillard requires for each node c in the digraph D the length $st(c, D)$ of the longest path up to and without c as well as the length $rt(c, D)$ of the longest path starting from and including c. After a step to a neighbour the graph changes and these lengths have to be recomputed. However, the computations of these longest paths are the most time consuming part of the algorithm. Profiling reveals that these two operations consume up to 90 percent of the total computation time. To reduce the computation time for the Taillard lower bound we introduce the following "bow-tie algorithm". Consider a graph D in which an arc (a, b) is reversed, resulting in the graph $D' = rev(D, a, b)$. A simple reduction can be obtained by noting that the values of $rt(c, D)$ and $rt(c, D')$ are equal for certain values of c. To be more precise, let V_{rt} be the set of vertices of D' from which there exists a path through a. Then $V \setminus V_{rt}$ is the set of vertices from which vertex a (and hence b) cannot be reached (in D'), see also Figure 25.1. So all paths in D' starting from a vertex $c \in V \setminus V_{rt}$ are also present in D. Hence $rt(c, D) = rt(c, D')$ for all $c \in V \setminus V_{rt}$. This yields the following 2-step method to update rt after going from D to D':

(i) compute the set V_{rt},

(ii) recompute $rt(c, D')$ only for $c \in V_{rt}$.

Both steps together take about $2|V_{rt}|$ elementary iterations. Computing st in a similar way also requires $2|V_{st}|$ iterations. Empirical analyses reveal that the bow-tie algorithm yields a reduction in running time of 35 to 40 percent.

Basic tabu search algorithm

The abstract code of a tabu search algorithm based on the ingredients given above is given in Figure 25.2. The procedures *Compute_rt* and *Compute_st* compute the distances rt and st as follows: if $It = 0$ the distances are computed with the standard longest path algorithm, if $It > 0$, the distances are recomputed with the Bow-tie algorithm, as described above. The neighbourhood search is performed by the procedure *Propose*, as described earlier. tl is the tabu list, with maximum length T. After a reverse of an edge (m_0, m_1) only m_1 is added to the tabu list, thus making all moves of the form (m_1, x) tabu.

Intensification

In order to make sure that our tabu search algorithm searches a large part of the solution space, we use intensification, i.e., we keep record of good solutions such that after termination of the basic tabu search algorithm, a restart in a different direction can be done. Of course, not only good solutions must be

```
proc Tabu_Search(var D);
begin var MaxIt, T, D_opt, tl, It, st, rt, m;
 MaxIt := "max. number of iterations"; T := "max. tabu list length";
 D := "some initial graph";
 D_opt := D; It := 0; tl := [ ]; globmin := false;
 while It < MaxIt and not globmin do
 begin
    Compute_rt(D, rt);
    Compute_st(D, st);
    Propose(D, rt, st, tl, globmin, m);
    (* if not globmin then m = (m0, m1) is the proposed edge to be reversed *)
    if not globmin then
    begin
          D := rev(D, m0, m1);
          tl := [m1] ⊕T tl;
          if f(D) < f(D_opt) then D_opt := D
    end;
    It := It + 1
 end;
 D := D_opt
end
```

Figure 25.2 Basic tabu search algorithm for the job shop scheduling problem.

stored, but also the corresponding tabu list. Moreover, when restarting the search from such a solution, the first step must be towards a not yet visited neighbour. This means that with each solution also the set of not yet used moves must be stored. So in fact triples $\langle D,\ tl,\ W\rangle$ are stored, where D is the digraph of a solution, tl is the current tabu list, and W are the moves to neighbours of D that have not yet been visited. This form of intensification has also been used by Nowicki and Smutnicki [9].

Results

We applied our new algorithm called TSI to the benchmark set that is introduced by Vaessens [12]. This benchmark set contains the *hard-to-solve* Lawrence-instances [7], and two of the easier ones, and the famous 10×10 instance ft10 of [4]. The algorithm is executed on an Intel Pentium 90 processor. We used the following parameter setting:

- $MaxIt$ = 5000, if current solution is the best so far; $MaxIt = 5000 - r * 1000$, if already restarts from all neighbours of r solutions have been done, and no new solution has been stored for intensification.

- Tabu tenure T: after each $T + 2$ iterations: $T :=$"Mean size of neighbourhood"$-1 + random(3)$.

Problem	Makespan: Lower bound[a]	Found by TSI[b]: Best	Avg.	Worst	CPU[sec.]: Termination[c]: Avg.	Worst	Up to makespan[d]: Avg.	Worst	Deviation[%][e]
10 jobs, 5 machines									
la02	655	655	655	655	4.9	7	0.5	2	0.00
10 jobs, 10 machines									
ft10	930	930	934	949	35.1	57	16.0	39	0.00
la19	842	842	843	849	24.0	32	9.7	22	0.00
15 jobs, 10 machines									
la21	1046	1048	1053	1055	42.9	61	22.1	42	0.19
la24	935	939	944	944	45.1	61	19.9	26	0.43
la25	977	977	980	984	49.0	79	29.6	65	0.00
15 jobs, 15 machines									
la36	1268	1268	1277	1278	49.6	83	18.6	46	0.00
la37	1397	1403	1414	1422	85.6	144	41.5	76	0.43
la38	1196	1196	1206	1215	96.6	161	46.1	96	0.00
la39	1233	1233	1238	1243	94.0	117	48.5	75	0.00
la40	1222	1228	1234	1299	62.7	101	22.0	78	0.49
20 jobs, 10 machines									
la27	1235	1241	1255	1270	51.2	75	23.7	42	0.49
la29	(1142)	1163	1175	1204	64.3	101	32.3	43	1.85

Table 25.1 The results of tabu search with intensification (TSI).

[a]The lower bound of the schedule length. For all instances except la29 this lower bound equals the optimum. [b]The solutions found by TSI: the best solution, the average solution (Avg.) and the worst solution, over 10 runs. [c]The average and the worst termination time of the algorithm over 10 runs. [d]The average and worst time to find the best makespan, over 10 runs. [e]The deviation of the best found makespan with respect to the lower bound. The average deviation over all instances is 0.30%.

- Maximum length backup list (for restarts): 5.

The results are given in Table 25.1. Table 25.2 compares the results of TSI with three other algorithms. The table shows that TSI finds equal quality or even better solutions than TSAB, GLS and SA1 for the la02, ft10, la19, la25, la36, la38, la39 and la40 instances in a reasonable computing time. Of course the computation times cannot be compared directly since different machines were used.

25.3 A PROBABILISTIC ANALYSIS OF PARALLEL INDEPENDENT WALKS

A parallel version of a local search algorithm may be applied in situations where the results of a sequential algorithm are insufficient. The use of a parallel algorithm may then lead to a shorter running time or a solution with a better quality. A very simple parallel version of a local search algorithm is the parallel, independent execution on P processors of a sequential algorithm. Of course, this parallelization method is only useful if the underlying sequential algorithm is nondeterministic. Many local search algorithms, including the one

Problem	Makespan					CPU[sec.]				
	OPT^{a}	TSAB^{b1}	GLS^{b4}	SA1^{m5}	TSI^{b10}	TSAB^{c}	GLS^{d}	SA1^{e}	TSI^{f}	TSI^{g}
10 jobs, 5 machines										
la02	655	655	655	669	655	8	33	19	49	7
10 jobs, 10 machines										
ft10	930	930	930	969	930	30	153	99	351	57
la19	842	842	842	855	842	60	134	94	240	32
15 jobs, 10 machines										
la21	1046	1055	1047	1083	1048	21	222	243	429	61
la24	935	948	938	962	939	184	243	235	451	61
la25	977	988	982	1003	977	155	330	255	490	79
15 jobs, 15 machines										
la36	1268	1275	1269	1307	1268	623	455	602	496	83
la37	1397	1422	1400	1440	1403	443	268	636	856	144
la38	1196	1209	1208	1235	1196	165	464	636	966	161
la39	1233	1235	1233	1258	1233	325	577	592	940	117
la40	1222	1234	1233	1256	1228	322	367	597	627	101
20 jobs, 10 machines										
la27	1235	1259	1236	1282	1241	66	435	492	512	75
la29	(1153)	1164	1157	1233	1163	493	627	471	643	101

Table 25.2 Comparison of TSI with three other algorithms, namely a tabu search algorithm by Nowicki and Smutnicki [9] (TSAB), a variable-depth algorithm by Balas and Vazacopoulos [2] (GLS), and a simulated annealing algorithm by Aarts, Van Laarhoven, Lenstra and Ulder [1] (SA1).

A superscript b followed by a number x, indicates that the reported makespan is the best one found after x runs; the computation time is the total time over all runs. The superscript m followed by a number x, indicates that the makespan is the mean over x runs; the computation time is the average over these runs. The TSAB^{b3} algorithm in [9] gives even better results than TSAB^{b1}, but no running times are given.

^{a}The best known upper bound. ^{c}The CPU-time on an Intel 386 DX. ^{d}The total CPU-time on a Sparc 330. ^{e}The mean CPU-time on a VAX 8650. ^{f}The total CPU-time on an Intel Pentium 90. ^{g}The worst computation time over 10 runs.

described in the previous section, have this property. Here we describe a simple probabilistic model that can be used to predict the speed-up of such a parallel algorithm.

Speed-up

A general discussion of the speed-up of parallel programs can be found in Foster [5]. For a correct definition of speed-up of a parallel local search algorithm, it is necessary that the sequential and the parallel algorithm achieve the same quality solutions. Usually this is done by requiring that a solution with cost at most some prescribed bound must be found. Then the speed-up could be defined as the quotient of the computing times of the sequential and parallel algorithm, see e.g., [5]. The definition of speed-up for parallel local search algorithms is more complicated by: i) the nondeterminism of most local search algorithms, ii) the fact that an algorithm may *fail*, i.e., it does not find a solution with required cost at all. To deal with these problems we introduce for an algorithm (statistical) distributions of computing times of successful and failing runs as well as the *success rate*, i.e., the probability that execution leads to a solution with the required quality.

The success rates of the sequential and parallel algorithm (with P processors) will be denoted by α and β_P, respectively. The distribution of the successful computations for the sequential and parallel algorithm (with P processors) will be denoted by G and H_P, respectively. So, for instance $H_P(x)$ is the probability that a successful run of the parallel algorithm with P processors terminates with a computing time of at most x. Similarly the distributions of the failing computations for the sequential and parallel algorithm (with P processors) will be denoted by K respectively L_P.

The parameters describing the sequential algorithm, i.e., α, G and K can again easily be found empirically by performing a large number of runs. Now we try to derive the corresponding parameters of the parallel algorithm. The parallel algorithm fails only if *all* of the P independent runs fail. Hence

$$\beta_P = 1 - (1 - \alpha)^P \; . \tag{25.1}$$

Next consider the distribution H_P of the computing time of the successful runs of the parallel algorithm. The probability $\gamma_{P,q}$ that q of the P processors of the parallel algorithm are successful is trivially given by

$$\gamma_{P,q} = \binom{P}{q} \alpha^q (1 - \alpha)^{(P-q)} \; .$$

The computing time in the case that $q \geq 1$ processors are successful is the *minimum* of the computing time of these q processors. It is a simple result from order statistics (see e.g., David [3]) that the minimum of q independent random variables, each with distribution $G(x)$, has as distribution $1-(1-G(x))^q$. The parallel algorithm is successful if at least one of its processors is successful, hence

$$H_P(x) = \sum_{q=1}^{P} \frac{\gamma_{P,q}}{\beta_P} (1 - (1 - G(x))^q) \; .$$

We use the ta31 instance as example. The best known upper bound for this instance is 1766. We require that the algorithms find a solution with cost at most 1783. For the used parameter setting this yields for the sequential algorithm a success rate of $\alpha = 0.255$. From (25.1) we see that the success rate β_P of the parallel algorithm increases strongly as a function of P, for instance $\beta_5 = 0.771$, $\beta_{10} = 0.947$ and $\beta_{20} = 0.997$. The densities g and h_P corresponding to G and H_P are given in Figure 25.3, also for $P = 5, 10, 20$. From that figure we can see that the computing time of the successful runs clearly decreases as the number of processors P increases.

Next we consider the distribution L_P of the computing time of the failing runs of the parallel algorithm. In this case *all* processors fail to find a solution with the required cost, consequently the computing time of the parallel algorithm is the *maximum* of the computing times of all the P processors. Hence (see e.g., David [3])

$$L_P(x) = K(x)^P \ .$$

For the ta31 instance the densities k and l_P corresponding to K and L_P are given in Figure 25.4 for $P = 5, 10, 20$. It is clear from the figure that the computing time of the failing runs of the parallel algorithm increases with the number of processors P. This is consistent with the fact that, for a failing run, the computing time of the parallel algorithm is the maximum of the computing times of the P processors.

Since both the sequential and the parallel algorithm have a probability that a run does not yield the required solution, the definition of the speed-up requires some care. Let E_G and E_K be the average computing times of a successful and failing run, respectively, of the sequential algorithm, i.e., E_G (E_K) is the

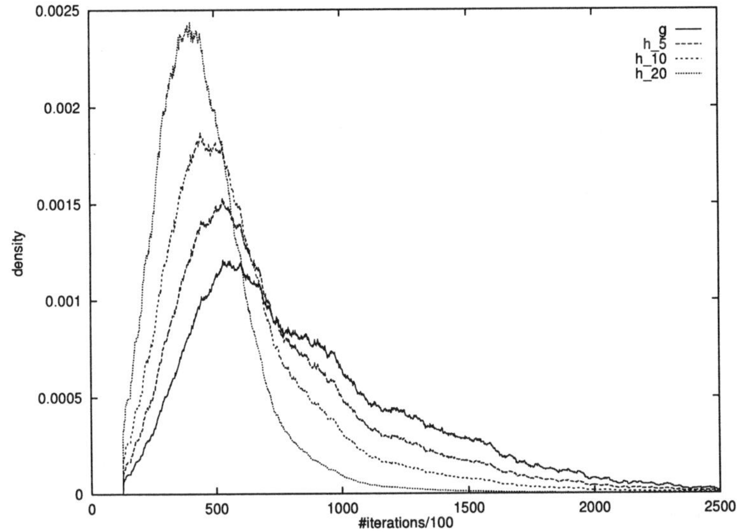

Figure 25.3 Densities g, h_5, h_{10} and h_{20}.

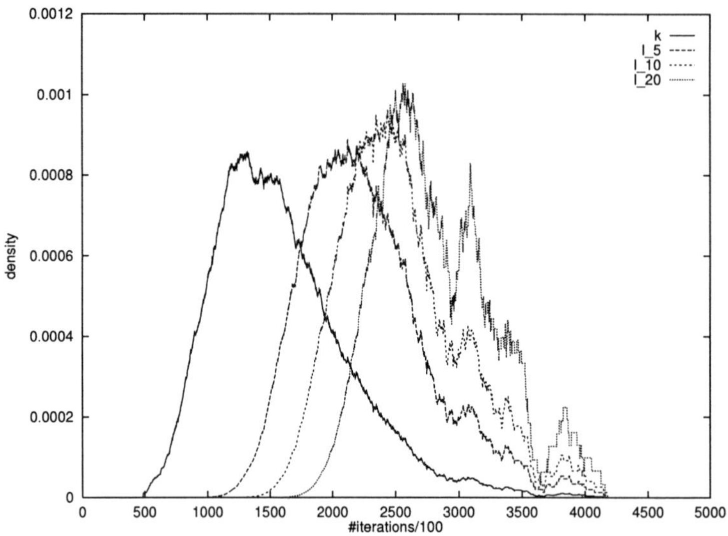

Figure 25.4 Densities k, l_5, l_{10} and l_{20}.

expectation of a random variable that has distribution G (K). Now consider a large number of runs of the sequential algorithm, taking totally a time T. If n_s and n_f are the successful and failing runs, respectively, then

$$T = n_s E_G + n_f E_K \text{ with } \alpha = \frac{n_s}{n_s + n_f} \ .$$

Hence the *average* time T_{seq} needed for one successful run of the sequential algorithm is given by

$$T_{seq} = E_G + \frac{1-\alpha}{\alpha} E_K \ .$$

A similar result holds for the parallel algorithm. If $E_{H,P}$ and $E_{L,P}$ denote the average computing times of a successful and failing run, respectively, of the parallel algorithm with P processes, the *average* time $T_{par,P}$ needed for one successful run of the sequential algorithm is given by

$$T_{par,P} = E_{H,P} + \frac{1-\beta_P}{\beta_P} E_{L,P} \ .$$

The speed-up now can be defined as

$$SP_P = \frac{T_{seq}}{T_{par,P}} \ .$$

In Figure 25.5 the speed-up obtained in this (theoretical) way is given for the ta11, ta21 and ta31 instances. That figure also gives the empirically found speed-up. For these three instances we required solutions with cost at most

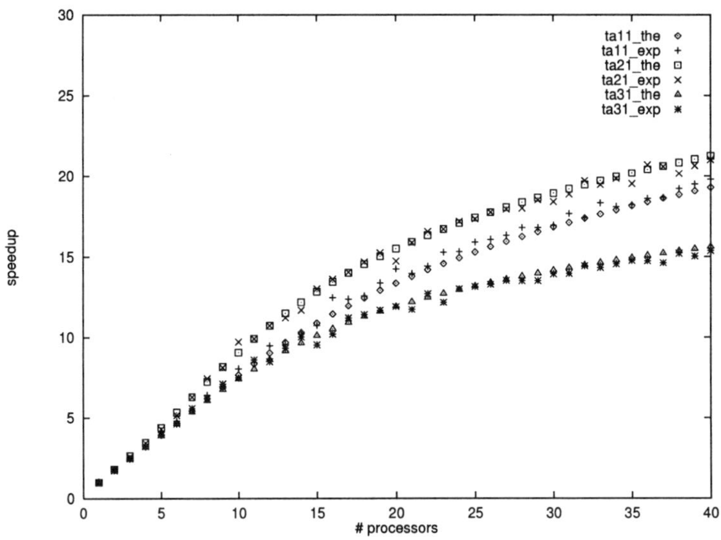

Figure 25.5 Theoretical and empirical speed-up for ta11, ta21, ta31 instances.

1391, 1679 and 1783, yielding success rates of 0.244, 0.289 and 0.255, respectively. Clearly our simple model gives a good description of the possible speed-up. It does not seem useful to use more than about 20-30 processors for this parallel algorithm.

It follows immediately from (25.1) that the success rate of the parallel algorithm tends quickly to 1 if the number of processors increases. In fact this high success rate of the parallel algorithm is an important reason for the speed-up, and not only the slowly decreasing running time of the successful parallel runs. This can also be seen from Figure 25.6, where we give the speed-up for ta31 when requiring solutions with cost of at most 1%, 2% and 3% over the upper bound of 1766. The resulting success rates of the sequential algorithm were 0.255, 0.616 and 0.836, respectively. Clearly the speed-up decreases when the success rate of the sequential algorithm tends to 1.

References

[1] E.H.L. Aarts, P.J.M. van Laarhoven, J.K. Lenstra, and N.J.L. Ulder. A computational study of local search algorithms for job shop scheduling. *ORSA Journal on Computing*, 6:118–125, 1995.

[2] E. Balas and A. Vazacopoulos. Guided local search with shifting bottleneck for job shop scheduling. Management science research report #MSRR-609, Graduate School of Industrial Administration, Carnegie-Mellon University, Pittsburgh, Pennsylvania, 1994.

[3] H.A. David. *Order Statistics.* Wiley, 1970.

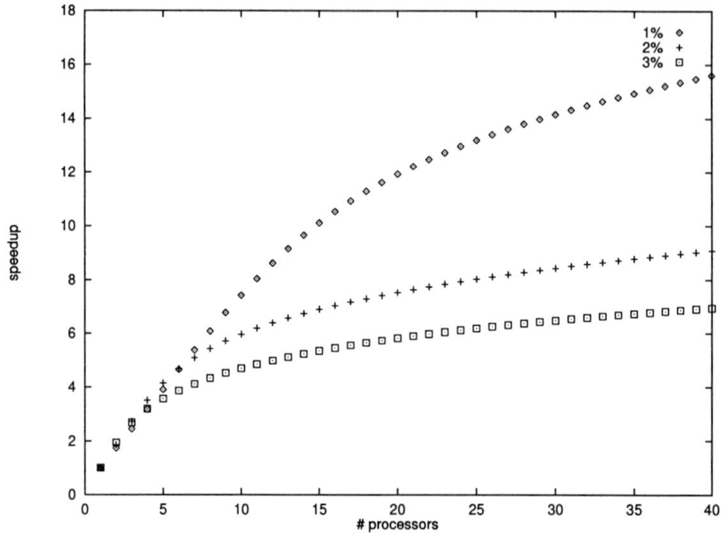

Figure 25.6 Speed-up for ta31 for sequential success rates 0.255 (1%), 0.616 (2%) and 0.836 (3%).

[4] H. Fisher and G.L. Thompson. Probabilistic learning combinations of local job-shop scheduling rules. In J.F. Muth and G.L. Thompson, editors, *Industrial Scheduling.* Prentice Hall, Englewood Cliffs, 1963.

[5] I.T. Foster. *Designing and Building Parallel Programs.* Addison-Wesley, 1995.

[6] E.L. Lawler, J.K. Lenstra, A.H.G. Rinnooy Kan, and D.B. Shmoys. Sequencing and scheduling: Algorithms and complexity. In S.C. Graves, A.H.G. Rinnooy Kan, and P. Zipkin, editors, *Handbooks in Operations Research and Management Science 4.* North-Holland, 1993.

[7] S. Lawrence. *Resource Constrained Project Scheduling: An Experimental Investigation of Heuristic Scheduling Techniques (Supplement).* Graduate School of Industrial Administration, Carnegie Mellon University, Pittsburgh, 1984.

[8] H. Matsuo, C.J. Suh, and R.S. Sullivan. A controlled search simulated annealing method for the general job shop scheduling problem. Working Paper 03-04-88, Graduate School of Business, The University of Texas at Austin, 1988.

[9] E. Nowicki and C. Smutnicki. A fast taboo search algorithm for the job shop problem. *Management Science*, 42:797–813, 1996.

[10] B. Roy and B. Sussmann. Les problèmes d'ordonnancement avec contraintes disjonctives. Note DS 9 bis, SEMA, Paris, France, 1964.

[11] E. Taillard. Parallel taboo search techniques for the job shop scheduling problem. *ORSA Journal on Computing*, 6:108–117, 1994.

[12] R.J.M. Vaessens. *Generalized Job Shop Scheduling: Complexity and Local Search.* PhD thesis, Eindhoven University, 1995.

[13] P.J.M. van Laarhoven, E.H.L. Aarts, and J.K. Lenstra. Job shop scheduling by simulated annealing. *Operations Research*, 40:113–125, 1992.

[14] M.G.A. Verhoeven and E.H.L. Aarts. Parallel local search and job shop scheduling. In A. Ferreira and J.D.P. Rolim, editors, *Parallel Algorithms for Irregular Problems: State ot the Art.* Kluwer, 1995.

26 AN EXPERIMENTAL STUDY OF SYSTEMIC BEHAVIOR OF COOPERATIVE SEARCH ALGORITHMS

Michel Toulouse[1,2], Teodor Gabriel Crainic[2,3] and Brunilde Sansó [3,4]

[1] School of Computer Science, University of Oklahoma
200 Felgar Street, Room 114, Norman, OK, USA.
toulouse@cs.ou.edu

[2] Centre de Recherche sur les Transports, Université de Montréal
C.P. 6128, succ. Centre-ville, Montréal, Québec, Canada H3C 3J7.
theo@crt.umontreal.ca, bruni@crt.umontreal.ca

[3] Dept. des Sciences Administratives, Université du Québec à Montréal
Montréal, Québec, Canada.

[4] École Polytechnique, Université de Montréal
Montréal, Québec, Canada.

Abstract: Parallel and distributed computer systems are increasingly used to find good solutions to difficult combinatorial optimization problems. An interesting and often used approach consists of executing concurrently different search methods that exchange information gathered in previously explored regions of the solution space. Cooperation among sequential programs strongly affects how the solution space is explored and it is hoped that it improves the efficiency of the search. A formal representation of these cooperative algorithms based on a discrete-time dynamical system model may be used to describe how the search behavior of cooperative programs depends on systems of complex and correlated interactions. We define *systemic cooperation* as the search behavior for programs interacting within the same system of correlated interactions, and show experimentally the importance of the phenomenon and its independence of specific search design parameters.

26.1 INTRODUCTION

Search methods use data on previously explored regions of the solution space as heuristics to overcome the inherent inefficiencies of an exhaustive search. For example, branch and bound methods keep track of the best solution to reduce the number of solutions that have to be explored. Metaheuristics such as tabu search keep track of different attributes of previously visited solutions to prevent local search from being trapped into local optima, and to define a consistent long-term search behavior. Genetic algorithms use the fitness of individuals in the population to bias the random selection of the parents.

The search history is increasingly becoming a central tool for developing parallelization strategies for search methods. Several instantiations of the same or different search methods are executed concurrently. At run time, the data gathered by one instantiation is not only used by the generating search thread, but is also shared (often after being suitably modified) with other concurrent instantiations. In the literature, these parallel algorithms are referred to as cooperative searches [3, 4, 23, 24, 40], multiple interacting walks [46] or population approaches [29]. Procedures such as coarse-grained Parallel Genetic Algorithms (PGA) [18, 33, 36], fine-grained PGAs [17, 27, 31], Memetic Algorithms [6, 28, 30], a number of parallel branch and bound [12] and tabu search (TS) algorithms [9, 8], error algorithms in annealing procedures [2, 10, 22, 25], and special purpose cooperative searches [4, 23], are examples of parallel search methods based on cooperation among several sequential search programs defined by the sharing of gathered information.

The rationale behind sharing information is simple: since search methods require heuristic information to ensure good performances, it is assumed that when programs have access to the best information available, it speeds up the exploration of the solution space. This objective is generally not too difficult to achieve at the level of each individual search. Each program may use the information it receives from other programs in the same way it uses information gathered locally to guide its exploration of the solution space. If one assumes that this is the only impact that shared information has on cooperating search programs then, under this assumption, the search behavior of cooperative algorithms is the sum of the search behaviors of the programs composing the parallel procedure and one may expect to speed up the search process from the sharing of information. Yet, several researchers have noted that cooperation among programs "was more than just hardware acceleration" (Cohoon et al. [5]). Tanese [36, 37], Jog and Gucht [26], Cohoon et al. [5], and Munetomo, Takai and Sato [32], among others, observed that cooperation not only produces speed-up, but also modifies the search behavior of the cooperating programs. For example, studies of the migration operators in coarse-grained PGA show that the frequency and the number of individuals that migrated between sub-populations affects the convergence properties of the PGA [37]. Similarly, studies of the population structure of fine-grained PGA have found that the population structure affects the niche formation and the spread of dominant individuals by the diffusion process [19].

There is empirical evidence scattered throughout the literature that sharing information has a cumulative impact on the search behavior among cooperating programs. Unfortunately, this issue has not always been properly addressed when designing cooperative algorithms: In fact, it has been addressed often from the narrow perspective of the search method making up the cooperative procedure. In this paper, we address this issue in a broader perspective in order to obtain more general and useful conclusions for the future design of this type of parallel algorithms.

Cooperative procedures are dynamical systems [27, 47, 48], where each search program is a subsystem and the sharing of information provides the interactions among the subsystems. We use a formal representation of cooperative search algorithms as discrete-time dynamical systems which, at each iteration, makes use of and depends on a global data structure that stores the information gathered by all the search programs. We use this model to describe how the search behavior of cooperating programs is determined by systems of complex and correlated interactions that build the cumulative impact of shared information on cooperating programs. We define as *systemic cooperation* the search behavior of programs interacting within such systems of correlated interactions. We experiment by using cooperative procedures with the same set of programs, the same initial conditions, and the same set of search parameter values, which, nevertheless have a completely different search pattern of the solution space. We prove that these divergences are related to the occurrence and the structure of systemic cooperation.

Our research shows that the cumulative impact of interactions among search programs controls the evolution of cooperative search in a similar way the interactions among components of other computational dynamical systems (e.g., artificial neural networks and cellular automata) control the evolution of these systems. These observations are critical to our understanding of how we can build more efficient cooperative algorithms. Today, it is widely assumed that initial conditions and search parameters of sequential programs define the search behavior of cooperative algorithms. The results of this paper show that on the contrary, when cooperation increases among programs, systemic cooperation becomes the main factor affecting the search behavior of cooperating programs, rather than the information exchanged or the search parameters of the individual programs. From a practical point of view, this means that a more efficient exploration of the solution space may result from cooperative algorithms if the design of the information sharing mechanism accounts for its impact on the search behavior. In particular, there exists a large body of concepts from dynamical systems theory (one may use the attractors of the system for example) that may help us in our quest to obtain more controlled interactions among cooperating search programs.

An important contribution of this work consists in the controlled and systematic experimentation that goes beyond intuition to show that shared information has a cumulative impact on the search behavior of cooperating programs. This paper also makes two important methodological contributions to the research on cooperative algorithms. First, we introduce a new perspective to study the impact of interactions among cooperating programs. All previous studies focused on the control structure of the sequential programs. In our research, we infer results on the control structure of cooperating programs by observing the evolution pattern of shared data. This approach is commonly used in computer science where data access patterns are studied to obtain more efficient execution of programs or to derive the conditions for the correct execution of parallel algorithms where processes share the content of some variables. By modeling the state transitions of a global memory (the shared data), we can explain and prove results about the global behavior of parallel algorithms such as cooperative search. Second, the notion of systemic cooperation is new to the literature on cooperative algorithms. Cooperation is not an instantaneous phenomenon consisting of a single exchange of information that occurs among two or more search programs, as it is almost universally described in the literature on cooperative algorithms. Cooperation is a systemic phenomenon, which has a time duration (across several iterations of a search program) and a space extension (systemic cooperations expand across several search programs) and it involves the exchange of several messages containing a broad range of correlated data. Shared data comes together through systemic cooperation phenomena and yields powerful attractors which affect globally the trajectory of several programs. The study of the time and space evolution of data patterns and how to combine the aggregate as part of an unique systemic cooperation has already provided us with very useful observations on how cooperation affects the control structure of search programs [42]. These observations have resulted in significant improvements in the design of cooperative algorithms [43].

This paper is organized as follows. In Section 26.2, we briefly examine cooperative parallel search, and recall the formal description of cooperative algorithms and the definition of systemic cooperation. The experimental design and results are presented Section 26.3, while Section 26.4 is dedicated to a brief discussion of their meaning and impact. Conclusions are presented in Section 26.5.

26.2 COOPERATIVE SEARCH ALGORITHMS

The first step in the design of a cooperative search algorithm is based on a parallelization strategy referred to as *independent searches* [1, 8, 34, 35]. It consists of choosing a set of sequential programs, exact or heuristic, that will become concurrent independent tasks, to be scheduled on the processors of a parallel or distributed computer system. This approach offers a large degree of parallelism, which is often not the case with standard functional or data decomposition techniques.

The next step consists of defining how the sequential programs will cooperate. A common approach to instantiate cooperation is based on sharing gathered information [4, 9, 8, 11, 18, 17, 20, 21, 23, 27, 31, 33, 36, 45]. The designer must specify the information to be exchanged, the programs between which the information will be exchanged, the frequency of the exchange among the sequential programs -see [40]-, as well as the role of the external information in each individual search thread.

Intuitively, cooperative algorithms appear simple to design. It is also easy to hypothesize how sharing information among programs may improve their search behavior. For example, if several bounds are computed by concurrent branch and bound programs, it would appear that the speed-up of such a parallel procedure would improve if the best bounds among those available are broadcast to several programs. For metaheuristics such as tabu search, the programs that find good solutions and share the corresponding information will help other search programs with less suitable initial conditions or search parameters. For evolutionary approaches, the inclusion of fitter individuals in a population means that schemata with higher fitness will be sampled, therefore introducing building blocks of higher value in the receiving populations.

However, cooperative algorithms only appear to be simple. Once programs borrow information from each other, the search behavior of each program becomes a function of the search pattern of all other programs, making the search trajectory quickly untraceable. In this paper, we argue that even if search programs are designed to solve an optimization problem, one should focus on programs as components of a system. From this perspective, it is easier to see that cooperative algorithms are complex systems of dynamically interacting components. This characterization then indicates the type of formal tools required to develop a more accurate understanding of cooperative search algorithms.

In [41], we formalized the concept of cooperative search as the evolution of a discrete-time dynamical system. The full model displays the ability to account for the asynchronous behavior of cooperative algorithms. Before proceeding to the experimental part of the paper, we briefly recall the main points of this development.

Let ψ be an p-dimensional state vector, and f some iterative operator. A discrete-time dynamical system has the structure

$$\psi(k+1) = f(\psi(k)), \quad k = 0, 1, 2, \dots, \tag{26.1}$$

where the state vector $\psi(k+1)$ depends only on vector $\psi(k)$ and the operator f (we assume the system has no input, and f is considered to be invariant over k). If ψ and f are separable into p components, we can then rewrite (26.1) as

$$\psi_i(k+1) = f_i(\psi_1(k), \psi_2(k), \dots, \psi_p(k)) \tag{26.2}$$

where $\psi_i(k)$ is the i-th component of $\psi(k)$, and f_i is the i-th component of operator f.

It is rather easy to fit cooperative searches into an iterative scheme such as (26.1). To simplify the presentation, we assume that the cooperative search is based on p sequential tabu search programs $P_0, P_1, \dots, P_{p-1}$. Program P_i then becomes the f_i component of the operator $f = \bigcup_{i=0}^{p-1} P_i$. Similarly, the solutions and the tabu memories of program P_i may be defined as the i-th component of the vector ψ. If no information is shared among the p search programs, equation (26.2) becomes $\psi_i(k+1) = f_i(\psi_i(k))$ for $i = 0, 1, \dots, p-1$, and these equations do not define a dynamical system. However, information sharing implies interaction among the equations in (26.2) by assuming that program P_i at iteration k depends on the data updated at iteration $k-1$ by the programs sharing information with P_i. In this case, the set of tabu search programs is a truly dynamical system where the subsystems P_i interact with each other.

Assume $min\{c(x)|x \in X\}$ is the problem to be solved by cooperative tabu search. Let x be the current solution, $N(x)$ and h the corresponding neighborhood and set of tabu lists and memories, respectively. The data structure h stores the information gathered on the solution space by a tabu program, which essentially consists of the attributes of solutions previously explored and used to guide the tabu search. Then, $x' \in N(x)$ is the solution selected to become the current one at the next iteration given the contents of h (for a more comprehensive description of tabu search see, for example, [13, 14, 15, 16]). Let k be the index of the iterative steps of the sequential TS method. The state vector ψ has two components: the current solution visited in the solution space, and the information gathered on the solution space. Therefore, $\psi(k) = (x(k), h(k))$. At each iteration, the operator f must update the two components of its state vector. Consequently, f is a tuple (s, v) of functions. The function s represents the search strategy of a tabu program, and takes as arguments $h(k)$ and $N(x(k))$, the neighborhood of the current solution $x(k)$. This function returns the current solution $x(k+1)$ of iteration $k+1$. The function v takes the values $x(k+1)$ and $h(k)$ as arguments, and returns $h(k+1)$. The function v updates the data structure $h(k)$ according to the attributes of the new solution $x' = x(k+1)$ and the state $h(k)$ of the tabu memories at iteration k. At each iteration, a subset $\mu \subseteq \{0, 1, 2, \ldots, m-1\}$ of the $h(k)$ fields are updated with the attributes of solution $x(k+1)$.

For the set of p sequential tabu search programs $CP = \{P_0, P_1, \ldots, P_{p-1}\}$, the set of search strategies is $s_0, s_1, \ldots, s_{p-1}$, where $s_i \neq s_j$, if $i \neq j$, (i.e., each TS program applies a different search strategy), x_i is the current solution of program P_i, and h_i is the current state of the tabu memories of program P_i. When the P_i programs do not share gathered information among each other, the choice of the solution x'_i in the neighborhood $N(x_i)$ of the current solution x_i of program P_i at iteration k is given by $x'_i(k) = s_i(N(x_i(k)), h_i(k))$, while the update of the tabu memories is given by $h_i(k+1) = v_i(x'_i(k), h_i(k))$. The value of $h_i(k+1)$ corresponds to the new state of the tabu memories of program P_i at iteration $k+1$, which results from updating the fields of $h_i(k)$ at iteration k using attributes of solution $x'_i(k)$. Together, the two equations are equivalent to the system in (26.2), since x_i and h_i are components of ψ_i, and s_i and v_i are components of f_i.

When interactions among sequential programs are allowed, the level of information available when choosing the new current solution x' is increased and the update of the current solution and memories of a TS thread may depend upon the history of other, interacting, threads. The previous equations have to be modified to account for this. Additional notation is required.

Let $x_i(0)$ be the initial solution of program P_i, and $\mathcal{S} = (\{\gamma(t)\}, \{\tau(t)\})$ be the schedule corresponding to operator f and starting with $x(0) = \bigcup_{i=0}^{p-1} x_i(0)$. The set $\{\gamma(t)\}$ is a sequence of subsets of $\{0, 1, \ldots, p-1\}$, and $\{\tau(t)\}$ is a sequence of elements of $\mathcal{N}^{p \times p}$ satisfying $0 \leq \tau_j^i < t$, for all $t = 1, 2, \ldots$ and $i, j = 0, 1, \ldots, p-1$ (see [44]). Since we make no assumption about the synchronization of procedures in CP, we need a function which returns k, the iteration index of program P_i at time unit t. This is given by $\tau_i^i(t)$. The notation $x_j(\tau_i^j(t))$ indicates the value of solution x_j at iteration $\tau_i^j(t)$ of program P_j which is used to determine $x_i'(k)$ of program P_i. Similarly, $h_j(\tau_i^j(t))$ indicates the state of memories h_j at iteration $\tau_i^j(t)$ of program P_j, which is used to find $h_i(k+1)$ of program P_i. Let $\phi_i \subset \{0, 1, \ldots, p-1\}$ be a subset of the indices of the sequential programs $P_0, P_1, \ldots, P_{p-1}$. For each sequential program P_i at iteration k, the choice of the solution x_i' in the neighborhood $N(x_i)$ of the current solution x_i is given by:

$$x_i'(k) = s_i(N(x_i(k)), (h_i(k), \{h_j(\tau_i^j(t))\})) \text{ where } j \in \phi_i \tag{26.3}$$

In equation (26.3), k corresponds to the k-th iteration of program P_i. This value depends on t, and is given by $k = \tau_i^i(t)$. Program P_i finds the new solution x_i' in the neighborhood $N(x_i)$ depending on the state $h_i(k)$ of its tabu memories and the state of the tabu memories represented by $\bigcup_{(j \in \phi)} h_j(\tau_i^j(t))$. If information is exchanged among the search programs, the state vector $\psi(t) = \bigcup_{i=0}^{p-1} \psi_i(t)$ of a cooperative procedure can then be found by solving the following iterative map:

$$x_i'(t) = \begin{cases} x_i'(t-1) & \text{if } i \notin \gamma(t) \\ s_i(N(x_i(\tau_i^i(t))), (h_i(\tau_i^i(t)), \{h_j(\tau_i^j(t))\})) & \text{if } i \in \gamma(t) \text{ and } j \in \phi_i \end{cases} \tag{26.4}$$

$$h_i(t) = \begin{cases} h_i(t-1) & \text{if } i \notin \gamma(t) \\ v_i(x_i'(\tau_i^i(t)), h_i(\tau_i^i(t))) & \text{if } i \in \gamma(t) \end{cases} \tag{26.5}$$

where $i \in \gamma(t)$, if program P_i has completely finished executing an iteration at time t. The set of equations (26.4) and (26.5) is a discrete-time dynamical system representation of a cooperative search. Each subsystem i has its own internal dynamics given by the search strategy of program P_i, and interacts with other subsystems by sharing the virtual memory $h(t)$.

By using this model, one can visualize how information sharing affects the state of a particular component ψ_i, which then propagates to other components of the state vector ψ, as well as how information exchanges among programs self-organizes into systemic cooperation. Information sharing among search programs as described in equation (26.3), does not always have an impact on the search behavior of programs. However, the following definition focuses on a particular form of information sharing:

Definition 1 *The sharing of information among two programs P_i and P_j is said to be an* **effective interaction** *for program P_i at iteration k, if and only* *if the information obtained by P_i from tabu memories h_j of program P_j affects the choice of the solution x'_i at iteration k, i.e. if and only if $x'_i(k) \neq \bar{x}'_i(k)$.*

Clearly, an effective interaction affects the choice of the solution $x_i(k+1)$, and therefore the search path of program P_i at iteration $k+1$. Let us assume that $\beta_i(k+1)$ is the current solution, $\beta'_i(k+1)$ is the new solution, and $h_i^\beta(k+1)$ is the state of the tabu memories of program P_i at the end of iteration $k+1$, after program P_i has performed an effective interaction with another program at iteration k; thus $\beta_i(k+1) = \bar{x}'_i(k)$. Let us also assume that $\alpha_i(k+1)$ is the current solution, $\alpha'_i(k+1)$ is the new solution, and $h_i^\alpha(k+1)$ is the state of the tabu memories of program P_i at the end of iteration $k+1$, given that program P_i has not performed an effective interaction at iteration k; thus $\alpha_i(k+1) = x'_i(k)$. There are two possible conditions that may cause the current new solution $\beta'_i(k+1)$ to be different from the solution $\alpha'_i(k+1)$ at iteration $k+1$:

1. $\alpha'_i(k+1) \notin N(\beta_i(k+1))$, or
2. $\alpha'_i(k+1) \in N(\beta_i(k+1))$ but $h_i(k+1) \neq \overline{h}_i(k+1)$ and $\alpha'_i(k+1)$ are evaluated as not being the best solution in the neighborhood $N(\beta_i(k+1))$ according to the information stored in $\overline{h}_i(k+1)$.

Figure 26.1 illustrates how a single effective interaction at iteration k of a program P_i can change the search line $\psi_i(k+1) = f_i(\psi_i(k))$, $\forall k \in [\tau_i^i(t-l), \tau_i^i(t-l+1), \ldots, \tau_i^i(t)]$, of program P_i, and therefore, the state of the memory component h_i (for a formal demonstration, see [41]). There are three potential outcomes of an effective interaction on the sequence ψ_i. In the first case, at iteration k of program P_i, an effective interaction chooses solution $\beta_i(k+1)$ as the current solution for iteration $k+1$, but at iteration $k+2$, the solution $\alpha'_i(k+1)$ is in the neighborhood of $\beta_i(k+1)$; the state of the memory $\overline{h}_i(k+1) = h_i(k+1)$, or the fields updated by $\beta_i(k+1)$ had no impact on the choice of the new current solution. The search path returns to its α-course in the solution space, that is, $\beta'_i(k+1) = \alpha'_i(k+1)$. In this case, the effective interaction changes the path of program P_i at iteration k, but $\beta_i(k+1)$ has no further impact on its search behavior. In the second case, an effective interaction attracts the search path to a new region of the solution space for a few iterations, but eventually resumes its original course after l iterations, that is, $\beta_{i,k+l} = \alpha_{i,k+l}$. In the third case, interference with the update pattern completely modifies the search behavior of program P_i.

The consequences of effective interactions on the state of the memory component h_i are also noteworthy. For each iteration where the sequence ψ_i is performed in the β-path of program P_i, the update $\overline{h}_i(k+2) = \upsilon_i(\bar{x}'_i(k+1), h_i(k+1))$ will store attributes of the β-solutions from the β-path in the memories of program P_i. Storing this information in the program memories will either overwrite fields which have been previously updated by other attributes of β-solutions, or they will overwrite fields which have been updated by attributes from α-solutions. It is then possible for the content of the tabu

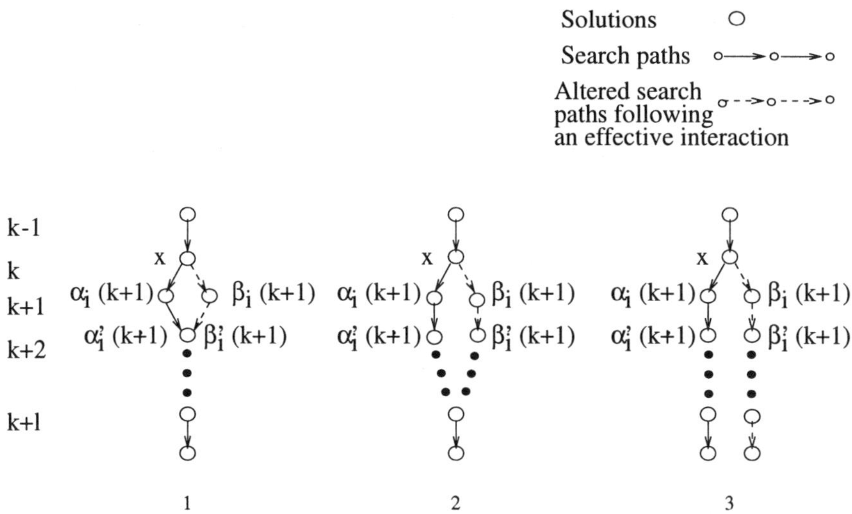

Figure 26.1 Potential outcomes following an effective interaction

memories of program P_i to be completely overwritten by attributes of solutions visited in the β-path. Clearly, an effective interaction affects the search trajectory of a sequential search program part of the cooperative procedure. From a formal perspective, an effective interaction modifies the trajectory of the dynamical system by changing the state component ψ_i of the global state vector ψ.

Furthermore, an effective interaction in sub-system f_i may change the state of other components ψ_y, $y \neq i$, of the dynamical system described by (26.4) and (26.5). Indeed, the modification of the search trajectory due to an effective interaction between two processes will often generate another interaction between one of the two processes with a third. These are identified as *correlated effective interactions.* We have then

Definition 2 *A* **systemic cooperation** *is a set B of β-paths in an interval $[t-l, t]$ such that, for any β-path $\in B$ and any point in the interval $[t-l, t]$, the system of equations (26.4) and (26.5) can be satisfied by a sequence of interactions $\tilde{h}_i^j$ involving only β-paths $\in B$, and leading to the initial independent effective interaction at point $(t-l)$.*

A systemic cooperation is therefore the search behavior of a subset of cooperating programs which is characterized by the execution of β-paths linked together by correlated interactions. Systemic cooperations are an example of the cumulative impact that information sharing has on the search behavior of cooperating programs.

26.3 EXPERIMENTATION

We introduce an experiment aimed to show the existence of systemic cooperations in cooperative algorithms, and their impact on the search behavior of cooperating programs. We use two cooperative procedures that have been designed to display a high level of correlated effective interactions. All the tests are performed on the location-allocation problem introduced in [7]. There are two types of decision variables in this problem: design variables, which are Boolean (0/1), and flow variables, which are continuous. Tabu search will run in the space defined by the design variables, therefore the problem solution space is given by $X^n \subseteq B^n$. Each sequential program is a tabu search algorithm similar to the one described in [7]. Tests are performed using a network of SUN Sparc stations, therefore the cooperative procedure CP is designed for a distributed memory system where gathered information is shared using the message passing paradigm. Exchanges among processes use our own communication software, which is stream-based using the TLI protocol interface and coded in C. The tabu search programs are coded in Fortran-77.

26.3.1 Design of the cooperative procedures

Let M be the set of possible messages that can be exchanged among programs. We define $m_s \in M$ as a message sent from program P_i to program P_j, and $m_r \in M$, a message received by program P_i from program P_j. Each tabu program P_i stores information on several attributes of the solutions visited during the exploration of the solution space. However, only the following attribute will be shared: given that the optimization problem in [7] is a minimization problem, at each iteration k, the solution $x'(k)$ is made available for sharing if $x'(k) < x(k)$. Information available for sharing is exchanged synchronously among programs at every tabu iteration. We define the special empty message m_0 ($\emptyset$) to be used by a program P_i, when no information is available to send to program P_j. If $x'(k) < x(k)$, at iteration $k+1$ a message is sent by P_i which contains the solution $x'(k)$; otherwise, P_i has no valuable information to share and message m_0 is sent.

Each program P_i executes the following operations at each iteration k:

- Find α, the best solution in the neighborhood $N(x)$ of the current solution,
- Compare α with β, the contents of message m_r received at iteration $k-1$,
- Choose the best solution among α and β as $x'(k)$, the current solution for iteration $k+1$.

If the contents of message m_r are chosen, then an effective interaction occurs for program P_i. Otherwise, if the solution in m_r is not chosen, or if a m_0 message was received at iteration $k-1$, the program proceeds according to information gathered in its own tabu memories.

The communication structure of CP has a low diameter; a program sends its information available for sharing to only one other program (a high diameter communication structure is one of a complete graph where a program can send information to every other program in one broadcast, see [39] for details). The communication structure indicates between which programs an exchange of information is to take place at each iteration. For this experimentation, the communication structure is given by the following model: $C(P_i, k) = P_{(i+k) \, mod \, p}$, where P_i is the source program, $P_{(i+k) \, mod \, p}$ is the destination program of the message sent by program P_i at iteration k, and p is the number of cooperating programs in the cooperative procedure. Special provision is made for the case where $P_{(i+k) \, mod \, p} = P_i$ (a program cannot send a message to itself).

To facilitate the comparison of the experimental results, each solution and message were mapped to a set of integers ranging from 1 to 800 on a first-come, first-served basis. Therefore, the values in the tables indicate only if solutions and messages are different; they do not refer to the objective values of solutions.

Table 26.1.A displays the search path of four tabu search programs for the first 50 iterations. Each program uses a different search strategy and shares information according to the rules defined for CP. Table 26.1.B focuses on the search program P_0, and shows the update pattern of variables $x(k)$ and $x'(k)$, as well as the messages sent (m_s) and received (m_r) by P_0 at each iteration.

Table 26.2 presents similar information, except that it corresponds to a modified version of CP, called $\overline{CP}$. $\overline{CP}$ is not a fully cooperative procedure; the program P_0 has been altered such that it cannot read the messages received from other programs. Therefore program P_0 will never execute an effective interaction with other programs. The search path of P_0 in $\overline{CP}$ is determined only by its initial solution and search parameters. Notice however, that although P_0 does not read its messages, it still continues to send messages to other programs using the same communication structure as defined in this section. Consequently, the other programs can still realize effective interactions based on information from P_0.

26.3.2 *Impact of shared gathered information on* P_0

Columns $x(k)$ in Tables 26.1.B and 26.2.B show the search path of program P_0 for the cooperative procedures CP and $\overline{CP}$, respectively. In Table 26.2.B, the program P_0 executes an α-path for all 50 iterations, since it does not perform any effective interaction. On the other hand, in Table 26.1.B, eight effective interactions are performed by program P_0 during the 50 iterations. As can be seen, these effective interactions using gathered information from P_1, P_2 and P_3, have a significant impact on the search behavior of program P_0 (resulting from the execution of several β-paths).

Given that $P_0 \in CP$ has executed several β-paths, according to the model presented in Section 26.2, P_0 should have stored attributes from β-solutions, and this information should be different from that stored by $P_0 \in \overline{CP}$. This hypothesis can be verified by examining the column of sent messages (m_s) in Tables 26.1.B and 26.2.B, where the two sets of messages are totally different.

	A: programs 0 to 3					B: program P_0			
it	P_0	P_1	P_2	P_3	it	$x(k)$	$x'(k)$	m_s	m_r
1	1	2	3	4	1	1	5	∅	∅
2	5	6	7	8	2	5	9	∅	6
3	6	7	8	12	3	6	13	6	∅
4	13	6	7	8	4	13	17	13	7
5	7	8	13	6	5	7	14	7	∅
6	14	21	7	8	6	14	24	14	∅
7	24	14	21	27	7	24	28	24	∅
8	28	21	24	31	8	28	32	∅	∅
9	32	33	7	8	9	32	36	32	∅
10	36	32	38	39	10	36	40	36	∅
11	40	21	36	27	11	40	42	∅	∅
12	36	43	44	45	12	36	46	∅	∅
13	46	15	47	48	13	46	49	∅	∅
14	49	50	38	15	14	49	52	49	∅
15	52	53	26	49	15	52	54	52	∅
16	54	52	53	27	16	54	56	∅	∅
17	56	57	7	8	17	56	59	∅	57
18	57	60	61	12	18	57	53	57	12
19	12	57	63	61	19	12	65	∅	63
20	63	61	67	68	20	63	69	∅	∅
21	57	67	57	61	21	57	53	∅	∅
22	53	71	72	57	22	53	60	∅	∅
23	60	74	57	71	23	60	66	∅	∅
24	66	76	71	76	24	66	53	∅	76
25	76	77	74	78	25	76	60	∅	74
26	74	79	80	74	26	74	62	∅	∅
27	62	76	81	76	27	62	82	∅	∅
28	82	83	74	85	28	82	66	∅	∅
29	57	74	86	71	29	57	67	∅	∅
30	67	76	71	74	30	67	87	67	∅
31	87	67	78	89	31	87	90	∅	∅
32	67	71	77	77	32	67	92	∅	∅
33	92	77	93	76	33	92	94	∅	∅
34	94	95	71	74	34	94	96	∅	71
35	67	71	76	89	35	67	97	∅	∅
36	97	79	74	79	36	97	98	97	∅
37	98	97	89	76	37	98	99	∅	∅
38	99	57	79	100	38	99	101	∅	∅
39	101	74	76	71	39	101	67	∅	∅
40	67	76	74	74	40	67	98	∅	∅
41	98	79	89	80	41	98	99	∅	∅
42	99	74	77	71	42	99	103	∅	∅
43	103	71	101	79	43	103	98	∅	∅
44	98	97	71	74	44	98	105	∅	97
45	97	104	79	57	45	97	106	∅	∅
46	97	107	85	76	46	97	98	∅	∅
47	98	108	71	107	47	98	110	∅	∅
48	110	111	74	89	48	110	112	∅	∅
49	97	113	111	77	49	97	99	∅	∅
50	99	111	77	71	50	99	98	∅	∅

Table 26.1 The cooperative procedure CP using four programs

The differences in columns $x(k)$ and m_s in Tables 26.1.B and 26.2.B demonstrate that effective interactions have a strong influence on both the search path and the gathered information made available for sharing. It is proof that changes to component h_0 of the virtual shared memory are increasing in magnitude.

26.3.3 *Identifying systemic cooperations*

We use values of m_r from Table 26.1.B and Table 26.2.B to prove the existence of a transfer of β-solution attributes among β-paths of the cooperative procedures, therefore proving systemic cooperation takes place among cooperating programs.

The variable m_r represents the messages received by program P_0 from the other programs. Recall that procedures CP and $\overline{CP}$ are identical, except for the reading of messages by program P_0. Therefore, changes in the search behavior of program P_0 between $\overline{CP}$ to CP are created due to the fact that P_0 reads its messages and performs some effective interactions. According to our definition in the last section, a program P_i performs an independent effective interaction if it uses α-solution attributes from program P_j to determine the next current

	A: programs 0 to 3					B: program P_0			
it	P_0	P_1	P_2	P_3	it	$x(k)$	$x'(k)$	m_s	m_r
1	1	2	3	4	1	1	5	0	0
2	5	6	7	8	2	5	9	∅	6
3	9	7	8	12	3	9	202	9	∅
4	202	9	7	8	4	202	203	202	7
5	203	8	202	9	5	203	205	203	∅
6	205	206	7	8	6	205	209	205	∅
7	209	205	206	51	7	209	210	209	∅
8	210	211	209	205	8	210	214	210	∅
9	214	205	7	8	9	214	218	214	∅
10	218	214	220	58	10	218	221	∅	∅
11	221	222	47	223	11	221	224	221	∅
12	224	205	38	221	12	224	226	224	∅
13	226	224	228	229	13	226	230	226	∅
14	230	219	226	27	14	230	231	∅	∅
15	226	232	220	51	15	226	233	∅	∅
16	233	234	38	45	16	233	235	∅	∅
17	235	217	7	8	17	235	236	∅	∅
18	226	227	237	238	18	226	239	∅	238
19	239	219	240	237	19	239	242	∅	240
20	242	205	243	244	20	242	233	∅	∅
21	233	243	245	237	21	233	247	∅	∅
22	247	248	243	250	22	247	239	∅	∅
23	239	251	252	248	23	239	242	∅	251
24	242	252	255	256	24	242	233	∅	256
25	233	251	258	255	25	233	247	∅	∅
26	226	260	261	262	26	226	230	∅	∅
27	230	263	255	256	27	230	266	230	∅
28	266	230	263	269	28	266	270	266	263
29	270	269	266	272	29	270	273	∅	∅
30	266	263	274	275	30	266	276	∅	∅
31	276	277	278	279	31	276	280	∅	278
32	280	267	281	282	32	280	283	∅	∅
33	266	260	282	285	33	266	286	∅	∅
34	286	277	263	279	34	286	276	∅	∅
35	276	278	287	288	35	276	289	∅	278
36	289	281	290	291	36	289	292	289	∅
37	292	289	281	285	37	292	293	292	∅
38	293	263	292	279	38	293	294	∅	∅
39	294	287	263	288	39	294	295	∅	∅
40	295	281	296	291	40	295	289	∅	∅
41	289	284	278	298	41	289	293	∅	∅
42	293	263	284	282	42	293	294	∅	∅
43	294	290	300	285	43	294	295	∅	∅
44	295	301	278	303	44	295	289	∅	∅
45	289	278	287	282	45	289	306	∅	∅
46	306	281	281	279	46	306	307	306	∅
47	307	308	306	285	47	307	309	∅	308
48	306	310	263	311	48	306	312	∅	∅
49	312	313	290	288	49	312	314	∅	∅
50	314	308	281	279	50	314	306	∅	∅

Table 26.2 The cooperative procedure $\overline{CP}$ using four programs

solution in its search path. If there were only independent effective interactions in the cooperative procedure CP (and therefore no systemic cooperation), the changes to the search behavior of program P_0 when going from $\overline{CP}$ to CP, should not affect the search behavior of the three other programs P_1, P_2 and P_3 (since these programs are supposed to pick up only α-solution attributes from P_0 when there is no systemic cooperation). But, in fact, as one can observe by comparing Table 26.2.A and Table 26.1.A, the programs in the two cooperative procedures do not explore the same regions of the solution space, except for the first three iterations. There is a dramatic difference in the search behavior of the sequential programs between the two cooperative procedures. The only possible way for these differences to occur, is for some β-solution attributes from P_0 to be picked up by programs P_1, P_2 and P_3 once P_0 has started to read its messages. Therefore, correlated effective interactions take place among sequential programs.

The existence of correlated interactions prove that systemic cooperations occur in CP. These systemic cooperations include at least two effective interactions (one performed by P_0 and one executed by either P_1, P_2 or P_3 using β-solution attributes from P_0). However, observing the results of these two cooperative procedures, we can show that the systemic cooperation structures are in fact more complex. For example, column m_r for P_0: this column is different for CP and $\overline{CP}$. The only way for this to happen, is for at least one program among P_i, $i \neq 0$, to have passed some of its β-values to P_0, therefore creating a chain of at least three effective interactions. There are very complex systemic cooperation structures in CP. However, the formal demonstration of this last statement goes beyond the scope of this paper (see [38]).

26.4 DISCUSSION

The tests presented in the previous section have been designed to show that the notion of systemic cooperation is experimentally sound. The cooperative procedures CP and $\overline{CP}$ are both built from the same set of sequential programs; they both use the same set of search strategies and each search program has the same initial solution in both CP and $\overline{CP}$. Nevertheless, they explore the solution space in a very different manner. The origin of these differences in the search behavior of the two cooperative procedures can be traced to the use of heuristic information by program P_0. The tabu search method is a local search procedure using information gathered on the solution space to guide its long-term search behavior. In procedures $\overline{CP}$ and CP, program P_0 does not use the same gathered information and therefore, its long-term search behavior is different. Because a cooperative procedure is a dynamical system, changes in one subsystem such as program P_0, affect the behavior of the other subsystems. Most models in the literature on cooperative algorithms assume that the sharing of information only has a local impact and therefore that the length of the propagation is limited to a single interaction [3, 23, 24]. We have proven that, on the contrary, cooperation is a systemic phenomenon which takes place among several programs and lasts for many iterations, therefore shared information has a cumulative impact on search programs.

The experiments show that a single systemic cooperation can have a significant impact on the search behavior of cooperating programs. As shown in [41], changes in the communication structure, the number of programs or the frequency of information exchanges directly induce the occurrence and the structure of the systemic cooperations. These kinds of observations raise some concern about the conclusions that have been drawn from experimental results published in the literature on cooperative algorithms, particularly when there is extensive cooperation among programs. Since in most cases these conclusions have been drawn without taking into account the profound impact of cooperation and how it is related to parameters such as number of programs, the communication structure or the frequency of information exchange, the possibility exists that diametrically different conclusions would result by simply modifying these parameters. Could the results possibly have been biased by these parameters of the cooperative procedures?

The iterative map of equations (26.4) and (26.5) of the dynamical system model aims to trace the impact of the complex interactions among the search programs from a data structure perspective. Equation (26.5) describes the data which are shared among the search programs in term of a lattice $h(t)$

which is updated according to the state of the search (current solutions) of currently interacting processes. Equation (26.4) provides a formal description of the evolution of the cooperative search also in terms of a lattice $x(t)$ which is updated according to the search strategies and the contents of the lattice of equation (26.5). This model is quite simple but it provides useful relations. Using this model we have been able to obtain a formal definition of the concept of systemic cooperation, which has been proved experimentally sound in the previous section. Based on this concept, the result of the present study shows not so much that interactions change the path of cooperating programs but rather that these interactions add up and have a cumulative impact on search programs. Actually, the present study singles out the control of the interactions among the search programs as a means to improve the design of cooperative algorithms. For example, equation (26.5) shows a dependency of the lattice $x(t)$ on the lattice $h(t)$. The study of this particular dependency has made the analysis of the propagation of blocks of data in the part of the lattice $x(t)$ related to a systemic cooperation possible. We found a direct relation between these blocks and the Hamming distance between the components x_i of the lattice and also the evolution of the Shannon entropy of $x(t)$ [42]. These small blocks are important in controlling the interactions among cooperating programs. In a recent project we have designed new operators that act more actively on these blocks by treating them as "partial solutions". We have then been able to come up with an improved design for cooperative algorithms which uses these "building blocks" to delimit regions of the solution space to be explored by the search programs [43]. This is an example of how the kind of model and concepts presented in this paper can be helpful in designing better cooperative algorithms.

26.5 CONCLUSION

Cooperative algorithms are parallel search algorithms based on the coupling of previously disjoint sequential search programs. In most cases, models of the design and analysis of cooperative algorithms have focused on sequential programs and aimed to optimize the use of information shared among programs based on the logic of the search strategies. The validity of this kind of approach to cooperative algorithms was questioned in this paper. Because of their underlying assumptions, we think these models cannot account for a number of fundamental aspects of the behavior of cooperating programs and may derive sub-optimal designs of cooperative algorithms.

In this paper we briefly presented a model of cooperative search which makes no simplifying assumptions about the interdependencies among cooperating programs. Although, for the sake of clarity and simplicity, our formulation was based on a particular search method, the observations and conclusions of this formulation can be extended to the behavior of cooperating programs in general. We described cooperating programs, which, in terms of the logic of their search strategy, make the best possible use of the information shared among them. Nevertheless, we observed a considerable variation in the search pattern of these cooperating programs, although their search parameters and initial condition did not change. This happened because the complex dependencies among cooperating programs support the emergence of "systemic cooperations", which is the search behavior of programs interacting within the same system of correlated interactions. Search strategies control the use of incoming information resulting from a particular interaction, but they have little control over the occurrences of interactions and their organization into systemic cooperations. Instead, the number of programs, the communication structure, and the frequency of the exchange have a stronger influence on how cooperation is organized.

Cooperative procedures are dynamical systems, and it is clear from our experimentation that the interactions among the subsystems can have a tremendous impact on the search behavior of cooperating programs. But at this point, we do not understand the existing relationships between these dynamics and the performance of cooperating programs. There are no formal relationships that describe how the state of a particular search program can depend on the state of other cooperating programs. However, nothing prevents us from creating cooperating programs that are based on formal interactions. We think this is a promising approach for the study of cooperative algorithms. For example, rather than exchanging information on the solution space, we can transform the state of the cooperating programs according to formal relations. One may study cooperative algorithms based on simulations of the interactions among cooperating programs using known formal dynamical systems such as cellular automata. This is the direction of our future work.

Acknowledgements

This research has been supported by grants from the Fonds F.C.A.R. of the Province of Québec, and the Natural Sciences and Engineering Research Council of Canada.

References

[1] R. Battiti and G. Tecchiolli. Parallel Biased Search for Combinatorial Optimization: Genetic Algorithms and TABU. *Microprocessors and Microsystems*, 16(7):351–367, 1992.

[2] N. Boissin and J.L. Lutton. A Parallel Simulated Annealing Algorithm. *Parallel Computing*, 19(8):859–872, 1993.

[3] S.H. Clearwater, T. Hogg, and B.A. Huberman. Cooperative Problem Solving. In B.A. Huberman, editor, *Computation: The Micro and the Macro View*, pages 33–70. World Scientific, 1992.

[4] S.H. Clearwater, B.A. Huberman, and T. Hogg. Cooperative Solution of Constraint Satisfaction Problems. *Science*, 254:1181–1183, 1991.

[5] J. Cohoon, S. Hedge, W. Martin, and D. Richards. Punctuated Equilibria: A Parallel Genetic Algorithm. In J.J. Grefenstette, editor, *Proceedings of the Second International Conference on Genetic Algorithms and their Applications*, pages 148–154. Lawrence Erlbaum Associates Publishers, 1987.

[6] D. Costa. An Evolutionary Tabu Search Algorithm and the NHL Scheduling Problem. *INFOR*, 33(3):161–178, 1995.

[7] T.G. Crainic, M. Gendreau, P. Soriano, and M. Toulouse. A Tabu Search Procedure for Multicommodity Location/Allocation with Balancing Requirements. *Annals of Operations Research*, 41:359–383, 1993.

[8] T.G. Crainic, M. Toulouse, and M. Gendreau. Synchronous Tabu Search Parallelization Strategies for Multicommodity Location-Allocation with Balancing Requirements. *OR Spektrum*, 17(2/3):113–123, 1995.

[9] T.G. Crainic, M. Toulouse, and M. Gendreau. Parallel Asynchronous Tabu Search for Multicommodity Location-Allocation with Balancing Requirements. *Annals of Operations Research*, 63:277–299, 1995.

[10] F. Damera-Rogers, S. Kirkpatrick, and V.A. Norton. Parallel Algorithms for Chip Placement by Simulated Annealing. *IBM Journal of Research and Development*, 31:391–402, 1987.

[11] P.S. de Souza. *Asynchronous Organization for Multi-Algorithm Problems.* PhD thesis, Carnegie Mellon University, 1996.

[12] B. Gendron and T.G. Crainic. Parallel Branch-and-Bound Algorithms: Survey and Synthesis. *Operations Research*, 42(6):1042–1066, 1994.

[13] F. Glover. Tabu Search – Part I. *ORSA Journal on Computing*, 1(3):190–206, 1989.

[14] F. Glover. Tabu Search – Part II. *ORSA Journal on Computing*, 2(1):4–32, 1990.

[15] F. Glover. Tabu Search and Adaptive Memory Programming – Advances, Applications and Challenges. In R.S. Barr, R.V. Helgason, and J. Kennington, editors, *Interfaces in Computer Science and Operations Research*, pages 1–75. Kluwer, Norwell, MA, 1996.

[16] F. Glover and M. Laguna. Tabu Search. In C.R. Reeves, editor, *Modern Heuristic Techniques for Combinatorial Problems*, pages 70–150. Blackwell, Oxford, 1993.

[17] V. Gordon, D. Whitley, and A. Bohm. Dataflow Parallelism in Genetic Algorithms. In R. Männer and B. Manderick, editors, *Parallel Problem Solving from Nature II*, pages 533–542. Elsevier Science Publishers B.V., 1992.

[18] V.S. Gordon and D. Whitley. Serial and Parallel Genetic Algorithms as Function Optimizers. In S. Forrest, editor, *Proceedings of the Fifth International Conference on Genetic Algorithms*, pages 177–183. Morgan Kaufmann Publishers, 1993.

[19] M. Gorges-Schleuter. Comparison of Local Mating Strategies in Massively Parallel Genetic Algorithms. In R. Männer and B. Manderick, editors, *Parallel Problem Solving from Nature II*, pages 553–562. Elsevier Science Publishers B.V., 1992.

[20] C. Graffigne. Parallel Annealing by Periodically Interacting Multiple Searches: an Experimental Study. In Robert Azencott, editor, *Simulated Annealing Parallelization Techniques*, pages 47–79. John Wiley & Sons Inc., 1992.

[21] D.R. Greening. Parallel Simulated Annealing Techniques. *Physica D*, 42:293–306, 1990.

[22] D.R. Greening and F. Darema. Rectangular Spatial Decomposition Methods for Parallel Simulated Annealing. In *Proceedings of the International Conference on Supercomputing*, pages 295–302, 1989.

[23] T. Hogg and C. Williams. Solving the Really Hard Problems with Cooperative Search. In *Proceedings of the 11th National Conference on Artificial intelligence (AAAI93)*, pages 231–236. AAAI Press, 1993.

[24] B.A. Huberman. The Performance of Cooperative Processes. *Physica D*, 42:38–47, 1990.

[25] R. Jayaraman and F. Darma. Error Tolerance in Parallel Simulated Annealing Technique. In *Proceedings of the IEEE International Conference on Computer-Aided Design: ICCAD-88*, pages 545–548, IEEE Computer Society Press, Washington, DC, 1988.

[26] P. Jog and D. Gucht. Parallelisation of Probabilistic Sequential Search Algorithms. In J.J. Grefenstette, editor, *Proceedings of the Second International Conference on Genetic Algorithms and their Applications*, pages 170–176. Lawrence Erlbaum Associates Publishers, 1987.

[27] B. Manderick and P. Spiessens. Fine-Grained Parallel Genetic Algorithm. In J. Schaffer, editor, *Proceedings of the Third International Conference on Genetic Algorithms*, pages 416–421. Morgan Kaufmann Publishers, 1989.

[28] P. Moscato. On Evolution, Search, Optimization, Genetic Algorithms and Martial Arts: Towards Memetic Algorithms. Publication Report 790, Caltech Concurrent Computation Program, 1989.

[29] P. Moscato. An Introduction to Population Approaches for Optimization and Hierarchical Objective Functions: A Discussion on the Role of Tabu Search. *Annals of Operations Research*, 41:85–121, 1993.

[30] P. Moscato and M.G. Norman. A 'Memetic' Approach for the Traveling Salesman Problem. Implementation of a Computational Ecology for Combinatorial Optimization on Message-Passing Systems. In M. Valero, E. Onate, M. Jane, J.L. Larriba and B. Suarez, editors, *Parallel Computing and Transputer Applications*, pages 187–194. IOS Press, Amsterdam, 1992.

[31] H. Mühlenbein, M. Gorges-Schleuter, and O. Krämer. New Solutions to the Mapping Problem of Parallel Systems - the Evolution Approach. *Parallel Computing*, 6:269–279, 1987.

[32] M. Munetomo, Y. Takai, and Y. Sato. An Efficient Migration Scheme for Subpopulation-Based Asynchronous Parallel Genetic Algorithms. In S. Forrest, editor, *Proceedings of the Fifth International Conference on Genetic Algorithms*, pages 649–658. Morgan Kaufmann Publishers, 1993.

[33] C. Pettey, M. Leuze, and J. Grefenstette. A Parallel Genetic Algorithm. In J.J. Grefenstette, editor, *Proceedings of the Second International Conference on Genetic Algorithms and their Applications*, pages 155–161. Lawrence Erlbaum Associates Publishers, 1987.

[34] E. Taillard. *Recherches Itératives Dirigées Parallèles.* PhD thesis, École Polytechnique Fédérale de Lausanne, 1993.

[35] E. Taillard. Parallel Taboo Search Techniques for the Job Shop Scheduling Problem. *ORSA Journal on Computing*, 6(2):108–117, 1994.

[36] R. Tanese. Parallel Genetic Algorithm for a Hypercube. In J.J. Grefenstette, editor, *Proceedings of the Second International Conference on Genetic Algorithms and their Applications*, pages 177–183. Lawrence Erlbaum Associates Publishers, 1987.

[37] R. Tanese. Distributed Genetic Algorithms. In J.D. Schaffer, editor, *Proceedings of the Third International Conference on Genetic Algorithms and their Applications*, pages 434–440. Morgan Kaufmann Publishers, 1989.

[38] M. Toulouse. *Heuristiques Parallèles de Recherche sans Contrôle Global Explicite.* PhD thesis, École Polytechnique, Université de Montréal, 1996.

[39] M. Toulouse, C. Crainic, and M. Gendreau. Issues in Designing Parallel and Distributed Search Algorithms for Discrete Optimization Problems. Publication CRT-96-36, Centre de recherche sur les transports, Université de Montréal, 1996.

[40] M. Toulouse, T. Crainic, and M. Gendreau. Communication Issues in Designing Cooperative Multi-Thread Parallel Searches. In I.H. Osman and J.P. Kelly, editors, *Meta-Heuristics: Theory & Applications*, pages 501–522. Kluwer Academic Publishers, 1996.

[41] M. Toulouse, T.G. Crainic, and B. Sansó. Systemic Behavior of Cooperative Search Algorithms. Publication CRT-97-55, Centre de recherche sur les transports, Université de Montréal, Montréal, QC, Canada, 1997.

[42] M. Toulouse, T.G. Crainic, B. Sansó, and K. Thulasiraman. Self-Organization and Optimization in Cooperative Tabu Search Algorithms. Publication CRT-98-34, Centre de recherche sur les transports, Université de Montréal, Montréal, QC, Canada, 1998.

[43] M. Toulouse, F. Glover, and K. Thulasiraman. Message Enhanced Self-Organization: A Multi-Scale Cooperative Search with an Application to Graph Partitioning. Publication, School of Computer Science, University of Oklahoma, Norman, OK, 1998.

[44] A. Üresin and M. Dubois. Asynchronous Iterative Algorithms: Models and Convergence. In L. Kronsjö and D. Shumsheruddin, editors, *Advances in Parallel Algorithms*, pages 302–342. John Wiley & Sons, Inc., 1992.

[45] M.G.A. Verhoeven. *Parallel Local Search.* PhD thesis, Eindhoven University of Technology, 1996.

[46] M.G.A. Verhoeven and E.H.L. Aarts. Parallel Local Search. *Journal of Heuristics*, 1(1):43–65, 1995.

[47] M.D. Vose and G.E. Liepins. Punctuated Equilibria in Genetic Search. *Complex Systems*, 5:31–44, 1991.

[48] D. Whitley. Applications of Modern Heuristic Methods. In V.J. Rayward-Smith, editor, *Proceedings of the Fifth International Conference on Genetic Algorithms*, pages 55–67. Alfred Waller, 1995.

27 A HOPFIELD-TANK NEURAL NETWORK MODEL FOR THE GENERALIZED TRAVELING SALESMAN PROBLEM

Ricardo Andresol, Michel Gendreau and Jean-Yves Potvin

Centre de recherche sur les transports
Département d'informatique et de recherche opérationnelle
Montréal, Québec, Canada H3C 3J7.
michelg@crt.umontreal.ca, potvin@iro.umontreal.ca

Abstract: The Generalized Traveling Salesman Problem consists of determining a shortest tour passing through each of several clusters of vertices. A version of this problem is examined where the distance matrix is symmetric and where exactly one vertex is visited in each cluster. A Hopfield-Tank neural network model is developed for solving this problem. Numerical results are presented for problems with up to 50 vertices and 10 clusters with either Euclidean or randomly generated distance matrices. A comparison with a recent heuristic shows that the neural network is competitive with respect to solution quality. However, it is computationally more expensive and is restricted to relatively small problems as it often fails to converge to feasible solutions when the search space of possible configurations is too large.

27.1 INTRODUCTION

The Generalized Traveling Salesman Problem (GTSP) extends the classical Traveling Salesman Problem (TSP) in the following way: the set of vertices is partitioned into a number of clusters and a least cost tour passing through each cluster must be found. Applications arise in many areas such as location-routing, computer design, loop material flow system design and postal box collection [5].

Different versions of the GTSP exist. In this paper, we consider a version of the problem where the distance matrix is symmetric and where exactly one vertex must be visited in each cluster. More formally, let $G = (V, E)$ be an undirected graph where $V = \{v_1, \ldots, v_n\}$ is the vertex set and $E = \{(v_i, v_j) : v_i, v_j \in V, i < j\}$ is the edge set. The set V is partitioned into m clusters $V_1, \ldots, V_m$ and a non negative distance d_{ij} is associated with each edge (v_i, v_j). The problem is then to find a shortest cycle that contains exactly one vertex in each cluster. Exact algorithms to solve this problem, using branch-and-cut or dynamic programming, are found in [1, 4, 11, 12]. Due to their exponential nature, however, these approaches are restricted to relatively small-sized instances. A composite heuristic, developed by Renaud and Boctor, is also reported in [9]. Finally, Laporte and Semet [6], based on a previous work for the asymmetric problem [7], have shown how a symmetric GTSP can be transformed into an equivalent symmetric TSP. These transformations, however, induce a fair amount of degeneracy and can hardly be exploited in practice.

In this paper, we present an alternative heuristic approach to the GTSP, based on a Hopfield-Tank neural network model [3]. Although the latter has been widely applied to the TSP [8], we are not aware of any application for the GTSP (see [13], however, for an application to another variant of the TSP, know as the orienteering problem). The Hopfield-Tank model looks like an interesting problem-solving approach here since (1) the original model developed for the TSP can be easily extended to the GTSP, (2) for the same number of vertices, the GTSP model is smaller than its TSP counterpart and (3) the model can be applied to both geometric and non-geometric instances.

In the following Section 27.2 we introduce the application of the Hopfield-Tank model to the TSP. Then, in Section 27.3 we explain how this model can be extended to the GTSP. In Section 27.4 computational results obtained on problems with Euclidean and randomly generated distance matrices are presented. Finally, the conclusion follows in Section 27.5.

27.2 THE HOPFIELD-TANK MODEL

The original Hopfield model [2] is a fully interconnected network of binary units with symmetric weighted connections. The connections' weights are not learned but are defined a priori from problem data, e.g., the distance matrix. Starting from some arbitrarily chosen initial configuration, where the activation level of each unit is set to 0 or 1, the network evolves by updating the activation level of each unit in turn. That is, an activated unit can be turned off and conversely. The update rule for any given unit i involves all units it is connected to, as well as the associated connection weights, that is:

$$\begin{array}{l} \text{if } \sum_j T_{ij} V_j < \theta_i, \text{ set } V_i \text{ to } 0, \\ \text{if } \sum_j T_{ij} V_j > \theta_i, \text{ set } V_i \text{ to } 1, \\ \text{if } \sum_j T_{ij} V_j = \theta_i, \text{ do nothing,} \end{array}$$

where V_i is the activation level of unit i, T_{ij} is the connection weight between units i and j and θ_i is the threshold of unit i. Through this update process,

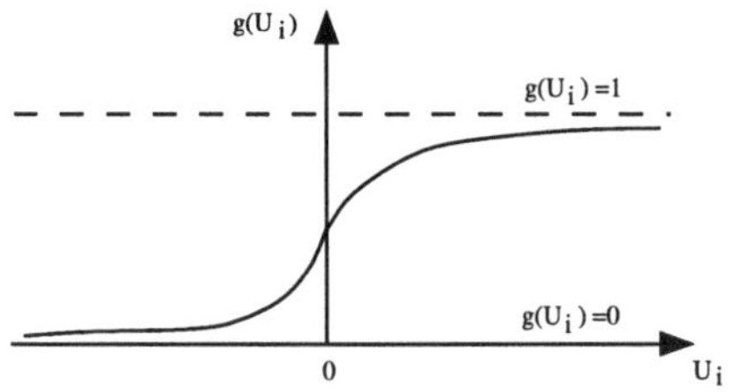

Figure 27.1 Sigmoidal transformation.

various configurations are explored until the network settles in a stable configuration where the activation levels do not change anymore. The network's behavior can be characterized through the following energy function:

$$E = -\frac{1}{2}\sum_i \sum_j T_{ij} V_i V_j + \sum_i \theta_i V_i. \tag{27.1}$$

Note that this is a function of the units' activation levels, since the weights and thresholds are fixed. The function is such that the energy of any new configuration produced through the update rule is always lower than or equal to the energy of the previous configuration. Hence, the network settles in a local minimum of the energy function.

The Hopfield-Tank neural network model is an extension of the previous model. In this case, a unit's activation level can be any value between 0 and 1. For each unit i, this value is continuously updated over time according to the following differential equation:

$$\frac{dU_i}{dt} = \sum_j T_{ij} V_j + I_i - U_i \tag{27.2}$$

where U_i is the total input to unit i and I_i is the external input or input bias of unit i (which replaces the threshold θ_i of binary units). Typically, the unit's activation level corresponds to a sigmoidal transformation of its input, namely:

$$V_i = g\left(U_i\right) = \frac{1}{\left(1 + e^{\frac{-U_i}{U_0}}\right)}. \tag{27.3}$$

The sigmoidal transformation always produces activation levels between 0 and 1 (see Figure 27.1). The constant U_0 controls the shape of the function and a "flat" curve converging slowly towards 0 and 1 can be obtained by setting U_0 to a large value.

The energy function of the Hopfield-Tank model, which is similar to equation (27.1), is:

$$E = -\frac{1}{2}\sum_i \sum_j T_{ij} V_i V_j - \sum_i I_i V_i + \int_0^{V_i} g^{-1}(x)dx. \tag{27.4}$$

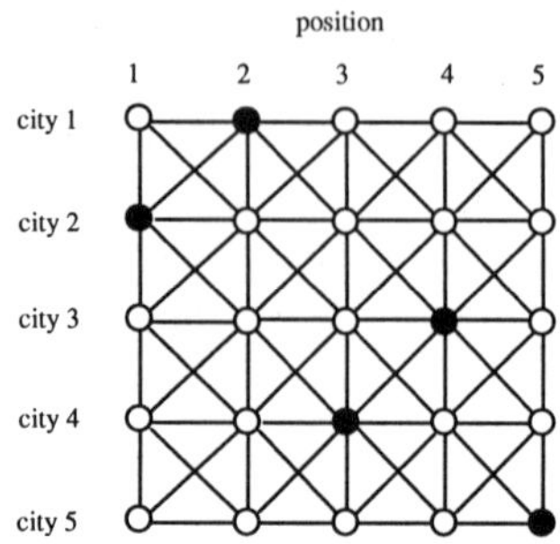

Figure 27.2 Hopfield-Tank model for the TSP

Note that $dU_i/dt = -\partial E/dV_i$. Thus, the model performs a gradient descent in the configuration space with respect to the energy function E. When the model reaches a stable state, $dU_i/dt = 0$ and the input to any unit i is simply:

$$U_i = \sum_j T_{ij} V_j + I_i \tag{27.5}$$

In practice, the continuous evolution of the model is approximated using discrete time intervals Δt. That is, starting with initial values $\{U_i(0)\}$ at time $t = 0$, the values at time $t + \Delta t$ are derived from the values at time t through equation (27.6), a discrete approximation of equation (27.2), until the model stabilizes.

$$\frac{U_i(t + \Delta t) - U_i(t)}{\Delta t} = \sum_j T_{ij} V_j(t) + I_i - U_i(t). \tag{27.6}$$

Figure 27.2 shows a Hopfield-Tank model for a TSP defined over $n = 5$ cities. In this case, it is a 5×5 matrix of units. Each row corresponds to a particular city and each column to a particular position in the tour. The black nodes correspond to activated units (i.e., units with activation levels sufficiently close to 1) that encode the current solution, namely the tour 2-1-4-3-5. Note that only a few connections are shown in Figure 27.2, since the full model includes a connection between each pair of units.

The energy function suggested in [3] for solving a TSP of size n with such a network architecture is:

$$\begin{aligned} E \;=\; & \frac{A}{2} \sum_X \sum_i \sum_{j \neq i} V_{Xi} V_{Xj} + \frac{B}{2} \sum_i \sum_X \sum_{Y \neq X} V_{Xi} V_{Yi} + \\ & \frac{C}{2} \left(\sum_X \sum_i V_{Xi} - n \right)^2 + \frac{D}{2} \sum_X \sum_{Y \neq X} \sum_i d_{XY} V_{Xi} \left(V_{Yi+1} + V_{Yi-1} \right). \end{aligned} \tag{27.7}$$

In this formula, V_{Xi} denotes the activation level of the unit in row (city) X and column (position) i while A, B, C and D are weighting parameters for each component of the sum. Note that the connections' weights and input biases of the Hopfield-Tank model can be obtained by observing that weights are associated with quadratic terms while input biases are associated with linear terms in equation (27.4). The first three components of the energy function are aimed at penalizing infeasible configurations: multiple activated units in any given row (cities visited more than once), multiple activated units in any given column (many cities visited at the same position in the tour) and a total number of activated units that is different from n, the number of cities. The fourth component favors short tours over longer ones (i.e., to minimize the energy function, the model will tend to set V_{Xi} and V_{Yi+1} (or V_{Yi-1}) to one, thus creating a solution where cities X and Y are visited consecutively, when the distance d_{XY} between the two cities is small). Large A, B and C values heavily penalize infeasible configurations, thus driving the model towards feasible TSP solutions, but not necessarily good ones, while a large D value emphasizes solution quality at the expense of feasibility. The next section will describe an extension to this basic model to address the GTSP.

27.3 AN EXTENSION FOR THE GTSP

For a GTSP defined over n cities and m clusters, the network representation is similar to the one shown in Figure 27.2. However, the matrix is $n \times m$ rather than $n \times n$. That is, there is a row for each city and a column for each cluster. Since m is typically much smaller than n, the configuration space is substantially reduced.

The energy function of the Hopfield-Tank model is then modified as follows:

$$\begin{aligned} E \;=\; & \frac{A}{2}\sum_i \left(\sum_X V_{Xi} - 1\right)^2 + \frac{B}{2}\sum_k \left(\sum_X \sum_i \delta_{Xk} V_{Xi} - 1\right)^2 + \\ & \frac{C}{2}\sum_X \sum_{Y \neq X} \sum_i d_{XY} V_{Xi} \left(V_{Yi+1} + V_{Yi-1}\right), \end{aligned} \tag{27.8}$$

where the summations over the positions run from $i = 1$ to m and where $\delta_{Xk} = 1$ if city X is in cluster k, 0 otherwise. The first component of this energy function penalizes columns with either no activated unit or more than one activated unit. The second component is added to the energy function to account for the specific nature of the GTSP where exactly one city must be visited in each cluster. That is, for any given cluster k, the summation over the rows that correspond to cities in this cluster must be equal to one, otherwise a penalty is incurred. The third component is the same as the one for the TSP and favors short tours over longer ones. Note that configurations with two or more activated units in the same row (i.e., same city visited more than once) are penalized through the second component. The first component is necessary to penalize configurations where two or more cities in different clusters are visited at the same position. The evolution of this Hopfield-Tank model can now be summarized as:

1. Initialize the activation level of each unit to $1/n$ plus or minus a small random perturbation; evaluate the energy E_0; set t to 0.

2. Update the activation level of each unit at time $t + \Delta t$ using equations (27.3) and (27.6).

3. Evaluate the energy $E_{t+\Delta t}$ of the new configuration

4. If $E_t - E_{t+\Delta t} \geq \Delta E_c$ then set t to $t + \Delta t$ and go to step 2.

5. Decode the configuration into a GTSP solution; to do this, a unit is on (activated) if its activation level is greater than or equal to some threshold value.

6. If the solution is feasible compute its total distance.

The initial activation levels are such that the total activation level over the $n \times m$ units is approximately equal to m, the desired number of activated units. The model then evolves until the variation of energy between two consecutive iterations is less than a critical ΔE_c value. At the end, the configuration of the network is decoded into a solution (hopefully, feasible) of the GTSP. In the computational results of Section 27.4, ΔE_c is set to 10^{-6} and the threshold value for deciding if a unit is activated is set to 0.5.

As already reported for the TSP [14], appropriately setting the penalty parameters A, B and C in equation (27.8) is a pretty difficult task. The network's behavior is very sensitive to those values which must be tuned to each problem. Without a proper setting, large variations in the units' activation levels, due to large variations in their input, are observed in the first iterations of the algorithm. In the process, many units quickly saturate at activation levels close to 1 and stay there forever (i.e., those units lie on the flat portion of the sigmoid curve shown in Figure 27.1). Consequently, the model "freezes" too quickly and settles in bad local optima.

To smooth up the network's evolution, large variations to the units' activation levels were forbidden through a ΔV_{max} parameter. That is, ΔV_i was set to ΔV_{max} whenever $\Delta V_i > \Delta V_{max}$. In the computational results of Section 27.4, ΔV_{max} was set to 0.1. This simple modification stabilized the network's evolution over a large range of values of A, B and C. It was thus possible to restrict ourselves to only a few different settings and to avoid a fine tuning of these values to each problem.

27.4 COMPUTATIONAL RESULTS

In this section, the heuristic of Renaud and Boctor [9] for solving the GTSP is briefly introduced. Then, the Hopfield-Tank model is compared with this heuristic on different sets of tests problems.

		GI^2		Hopfield-Tank					
n	m	Length	Comp. time	% fsb.	Length avg.	Length best	A,B	C	Comp. time
12	4	123	0.1	100	124	120	2	3	1.0
12	6	143	0.1	90	151	143	2	3	1.8
24	4	70	0.1	100	74	70	2	3	6.5
24	6	97	0.3	95	107	97	2	3	14.8
24	8	169	0.8	90	189	178	2	3	28.2
24	12	241	1.5	85	268	245	2	3	58.5
32	4	98	0.4	100	95	88	2	3	11.3
32	8	129	1.6	95	129	120	2	3	42.1
32	16	262	2.5	80	285	264	2	3	315.0
50	5	80	0.8	90	78	69	2	2	410.0
50	10	143	2.5	85	148	141	2	2	1310.0

Table 27.1 Numerical results on Euclidean problems

27.4.1 An alternative heuristic

For comparison purposes, we implemented the Generalized Initialization and Insertion (GI^2) heuristic of Renaud and Boctor. The two phases of this heuristic are the following.

Initialization phase: The initialization phase first selects a vertex in each cluster. For each cluster k, the selected vertex is the one that minimizes the sum of the distances to vertices that are not in k. An enveloping subtour that contains some, but not necessarily all, selected vertices is then constructed (see [10] for a detailed description of this procedure). Note that for non-geometric instances based on randomly generated distance matrices, the enveloping subtour heuristic cannot be applied. In this case, the authors suggest the "maximum triangle" rule which constructs an initial subtour with the three vertices that are the farthest apart.

Insertion phase: Clusters that are not included in the initial subtour are inserted through a cheapest insertion procedure. That is, for each vertex in clusters that are not yet included, the least cost insertion between two consecutive clusters is computed and the vertex with the overall lowest cost is inserted at its best insertion place. The procedure is then repeated with the new subtour until all clusters are done.

27.4.2 Numerical experiments

To test the Hopfield-Tank model, symmetric GTSPs with Euclidean or random (uniform) distance matrices were generated with up to $n = 50$ vertices and $m = 10$ clusters, with n/m vertices in each cluster. The code was written in C and the tests were run on a Sun Sparc 10 workstation. All results are based on pure tour construction heuristics. That is, postoptimization procedures, such as those reported in [9], have not been used to improve the solutions produced either by the Hopfield-Tank model or the GI^2 heuristic.

		GI^2		Hopfield-Tank					
n	m	Length	Comp. time	% fsb.	Length avg.	Length best	A,B	C	Comp. time
12	4	109	0.1	100	107	102	2	3	1.0
12	6	152	0.1	95	150	143	2	3	1.8
24	4	34	0.1	100	36	34	2	3	9.6
24	6	80	0.3	95	78	74	2	3	10.2
24	8	118	0.8	95	115	109	2	3	28.5
24	12	159	1.4	90	160	152	2	3	51.0
32	4	34	0.4	100	37	34	2	3	9.3
32	8	61	1.8	90	68	56	2	3	65.3
32	16	189	3.0	85	192	178	2	3	335.0
50	5	36	1.4	95	42	36	2	2	245.0
50	10	89	4.5	90	93	90	2	2	1075.0

Table 27.2 Numerical results on random problems

Tables 27.1 and 27.2 show the results obtained with the Hopfield-Tank model and the GI^2 heuristic on our test problems. For GI^2, we show the tour length (*Length*) as well as the computation time in seconds (*Comp. time*). The Hopfield-Tank model was run 20 times on each problem, starting from different initial configurations. The tables indicate the percentage of feasible configurations obtained over the 20 runs (*% fsb.*), the average tour length over feasible instances (*Length avg.*), the best feasible tour length (*Length best*), the values of the A, B and C parameters and the average computation time in seconds (*Comp. time*).

It is worth noting that the Hopfield-Tank converged to feasible configurations in 85% to 100% of the runs. This is opposed to results previously reported for TSPs of equivalent size [8], where the convergence rate is much lower. This increased stability is due to a reduced search space, since the matrix of units is $n \times m$ for the GTSP, as opposed to $n \times n$ for the TSP, where m is typically much smaller than n. The results in Tables 27.1 and 27.2 show that the percentage of feasible runs tends to decrease when the number of clusters increases (for the same number of vertices) due to the growth of the associated neural network model. As mentioned in the previous section, putting a limit on the variation to the units' activation levels was also very helpful to stabilize the network.

The neural network model is quite competitive with the GI^2 heuristic with respect to solution quality, particularly on problems with random distance matrices. The Hopfield-Tank model is well adapted to both types of problems, because it does not rely on any geometric intuition. This is not the case for the GI^2 heuristic whose initialization phase must be modified to a simple "ad hoc" triangular rule for non-Euclidean problems. On the other hand, the GI^2 heuristic is much faster. For example, on the largest problems with $n = 50$ and $m = 10$, the latter heuristic takes only a few seconds, while the neural network model runs for about 20 minutes. We observed that the Hopfield-Tank model becomes very computationally expensive and quite unstable beyond this limit. It would probably be possible to solve larger problems, but a fine tuning of the model on each instance would then be required.

The Hopfield-Tank model performs better on the GTSP, as compared to previous results reported for the TSP [8], due to the partition of the set of vertices into clusters that lead to a smaller model: a greater stability and a better solution quality are thus achieved. However, the limitations are still important as the model remains computationally expensive and restricted to small-sized problem instances. Note, in particular, that optimal solutions are reported by Fischetti et al. [1] on larger instances with 51 to 442 vertices using a branch-and-cut algorithm. The average computation time for solving these problems is 3821.2 seconds, with a maximum of 58770.5 seconds, on a HP 9000 series 700 Apollo.

27.5 CONCLUSION

A Hopfield-Tank model was developed to solve a symmetric version of the GTSP. Through the partition of the initial set of vertices into clusters, a smaller neural network model is obtained. Thus, feasible solutions to instances with up to 50 vertices and 10 clusters were easily obtained. Beyond this limit, however, the model becomes unstable and would require a fine tuning of its parameters. Although competitive with a recent heuristic on small-sized instances, this work should be put in perspective as optimal solutions to some larger instances are reported in [1] using a branch-and-cut algorithm.

Finally, it is worth noting that the development of Hopfield-Tank neural networks to solve combinatorial optimization problems is still an active area of research, as these models are designed to run on specialized parallel hardware, known as "neurocomputers". Such devices, which are currently being developed, could lead to electronic implementations that are many orders of magnitude faster than the usual computer simulations (see, e.g., Chapter 9 in [15]).

Acknowledgements

Financial support for this work was provided by the Natural Sciences and Engineering Research Council of Canada (NSERC) and by the Quebec Fonds pour la Formation de Chercheurs et l'Aide à la Recherche (FCAR). This support is gratefully acknowledged.

References

[1] M. Fischetti, J.J. Salazar and P. Toth. A Branch-and-Cut Algorithm for the Symmetric Generalized Traveling Salesman Problem. Working paper, DEIS, University of Bologna, 1994.

[2] J.J. Hopfield. Neural Networks and Physical Systems with Emergent Collective Computational Abilities. In *Proceedings of the National Academy of Sciences 79*, 2554–2558, 1982.

[3] J.J. Hopfield and D.W. Tank. Neural Computation of Decisions in Opti-

mization Problems. *Biological Cybernetics 42*, 141–152, 1985.

[4] G. Laporte and Y. Nobert. Generalized Traveling Salesman Problem through n Sets of Nodes: An Integer Programming Approach. *INFOR 21*, 61–75, 1983.

[5] G. Laporte, A. Asef-Vaziri and C. Sriskandarajah. Some Applications of the Generalized Traveling Salesman Problem. Technical report CRT-95-59, Centre de recherche sur les transports, Université de Montréal, 1995.

[6] G. Laporte and F. Semet. A Transformation Procedure for the Symmetric Generalized Traveling Salesman Problem. Technical report CRT-96-16, Centre de recherche sur les transports, Université de Montréal, 1996.

[7] J.E. Noon and J.C. Bean. An Efficient Transformation of the Generalized Traveling Salesman Problem. *INFOR 31*, 39–44, 1993.

[8] J.Y. Potvin. The Traveling Salesman Problem: A Neural Network Perspective. *ORSA Journal on Computing 5*, 328–348, 1993.

[9] J. Renaud and F.F. Boctor. An Efficient Composite Heuristic for the Symmetric Generalized Traveling Salesman Problem. Notes de recherche, Télé-Université, Université du Québec, 1996.

[10] J. Renaud, F.F. Boctor and G. Laporte. A Fast Composite Heuristic for the Symmetric Traveling Salesman Problem. *INFORMS Journal on Computing 8*, 134–143, 1996.

[11] J.P. Saksena. Mathematical Model of Scheduling Clients through Welfare Agencies. *CORS Journal 8*, 185–200, 1970.

[12] S.S. Srivastava, S. Kumar, R.C. Garg and P. Sen. Generalized Travelling Salesman Problem through n Sets of Nodes. *CORS Journal 7*, 97–101, 1969.

[13] Q. Wang, X. Sun, B.L. Golden and J. Jia. Using Artificial Neural Networks to solve the Orienteering Problem. *Annals of OR 61*, 111–120, 1995.

[14] G.V. Wilson and G.S. Pawley. On the Stability of the Traveling Salesman Problem Algorithm of Hopfield and Tank. *Biological Cybernetics 58*, 63–70, 1988.

[15] J.M. Zurada. *Artificial Neural Systems.* West Publishing Company, St-Paul, MN, 1992.

28 GENERALIZED CYBERNETIC OPTIMIZATION: SOLVING CONTINUOUS VARIABLE PROBLEMS

Mark A. Fleischer

Department of Engineering Management
Old Dominion University, Norfolk, VA 23529-0248, USA.
mfleisch@odu.edu

Abstract: *Cybernetic optimization by simulated annealing* (COSA) is a method of parallel processing for solving discrete optimization problems. This paper extends the theory of COSA to the continuous domain. This is done by applying the concept of *probabilistic feedback control* to the generation of candidate solutions in continuous variable problems. Three general principles of candidate generation are presented leading to the formulation of the *candidate generation criterion* and its theoretical implications. A practical method of generating candidate solutions is also presented in which the generation of candidate solutions is achieved by making the magnitude of the expected step size to candidate solutions functionally dependent on the proximity of the parallel processors in the solution space. Experimental results show that this method of generating candidate solutions accelerates the convergence of the parallel processors to the global optimum.

28.1 INTRODUCTION

Cybernetic optimization by simulated annealing (COSA) is a method of parallel processing that uses a network of self-similar, self-reflexive, and self-referential processors that accelerates the convergence of simulated annealing (SA) to the global optima (see [6]). This is done by linking processors in a *probabilistic feedback control* (PFC) network and using negative feedback to improve the search mechanism of individual processors.

Applying the COSA concept to continuous variable problems raises several new issues, problems and possibilities regarding candidate selection and thus motivates further development of the COSA concept. This paper extends the theory and application of COSA by developing a second form of PFC based on the generation of candidate solutions. Theoretical and experimental results are presented demonstrating the application of these concepts to the continuous domain.

The basic elements of COSA are described in Section 28.2. Implementation issues for the continuous domain are described in Section 28.3. The concept of PFC by candidate generation is discussed in Section 28.4. Experimental methodology and results are explored in Section 28.5. Section 28.6 summarizes these results.

28.2 BACKGROUND AND MOTIVATION

Definitions:

s the number of solution states in a combinatorial optimization problem (COP).

$S_m^{[k]}$ a random solution state associated with processor m at time index k.

i a state in a discrete solution space.

f_i the objective function value of solution i. When discussed in association with a particular processor m, then this is denoted as f_{i_m}.

$i_1 i_2 \cdots i_p$ the solution in a discrete product space spanned by p processors.

$f_{i_1 i_2 \cdots i_p} \equiv \sum_{m=1}^{p} f_{i_m}$ the objective function value in the product space spanned by p processors.

$x_{[i,m]}^{[k]}$ the value of the ith decision variable in the continuous domain with $i = 1, 2, \ldots, n$ obtained by processor m at time index k.

$x'_{[i,m]}$ the value of the ith element of the candidate solution vector for processor m.

$\mathbf{x}_m^{[k]} \equiv \left\{ x_{[1,m]}^{[k]}, x_{[2,m]}^{[k]}, \ldots, x_{[n,m]}^{[k]} \right\}$ an n-tuple or vector corresponding to the solution obtained by processor m at time index k. When referring just to some point in the configuration space, the superscripts are dropped, i.e., $\mathbf{x}_m$.

$\mathbf{x}^*$ the optimal solution.

$\mathbf{X} \equiv \mathbf{x}_1, \ldots, \mathbf{x}_p$ a set of p, n-tuples corresponding to the solution in the product space spanned by p processors in a continuous domain.

$f_{\mathbf{X}} \equiv \sum_{m=1}^{p} f_{\mathbf{x}_m}$ the objective function value in the product space.

$F_m^{[k]}$ the random objective function value associated with processor m at time index k produced in a PFC network. Similarly, $G_m^{[k]}$ produced in a network without PFC.

$d(\mathbf{x}_m, \mathbf{x}_{m+1})$ a metric on the domain of $\mathbf{x}$ associated with the solutions of processors m and $m+1$.

$d_\mathbf{X}$ the distance metric associated with system state $\mathbf{X}$ such that for all $d_\mathbf{X} \geq 0$ and $d_\mathbf{X} = 0$ if and only if all processors have coincident solutions. For example: $d_\mathbf{X} \equiv \sum_{m=1}^{p-1} d(\mathbf{x}_m, \mathbf{x}_{m+1}) + d(\mathbf{x}_p, \mathbf{x}_1)$.

$\Delta d_{\mathbf{YX}} = d_\mathbf{Y} - d_\mathbf{X}$ and $\Delta d^+_{\mathbf{YX}} = \max\{0, \Delta d_{\mathbf{YX}}\}$.

28.2.1 Parallelism and SA

There is a natural, simple and beautiful parallelism inherent in SA. The exponential form of the Metropolis Acceptance criterion [13] readily lends itself to *scalability*— products of probabilities can be described simply by the addition of exponents. Indeed, many of the important equations governing SA remain the same when described in the context of independent parallel processors. For example, the stationary probability in SA, known as the Boltzmann Distribution [1], is given as

$$\pi_i(t) = \frac{e^{-f_i/t}}{\sum_{i=1}^{s} e^{-f_i/t}}. \tag{28.1}$$

The joint stationary probability for a system of p independent processors is

$$\begin{aligned} \pi_{i_1\cdots i_p}(t) &= \prod_{m=1}^{p} \pi_{i_m}(t) = \prod_{m=1}^{p} \frac{e^{-f_{i_m}/t}}{\sum_{i_m=1}^{s} e^{-f_{i_m}/t}} \\ &= \frac{e^{-(\sum_{m=1}^{p} f_{i_m})/t}}{\prod_{m=1}^{p}\sum_{i_m=1}^{s} e^{-f_{i_m}/t}} = \frac{e^{-f_{i_1\cdots i_p}/t}}{\sum_{i_1\cdots i_p}^{s^p} e^{-f_{i_1\cdots i_p}/t}} \end{aligned} \tag{28.2}$$

since for exponentials, the product of the sums is equal to the sum of the products.

As (28.1) and (28.2) suggest, it is straightforward to describe the behavior of parallel SA simply by changing the form or meaning of the subscripts; instead of solutions i, there are *system* solutions $i_1 i_2 \ldots i_p$ in the discrete case, or $\mathbf{x}_1, \mathbf{x}_2, \ldots, \mathbf{x}_p$ in the continuous case. In a very real sense, one processor is the same as many parallel and independent processors [6] (see also [2]). For this parallelism to be of benefit, however, the processors must be linked in an advantageous manner, so that useful information is communicated between them. This linkage of parallel processors is what COSA is all about.

28.2.2 The Fundamental Theorem of COSA

The COSA concept utilizes the ergodicity of SA to provide a meaningful feedback control signal for SA: multiple processors, each running the SA algorithm on the same or related problem, converge in probability to the optima and therefore the distance between the solutions of these processors converges to zero. It is therefore possible to make a probabilistic inference of how close a processor is to the optimum by measuring its proximity to the solutions (or objective function values) of another processor. Each processor therefore gauges its progress toward the optimum by this proximity measure. This idea is given mathematical form in the fundamental theorem of COSA from [6]:

Theorem 1 of COSA [6]: *Let $X_m^{[k]}$ represent a nonnegative random variable at time index k associated with some process m such that $X_m^{[k]} \overset{Prob.}{\longrightarrow} 0$ as $k \longrightarrow \infty$. Let $Z^{[k]}$ represent some metric between these processes such that $Z^{[k]} = 0$ if and only if all processes have the same value and $Z^{[k]} \overset{Prob.}{\longrightarrow} 0$ as $k \longrightarrow \infty$. Then for k sufficiently large,*

$$\Pr\{X_m^{[k]} = 0 \mid Z^{[k]} = 0\} > \Pr\{X_m^{[k]} = 0\} \tag{28.3}$$

This theorem is applicable to a system of multiple processors running SA by letting $X_m^{[k]}$ represent a metric between an optimal value, and the random solution $S_m^{[k]}$; *e.g.*$X_m^{[k]} = d(S_m^{[k]}, \mathbf{x}^*)$ and $Z^{[k]} = d(S_m^{[k]}, S_{m+1}^{[k]})$, a metric *between* the solutions of two processors. This suggests a benefit in *inducing* an increase in this proximity.

Fleischer [6] describes this proximity measure in terms of a *probabilistic measure of quality* (PMQ) that serves as an error signal used to modulate the temperature control parameter. In general, the smaller the distance between the processors' solutions, the lower the temperature parameter; the larger the distance, the higher the temperature parameter. Fleischer [6] bases this feedback mechanism on information theoretic concepts to show how modulating the temperature effectively modulates the entropy of the search process.

28.3 APPLICATION OF COSA TO THE CONTINUOUS DOMAIN

SA has been both theoretically and practically applied to continuous variable problems [3, 4]. Bohachevsky [4] applied SA to the following 2- variable objective function (see Figure 28.1).

$$f(x, y) = x^2 + 2y^2 - 0.3\cos(3\pi x) - 0.4\cos(4\pi y) + 0.7 \tag{28.4}$$

This objective function has many local minima with the global optimum at point $(0,0)$ and is eminently suited to serve as a test- bed for application of COSA to continuous, nonlinear problems.

A number of approaches have been used in proving SA's convergence in the continuous domain. Fleischer [5, 8] uses a scaling property inherent in SA to prove convergence to an arbitrary neighborhood of the optimum. Bélisle [3] also showed convergence for a class of SA algorithms in the continuous domain. These proofs preserve those aspects of SA upon which COSA depends in the continuous domain.

28.3.1 The Probabilistic Measure of Quality

Central to the COSA concept is the manner in which PMQs are defined and used to modulate the temperature control parameter. In [6], the PMQ corre-

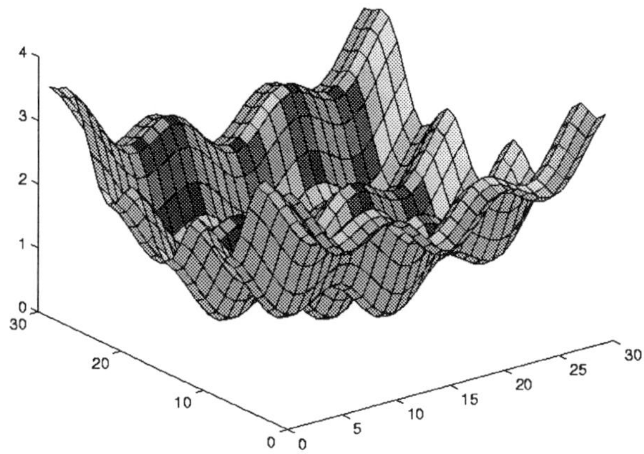

Figure 28.1 A Continuous, 2-Variable Configuration Space

sponds to the shortest distance between solutions in the underlying graph or solution space of the COP being solved. For continuous variable problems, the Euclidean distance between the current solutions of neighboring processors is an obvious choice. Thus,

$$PMQ_m^{[k]} = d(\mathbf{x}_m^{[k]}, \mathbf{x}_{m+1}^{[k]}) = \|\mathbf{x}_m^{[k]} - \mathbf{x}_{m+1}^{[k]}\| \tag{28.5}$$

Other definitions are possible. For example,

$$PMQ_m^{[k]} = \left| f_{\mathbf{x}_m^{[k]}} - f_{\mathbf{x}_{m+1}^{[k]}} \right| \tag{28.6}$$

Note that these PMQs are minimized when processors m and $m+1$ are coincident. (see [6, p. 230]). The nature of the domain may suggest the type that is most appropriate. If the domain is smooth and allows one to utilize Glover's Proximate Optimality Principle [11] then (28.5) may be appropriate. An advantage of (28.6) is that it avoids problems associated with multiple optima.

28.3.2 *The Quality Coefficient and the System Gain*

Define the *base* temperature at time index k as t_k, the *effective* temperature for processor m at time index k as t'_m, and relate these via the *quality coefficient* $q_m^{[k]}$ for processor m at time index k as in [6]. Thus,

$$t'_m = \left(\frac{PMQ_m^{[k]}}{PMQ_m^{[0]}} + \theta \right) t_k = q_m^{[k]} t_k \tag{28.7}$$

where $q_m^{[k]} = (\frac{PMQ_m^{[k]}}{PMQ_m^{[0]}} + \theta)$. $\theta > 0$ ensures that neighboring processors do not couple. See [5, 6] for details.

The effective temperature is used as the temperature parameter in standard SA but fluctuates in accordance with the value of $PMQ_m^{[k]}$. The value of $PMQ_m^{[0]}$ affects the *system gain*; i.e., the magnitude of temperature fluctuations [6, p.235]. This value is determined by the *gain setting* and set by initialization routines (see the COSA algorithm in Section 28.4.4).

28.4 PFC BY CANDIDATE GENERATION

The effectiveness of SA often depends on how candidate solutions are generated [7]. For COPs, the generation of candidate solutions is often fast, efficient, and typically suggested by the problem itself [1, 5]. Usually, there are limited choices for the neighborhood from which these candidate solutions are generated [1, Ch. 5].

Continuous variable problems provide a very different framework from which to study SA. In general, the entire domain may be considered as potential candidate solutions since there usually is no obvious neighborhood structure. This raises a number of possible advantages and problems. Bohachevsky [4] describes these problems and issues regarding the generation of candidate solutions. In their problem, minimizing (28.4), they used a fixed step size which can preclude getting very close to the optimum. Smaller step sizes permit getting closer, but slow down convergence as more steps are needed.

The continuous domain however offers an intriguing way around these difficulties. Rather than fixing the step size based on some compromise between speed and accuracy, the COSA concept provides a method for *intelligently varying the step size.* An appropriate method for doing this is suggested by the fundamental theorem of COSA—biasing the generation of candidate solutions so as to increase the proximity of the current solutions. In this section, a second form of PFC is proposed that incorporates such a bias. First, the intuition behind these ideas is described.

28.4.1 The Intuition Behind Candidate Generation Using PFC

The same logic that justifies the way the PMQ affects the temperature parameter (and the entropy of the search) is applicable to the generation of candidate solutions:

> If the current solutions of two or more processors converge in probability to the optimum, then these current solutions will be close to one another an increasing proportion of time. Thus, if two or more processors have current solutions that are close to each other, then it is more likely that one or more are close to the optimum. Under these circumstances it makes sense to generate candidate solutions that *tend* to be closer to the current solutions by making smaller step sizes. When these processors are far apart, larger step sizes should be made.

28.4.2 General Principles of COSA Search in the Continuous Domain

The following general principles derive from implications of the fundamental theorem when applied to the continuous domain.

General Principles for Candidate Generation

Principle I: candidate solutions must be randomly generated. This is based on the fact that the values of $X_m^{[k]}$ and $Z^{[k]}$ in (28.3) are random variables.

Principle II: all solutions should be reachable in a single step as the domain is continuous. Furthermore, it permits intelligently varying the step sizes as Principle III requires.

Principle III: the magnitude of the jumps (to the candidate solutions) must be proportional to, or functionally dependent on, some PMQ that induces processors to produce solutions close to each other in a manner consistent with negative feedback control.

These general principles provide a guide in developing theoretical and practical methods of generating candidate solutions in a PFC network.

28.4.3 The Candidate Generation Criterion

In this section, a theoretical construct referred to as the *candidate generation criterion*, similar in spirit to the Metropolis Acceptance Criterion used in standard SA, is articulated. This criterion is an *idealized* method of searching the configuration space that abides by the three principles above and directly demonstrates the applicability of the fundamental theorem of COSA in (28.3). The following definitions are needed:

Definitions:

$\omega_{\mathbf{X}}(t)$ the stationary probability of system state $\mathbf{X}$ when the metric $d_{\mathbf{X}}$ is incorporated into the objective function.

$\pi_{\mathbf{X}}(t)$ the stationary probability of system state $\mathbf{X}$ in traditional SA (without PFC).

$G_{\mathbf{YX}}$ the probability of generating a *provisional* candidate solution $\mathbf{Y}$ given the current solution $\mathbf{X}$.

A provisional candidate solution $\mathbf{Y}$ is either accepted as the *candidate* solution (in the product space), or it is rejected based on the following *candidate generation criterion.*

$$\Pr\left\{\begin{matrix}\text{Accept Candidate}\\ \text{Provisionally}\end{matrix}\right\} = C_{\mathbf{YX}} = \begin{cases} e^{-\Delta d_{\mathbf{YX}}/t} & \text{if } \Delta d_{\mathbf{YX}} > 0 \\ 1 & \text{if } \Delta d_{\mathbf{YX}} \le 0 \end{cases} \tag{28.8}$$

It follows from (28.8) that the provisional candidate acceptance probability $C_{\mathbf{YX}} = e^{-\Delta d^+_{\mathbf{YX}}/t}$, and the transition probability for all system states $\mathbf{Y} \neq \mathbf{X}$ is

$$p_{\mathbf{XY}}(t) \quad = \quad G_{\mathbf{YX}}\, C_{\mathbf{YX}}\, A_{\mathbf{YX}} = G_{\mathbf{YX}}\, e^{-\Delta d^+_{\mathbf{YX}}/t}\, e^{-\Delta f^+_{\mathbf{YX}}/t}. \tag{28.9}$$

Define

$$p_{\mathbf{XX}}(t) = 1 - \sum_{\mathbf{Y}} p_{\mathbf{XY}}(t). \tag{28.10}$$

Using this candidate generation criterion, the transition probability is composed of three distinct elements: 1) the generation probability of a *provisional* candidate solution, 2) the acceptance probability of the provisional candidate as the candidate, and, 3) the acceptance probability of the candidate solution as the current solution.

This criterion implements the three principles outlined above. For Principle I; candidate solutions are produced in a two step process involving the random generation of a provisional candidate solution. For Principle II; $G_{\mathbf{YX}}$ is a symmetric function and the generation of provisional candidates is based on a uniform distribution over the entire configuration space. For Principle III; the value of $C_{\mathbf{YX}}$ is functionally dependant on the value of $d_{\mathbf{X}}$ and, with probability 1, a provisional candidate is accepted as the candidate solution if the distance metric is *reduced* while increases in the distance metric are permitted with a probability less than 1.

The following theorem and corollary show how the transition probabilities (28.9) and (28.10) affect the system stationary probability consistent with the conditional probability in (28.3).

Theorem 1 *Given p processors in a PFC network and system configuration space spanned by $\mathbf{X}$, let the transition probability of parallel SA be defined as in (28.9) and (28.10). Then, given a symmetric function $G_{\mathbf{XY}} = G_{\mathbf{YX}}$ and for all temperatures $t > 0$, the stationary probability of the system of p processors in the product space spanned by $\mathbf{X}$ is*

$$\omega_{\mathbf{X}}(t) = \frac{e^{-(f_{\mathbf{X}}+d_{\mathbf{X}})/t}}{\sum_{\mathbf{X}} e^{-(f_{\mathbf{X}}+d_{\mathbf{X}})/t}}$$

Proof:

It is sufficient to show that $\omega_{\mathbf{X}}(t)$ is the steady-state probability given the transition probabilities in (28.9) and (28.10) by showing that the detailed and global balance equations are maintained [1, p.38]. Thus, we must show that for all $\mathbf{X} \neq \mathbf{Y}$

$$\omega_{\mathbf{X}}(t)\; p_{\mathbf{XY}}(t) = \omega_{\mathbf{Y}}(t)\; p_{\mathbf{YX}}(t) \tag{28.11}$$

and for all $\mathbf{X}$

$$\omega_{\mathbf{X}}(t) = \sum_{\mathbf{Y}} \omega_{\mathbf{Y}}(t)\; p_{\mathbf{YX}}(t). \tag{28.12}$$

To show (28.11), note that

$$\frac{\omega_{\mathbf{X}}(t)}{\omega_{\mathbf{Y}}(t)} = \frac{e^{-(f_{\mathbf{X}}+d_{\mathbf{X}})/t}}{e^{-(f_{\mathbf{Y}}+d_{\mathbf{Y}})/t}} = e^{-[(f_{\mathbf{X}}+d_{\mathbf{X}})-(f_{\mathbf{Y}}+d_{\mathbf{Y}})]/t} = e^{-\Delta f_{\mathbf{XY}}/t} \cdot e^{-\Delta d_{\mathbf{XY}}/t} \tag{28.13}$$

Note that for all $\mathbf{X}, \mathbf{Y}$

$$e^{-\Delta f_{\mathbf{XY}}/t} = \frac{e^{-\Delta f^+_{\mathbf{XY}}/t}}{e^{-\Delta f^+_{\mathbf{YX}}/t}} \tag{28.14}$$

and similarly for $e^{-\Delta d_{\mathbf{XY}}/t}$. Substituting (28.14) for $e^{-\Delta f_{\mathbf{XY}}/t}$ and $e^{-\Delta d_{\mathbf{XY}}/t}$ into (28.13) gives

$$\frac{\omega_{\mathbf{X}}(t)}{\omega_{\mathbf{Y}}(t)} = \frac{e^{-\Delta d^+_{\mathbf{XY}}/t}\, e^{-\Delta f^+_{\mathbf{XY}}/t}}{e^{-\Delta d^+_{\mathbf{YX}}/t}\, e^{-\Delta f^+_{\mathbf{YX}}/t}} = \frac{G_{\mathbf{XY}} e^{-\Delta d^+_{\mathbf{XY}}/t}\, e^{-\Delta f^+_{\mathbf{XY}}/t}}{G_{\mathbf{YX}} e^{-\Delta d^+_{\mathbf{YX}}/t}\, e^{-\Delta f^+_{\mathbf{YX}}/t}} = \frac{p_{\mathbf{YX}}}{p_{\mathbf{XY}}}$$

since $G_{\mathbf{XY}} = G_{\mathbf{YX}}$, and (28.11) follows.

To show global balance, note that the matrix of system states is stochastic by virtue of (28.10). Therefore, $\sum_{\mathbf{Y}} p_{\mathbf{XY}}(t) = 1$ and

$$\omega_{\mathbf{X}}(t) = \omega_{\mathbf{X}}(t) \sum_{\mathbf{Y}} p_{\mathbf{XY}}(t) = \sum_{\mathbf{Y}} \omega_{\mathbf{X}}(t)\, p_{\mathbf{XY}}(t) = \sum_{\mathbf{Y}} \omega_{\mathbf{Y}}(t)\, p_{\mathbf{YX}}(t) \tag{28.15}$$

from detailed balance, and the proof is complete. See [1, 14] for similar derivations. □

Corollary to Theorem 1:

Given the same configuration space, let $\mathbf{X}_0 = \{\mathbf{X} : d_{\mathbf{X}} = 0\}$, *then for all* $X' \in \mathbf{X}_0$ *and* $t > 0$,

$$\omega_{X'}(t) > \pi_{X'}(t)$$

Proof:

Since $d_{\mathbf{X}} > 0$ for any $\mathbf{X} \notin \mathbf{X}_0$, then

$$1 > e^{-d_{\mathbf{X}}/t} \tag{28.16}$$

and for any $\mathbf{X} \in \mathbf{X}_0$

$$1 = e^{-d_{\mathbf{X}}/t}. \tag{28.17}$$

Multiplying both sides of (28.16) by $e^{-f_{\mathbf{X}}/t}$ and summing over $\mathbf{X} \setminus \mathbf{X}_0$

$$\sum_{\mathbf{X}\setminus\mathbf{X}_0} e^{-f_{\mathbf{X}}/t} > \sum_{\mathbf{X}\setminus\mathbf{X}_0} e^{-(d_{\mathbf{X}}+f_{\mathbf{X}})/t} \tag{28.18}$$

Multiplying both sides of (28.17) by $e^{-f_{\mathbf{X}}/t}$ and summing over $\mathbf{X}_0$

$$\sum_{\mathbf{X}_0} e^{-f_{\mathbf{X}}/t} = \sum_{\mathbf{X}_0} e^{-(d_{\mathbf{X}}+f_{\mathbf{X}})/t} \tag{28.19}$$

Adding (28.18) and (28.19), we obtain the strict inequality

$$\sum_{\mathbf{X}} e^{-f_{\mathbf{X}}/t} > \sum_{\mathbf{X}} e^{-(d_{\mathbf{X}}+f_{\mathbf{X}})/t} \tag{28.20}$$

Multiplying both sides of (28.20) by $e^{-(d_{X'}+f_{X'})/t}$ leads to

$$e^{-(d_{X'}+f_{X'})/t} \sum_{\mathbf{X}} e^{-f_{\mathbf{X}}/t} > e^{-(d_{X'}+f_{X'})/t} \sum_{\mathbf{X}} e^{-(d_{\mathbf{X}}+f_{\mathbf{X}})/t} \tag{28.21}$$

Dividing both sides of (28.21) by the summations (both positive), leads to

$$\frac{e^{-(d_{X'}+f_{X'})/t}}{\sum_{\mathbf{X}} e^{-(d_{\mathbf{X}}+f_{\mathbf{X}})/t}} > \frac{e^{-(d_{X'}+f_{X'})/t}}{\sum_{\mathbf{X}} e^{-f_{\mathbf{X}}/t}} = \frac{e^{-f_{X'}/t}}{\sum_{\mathbf{X}} e^{-f_{\mathbf{X}}/t}} = \pi_{X'}(t),$$

since $d_{X'} = 0$. Therefore, $\omega_{X'}(t) > \pi_{X'}(t)$ and the proof is complete. □

28.4.4 Non-Monotonic Candidate Generation Using PFC

In this section a practical method of generating candidate solutions based on the above principles is described. This method is somewhat different than the one embodied in the candidate generation criterion. To develop another method, it is essential that each all three principles are utilized as they work well only when used together. For instance, simply biasing the candidate solutions in some fashion to get the solutions close to each other by taking smaller step sizes monotonically makes it impossible to reach any system solution in a single step (Principle II). Indeed, the experimental evidence in Section 28.5 indicates that using such monotonic step sizes is inferior to the non- monotonic step sizes suggested by Principle III (see [10, 15] for other discussions on non-monotonicity).

So why bother develop another method? Afterall, the candidate generation criterion has the value of simplicity and clearly embodies all three principles. Moreover, its effects on the stationary probabilities are understood from Theorem 1.

Unfortunately, actually implementing the candidate generation criterion is problematic, because there is an inherent inefficiency that stems from use of an acceptance-rejection method. This can lead to repeated attempts to generate a single random variable or, in this case, a single *candidate* solution (see *e.g.* [12, p. 478]). While the SA algorithm has always been based on this idea—moving from a current solution depending upon the acceptance of the candidate solution—this criterion adds yet another similar hurdle. The Metropolis Acceptance criterion is applied only after a provisional candidate solution is accepted as the candidate solution. It would be advantageous, therefore, to find a method of satisfying the three principles that underpin the candidate generation criterion, but without the baggage of the acceptance-rejection method implicit in the criterion itself.

The approach used here is to have each processor generate a candidate solution by making random sized jumps from the current solution. To make the size of these jumps functionally dependent on the PMQ (required by Principle III), the bilateral exponential density function

$$f(x) = \begin{cases} \frac{\lambda}{2}e^{-\lambda x} & \text{for } x > 0 \\ \frac{\lambda}{2}e^{\lambda x} & \text{for } x \leq 0 \end{cases} \tag{28.22}$$

is used with parameter $\lambda = 1/q_m^{[k]} t_k$. Thus, *the scale of the jumps from current solutions is based on the current size of the distance metric.* This approach combines PFC by candidate generation *and* temperature modulation. It also abides by all three principles and produces candidate solutions in $\mathcal{O}(1)$ time as all processors generate candidate solutions concurrently.

Figures 28.2*a* and *b* illustrate this approach where the current solutions of four processors are indicated by solid circles and random jumps to candidate solutions are indicated by arrows of varying lengths.

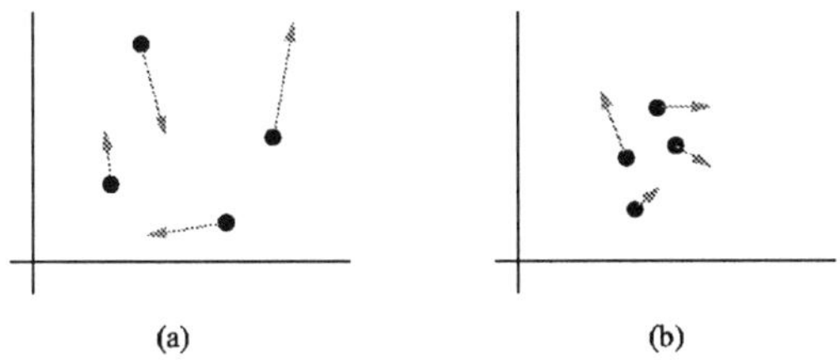

Figure 28.2 PFC by Candidate Generation

Note how the distance between the processors in Figure 28.2*a* is large relative to those in Figure 28.2*b*. Consequently, the jumps to candidate solutions in Figure 28.2*a* *tend* to be larger than those in Figure 28.2*b*.

In the following C-like algorithm, processors are initialized in the following manner: Processor 1 is initialized from a uniform distribution over the feasible region. The next processor is initialized by adding a random distance D to each vector element where D is generated using (28.22) with λ equal to the inverse of the *gain setting* parameter. Higher gain settings tend to produce initial solutions with processors farther apart resulting in lower *system gains* (see Section 28.3.2). The opposite holds for low gain settings which lead to high system gains. Candidate solutions are generated by adding a random distance D to each vector element where D is also generated using (28.22) but with $\lambda = 1/q_m^{[k]} t_k$.

The COSA Algorithm

Define the following function:
RandomDistance(processor m, λ):
{ For $i = 1$ to n (n is the number of variables)
  Generate D using (28.22)
  $x'_{[i,m]} = x^{[k]}_{[i,m]} + D$
  if $x'_{[i,m]} < a$ or $> b$, then $x'_{[i,m]} = a$ or b, respectively.
 next i
 return($\mathbf{x}'_m$);
}
Initialize: obtain initial solution $\mathbf{x}_m^{[0]}$ for processors 1 to p.
 initialize solution $\mathbf{x}_1^{[0]}$ using a uniform distribution over the feasible region.
 For $m = 1$ to $p - 1$
  $\mathbf{x}_{m+1}^{[0]} =$ RandomDistance(m, 1/gain_setting);
 next m;
 For $m = 1$ to $p - 1$: $PMQ_m^{[0]} = d(\mathbf{x}_m^{[0]}, \mathbf{x}_{m+1}^{[0]})$ (see (28.5) and (28.6))
For $k = 1$ to l for each processor m
 Calc Quality Coefficient:
  $PMQ_m^{[k]} = d(\mathbf{x}_m^{[k]}, \mathbf{x}_{m+1}^{[k]})$
  $q_m^{[k]} = PMQ_m^{[k]} / PMQ_m^{[0]}$
 Candgen:
  $\mathbf{x}'_m$ =RandomDistance(m, $1/q_m^{[k]} t_k$);

Metropolis Accept:

if $f(\mathbf{x}'_m) \le f(\mathbf{x}_m^{[k]})$
    then $\mathbf{x}_m^{[k+1]} = \mathbf{x}'_m$
else Generate $R \sim U(0,1)$
if $R \le e^{-(f(\mathbf{x}'_m)-f(\mathbf{x}_m^{[k]}))/q_m^{[k]} t_k}$
    then $\mathbf{x}_m^{[k+1]} = \mathbf{x}'_m$
else $\mathbf{x}_m^{[k+1]} = \mathbf{x}_m^{[k]}$

next k

This algorithm follows the intuition in Section 28.4.1 and the three principles closely: the smaller the quality coefficient, the smaller the expected value of the magnitude of the jumps to candidate solutions. Step sizes are therefore non-monotonic due to their randomness as is the *expected step size* due to the random nature of the quality coefficient in the bilateral exponential distribution.

28.5 EXPERIMENTAL METHODOLOGY AND RESULTS

This section describes the motivation and methodology behind several experiments designed to test whether the COSA algorithm accelerates the convergence of a system of processors to the global optimum. It bears emphasis that the purpose here is not to demonstrate a practical implementation of COSA. That is an area of future research. The purpose here is strictly limited to testing the efficacy of PFC by candidate generation as implemented in the algorithm in Section 28.4.4.

28.5.1 Measuring Convergence

Testing the efficacy of this algorithm is best done by comparing its performance to the performance of a similar system of processors that do not use PFC by candidate generation. But how shall we measure performance? In practical implementations of SA, one uses the historical best value obtained in the search because such implementations have the purpose of actually obtaining good solutions. Here, a different measure of performance is required; one that reflects the finite-time convergence of the associated Markov chain that embodies SA and COSA. Several approaches are possible.

One approach for measuring convergence is to evaluate a metric between the state vector of the system at finite-time and the limiting value of the state vector. Unfortunately, for inhomogeneous Markov chains such as SA, analytically determining this state vector at some finite-time has not been possible [14]. This task is made even more difficult when the form of the stationary probability of a PFC system with temperature modulation has not been determined (this was attemped, incorrectly, in [6] where this writer's steady-state analysis was flawed. Although it showed the inherent parallelism in the transition probabilities, as described here in Section 28.2.1, it incorrectly reported the stationary distribution).

Because the state probability vector at some fixed time index is virtually impossible to determine, a simulation—actually running the algorithm using multiple processors—must be performed and the measures of performance that reflect the degree of convergence then evaluated.

One suitable and convenient measure associated with SA are objective function values. Like the state vector, objective function values converge in probability to the optimum value from above (for minimization problems). With p processors, some measure associated with p objective function values is therefore appropriate. Define the following measures of performance for replication $r = \{1, \ldots, R\}$ of a process, each l iterations long, in a system with p processors with PFC and without PFC, respectively: $F_{\min}^{[l,r]} = \min\{F_1^{[l]}, F_2^{[l]}, \ldots, F_p^{[l]}\}, G_{\min}^{[l,r]} = \min\{G_1^{[l]}, G_2^{[l]}, \ldots, G_p^{[l]}\}$. The values $F_{\min}^{[l,r]}$ and $G_{\min}^{[l,r]}$ are the extreme order statistics of the objective function values of the p processors after l iterations of the algorithm in the rth experimental replication. In other words, they constitute the best objective function values among the p processors at the end of each execution of the algorithm.

Although we could use the average of the objective function values of the p processors as the measure of performance to estimate the expectation value of the objective function (which also converges to the optimum value), the extreme values have some advantages; it is a biased estimate of the expected objective function value and bounded by the optimal value. Consequently, its variance should be smaller thus providing a more powerful test statistic [9]. This same extreme order statistic is used in the non-PFC systems. Thus, comparisons provide a way of assessing performance differences.

28.5.2 Finite-Time Considerations

The finite-time behavior of SA poses an additional methodological issue insofar as statistical tests are concerned. This problem stems from the asymptotic behavior of SA and our desire to test rates of convergence—running the algorithm too long leads to convergence whether we use PFC or not. If we wait until both systems have converged to the optima, as both surely will, then it becomes impossible to tell which system got closer to the optimum in less time making it difficult to discern performance differences. Running the algorithm for too many iterations therefore causes a dilution of statistical significance. Thus, to combat this problem, PFC and non-PFC systems are run for a small enough number of iterations ($l = 50$) so that statistical differences become discernable.

28.5.3 Statistical Analysis

To estimate the expectation values of $F_{\min}^{[l,r]}$ and $G_{\min}^{[l,r]}$, the following statistics based on $R = 30$ experimental replications were calculated:

$$\overline{F}_{\min} = \frac{\sum_{r=1}^{R} F_{\min}^{[l,r]}}{R}, \qquad \overline{G}_{\min} = \frac{\sum_{r=1}^{R} G_{\min}^{[l,r]}}{R} \tag{28.23}$$

A simple hypothesis test was used:

$$H_0 : E\left\{F_{\min}^{[l,r]}\right\} = E\left\{G_{\min}^{[l,r]}\right\}$$

and tested using the following test statistic:

$$z = \frac{\overline{F}_{\min} - \overline{G}_{\min}}{\sqrt{\frac{s_{\mathrm{F}}^2}{R} + \frac{s_{\mathrm{G}}^2}{R}}} \tag{28.24}$$

where the estimated variance for PFC and non-PFC systems are s_{F}^2 and s_{G}^2, respectively. The of experiments were performed on a PowerMac 8100 using the MkLinux operating system.

28.5.4 Parameters and Initialization

Six sets of experiments were run using $p = 5$ and 10 processor systems at each of three gain settings: 1, 5, and 10. Each decision variable was bounded between $a = -10$ and $b = 10$. The cooling schedule of the system base temperature monotonically decreased based on the cooling schedule $t_k = \frac{\gamma}{\log(c+k+1)}$ where the parameters γ and c were such that $t_1 = 10$ and $t_{50} = 1$ (see [5, 7]). Experiments in corresponding sets (with PFC and without PFC using the same numbers of processors and gain settings) used identical random number streams and identical initial solutions.

In the experiments using PFC, the PMQ was based on (28.5). In the experiments without PFC, everything was done exactly the same way as described above except that the quality coefficients were set to 1, i.e., $q_m^{[k]} = 1$ for all k and m. Thus, the effective temperature was t_k, hence, monotonic and candidate solutions were generated using bilateral exponentially distributed jumps with parameter $1/t_k$, hence, also monotonic.

28.5.5 Experimental Results and Observations

Number of Processors	Gain Setting	Objective Function Values: With PFC $\overline{F}_{\min}$	With PFC s_{F}^2	No PFC $\overline{G}_{\min}$	No PFC s_{G}^2	z-value
5	1	0.683	0.303	0.541	0.057	1.290
5	5	0.212	0.057	0.536	0.536	-2.300
5	10	0.232	0.090	0.541	0.057	-4.419
10	1	0.488	0.560	0.388	0.040	0.709
10	5	0.130	0.027	0.324	0.047	-3.899
10	10	0.110	0.020	0.438	0.062	-6.261

Table 28.1 Efficacy of PFC by Candidate Generation

Table 28.1 reports the results of these experiments and permits a number of observations. First, the obvious one which conforms to common-sense: for any particular gain setting, the more processors there are, the lower the values of $F_{\min}^{[l,r]}$ and $G_{\min}^{[l,r]}$. This demonstrates the value of scatter search—the more initial solutions and processors employed, the more likely it is that better solutions will be obtained.

More significant is the relationship between the sets of experiments. Notice that for the gain setting of 1 corresponding to the highest system gain, the use of PFC degrades the performance of the processors compared to the monotonic, non-PFC experiments. This holds true for both 5 and 10 processor systems. In this case, the use of PFC creates an error *amplifying* system which overcorrects. When the gain setting is increased to 5 and 10, corresponding to smaller system gains, the use of PFC yields superior performance showing increasing statistical significance with z-values ranging from -2.3 to -6.2 for 5 and 10 processor systems, respectively.

28.6 CONCLUSION

This paper has examined the issues, theory, and possible implementation schemes for applying the COSA concept to continuous variable problems. The issue of candidate generation in the continuous domain was addressed by articulating three general principles of PFC by candidate generation. A candidate generation criterion was defined that provided an idealized method for implementing these principles. It also permitted a theoretical examination of the effects of these principles on the system stationary probabilities and thereby demonstrated the benefits of inducing processors to get close together per the fundamental theorem of COSA.

These principles were used to develop a practical implementation of the COSA concept for continuous variable problems. A number of experiments were performed using this implementation and the resulting experimental evidence was consistent with theory. Larger numbers of processors improved the performance of the network. Different gain settings affected performance in predictable ways: high system gains degraded performance compared to the monotonic systems while lower system gains showed statistically significant benefits from using non-monotonic candidate generation.

The use of self-similar SA processors in a PFC network adds intelligence to random search. Like the two hemispheres of the brain, pairs of processors are "conscious" of each other, continuously gauging and improving each other's performance. Each tending toward its ultimate goal by virtue of SAs ergodicity, yet each benefitting from information from other similar processors.

References

[1] E.E. Aarts and J. Korst. *Simulated Annealing and Boltzmann Machines: A Stochastic Approach to Combinatorial Optimization and Neural Computing.* Wiley, 1989.

[2] R. Azencott. *Simulated Annealing: Parallelization Techniques*, chapter Ch.

4 Parallel Simulated Annealing: An Overview of Basic Techniques. Wiley, 1992.

[3] C.J.P. Bélisle. Convergence theorems for a class of simulated annealing algorithms on $\mathcal{R}^d$. *J. of Applied Probability*, 29:885–895, 1992.

[4] I. Bohachevsky, M. Johnson, and M. Stein. Generalized simulated annealing for function optimization. *Technometrics*, 28:209–217, 1986.

[5] M.A. Fleischer. *Assessing the Performance of the Simulated Annealing Algorithm Using Information Theory.* Doctoral Dissertation, Case Western Reserve University, Cleveland, Ohio, 1993.

[6] M.A. Fleischer. Cybernetic optimization by simulated annealing: Accelerating convergence by parallel processing and probabilistic feedback control. Journal of Heuristics, 1:225–246, 1996.

[7] M.A. Fleischer and S.H. Jacobson. Information theory and the finite-time performance of the simulated annealing algorithm: Experimental results. Technical Report, 1994.

[8] M.A. Fleischer and S.H. Jacobson. Using scaling properties in simulated annealing to prove convergence in the continuous domain. Technical Report, 1996.

[9] J. Galambos. *The Asymptotic Theory of Extreme Order Statistics.* Wiley, 1978.

[10] F.W. Glover. Future paths for integer programming and links to artificial intelligence. *Computers & Operations Research*, 13:533–549, 1986.

[11] F.W. Glover. Tabu thresholding: Improved search by nonmonotonic trajectories. *INFORMS J. on Computing*, 7:426–442, 1995.

[12] A.M. Law and W.D. Kelton. *Simulation Modeling and Analysis, 2nd Ed.* McGraw-Hill, 1991.

[13] N.A. Metropolis, A. Teller A. Rosenbluth, M. Rosenbluth, and E. Teller. Equation of state calculations by fast computing machines. *J. of Chemical Physics*, 21:1087–1092, 1953.

[14] D. Mitra, F. Romeo, and A. Sangiovanni-Vincentelli. Convergence and finite-time behavior of simulated annealing. *Advances in Applied Probability*, 18:747–771, 1986.

[15] I. Osman. Metastrategy simulated annealing and tabu search algorithms for the vehicle routing problem. *Annals of OR*, 41:421–451, 1993.

29 SOLVING THE PROGRESSIVE PARTY PROBLEM BY LOCAL SEARCH

Philippe Galinier and Jin-Kao Hao

LGI2P, EMA-EERIE, Parc Scientifique Georges Besse, F-30000 Nimes, France.
galinier@eerie.fr, hao@eerie.fr

Abstract: The Progressive Party Problem (PPP) is a complex, constrained combinatorial optimization problem. The goal of the problem consists of finding assignments of resources to variables for a given number of time periods while satisfying a set of multiple constraints. So far, two quite different approaches have been tried to solve the problem: Integer Linear Programming and Constraint Programming. All attempts with Integer Linear Programming failed, while Constraint Programming obtained better results. In this work, we present a third approach based on Local Search. We show that this approach gives better results than the previous ones and constitutes a very effective alternative to solve the PPP. We investigate different techniques for solving the problem, in particular issues related to search space, cost function and neighborhood functions.

29.1 INTRODUCTION

The Progressive Party Problem is a complex, constrained combinatorial problem. It consists in assigning resources to several variables for a given number of time periods while satisfying a set of multiple constraints. The PPP is interesting because it is known to be extremely hard for some mathematical method and remains unsolved. Moreover, the problem possesses many non trivial, heterogeneous constraints.

The problem has been proposed by a yacht club in order to organize a *party* taking place during several successive time periods [2, 8]. There are some host boats that receive the crews of other boats. For each time period, each guest crew must visit one of the host boats while respecting the following constraints:

- Each guest crew moves to a different host at each time period.
- Two crews meet each other at most once.

- The capacities of the host boats must be respected: for each time period and each host boat, the sum of the sizes of the crews who visit this boat during this time period cannot be greater than the capacity of the host.

Given these constraints, the problem consists in finding an assignment of the boats to the guest crews that respects all the constraints for the maximal possible number of time periods. There exists in the literature a well-known instance coming directly from the initial organization problem. Moreover, the basic version of the problem was to find, for this instance, a consistent assignment for a number of time periods fixed to six. So far, two different approaches have been tried to solve the above mentioned benchmark instance of the problem: Integer Linear Programming (ILP) [2] and Constraint Programming (CP) [8]. All attempts with Integer Linear Programming failed since no solution was found for the basic problem of 6 time periods. On the contrary, CP proved to be a more successful approach because it found solutions not only for the basic problem, but also for 7 and 8 time periods[1]. However, it was unknown if a solution would exist for more time periods.

In this work, we investigate a third approach based on Local Search to tackle the PPP. To model the PPP, we follow the CP approach, *i.e.* we consider the PPP as a Constraint Satisfaction Problem (CSP) [9, 14]. Based on this modeling, we introduce a search space, a cost function and two different neighborhoods, and experiment them with some well-known local search meta-heuristics. We show that this approach allows us to find better results than previous methods for the benchmark instance. Indeed, we are able to find solutions for 9 time periods in several seconds of CPU time.

The paper is organized as follows. In the next section, we present an overview of solving constraint satisfaction problems using Local Search and we represent the PPP using the formalism CSP (Section 29.2). Then, we present the different components of our local search algorithms (Section 29.3) and computational results (Section 29.4). In the last section, we give some conclusions of the work.

29.2 LOCAL SEARCH FOR CSP

29.2.1 Constraint Satisfaction Problem (CSP)

A CSP [9, 14] is defined by a triplet $(X, \mathcal{D}, \mathcal{C})$ with:

- *a finite set X of n variables :* $X = \{x_1, \cdots, x_n\}$.
- *a set $\mathcal{D}$ of associated domains :* $\mathcal{D} = \{D_{x_1}, \cdots, D_{x_n}\}$. Each domain D_{x_i} specifies the finite set of possible values of the variable x_i.
- *a finite set $\mathcal{C}$ of p constraints :* $\mathcal{C} = \{C_1, \cdots, C_p\}$. Each constraint is defined on a set of variables and specifies which combinations of values are compatible for these variables.

[1]Better results were obtained very recently, see Section 29.4 for more discussions.

Given such a triplet, the problem consists in finding a complete assignment of the values to the variables that satisfies all the constraints: such an assignment is then said consistent. Since the set of all assignments (not necessarily consistent) is defined by the Cartesian product $D_{x_1} \times \cdots \times D_{x_n}$ of the domains, solving a CSP means to determine a particular assignment among a potentially huge number of possible assignments.

The CSP formalism is a powerful and general model. In fact, it can be used to model conveniently some well-known problems such as k-coloring and satisfiability as well as many practical applications related to resource assignments, planning or timetabling. As we will see below, the PPP is also easily represented as a CSP.

29.2.2 Local Search (LS)

LS, also called neighborhood search, constitutes a powerful approach for tackling hard optimization problems [12]: given a couple (S, f) with S being a finite set of configurations and f a cost function $f : S \rightarrow R$, to determine an element $s_* \in S$ such that $f(s_*) = \min_{s \in S} f(s)$. To solve an optimization problem, Local Search needs a so-called neighborhood function $N : S \rightarrow 2^S$ ($N(s) \subseteq S$ is called the neighborhood of $s \in S$). A Local Search algorithm begins with an initial configuration $s_0 \in S$ and then generates a series of configurations $(s_i)_{i \in \{0,1,\cdots\}}$ such that $\forall i \in \{0, 1, \cdots\}$, $s_{i+1} \in N(s_i)$. Well-known examples of LS methods include various descent methods, simulated annealing (SA) [7] and Tabu Search (TS) [4, 5]. The main difference among LS methods concerns the way of visiting the given neighborhood.

Several LS methods, called repair methods, have been developed to solve CSPs in Artificial Intelligence. One well-known example is the Min-conflict (MC) heuristic [10]. With this heuristic, one seeks for a best value for a given conflicting variable at each iteration[2]. This method corresponds in fact to a special descent method and cannot go beyond a local optimum. This method has been enforced by a noise strategy called *random-walk* [13]. Other extensions or similar methods are reported in [6, 11, 15]. Recently, more advanced LS methods such as Tabu Search are also used to solving CSPs [3].

In all these studies, a *configuration* is a complete assignment of values to variables and the *search space* is defined by the Cartesian product $D_{x_1} \times \cdots \times D_{x_n}$. Two configurations are *neighbors* if and only if they are different at the value of a single variable. The *cost* of a configuration is simply the number of violated constraints by the configuration (a cost of zero corresponds to a consistent assignment, thus a solution).

[2]A variable is said conflicting if it is involved in some unsatisfied constraints.

29.2.3 Formulation of the PPP as a Constraint Satisfaction Problem

Recall the PPP consists in finding an assignment of the boats to the guest crews that respects all the constraints for the maximal possible number of time periods. Let T be the number of time periods, G the number of guest crews and H the number of host boats. We use $c(g)$ to denote the size of a guest crew $g \in \{1, \cdots, G\}$, $C(h)$ the capacity of a host boat $h \in \{1, \cdots, H\}$ and $x_{g,t} \in \{1, \cdots, H\}$ the host boat visited by the crew g during the time period t. In the benchmark instance, there are $H = 13$ host boats and $G = 29$ guest crews. Table 29.1 indicates the capacities of the boats and the sizes of the crews.

capacity C	*#boats*	size c	*#crews*
10	2	7	1
9	1	6	1
8	6	5	3
7	1	4	8
6	1	3	2
4	2	2	14
	13		29

Table 29.1 Capacities of the boats and sizes of the crews

Note however that an obvious upper bound for T is 10. Indeed, there is a guest crew of size 7 and only 10 hosts have a capacity greater or equal to 7. Hence only this 10 hosts can be assigned to this guest crew in order to respect the capacity constraint. Moreover, this guest crew must visit a different host at each time period. Hence, the number of time periods cannot exceed 10.

We use P_n to denote the problem of finding a consistent assignment for a fixed number n of time periods in the benchmark instance of the problem. The instance is naturally solved by solving P_n for the different possible values of n (from 1 to 10). For a fixed value of n, P_n can be modeled as a CSP $(X, \mathcal{D}, \mathcal{C})$ as follows. The set of variables is naturally $X = \{x_{g,t}, 1 \leq g \leq G, 1 \leq t \leq T\}$ and all domains are equal to $D = \{1, 2, \cdots, H\}$: $\forall x \in X, D_x = D$. The constraint set $\mathcal{C}$ contains the following three different types of constraints:

- The first type of constraints imposes that each crew moves to a different host at each time period. For each g, $1 \leq g \leq G$, we impose the constraint: DIFF(g) $\Leftrightarrow x_{g,1}, x_{g,2}, \cdots, x_{g,T}$ are all different.

- The second type of constraints imposes that two crews meet at most once: for each pair of guest crews, the number of time periods when these two crews meet is 0 or 1. For each pair $\{g_1, g_2\}$, we impose the constraint: ONCE($\{g_1, g_2\}$) $\Leftrightarrow |\{t, 1 \leq t \leq T \mid x_{g_1,t} = x_{g_2,t}\}| \leq 1$.

- The third type of constraints imposes that the capacity of the host boats must be respected. For each time period t, $1 \leq t \leq T$ and each host boat h, $1 \leq h \leq H$, we impose the constraint: CAPA(h,t) $\Leftrightarrow \sum_{1 \leq g \leq G, x_{g,t}=h} c(g) \leq C(h)$

Note that although each DIFF constraint may be replaced by a set of $|T| * (|T|-1)/2$ binary constraints of difference, the constraints of types ONCE and CAPA cannot be defined simply using binary constraints.

29.3 SOLVING THE PPP WITH LOCAL SEARCH

We define below the components of our local search procedures that are the search space, the cost function, the neighborhood function and the meta-heuristic.

29.3.1 Search space

As presented above, a configuration is a complete assignment of the numbers in $D = \{1 \cdots H = 13\}$ to the variables of $X = \{x_{g,t}, 1 \leq g \leq G, 1 \leq t \leq T\}$: a configuration is a $G \times T$ table (G=29 and T = 6,7,8,9 or 10) whose elements are integers from $\{1 \cdots 13\}$. The search space is the set of all such assignments.

29.3.2 Cost function

As stated in Section 29.2, the cost of a configuration of a CSP is often defined as the number of violated constraints and one associates to each constraint C a penalty $f_C(s)$ which takes 1 or 0 according to whether the constraint is violated or not for the configuration s. This 0/1 penalty function works generally well for binary constraints, but is no more appropriate for n-ary constraints. In fact, this function cannot distinguish a "strongly" violated constraint from a "weakly" violated one. In the PPP, constraints are more complex, this is why we introduce a more informative multi-valued penalty function for each constraint.

Penalty for a constraint of type DIFF Recall that $DIFF(g)$ is satisfied if and only if $x_{g,1}, x_{g,2}, \cdots, x_{g,T}$ are all different. For a constraint $DIFF(g)$, we define the penalty as the number of pairs $\{x_{g,t_1}, x_{g,t_2}\}$ of variables having the same value. That is, for any DIFF constraint $C = DIFF(g)$ and any configuration s, $f_C(s) = |\{\{x_{g,t_1}, x_{g,t_2}\} \mid s(x_{g,t_1}) = s(x_{g,t_2})\}|$, where $s(x)$ denotes the value assigned to x in s.

Penalty for a constraint of type ONCE Given a constraint of type ONCE $C = ONCE(\{g_1, g_2\})$, we use $Meet_{g_1,g_2}$ to denote the number of time periods when these two crews meet: $Meet_{g_1,g_2}(s) = |\{t, 1 \leq t \leq T \mid s(x_{g_1,t}) = s(x_{g_2,t})\}|$. Recall that the constraint is satisfied for s if and only if $Meet_{g_1,g_2}(s) \leq 1$. If the constraint is violated ($Meet_{g_1,g_2} > 1$), we define the penalty f_C to be equal to $Meet_{g_1,g_2} - 1$ in order to incite the value of $Meet_{g_1,g_2}$ to decrease progressively until the constraint becomes satisfied (*i.e.* $Meet_{g_1,g_2} = 1$).

$$f_C(s) = \begin{cases} 0 & \text{if } Meet_{g_1,g_2}(s) \leq 1 \\ Meet_{g_1,g_2}(s) - 1 & \text{otherwise} \end{cases}$$

Penalty for a constraint of type CAPA Given a constraint of type CAPA $C = CAPA(h,t)$, let $\sigma_{h,t}$ be the overloading of the host h for the time period t: $\sigma_{h,t}(s) = C(h) - \sum_{1 \leq g \leq G, s(x_{g,t})=h} c(g)$. When $\sigma_{h,t} \leq 0$, the penalty is naturally 0 because the constraint is satisfied. The penalty is fixed to 1 if $\sigma_{h,t} = 1$ and increases progressively for higher values of $\sigma_{h,t}$, in order to incite $\sigma_{h,t}$ to decrease: $f_C = 1 + B * (\sigma_{h,t} - 1)$. We fixed $B = 1/4$ empirically.

$$f_C(s) = \begin{cases} 0 & \text{if } \sigma_{h,t}(s) \leq 0 \\ 1 + (\sigma_{h,t}(s) - 1)/4 & \text{otherwise} \end{cases}$$

Remark Note that the penalty $f_C(s)$ for a constraint C of type ONCE is exactly the minimum number of variables that must be changed in s in order to satisfy the constraint C. It is also roughly the case for the constraints of type CAPA, considering that the average size of a crew is about 4 (3.24 more precisely). This principle is applicable to other n-ary constraints in order to define an effective penalty function.

Cost function Finally, the cost function is the weighted sum of the penalty functions of all the constraints $C \in \mathcal{C}$:

$$\forall s \in S, f(s) = \sum_{C \in \mathcal{C}} p(C) * f_C(s)$$

where p is a weight function $p : \mathcal{C} \rightarrow R$.

The weight for each of three types of constraints is determined empirically. The best weights found are 2, 1 and 2 for DIFF, ONCE and CAPA respectively - more precisely, the weights are 4, 2 and 4, in order to have integer values for f.

29.3.3 Conflicting variables

Many search methods give a different role to a variable depending on whether it is in conflict or not. If a constraint C is satisfied, no variable is conflicting for this constraint. But if C is violated, some variables are conflicting. The notion of conflicting variable in the PPP is defined as follows.

For a DIFF constraint $C = DIFF(g)$, variable x_{g,t_1} is in conflict in a configuration s if and only if there exists another variable x_{g,t_2} ($t_1 \neq t_2$) such that $s(x_{g,t_1}) = s(x_{g,t_2})$.

For a constraint $C = ONCE(\{g_1, g_2\})$: variables $x_{g_1,t}$ and $x_{g_2,t}$ are conflicting for the constraint C if the constraint is violated ($Meet_{g_1,g_2}(s) > 1$) and $s(x_{g_1,t}) = s(x_{g_2,t})$.

For a constraint $C = CAPA(h,t)$: variable $x_{g,t}$ is conflicting for the constraint C if, first, the constraint is violated ($\sigma_{h,t}(s) > 0$) and, second, $s(x_{g,t}) = h$.

29.3.4 Definition of the neighborhood function

Recall that a neighborhood function is any function $N : S \to 2^S$. In order to simplify the presentation, we will use the notion of *move* to define a neighborhood. Applying a move m to a configuration $s \in S$ leads to a new configuration denoted by $s \oplus m$. Let $M(s)$ be the set of all possible moves which can be applied to s, then the neighborhood of s is defined by: $N(s) = \{s \oplus m \mid m \in M(s)\}$.

For the PPP, we introduce two quite different types of moves denoted by *OneMove* and *Swap*. From these two types of moves, we define two neighborhoods denoted by N_1 and N_2.

Neighborhood N_1 A move of type *OneMove* consists simply in changing the host boat affected to a given crew for a given period, *i.e.* the current value of a single variable x is replaced by a new one v. Such a move is denoted by the couple $(x, v) \in X \times D$ where $s(x) \neq v$. The neighborhood N_1 is defined from $M_1 = OneMove$.

- $M_1(s) = OneMove(s) = \{(x, v) \in X \times D \mid s(x) \neq v\}$,
- $s' = s \oplus (x, v) \Leftrightarrow \begin{cases} s'(x) = v \text{ and} \\ \forall y \in X - \{x\}, s'(y) = s(y) \end{cases}$
- $N_1(s) = \{s \oplus (x, v) \mid (x, v) \in M_1(s)\}$.

Neighborhood N_2 A *swap* move consists in reversing the host boats assigned to two crews g_1 and g_2 during the same time period t. It is denoted by a couple $(x_{g_1,t}, x_{g_2,t})$. Applying the move $(x_{g_1,t}, x_{g_2,t})$ to s consists in assigning the value $s(x_{g_2,t})$ to $x_{g_1,t}$ and inversely the value $s(x_{g_1,t})$ to $x_{g_2,t}$. The neighborhood N_2 combines the moves of *OneMove* and *Swap*.

- $Swap(s) = \{(x_{g_1,t}, x_{g_2,t}) \mid s(x_{g_1,t}) \neq s(x_{g_2,t})\}$,
- $s' = s \oplus (x_{g_1,t}, x_{g_2,t}) \Leftrightarrow \begin{cases} s'(x_{g_1,t}) = s(x_{g_2,t}) \text{ and} \\ s'(x_{g_2,t}) = s(x_{g_1,t}) \text{ and} \\ \forall y \in X - \{x_{g_1,t}, x_{g_2,t}\}, s'(y) = s(y) \end{cases}$
- $M_2(s) = OneMove(s) \cup Swap(s)$,
- $N_2(s) = \{s' = s \oplus m \mid m \in M_2(s)\}$.

Candidate list based on conflicting variables A common heuristic used to make the search more efficient is to restrict the choice of a move to the subset of $M(s)$ involving variables which are conflicting in s. Such an heuristic can be seen as a particular *candidate list strategy*. For the PPP, this conflict-based subset, denoted by $M_{1,CFL}(s)$ and $M_{2,CFL}(s)$ for the 2 neighborhoods, is defined as follows.

- $M_{1,CFL}(s) = \{(x, v) \in M_1(s) \mid x \text{ is conflicting in s }\}$
- $M_{2,CFL}(s) = \{(x_1, x_2) \in M_2(s) \mid x_1 \text{ or } x_2 \text{ is conflicting in s }\}$

Data : θ : parameter, N : neighborhood
Result : the best configuration found
begin
  generate a random configuration s
  while *not Stop-Condition* **do**
    choose randomly a move $m \in M$
    $\delta := f(s \oplus m) - f(s)$
    if $\delta \leq 0$ **then** $s := s \oplus m$
    else with probability $e^{-\delta/\theta}$ **do** $s := s \oplus m$
end

Figure 29.1 The Metropolis algorithm

29.3.5 Meta-heuristics tested

We experimented mainly two meta-heuristics: a *Metropolis algorithm* and a *Tabu Search* algorithm. Both algorithms use the above mentioned candidate list strategy.

Algorithm of Metropolis Metropolis algorithm is a simplified version of Simulated Annealing [7] using a constant value for its temperature parameter. The algorithm begins with an initial configuration in the search space S and then performs a series of iterations. At each iteration, a single neighbor (or equivalently a single move) of the current configuration is randomly chosen and then a probabilistic criterion is performed in order to decide if this neighbor is accepted or not. The principle of the Metropolis algorithm is to accept any move that does not increase the cost function, and to accept, in a controlled manner, deteriorating moves. So the algorithm is not trapped in local optima.

The method has one parameter θ called *temperature*: the higher the value of the temperature, the easier degradations are accepted (a zero temperature corresponds to a simple descent algorithm as all deteriorating moves are refused while an infinite temperature is a random walk in the search space as all moves are accepted).

More precisely, suppose that the move m has been chosen at the current iteration. Then, one considers the difference $\delta = f(s \oplus m) - f(s)$ that represents the effect of the chosen move on the cost function. If the move m does not increase the cost function ($\delta \leq 0$), then it is always accepted. If it does ($\delta > 0$), the move is accepted with a probability $e^{-\delta/\theta}$ so that a bigger deterioration is more rarely accepted.

Tabu algorithm A typical TS procedure [4] begins with an initial configuration in the search space S and then proceeds iteratively to visit a series of locally best configurations following the neighborhood. At each iteration, a *best* move m is applied to the current configuration s even if $s' = s \oplus m$ does not improve the current configuration in terms of the value of the cost function.

Data : k : tabu tenure; N : neighborhood
Result : the best configuration found
begin
  generate a random configuration s
  initialize the Tabu list to empty
  while *not Stop-Condition* **do**
    choose a best move m such that m is not tabu or satisfies the aspiration criterion
    introduce during k iterations the couple $< x, s(x) >$ in the Tabu list for all modified variables $x \in X$
    $s := s \oplus m$
end

Figure 29.2 The tabu search algorithm

This iterative process may suffer from cycling and get trapped in local optima. To avoid the problem, TS introduces the notion of *Tabu lists*. A tabu list is a special short term memory that maintains a selective history $\mathcal{H}$, composed of previously encountered solutions or more generally pertinent attributes of such solutions. A simple TS strategy based on this short term memory $\mathcal{H}$ consists in preventing solutions of $\mathcal{H}$ from being reconsidered for next k iterations (k, called tabu tenure, is problem dependent). Now, at each iteration, TS searches for a best neighbor from this dynamically modified neighborhood $N(\mathcal{H},s)$, instead of from $N(s)$ itself. Such a strategy prevents Tabu from being trapped in short term cycling and allows the search process to go beyond local optima. Moreover, note that Tabu restrictions may be overridden under certain conditions, called *aspiration criterion.*

A move for the PPP corresponds to changing the value of one variable (*OneMove*) or two variables (*Swap*). When a variable x is involved in a move, its current value v is replaced by a new one v'. At this time, the couple $< x, v >$ is classified tabu for the next k iterations and that means that the value v is not allowed to be re-assigned to x during this period. Nevertheless, a tabu move leading to a configuration better than the best configuration found so far is always accepted (*aspiration criterion*).

More precisely, a move $(x, v) \in OneMove$ is declared tabu iff the couple $< x, v >$ is tabu. A move $(x_{g_1,t}, x_{g_2,t}) \in Swap$ is declared tabu iff $< x_{g_1,t}, s(x_{g_2,t}) >$ or $< x_{g_2,t}, s(x_{g_1,t}) >$ is tabu.

Remarks: The algorithms stop if $f(s) = 0$ or if a given limit is reached concerning the time, the number of iterations or the number of moves.

We also tested a simple *Descent Method.* This algorithm chooses at each iteration a best possible neighbor that does not degrade the performance and hence it is easily trapped in a local optimum.

The efficiency of these algorithms is greatly influenced by the ability of finding quickly a best move at each iteration. For moves of type $OneMove$, we use a data structure that permits to find in constant time the performance of a given move. This data structure is updated each time a move is performed.

29.4 COMPUTATIONAL RESULTS

In this section, we first compare the best results of Local Search with the best results obtained by Integer Linear Programming (ILP) and Constraint Programming (CP) for the PPP. Then we present and analyze in detail the results obtained by our Local Search procedures (LS).

29.4.1 Comparison of LS, ILP and CP

Table 29.2 presents the best known results of the three methods ILP, CP and LS. The published results of CP are denoted by CP1. From the table, we see that ILP fails to solve any of P_6 to P_{10}. CP1 solves P_7 and P_8 in 27 and 28 minutes respectively (using a SPARCstation IPX), but fails to solve P_9 and P_{10}. Recently, new strategies using CP were reported leading to much better results (CP2) [1]. Indeed P_7 and P_8 are now solved in a few seconds (after several hundreds of backtracks) by CP2. The problem P_9 is also solved, but using several hours (and millions of backtracks). Using LS we solve the problem up to 9 time periods [3] and fail for 10 periods. More detailed results will be presented below.

problem	ILP	CP1	CP2	LS
P_6	fail	27 min.	a few sec.	$<$ 1 s.
P_7	fail	28 min.	a few sec.	$<$ 1 s.
P_8	fail	fail	a few sec.	1 s.
P_9	fail	fail	hours	4 s.
P_{10}	fail	fail	fail	fail

Table 29.2 Results of the ILP, CP and Local Search for the PPP

29.4.2 Results of Local Search

We tested the two LS algorithms presented in Section 29.3 (algorithm of Metropolis denoted by Mt and Tabu algorithm denoted by TS) and used the two neighborhoods N_1 and N_2. So we tested four different procedures (denoted by Mt-N_1, TS-N_1, Mt-N_2 and TS-N_2). With these four procedures, we tried to solve the increasingly difficult problems from P_6 up to P_{10}. All these four procedures solved P_9 (and easier problems) but none of them could solve P_{10}. Note however that we don't know if P_{10} has a solution. Our local search algorithms give repeatedly solutions violating a single constraint (of any of the

[3]The running time is obtained on a Sun ULTRA 1 (128 RAM, 134MHz). The LS algorithms are implemented in C++.

three types) for P_{10}. Table 29.3 presents the results obtained by these four procedures for P_7, P_8 and P_9.

Each procedure was run 20 times. For each problem, the left column indicates the average number of moves ($\#moves$) and of iterations ($\#iter$) and the right column the average computing time (in seconds). Recall that for Tabu, each iteration leads to a move (hence, $\#moves = \#iter$), while for Metropolis several iterations are needed to lead to a move. Each method has been run with different values of its parameter on the problem P_9. The best value found has been used for the two other problems. These values are $\theta = 130$ for Mt-N_1 and Mt-N_2, $k = 9$ for TS-N_1 and $k = 6$ for TS-N_2.

	P_7		P_8		P_9	
	$\#moves/\#iter$	time	$\#moves/\#iter$	time	$\#moves/\#iter$	time
Mt-N_1	1,228/38,438	1.6	2,635/111,628	4.3	24,458/2,064,295	66.4
TS-N_1	330	0.5	1,366	1.7	51,507	67.5
Mt-N_2	494/5,724	0.6	682/11,582	1.0	1,488/63,685	3.6
TS-N_2	110	2.0	171	3.3	347	6.5

Table 29.3 Results of Local Search procedures for solving the PPP

From Table 29.3, we first note that all methods can solve the problems P_7 and P_8 very easily (in less than 5 seconds and 3,000 moves). P_9 can be solved very easily using the enlarged neighborhood N_2. Using N_1 to solve P_9 needs more effort (more than one minute for Mt-N_1 and TS-N_1). Moreover, we observe that the results obtained do not strongly depend on the method used since the number of moves and computing time are often similar when the two methods are used to solve the same problem using the same neighborhood.

method	problem	success rate	method	problem	success rate
Desc-N_1	P_3	26	Desc-N_2	P_5	93
	P_4	9		P_6	52
	P_5	3		P_7	11
				P_8	1

Table 29.4 Results of the Descent Method

Table 29.4 shows results obtained by the two Descent procedures with the two different neighborhoods (denoted by Desc-N_1 and Desc-N_2). Each procedure is run 100 times for each problem. The table indicates the number of success for the 100 runs. Each run corresponds to a single execution of the algorithm (there is no retry) and the algorithm is stopped after 1,000 iterations without improvement of the cost function. We observe that the Descent Algorithm can solve the PPP up to 5 time periods with the neighborhood N_1 and up to 8 time periods with N_2. These results illustrate the power of Local Search for the problem in that a problem that a simple descent method can solve can be estimated very easy for Local Search.

29.4.3 Comparisons between the two neighborhoods

From the above results, we observe a big difference of performance with the two neighborhoods. We compare now precisely the performance of the two neighborhoods. Concerning the number of iterations of TS, we observe that the ratio between the two neighborhoods is of 1 to 3 for P_7 (330 moves using TS-N_1 and 110 moves using TS-N_2), but becomes 8 and 148 for P_8 and P_9. This means that this ratio increases when the problem becomes more difficult and, in particular, the ratio increases dramatically for P_9.

Concerning computing time, we can compute from Table 29.4 that an iteration using the enlarged neighborhood N_2 costs (for Tabu) 12 times more than using N_1. Although P_7 and P_8 are solved more quickly using the limited neighborhood, the advantage turns strongly to the enlarged neighborhood for the most difficult problem P_9.

Moreover N_2 allows a simple descent method to solve up to P_8 while N_1 cannot go beyond P_5. Therefore the enlarged neighborhood N_2 is more powerful than N_1 especially for solving difficult problems. Experiments on P_{10} confirm this remark, as the two algorithms TS and Mt can easily find solutions that violate a single constraint using N_2, but not using N_1. So we can hope that an improved procedure using N_2 will find a solution for P_{10}, if there exists one.

	[0..5[	[5..10[	[10..15[	[15..20[	[20..25[	[25..30[	[30..35[
$\psi_1(\%)$	1.7	1.4	15.6	37.3	42.3	42.8	52.0
$\psi_2(\%)$	10.3	41.1	51.2	60.6	67.1	67.9	69.5

Table 29.5 Percentage of configurations which are not local optima for N_1 and N_2

Now we present another measure in order to explain the power difference of the two neighborhoods. We observed for the PPP a well-known phenomenon: starting with a high value of the cost function, the cost value decreases very quickly at the beginning of the search and then oscillates when the search becomes more and more difficult. The reason of the initial fast decrease of the cost function is that most of configurations have at least one neighbor which has a smaller cost: these configurations are not local optima for the neighborhood considered. On the contrary, many configurations encountered at the end of the search are local optima. So, the percentage of local optima tends to be related to the difficulty of the search. To study more precisely this point, the following experiment was carried out.

We used a local search procedure to solve P_9. At each iteration we tested if the current configuration is a local optimum for the two neighborhoods N_1 and N_2. Using this information, we computed the percentages $\psi_1(f)$ and $\psi_2(f)$ of configurations of cost f that are not local optima for N_1 and N_2 respectively. Table 29.5 gives a summary of this experiment. The different columns represent the values of the cost function grouped in classes. For example, the first column means that 1.7% and 10.3% of the encountered configurations having a cost of 0, 1, 2, 3 or 4 are not local optima for N_1 and N_2 respectively.

For high values of f, i.e., at the beginning of the search, we observe that the percentage is high for both neighborhoods: for example, for $f \in [30, 35[$, this percentage is about 50% and 70% for N_1 and N_2 respectively. Hence it is generally easy to find a move that improves the current configuration, using indifferently one of the two neighborhoods. For smaller values of f, this percentage decreases dramatically for N_1, while it remains much higher for N_2. For example, for $f \in [0, 4[$, $\psi_1(f) = 1.7\%$ and $\psi_2(f) = 10.3\%$. Hence there are much more useful improving moves in N_2 than in N_1. These results are averages computed on 100 runs performed with the algorithm TS-N_2 on the problem P_9, but experiments carried out with the other LS procedures show similar properties.

Besides, we note that N_2 is about only two times larger than N_1: in P_9, the size of N_1 is $29*9*(13-1) = 3132$ while the size of N_2 is $3132+9*(29*28/2) = 3132 + 3654 = 6876$.

In summary, we see that the neighborhood N_2 contains many solutions of high quality that are not present in N_1 while its size is about only two times larger than N_1. This is probably an important factor that explains the observed difference of efficiency between the two neighborhoods.

29.5 CONCLUSIONS

In this paper we have presented a local search approach for the Progressive Party Problem. Based on a formulation of the PPP as a Constraint Satisfaction Problem, the proposed approach introduces two different neighborhoods and penalty-based cost function for handling complex and heterogeneous constraints.

The approach was tested on the available benchmark instance. Results were compared with previous ones obtained with Integer Linear Programming and Constraint Programming. Numerical experiments showed that both a simple tabu algorithm and a metropolis algorithm give better results since they solve the problem up to 9 time periods in only several seconds of cpu time. Even a descent is successful for up to 8 time periods if the swap neighborhood is used. Therefore, LS should be considered as one of the most competitive approaches for the PPP.

Until now, all reported studies on the PPP concern a particular instance arising form the initial party organization problem. It is now natural and interesting to know how different approaches will behave for other instances of the problem.

Acknowledgements

We would like to thank the referees of this paper for their useful comments.

References

[1] N. Beldiceanu, E. Bourreau, P. Chan and D. Rivreau, Partial Search

Strategy in CHIP, *presented at MIC'97*, Sophia-Antipolis, 1997.

[2] S. C. Brailsford, P. M. Hubbard and B. M. Smith, The Progressive Party Problem: A Difficult Problem of Combinatorial Optimization, *Computers and Operations Research*, 23:845-856, 1996.

[3] P. Galinier and J. K. Hao, Tabu Search for Maximal Constraint Satisfaction Problems, *Lecture Notes in Computer Science 1330*, pp196-208, 1997.

[4] F. Glover and M. Laguna, Tabu Search, *USA, Boston: Kluwer Academic Publishers*, 1997.

[5] P. Hansen and B. Jaumard, Algorithms for the Maximum Satisfiability Problem, *Computing*, 44:279-303, 1990.

[6] J. K. Hao and R. Dorne, Empirical Studies of Heuristic Local Search for Constraint Solving, *Lecture Notes in Computer Science 1118*, pp194-208, 1996.

[7] S. Kirkpatrick, C.D. Gelatt Jr. and M.P. Vecchi, Optimization by Simulated Annealing, *Science*, 220:671-680, 1983.

[8] B. M. Smith, S. C. Brailsford, P. M. Hubbard and H. P. Williams, The Progressive Party Problem: Integer Linear Programming and Constraint Programming Compared, *Constraints*, 1(1/2):119-138, 1996.

[9] A.K. Mackworth, Constraint Satisfaction, *in S.C. Shapiro (Ed.) Encyclopedia on Artificial Intelligence, John Wiley & Sons, NY*, 1987.

[10] S. Minton, M.D. Johnston and P. Laird, Minimizing Conflicts: a Heuristic Repair Method for Constraint Satisfaction and Scheduling Problems, *Artificial Intelligence*, 58(1-3):161-206, 1992.

[11] P. Morris, The Breakout Method for Escaping from Local Minima, *Proc. of AAAI-93*, pp40-45, 1993.

[12] C.H. Papadimitriou and K. Steiglitz, Combinatorial Optimization - Algorithms and Complexity, *Prentice Hall*, 1982.

[13] S. Selman, H.A. Kautz and B. Cohen, Noise strategies for improving local search, *Proc. of AAAI-94*, pp337-343, Seattle, WA, 1994.

[14] E. Tsang, Foundations of Constraint Satisfaction, *Academic Press*, 1993.

[15] N. Yugami, Y. Ohta and H. Hara, Improving Repair-based Constraint Satisfaction Methods by Value Propagation, *Proc. of AAAI-94*, pp344-349, Seattle, WA, 1994.

30 AN INTRODUCTION TO VARIABLE NEIGHBORHOOD SEARCH

Pierre Hansen and Nenad Mladenović

GERAD and École des Hautes Études Commerciales
3000 chemin de la Côte-Sainte-Catherine, Montréal, Canada H3T 2A7.
pierreh@crt.umontreal.ca, nenad@crt.umontreal.ca

Abstract: In this paper we examine a relatively unexplored approach to the design of heuristics, the guided change of neighborhood in the search process. Using systematically this idea and very little more, i.e., only a local search routine, leads to a new metaheuristic, which is widely applicable. We call this approach Variable Neighborhood Search (VNS).

30.1 INTRODUCTION

Local search methods for combinatorial and global optimization proceed by performing a sequence of local changes in an initial solution which improve each time the value of the objective function until a local optimum is found. That is, at each iteration an improved solution x' in the neighborhood $\mathcal{N}(x)$ of the current solution x is obtained, until no further improvement are found. In recent years, several metaheuristics have been proposed which extend in various ways this scheme and avoid being trapped in local optima with a poor value (see [67] for a multi-authored book-length survey and [64] for an extensive bibliography).

These metaheuristics, e.g. tabu search [31, 32, 34, 33, 35, 37], simulated annealing [50], genetic search [44], GRASP [28], variable depth search [54, 65], adaptive multi-start [9] have led to much improved results in many practical contexts. Usually, these results are better approximate solutions to large problem instances. But heuristics can also be used for other purposes. Important steps of exact algorithms can be performed in a heuristic way. For instance, solving the auxiliary problem of finding a column with negative reduced cost in column generation methods can be done heuristically. Only when no such column is detected need one turn to an exact algorithm. Moreover, while finding a good initial solution is often helpful in branch and bound algorithms, such a solution can be used to do much more than provide an initial bound for an optimality test, e.g. it can be used to stabilize a column generation scheme [20]. Then, heuristics can be used to explore theoretical questions. For instance, they can help substantially to suggest, disprove or give hints on how to prove conjectures in graph theory [14].

In this paper we examine a relatively unexplored approach to the design of heuristics: change of neighborhood in the search. Using systematically this idea and very little more, i.e., only a local search routine, leads to a new metaheuristic, which is widely applicable. We call this approach Variable Neighborhood Search (VNS). Contrary to other metaheuristics based on local search methods, VNS does not follow a trajectory but explores increasingly distant neighborhoods of the current incumbent solution, and jumps from this solution to a new one if and only if an improvement has been made. In this way often favorable characteristics of the incumbent solution, e.g. that many variables are already at their optimal value, will be kept and used to obtain promising neighboring solutions. Moreover, a local search routine is applied repeatedly to get from these neighboring solutions to local optima. This routine may also use several neighborhoods. Therefore, to construct different neighborhood structures and to perform a systematic search, one needs to have a way for finding the distance between any two solutions, i.e., one needs to supply the solution space with some metric (or quasi-metric) and then induce neighborhoods from it. In the application part of the next sections we answer this problem-specific question for each problem considered.

The remainder of the paper is organized as follows. The rules of the basic VNS and some easy extensions are given in the next section, followed by application to six classical combinatorial optimization problems. In Section 30.3 we develop a variable neighborhood decomposition search heuristic (VNDS) for solving very large problem instances and apply it to three of the six problems. In Section 30.4, it is shown how VNS can be applied in graph theory, an area where (contrary to graph optimization) metaheuristics do not appear to have been used before. Section 30.5 concludes the paper with a brief discussion of criteria for evaluating metaheuristics with application to VNS.

This paper being an introductory one, the focus is on basic ideas (we mention briefly a few other ones). Further developments of VNS, now under study, will be reported on later. These include possible hybrids of VNS and Tabu Search (TS) or other metaheuristics. Results of various papers are summarized and some new ones (on using 2-OPT for VNS in the traveling salesman problem, and on applying VNS and VNDS to minimum sum-of-squares clustering) are presented.

30.2 VNS HEURISTIC

In this section we give the rules of the basic VNS and apply them in solving the following six classical combinatorial optimization problems: (i) traveling salesman problem (TSP); (ii) TSP with back-hauls (TSPB); (iii) weighted maximum satisfiability problem (WMAXSAT); (iv) p-median problem (PM); (v) multisource Weber problem (MW) and (vi) minimum sum-of-squares clustering problem (MSSC).

Let us denote a finite set of pre-selected neighborhood structures with $\mathcal{N}_k$, $(k = 1, \ldots, k_{max})$, and with $\mathcal{N}_k(x)$ the set of solutions in the kth neighborhood of x. (Local search heuristics usually use one neighborhood structure, i.e., $k_{max} = 1$.) Our basic VNS heuristic comprises the following steps.

Initialization. Select the set of neighborhood structures $\mathcal{N}_k$, $k = 1, \ldots, k_{max}$, that will be used in the search; find an initial solution x; choose a stopping condition;

Repeat the following until the stopping condition is met:

(1) Set $k \leftarrow 1$; (2) Until $k = k_{max}$, repeat the following steps:

(a) *Shaking.* Generate a point x' at random from the kth neighborhood of x $(x' \in \mathcal{N}_k(x))$;

(b) *Local search.* Apply some local search method with x' as initial solution; denote with x'' the so obtained local optimum;

(c) *Move or not.* If this local optimum is better than the incumbent, move there $(x \leftarrow x'')$, and continue the search with $\mathcal{N}_1$ $(k \leftarrow 1)$; otherwise, set $k \leftarrow k + 1$;

Figure 30.1 Steps of the basic VNS.

The stopping condition may be e.g. maximum CPU time allowed, maximum number of iterations, or maximum number of iterations between two improvements. Often successive neighborhoods $\mathcal{N}_k$ will be nested. Note that point x' is generated at random in step 2a in order to avoid cycling, which might occur if any deterministic rule was used.

It is worth stressing the ease of implementation of both the basic version of VNS (with only one parameter k_{max}) and various simple extensions discussed below and inspired by TS [31, 32, 34, 37] or other metaheuristics. Step 2a is easy to program. For example, if $\mathcal{N}_k$ is obtained by k-interchanges of solution attributes, one need only add a few lines to an existing code for a local search method (step 2b).

As a local optimum within some neighborhood is not necessarily one within another, change of neighborhoods can be performed during the local search phase, too. In some cases, as when applying VNS to graph theory, the use of many neighborhoods in the local search is crucial. This search is then called Variable Neighborhood Descent (VND) and its steps are the following.

The basic VNS is in fact a descent, first improvement method. Without much additional effort it could be transformed into a descent-ascent method: in step 2c set also $x \leftarrow x''$ with some probability even if the solution is worse than the incumbent, and/or a best improvement method: make a move to the best neighborhood k^* among all k_{max} of them. Of course, the former variant is reminiscent of simulated annealing [50] and the latter of variable depth search [54, 65]. Other variants of the basic VNS could be:

(i) find solution x' in step 2a as the best among ℓ (a parameter) randomly generated solutions from the kth neighborhood;

Initialization. Select the set of neighborhood structures $\mathcal{N}'_k$, $k = 1, \ldots, k'_{max}$, that will be used in the descent; find an initial solution x (or apply the rules to a given x);

Repeat the following until no improvement is obtained:
(1) Set $k \leftarrow 1$; (2) Until $k = k'_{max}$, repeat the following steps:
(a) *Exploration of neighborhood.* Find the best neighbor x' of x ($x' \in \mathcal{N}'_k(x)$);
(b) *Move or not.* If the solution thus obtained x' is better than x, set $x \leftarrow x'$; otherwise, set $k \leftarrow k + 1$.

Figure 30.2 Steps of the basic VND.

(ii) introduce k_{min} and k_{step}, two parameters that control the change of neighborhood process, i.e., in the previous algorithm instead of $k \leftarrow 1$ set $k \leftarrow k_{min}$ and instead of $k \leftarrow k + 1$ set $k \leftarrow k + k_{step}$. Then intensification and diversification of the search is achieved in an easy and natural way: indeed, if k_{min} and/or k_{step} are set to some fairly large integer values, then the search continues in far away regions of the solution space, i.e., it is diversified; if k_{min} is small and $k_{step} = \lfloor \frac{k}{\ell} \rfloor + 1$ where $\lfloor\ b\ \rfloor$ is the largest integer not greater than b, then the search spends more time in the region close to the incumbent, i.e., it is intensified;

(iii) remove the local search step (2b). This variant, which we call Reduced VNS (RVNS), is useful for very large instances for which local search is costly. It is akin to a Monte-Carlo method, but more systematic. Its relationship to the Monte-Carlo method is the same as that of VNS to Multistart.

The second variant is reminiscent of some aspects of the strategic oscillation approach of TS [33, 34, 35], in the way it uses two parameters to guide movements in the search space. But there are also many differences, e.g., use of an oscillation guidance function and choice of a target level (or oscillation boundary) for this function versus consideration of local minima to define neighborhoods, creation of an oscillating behavior in the trajectory by generating "modified evaluations and rules of movement depending on the region navigated and the direction of search", versus descents from the points selected in the successive neighborhoods and jumps when a better solution than the incumbent has been found, etc.

When using more than one neighborhood structure in the search, as we do in VNS, the following problem specific questions have to be answered:
(i) what $\mathcal{N}_k$ should be used and how many of them?
(ii) what should be their order in the search?
(iii) what search strategy should be used in changing neighborhoods?

Furthermore, for each problem below, we must decide (iv) what local search routine will be used in step 2b. Development of the VNS heuristic for solving each particular problem is based on answers to questions (i) to (iv) above.

Therefore, after formulation of the problem in each of the subsections to follow, we develop our implementation of VNS by answering those questions.

30.2.1 Traveling Salesman Problem (TSP)

Given n cities with intercity distances, the traveling salesman problem (TSP) is to find a minimum cost tour x (i.e., a permutation of the cities which minimizes the sum of the n distances between adjacent cities in the tour). It is a classical NP-hard problem (for a recent survey of heuristics for the TSP, see [47]).

Probably the most popular heuristic for solving TSP is 2-OPT, where in turn two connections between cities in the current tour x are broken and these cities reconnected again in the only other way which gives a tour. Since 2-OPT is a descent local search heuristic, it stops in a local minimum. We first apply the basic VNS rules using 2-OPT (or a quicker variant, see below) as local search routine.

Let us denote with $\mathcal{S}$ the solution space (i.e., the set of all permutations) of the TSP ($x \in \mathcal{S}$). One may measure the distance ρ between two TSP tours simply in terms of their difference in form: if tours x_1 and x_2 differ in k links, then $\rho(x_1, x_2) = k$. It is clear that ρ is a metric on $\mathcal{S}$. The neighborhoods induced by the metric ρ are

$$\mathcal{N}_k(x) = \{x' \mid \rho(x, x') = k,\ x \in \mathcal{S}\}, k = 2, \ldots, n \tag{30.1}$$

that is the k-OPT neighborhoods [53]. (The number of tours in $\mathcal{N}_k(x)$ is not known and appears very difficult to determine, see [57] for some approximations.) We obtain a parameter free VNS heuristic by setting the single parameter k_{max} to n. Then all ingredients for application of the basic VNS algorithm (VNS-1 for short) are given. Another version (VNS-2), that uses a quicker 2-OPT in step 2b works as follows. Instead of deleting and reconnecting all possible pairs of links, the $O(n^2)$ neighborhood is reduced by neglecting the $r\%$ (a parameter) longest links from each city. Let (i, j) be the link to be deleted in the outer loop of 2-OPT. If city j does not belong to the $(1 - r)\%$ cities closest to city i, then the inner loop does not take place.

In Table 30.1 average results for random Euclidean problems over 100 trials for $n = 100, \ldots, 500$ and 10 trials for $n = 600, \ldots, 1000$ are reported. Each problem instance is first solved twice by 2-OPT and the best solution value kept. The CPU time spent for these two independent calls of 2-OPT is then used as stopping criterion for both VNS variants. The first column of Table 30.1 gives the size of the problem, the next three the average objective function values, while the last three columns report the average CPU time when the best solution was found by each method, (i.e., at the last iteration for 2-OPT). In VNS-2 the size of the 2-OPT neighborhood is reduced by taking $r = 40\%$ for $n = 100$, 60% for $n = 200$ and 80% for $n \geq 300$. Average improvements in value of 2.73% and 4.43% over the classical 2-OPT heuristic are obtained by VNS-1 and VNS-2 respectively. These improvements are significant when compared with those of other studies. For example a 0.5% average improvement for the 2.5-OPT heuristic [8] over the 2-OPT heuristic (on random Euclidean instances

	Best value found			% Improvement		CPU times		
n	2-OPT	VNS-1	VNS-2	VNS-1	VNS-2	2-OPT	VNS-1	VNS-2
100	825.69	817.55	811.95	0.99	1.66	0.25	0.18	0.17
200	1156.98	1143.19	1132.63	1.19	2.10	3.88	3.21	2.82
300	1409.24	1398.16	1376.76	0.79	2.30	12.12	10.29	9.35
400	1623.60	1602.59	1577.42	1.29	2.84	46.13	40.03	34.37
500	1812.08	1794.59	1756.26	0.96	3.07	110.64	99.57	91.00
600	1991.56	1959.76	1925.51	0.97	3.32	204.60	191.85	173.07
700	2134.86	2120.59	2089.33	0.67	2.13	347.77	307.93	259.06
800	2279.18	2242.11	2190.83	1.63	3.88	539.94	480.50	462.23
900	2547.43	2399.52	2342.01	5.81	8.06	699.33	656.96	624.74
1000	2918.10	2555.56	2483.95	12.42	14.88	891.61	844.88	792.88
Av.	1869.87	1803.36	1768.67	2.73	4.43	285.63	263.54	244.97

Table 30.1 TSP: average results for random Euclidean problems over 100 trials for $n = 100, \ldots, 500$ and 10 trials for $n = 600, \ldots, 1000$. Computing times in seconds CPU on a SUN SPARC 10, 135.5 Mips (as all other results in this paper).

as well) at the cost of a 30-40% increase in running time is reported in [47]. It is interesting to note that although the quicker version of 2-OPT does not seem to be efficient by itself (links with large length can hardly be removed), combined with a k-OPT *shaking* operation within VNS (large edges are deleted by shaking), surprisingly good results are obtained. It thus appears that not only the quality of the local search or the shaking process are important in VNS, but also their interrelation.

We then try our basic VNS with the GENIUS heuristic for the TSP developed by Gendreau *et al.* [29]. It is a sophisticated insertion followed by local deletion/insertion and correction procedure. The size of the neighborhood in GENIUS depends on a parameter p (the number of cities already in the tour closest to the city that is considered for possible deletion or insertion). We immediately get a set of neighborhood structures for VNS by denoting with $\mathcal{N}_p(x)$ all tours obtained by deletion/insertion with parameter value p. The details can be found in [62] where results on the same type of test problems (i.e., Euclidean instances) as reported in [29] are given. A 0.75% average improvement in value is obtained within a similar CPU time. Moreover, improvements are obtained for all problem sizes.

30.2.2 Traveling Salesman Problem with Back-hauls (TSPB)

In the traveling salesman problem with back-hauls (TSPB) customers (or cities) are divided into three disjoint sets: depot, line-haul and back-haul customers. Starting from the depot, a route must be designed such that all line-haul customers are visited contiguously before all back-haul customers. In fact, TSPB is a single-vehicle routing problem which can be reformulated as a TSP by adding large distances (or costs) between customers that belong to different subsets.

In Gendreau *et al.* [30] six heuristic methods for TSPB are proposed and compared. Their results indicate that, in terms of solution value, GENIUS (applied to the reformulated problem) is the best algorithm. In [62] GENIUS is compared with GENIUS based VNS on a series of random Euclidean instances, as designed and explained in [30]. A 0.4% average improvement in value is obtained with 30% increase in running time. Again, improvement is obtained for all problem sizes.

30.2.3 Weighted Maximum Satisfiability Problem (WMAXSAT)

A central problem in Artificial intelligence, Logic and Computational complexity is the Satisfiability problem. It was the first problem shown to be NP-complete [18]. Here we consider its extension, weighted maximum satisfiability (WMAXSAT) which can be expressed as follows: let $C_1, \ldots, C_m$ be m clauses, involving n Boolean variables $x_1, \ldots, x_n$. In addition, for each clause C_i, there is an associated nonnegative weight w_i. Determine an assignment of truth values to the n variables that maximizes the sum of the weights of the satisfied clauses. A mixed integer linear programming formulation of WMAXSAT is:

$$\max\ F(x,z) = \sum_{i=1}^{m} w_i z_i$$

subject to

$$\sum_{j \in I_i^+} x_j + \sum_{j \in I_i^-} (1 - x_j) \geq z_i,$$

$$x_j \in \{0,1\}, j = 1, \ldots, n$$

$$0 \leq z_i \leq 1, i = 1, \ldots, m,$$

where I_i^+ (resp. I_i^-) denotes the index set of the variables appearing unnegated (resp. negated) in clause C_i.

The simplest possible local search heuristic (*Greedy*) uses the neighborhood defined as the set of solutions obtained by complementation of a component x_j of x. Such a neighborhood has cardinality n. Best complementations are performed until no further improvement is possible. The solution space $\mathcal{S}$ consists of the 2^n n-dimensional binary vectors x ($\mathcal{S} = \{0,1\}^n$). The metric on $\mathcal{S}$ which we use is the Hamming metric ρ, where two vectors x_1 and x_2 are at distance k ($k \leq n$) if and only if they differ in k components ($\rho(x_1, x_2) = k$). This metric induces neighborhoods $\mathcal{N}_k$ as

$$x' \in \mathcal{N}_k(x) \Longleftrightarrow \rho(x, x') = k. \tag{30.2}$$

Note that the Greedy heuristic also uses the neighborhood induced by the Hamming metric (for $k = 1$).

The basic VNS from Figure 30.1 is applied for solving WMAXSAT in [39]. In step 2a shaking is done by picking a random point from $\mathcal{N}_k(x)$, where x is the current solution and $\mathcal{N}_k(x)$ is defined by (30.2). For step 2b, the Greedy heuristic is used. The method is compared with GRASP [69] and a simple TS (Steepest ascent mildest descent, [37]). Results of VNS and TS are roughly equal in quality and outperform GRASP: for instance, in the three series of problems derived from the SAT instance class *jnh* of the 2nd DIMACS Implementation Challenge [48], 42 problems out of 44 are solved exactly by VNS instead of 3 by GRASP and the average error is reduced from 0.053% to 0.0001% within the same CPU time.

30.2.4 p-Median Problem (PM)

Consider a set L of m potential locations for p facilities and a set U of given locations for n users. The p-median problem (PM) is to locate simultaneously the p facilities at locations of L in order to minimize the total transportation cost for satisfying the demand of the users, each supplied from its closest facility. PM can also be defined as a purely mathematical problem: given an $n \times m$ matrix D, select p columns of D in order that the sum of minimum coefficients in each line within these columns be smallest possible. This NP-hard problem [49], is a basic one in location theory (see [59] for an introduction to discrete location theory). Let $D = (d_{ij})$ be a $n \times m$ matrix of distances d_{ij} traveled (or costs incurred) for satisfying the demand of the user located at i from a facility located at j, for all $i \in U$ and $j \in L$. PM can be expressed concisely as

$$\min_{J} \sum_{i \in U} \min_{j \in J} d_{ij}.$$

where $J \subseteq L$ and $|J| = p$.

Let us denote with $\mathcal{S} = \{x \mid x = set\ of\ p\ (out\ of\ m)\ locations\ of\ L\}$ the solution space of the problem. We say that the distance between two solutions x_1 and x_2 ($x_1, x_2 \in \mathcal{S}$) is equal to k, if and only if they differ in k locations. Since $\mathcal{S}$ is a set of sets, a (symmetric) distance function ρ can be defined as

$$\rho(x_1, x_2) = |x_1 \setminus x_2| = |x_2 \setminus x_1|, \quad \forall x_1, x_2 \in \mathcal{S}. \tag{30.3}$$

It can easily be checked that ρ is a metric function in $\mathcal{S}$, thus, $\mathcal{S}$ is a metric space. The neighborhood structures $\mathcal{N}_k$ we use are induced by metric ρ as in (30.2), i.e., k locations of facilities ($k \leq p$) from the current solution are replaced by k others. Note that such moves have already been mentioned in the literature as λ-interchange mechanism [63] or k-substitution move [61]. The cardinality of $\mathcal{N}_k(x)$ is

$$|\mathcal{N}_k(x)| = \binom{p}{k}\binom{m-p}{k},$$

since k out of p facilities are dropped and k out of $m-p$ added to the solution. This number first increases then decreases with k. Sets $\mathcal{N}_k(x)$ are disjoint ($\mathcal{N}_k(x) \cup \mathcal{N}_l(x) = \emptyset, \forall k, l \neq k$), and their union, together with x, gives $\mathcal{S}$.

PM is the first problem solved by VNS, in [60]. Results of two versions are reported: in the first, step 2b of the basic VNS is omitted, while in the second, Maranzana's heuristic [58] is used as local descent; shaking is performed by substituting k facilities; k_{max} is set to 2. Computational results obtained by these two versions of VNS are compared with those given by *Interchange* [73] and TS [61] heuristics. (The *Interchange* heuristic is also known as *node substitution* if PM is expressed in terms of an underlying graph). When solving small problem instances (47 and 75 users), average errors of 1.63%, 0.21%, 0.24% and 0.06% for Interchange, TS, first and second VNS respectively are obtained.

Much better results in solving PM by VNS and TS are reported in [41], where an efficient implementation of *Fast Interchange* (FI) [75] descent is developed and used in both step 2b of the basic VNS and in TS. For VNS, the same metric ρ is used for both shaking ($k = 1, \ldots, k_{max}$) and local search ($k = 1$). In order to get a parameter free version of VNS, we set the value of the single parameter k_{max} to p. The basic VNS heuristic is compared with FI and two recent methods based on the TS approach: TS where the reverse elimination method is used for diversifying the search [74]; we denote it by TS-1; TS where the so-called 1-chain-substitution move is used in the search [61]; we denote it by TS-2. Results reported are favorable for the basic VNS. For example, for the first 25 test problems from [7] (as results for them only are reported in [74]), all but 2 test problems are solved exactly by VNS, with a total error of 0.07%; all but 3 are solved exactly by TS-2, with a total error of 0.57%; all but 4 are solved exactly by TS-1, with a total error of 0.78%. We next discuss briefly some possible reasons why this is so.

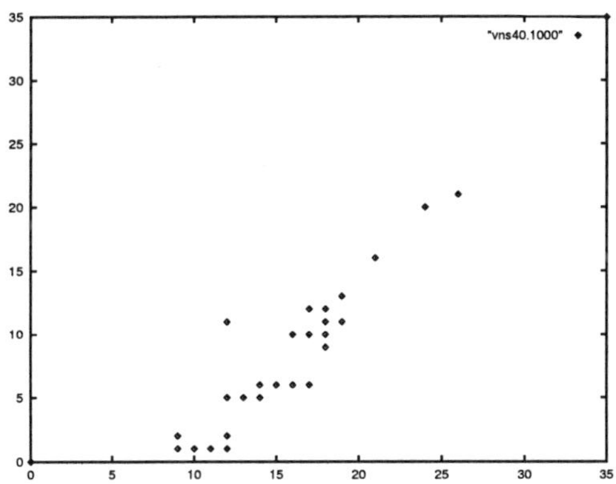

Figure 30.3 Local minima obtained after 1,000 iterations by VNS on PM test problem #40 ($n = 900$, $p = 90$); optimal solution has not been found.

All three methods use the same interchange neighborhoods and do the search in an attractive area of the solution space (in its "deep valley"), but the TS methods do not visit *real* but *pseudo* local minima defined by both the objective function and the forbidden moves. The solutions on the path from pseudo to real local minima might be tabu (i.e., in forbidden areas of the solution space). This happens with TS-2, for example, which puts in the Tabu List (TL) attributes of the solutions, not solutions themselves (to save memory and time, which is usual in TS). If the length of TL is small, a few solutions are

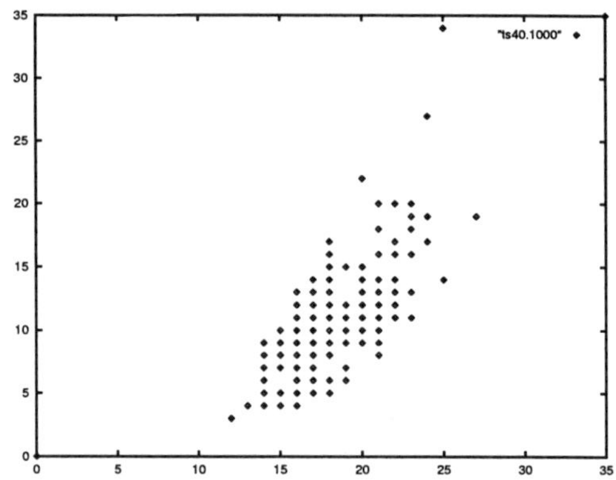

Figure 30.4 Pseudo local minima obtained after 1000 iterations by TS-2 on PM test problem #40 ($n = 900$, $p = 90$); optimal solution has not been found.

tabu, but probability of cycling is large. If TL length is large, many solutions from the neighborhood are forbidden. Note that the simple *Aspiration* function does usually not help much here, because it considers only solutions from the closest neighborhood, i.e., only one point from the path from pseudo to real local minimum; that path often contains several solutions and if at least one of them is tabu, it cannot be used to reach the real local minimum (which might itself be tabu). However, more sophisticated aspiration criteria do remove tabu restrictions when there is a direction of descent. For instance, "aspiration by strong admissibility" [33] overrules a tabu restriction when there are no non-tabu descent directions and an improving move (not due to an aspiration criterion other than the usual one) has taken place since the last non-improving move. Clearly, such a rule augments the danger of cycling. To the best of our knowledge, it has not been applied to PM, and no computational experience using it for any problem is reported in [33].

Solutions of the largest problem from ORLIB (#40, $n = 900$, $p = 90$) is illustrated on Figures 30.3 and 30.4, where local minima are generated by VNS and TS-2 respectively. This example is chosen because the optimal solution has not been found by either of these methods. Each point (r_i, s_i) on the graph of Figures 30.3 and 30.4 refers to one local minimum x_i obtained by the VNS or TS respectively, which were run 1000 times. The origin represents the known optimal solution. The r_i represent the distance between optimal solution and the local minimum ($r_i = \rho(x_{opt}, x_i)$), while the s_i correspond to the difference in their objective function values ($s_i = f(x_i) - f(x_{opt})$). It appears that: (i) the number of *real* local minima of VNS (represented in Figure 30.3) is less than the number of *pseudo* local minima (represented in Figure 30.4); (ii) solutions obtained by VNS are of better quality: with VNS four solutions whose objective function value is only 1 larger than the optimal value are found, while the best TS value is 3 larger than the optimal one.

30.2.5 *Multisource Weber Problem (MW)*

The multisource Weber (MW) problem (also known as continuous location-allocation problem) is the continuous counterpart of PM: instead of locating the p facilities at some locations of L, they can be located anywhere in the plane. MW can be formulated as follows [52]:

$$\min_{u_i, v_i, z_{ij}} \sum_{i=1}^{p} \sum_{j=1}^{n} z_{ij} \cdot w_j \cdot d_j(u_i, v_i)$$

subject to

$$\sum_{i=1}^{p} z_{ij} = 1, \quad j = 1, 2, \ldots, n,$$

$$z_{ij} \in [0, 1] \quad i = 1, 2, \ldots, p; \; j = 1, 2, \ldots, n,$$

where p facilities must be located to satisfy the demand of n users, u_i, v_i denote the coordinates of the ith facility, d_j the Euclidean distance from (u_i, v_i) to the jth user, w_j the demand (or weight) of the jth user and z_{ij} the fraction of this demand which is satisfied from the ith facility. It is easy to see that there is always an optimal solution with all $z_{ij} \in \{0, 1\}$, i.e., each user is satisfied from a single facility, which is the (or a) closest one.

A variable neighborhood concept for solving MW is first applied in [12], which systematically increases the number of exchanges (k customer-to-facility reallocations) of the H-type neighborhood [55] to expand the search radius about a local optimum. Another local search procedure is suggested in [13], which constructs its neighborhood as the set of points obtained by a given number of facility relocations. The simplest construction considers the relocation of a single facility to any customer location unoccupied by another facility. Using this neighborhood structure, several basic TS and VNS heuristics were developed and an extensive empirical study carried out to compare various heuristics - old, recent, and new - in a unified setting. The different methods (i.e., Genetic search, three TS variants, four VNS variants etc.) were compared on the basis of equivalent CPU times. Results of this study indicate that the VNS can be effectively used to obtain superior solutions. For instance on a series of 20 problems with 1060 users the average error (by comparison with best known solution) is of 0.02% only for the best VNS, while it can rise to more than 20% for some well-known heuristics of the literature. Average error for a Genetic algorithm was 1.27% and for the best TS 0.13%.

Note that the VNS, TS and other algorithms considered in this introductory paper do not make systematic use of knowledge on the distribution of local optima (or, as a rough proxy, on the set of local minima encountered during the search). Versions which do so will be studied later.

30.2.6 *Minimum Sum-of-Squares Clustering Problem (MSSC)*

Consider a set X of n entities in q-dimensional Euclidean space $\mathbb{R}^q$: $X = \{x_1, \ldots, x_n\}$, $x_j = (x_{1j}, \ldots, x_{qj}) \in \mathbb{R}^q$. The minimum sum-of-squares clustering (MSSC) problem is to find a partition of X into m disjoint subsets (clusters) C_i, such that the sum of squared distances from each entity to the centroid $\overline{x}_i$ of its cluster is minimum, i.e.,

$$\min \sum_{i=1}^{m} \sum_{\ell:x_\ell \in C_i} \|x_\ell - \overline{x}_i\|^2, \quad \overline{x}_i = \frac{1}{n_i} \sum_{\ell:x_\ell \in C_i} x_\ell, \tag{30.4}$$

where $\|.\|$ denotes the Euclidean norm and $n_i = |C_i|$ $(n_1 + \cdots + n_m = n)$.

Among many heuristics for MSSC (see [36] or [72] for surveys), the K-MEANS local search algorithm [46, 56] is the most popular. Given an initial partition, an entity x_j that belongs to the cluster C_ℓ in the current solution is assigned to some other cluster C_i, $i \neq \ell$. New centroids and objective function values are then updated by using simple formulae; a neighborhood of the current solution is defined by all possible such exchanges (i.e., for all i and j). A move is made if the best solution in the neighborhood is better than the current one. Otherwise, the procedure stops. Another popular heuristic is the so-called H-MEANS algorithm [2], which is very similar to Cooper's alternating heuristic [19] for the multisource Weber problem. A new descent local search heuristic, called J-MEANS, is proposed in [42], where the cluster centroid $\overline{x}_i$ is re-located at some entity which does not already coincide with a centroid. Since this move may be a large one and corresponds to n_i reallocations (or n_i K-MEANS moves), we refer to it as a *jump*, hence the name J-MEANS. Obviously, heuristic solutions obtained with the *jump* neighborhood could be improved by the H-MEANS and/or K-MEANS heuristics. Using them after J-MEANS gives a VND heuristic called J+H+K-MEANS. Metaheuristics previously applied to MSSC are: Simulated annealing [51], Genetic search [6], and TS [1].

In Table 30.2, two variants of VNS [42] are compared with multistart versions of K-MEANS, H-MEANS and H+K-MEANS local searches based on equivalent CPU time (150 sec.). Both VNS-1 and VNS-2 are parameter free, i.e., $k_{max} = m$. VNS-1 method uses K-MEANS neighborhoods (for step 2a of the basic VNS from Figure 30.1) and H+K-MEANS for step 2b. VNS-2 uses relocation neighborhoods and J+H+K-MEANS local search. The large difference in the quality of the solutions obtained by these two VNS versions can be explained by the effectiveness of the re-location (*jump*) neighborhood structure. A similar difference in efficiency between re-allocation and re-location neighborhoods for the Multisource Weber problem is reported in [13].

30.3 VARIABLE NEIGHBORHOOD DECOMPOSITION SEARCH

Effectiveness of local search based metaheuristics, i.e., time to find a near-optimal solution, largely depends on the size of the neighborhoods they use. For small and medium size problems, all descent local searches are very fast and metaheuristics usually use several times longer CPU times. But, for very large problem instances, even greedy heuristics can work very long. Consequently, the decomposition methods or parallel computing are needed if computing time is to remain moderate.

		% Deviation from best found									
	Best	K-MEANS		H-MEANS		H+K-MEANS		VNS-1		VNS-2	
m	found	Av.	Best	Av.	Best	Av.	Best	Av.	Best	Av.	Best
10	1754840264.9	0.19	0.19	0.01	0.00	0.01	0.00	0.14	0.00	0.14	0.04
20	791925963.7	2.89	0.00	1.87	1.29	1.31	0.46	3.50	1.84	0.74	0.01
30	482302357.1	9.34	6.68	11.77	9.01	8.20	5.79	10.64	5.16	0.86	0.00
40	342844809.0	15.50	12.04	20.01	15.28	16.86	11.53	15.95	9.64	0.81	0.00
50	256892529.0	27.16	24.57	35.60	30.59	28.66	17.14	29.95	13.50	1.42	0.00
60	199151542.6	35.79	32.53	44.59	33.56	36.21	30.00	35.19	20.50	0.59	0.00
70	159781533.1	44.28	33.08	56.63	47.90	47.39	38.89	43.85	25.59	0.78	0.00
80	130038918.6	53.15	46.64	62.34	50.77	56.69	43.98	51.08	35.07	0.75	0.00
90	111322621.7	56.12	48.94	63.94	51.38	54.40	48.38	44.94	28.17	0.73	0.00
100	97352045.7	60.41	54.74	46.21	46.21	35.95	35.95	42.99	28.00	1.16	0.00
110	86287804.2	60.69	52.78	59.92	49.73	41.44	40.79	43.97	23.76	1.34	0.00
120	76380389.5	62.90	54.00	62.32	52.66	48.96	41.28	38.58	33.56	1.02	0.00
130	68417681.6	65.91	50.73	54.66	38.95	42.34	24.64	38.46	26.30	0.67	0.00
140	61727504.5	62.16	49.82	53.05	45.51	36.00	36.00	30.85	22.04	1.43	0.00
150	56679822.6	66.06	55.05	47.82	40.74	33.43	26.88	25.41	20.05	1.34	0.00
160	52210995.2	59.37	53.16	41.74	34.88	30.85	25.61	25.83	19.32	0.70	0.00
Av. error		42.62	35.93	41.40	34.28	32.42	26.71	30.08	19.53	0.90	0.00

Table 30.2 MSSC for a 1060-entity problem: average and best results in 10 trials; stopping rule: 150 sec. for each trial.

In [43] a Variable Neighborhood Decomposition Search (VNDS) heuristic is proposed. This method follows a basic VNS scheme within a successive approximations decomposition method. For a given solution x, all but k attributes (or variables) are fixed in the local search phase. All possible such fixations define a neighborhood $\mathcal{N}_k(x)$. As in VNS, we start with a random solution x' from $\mathcal{N}_1(x)$. But, instead of performing local search in the whole solution space S with x' as a starting point, we solve a one-dimensional problem in the space of the unfixed variable that has been chosen at random. We then return a new value for this variable into the solution and compute (or update) the objective function value. The other steps are the same as in VNS: if the new solution is not better than the incumbent, we set $k = k+1$, i.e., we look for improvement in the subspace where all but two variables are fixed, etc.; otherwise a move has been made and we set $k \leftarrow 1$ again. The steps of the basic VNDS are the following:

Initialization. Select the set of neighborhood structures $\mathcal{N}_k$, $k = 1, \ldots, k_{max}$, that will be used in the search; find initial solution x; choose stopping condition;

Repeat the following until the stopping condition is met:

(1) Set $k \leftarrow 1$; (2) Until $k = k_{max}$, repeat the following steps:

(a) *Shaking.* Generate a point x' at random from the kth neighborhood of x ($x' \in \mathcal{N}_k(x)$); in other words, let y be a set of k solution attributes present in x' but not in x ($y = x' \setminus x$).

(b) *Local search.* Find the local optimum in the space of y either by inspection or by some heuristic; denote the best solution found with y' and with x'' the corresponding solution in the whole space S ($x'' = (x' \setminus y) \cup y'$);

(c) *Move or not.* If the solution thus obtained is better than the incumbent, move there ($x \leftarrow x''$), and continue the search with $\mathcal{N}_1$ ($k \leftarrow 1$); otherwise, set $k \leftarrow k+1$;

Figure 30.5 Steps of the basic VNDS.

Note that the only difference between the basic VNS and VNDS is in step 2b: instead of applying some local search method in the whole solution space $\mathcal{S}$ (starting from $x' \in \mathcal{N}_k(x)$), in VNDS we solve at each iteration a subproblem in some subspace $V_k \subseteq \mathcal{N}_k(x)$ with $x' \in V_k$. If a given local search heuristic is used for solving this subproblem (in step 2b of VNDS), then VNDS uses a single parameter, k_{max}. However, we can use some better heuristic, such as VNS, and then an additional problem specific parameter, say b, can be considered. Its aim is to strike a balance between the number of subproblems solved and the desired quality of the solution of each subproblem. Parameter b could represent, e.g., maximum CPU time allowed for solving each subproblem, or maximum size of the subproblem we are willing to solve. If the CPU time or the size of the subproblem exceeds its limit, we set $k \leftarrow 1$, i.e., we continue the decomposition by solving smaller subproblems. In this case we have in fact a recursive, two level, VNS heuristic. Note that it may be worthwhile to consider neighborhoods built from close smaller ones if there is a significant proximity relation between them. This happens in the examples next discussed.

30.3.1 VNDS for the p-Median Problem

The p-median problem has been defined in Section 30.2.4. Here we describe an application of VNDS to PM [43]. The algorithm for solving large PM is outlined in Figure 30.6.

	Objective values				CPU time			% Error		
p	VNS	FI	RVNS	VNDS	FI	RVNS	VNDS	FI	RVNS	VNDS
50	507809.5	510330.2	510216.4	507655.2	612.9	60.7	311.1	0.50	0.47	-0.03
100	354488.7	356005.1	356666.3	353255.2	1040.7	132.4	885.8	0.43	0.61	-0.35
150	281911.9	284159.0	283024.6	281772.1	1459.2	128.5	1432.4	0.80	0.39	-0.05
200	239086.4	240646.2	241355.6	238623.0	1943.6	107.6	1796.8	0.65	0.95	-0.19
250	209718.0	210612.9	210727.7	209343.3	2395.6	150.3	2189.7	0.43	0.48	-0.18
300	188142.3	189467.5	188709.3	187807.1	2583.4	130.6	1471.7	0.70	0.30	-0.18
350	171726.8	172668.5	172388.5	171009.3	2804.3	153.1	2270.3	0.55	0.39	-0.42
400	157910.1	158549.5	158805.0	157079.7	4083.3	158.7	3670.9	0.40	0.57	-0.53
450	146087.8	146727.2	147062.0	145449.0	4223.8	179.5	1652.7	0.44	0.67	-0.44
500	136081.7	136680.5	136665.0	135468.0	4649.3	209.7	2599.8	0.44	0.43	-0.45
Average					2579.6	141.1	1828.12	0.53	0.53	-0.28

Table 30.3 Results for PM problem with 3038 users; Methods: VNS - Basic VNS; FI - Fast Interchange; RVNS - Reduced VNS; VNDS - VN Decomposition search.

Results for a 3038-user problem [68] are given in Table 30.3. It is solved by the following heuristics: VNS - Basic VNS; FI - Fast Interchange; RVNS - Reduced VNS and VNDS - Variable neighborhood decomposition search. The CPU time FI spends in the search (t) is multiplied by five to get a stopping condition for basic VNS. We also use t as maximum time allowed for VNDS; the parameter b for VNDS is set to 300; the stopping rule for RVNS is $vr_{max} = 1000$ iterations. In the last three columns of Table 30.3, errors relative to the solution obtained by VNS are reported. It appears that: (i) solutions of similar quality are obtained by FI and RVNS, despite the fact that this last method uses no local search (0.53% error in both cases), but RVNS is more than 18 times faster

Initialization. (1) Rank by Algorithm 4 of [38], locations for potential facilities from L, i.e., for each potential facility, find its closest, second closest, etc. potential facility, with respect to the sum of distance criterion;
(2) Neighborhood structures $\mathcal{N}_k$, $k = 1, \ldots, k_{max}$ are induced by distance as defined in (2);
(3) Find initial solution x by Reduced VNS (RVNS): in the local search, a facility is added at random and the best deletion found; RVNS stops when the number of unsuccessful attempts at improvement is greater than a given value vr_{max}; use $k_{max} = 2$ in RVNS.
(4) Set $k_{max} = p$ and choose parameter b (i.e., the size of the subproblem we are willing to solve by VNS cannot have more than b users); choose stopping conditions for both VNDS and VNS (that will be used in step 2b below);

Repeat the following until the stopping condition is met:
(1) Set $k \leftarrow 1$; (2) Until $k = k_{max}$, repeat the following steps:
(a) Choose at random one facility from the solution x and take its $k-1$ closest medians among those that belong to the solution x. (Note that the ranking of facilities is done in the Initialization step); denote the set of k facilities with y; (note also that now $x' \in \mathcal{N}_k(x)$ can be obtained by replacing y with any other subset of k facilities $y' \subseteq L$ not being in the solution).
(b) Denote with U_k the set of users whose closest facilities belong to y. If $|U_k| \leq b$, use VNS to solve the k-median subproblem, with U_k as set of users; if $|U_k| > b$, solve the k-median problem by RVNS; denote the solution obtained with y'; replace y by y' in x' to get the new solution $x'' \in \mathcal{N}_k(x)$ ($x'' = (x' \setminus y) \cup y'$);
(c) If the solution thus obtained is better than the incumbent, move there ($x \leftarrow x''$), and continue the search with $\mathcal{N}_1$ ($k \leftarrow 1$); otherwise, set $k \leftarrow k + 1$;

Figure 30.6 VNDS for solving large PM.

then FI; (ii) VNDS outperforms FI within similar CPU times; (iii) VNDS is 0.28% better on average than basic VNS, using five times less CPU times.

	Objective values			CPU time			% Error		
p	FI	RVNS	VNDS	FI	RVNS	VNDS	FI	RVNS	VNDS
50	508786.1	508168.6	506392.5	635.9	61.0	322.2	0.47	0.35	0.00
100	354699.9	355068.2	352204.2	1234.4	157.3	1082.0	0.71	0.81	0.00
150	282809.2	281712.3	280733.1	1868.6	167.9	1403.0	0.74	0.35	0.00
200	238844.9	239434.1	237220.9	2792.8	165.8	2791.9	0.68	0.93	0.00
250	208446.4	208548.4	207383.4	3417.1	229.2	3019.2	0.51	0.56	0.00
300	186899.3	186319.0	185668.7	3598.9	192.9	3043.3	0.66	0.35	0.00
350	170461.2	170049.9	169056.9	3885.1	223.6	2815.4	0.83	0.59	0.00
400	156547.1	156931.4	155467.4	4463.3	180.3	4164.4	0.69	0.94	0.00
450	144994.1	145228.4	143961.8	4603.3	200.7	2276.2	0.72	0.88	0.00
500	135205.0	135027.9	133993.6	4954.4	237.6	3501.0	0.90	0.77	0.00
	Average			3145.4	181.6	2441.9	0.69	0.65	0.00

Table 30.4 Results for MW problem with 3038 users.

30.3.2 Multisource Weber Problem

The similarity between PM and MW has been discussed above. It is exploited in [40] where a solution of PM is first obtained and then single facility location problems (Weber problems) are solved for each set of users. For large instances of MW we use the same idea but solve PM by VNDS. Results are given in Table 30.4 for the same data as in Table 30.3 and similar conclusions to those for PM hold.

30.3.3 Minimum Sum-of-Squares Clustering Problem

p	Objective values			CPU time			% Error		
	FI	RVNS	VNDS	FI	RVNS	VNDS	FI	RVNS	VNDS
50	100156520	99284944	99020984	674.7	74.3	220.5	1.15	0.27	0.00
100	49243836	48803476	48027868	1132.3	152.8	1071.7	2.53	1.61	0.00
150	31485328	31479952	30963546	1675.9	153.0	1569.5	1.69	1.67	0.00
200	22663716	22419824	22116072	2124.2	159.9	1310.6	2.48	1.37	0.00
250	17117302	17047274	16844040	2954.3	182.1	2078.4	1.62	1.21	0.00
300	13665038	13622411	13497677	3151.0	229.3	1285.3	1.24	0.92	0.00
350	11356726	11306206	11169431	3760.3	230.9	2170.0	1.68	1.22	0.00
400	9662759	9675598	9532412	3446.1	165.0	2666.7	1.37	1.50	0.00
450	8355880	8382812	8240386	4151.7	242.6	336.1	1.40	1.73	0.00
500	7308130	7304776	7235066	4059.6	204.4	997.0	1.01	0.96	0.00
Average				2713.0	179.4	1370.6	1.62	1.25	0.00

Table 30.5 Results for MSSC with 3038 entities.

The same approach applies also for solving large MSSC. A matrix D for PM is obtained by computing squared distances between each pair of entities from X. Then clusters are found by VNDS (for the p-median problem) and their centroids and objective function value evaluated. A VNS phase follows. Results are given in Table 30.5 again for the same data as in Table 30.3. Conclusions are similar to those obtained above.

30.4 VNS FOR EXTREMAL GRAPHS

Computers can help graph theorists in many other ways than performing numerical calculations. Among such ways are: (i) computer assisted theorem proving (e.g. [3, 4, 5, 70]); (ii) automated or semi-automated theorem proving (e.g. [15]); (iii) automated or interactive generation, analysis and description of graphs (e.g. [10, 11, 16]); (iv) automated generation, checking and selection of conjectures (e.g.[23, 24, 25, 26, 27]).

In [14], conjectures in graph theory are viewed as combinatorial optimization problems on an infinite family of graphs (of which only those of moderate size will be explored). Then VNS can be applied. This leads to an automated system, called AutoGraphiX (or AGX for short) whose aim is to suggest, disprove or give hints on how to prove conjectures in graph theory.

30.4.1 AGX

Assume an objective function $f(G)$ depending on graph invariants $i_1(G)$, $i_2(G)$, $\ldots i_l(G)$ (e.g. order, size, radius, diameter, chromatic number, maximum or minimum degree...) is given. Let $\mathcal{G}_n$ denote the family of all graphs G with n

vertices (a parameter). The problem considered is then

$$\min_{G \in \mathcal{G}_n} (\max) f(G) \tag{30.5}$$

to which we apply VNS specialized to extremal graphs problems.

AGX comprises a VND component with many neighborhoods $\mathcal{N}_1(G)$, $\mathcal{N}_2(G)$ .... In the present version there are ten of them, defined by the following transformations on G: *deletion* of an edge; *addition* of an edge between non-adjacent vertices; *move* of an edge, i.e., deletion followed by addition, but not in the same position (these three neighborhoods are used in most applications); *detour:* removal of an edge and addition of two edges between endpoints of the deleted one and a vertex not adjacent to either of these endpoints; *short cut*, i.e., the reverse operation than detour; *2 opt:* remove two non adjacent edges and add two different non adjacent edges connecting the endpoints of the removed ones; *insert pending vertex:* remove the edge of a pending vertex, add edges between it and two adjacent vertices and remove the edge between them; *add pending vertex:* add an edge from one vertex to a new one; *delete vertex* of bounded degree and all edges adjacent to it.

It is up to the user to apply all transformations in turn or to select a subset of them, e.g. the less time-consuming ones. VNS itself uses a set of nested neighborhoods $\mathcal{N}'_1(G)$, $\mathcal{N}'_2(G), \ldots$ defined by the cardinality of the symmetric difference between the edge sets of two graphs $G = (V, E)$ and $G' = (V', E')$:

$$\rho(G, G') = |(E \setminus E') \cup (E' \setminus E)| \tag{30.6}$$

i.e., $\rho(G, G') = k$ if there are exactly k pairs of vertices adjacent in G and not in G' or adjacent in G' and not in G. Then

$$G' \in \mathcal{N}_k(G) \Longleftrightarrow \rho(G, G') = k. \tag{30.7}$$

In addition to these components, which follow the basic schemes of Figures 30.2 and 30.1, AGX comprises a routine for graph analysis which performs the following: (i) recognition of the class to which the graph G obtained belongs (if it is one of the series of well known classes such as path, star, tree, complete or bipartite graph, ...) (ii) computation of various invariants for G, (iii) visualization of G on the screen using X-Windows, with interactive modifications of position of vertices, (iv) subroutines for interactive modifications of G by deletion or addition of edges or vertices.

30.4.2 Uses of AGX

Variable Neighborhood Search can be used in graph theory for several purposes: (a) *Find a graph satisfying given constraints.* Let $i_1(G), i_2(G), \ldots, i_l(G)$ denote l invariants of G and $p_1, p_2, \ldots p_l$ values given to them. Consider the problem

$$\min_{G \in \mathcal{G}_n} f(G) = \sum_{k=1}^{l} |i_k(G) - p_k|. \tag{30.8}$$

Any graph such that $f(G) = 0$ satisfies these constraints. Note that constraints involving formulae on several invariants can be treated in a similar way. Expression (30.8) shows constrained problems can be reduced to unconstrained ones. An example of problem (30.8), posed by Seymour [71] and solved by AGX [14], is to find a cubic graph of diameter 3 and largest possible order.

(b) *Find optimal or near optimal values for an invariant subject to constraints.* Let $i_0(G)$ denote the objective function invariant and assume constraints expressed as above. Consider the problem

$$\min_{G \in \mathcal{G}_n} f(G) = i_0(G) + M \sum_{k=1}^{l} |i_k(G) - p_k| \tag{30.9}$$

where M is a constant sufficiently large to ensure that for any pair of graphs $G,\ G' \in \mathcal{G}_n$ such that $\sum_{k=1}^{l} |i_k(G) - p_k| = 0$ and $\sum_{k=1}^{l} |i_k(G') - p_k| > 0$, $f(G) < f(G')$. Maximum values are tackled in a similar way.

An example of problem (30.9), with a known answer, posed by Cvetković [17] as a test for AGX is to find trees with minimum or maximum first eigenvalue and $n = 2$ to 20 vertices. AGX recognized easily that they are paths and stars, respectively [14].

Note that some invariants take integer values and remain constant on large plateaus when G is modified a little at a time, i.e., replaced sequentially by a graph in the neighborhood $\mathcal{N}(G)$. It is then convenient to modify the objective function by adding a secondary objective, with a coefficient sufficiently small not to change the ordering of graphs according to the first objective, and influencing the transformations in the right direction.

(c) *Refute a conjecture.* Consider a conjecture $h(G) \leq g(G)$ where $h(G)$ and $g(G)$ are formulae depending on one or more invariants of G.
Consider the problem

$$\min_{G \in \mathcal{G}_n} f(G) = g(G) - h(G). \tag{30.10}$$

If a graph G for which $f(G) < 0$ is found, the conjecture is refuted.

(d) *Suggest a conjecture (or sharpen one).* This can be done in various ways, which usually use parameterization on n or other invariants $i_1(G), i_2(G)$, $i_3(G) \ldots i_l(G)$. For instance consider

$$\min_{G \in \mathcal{G}_n} f(G) = i_2(G) - i_1(G). \tag{30.11}$$

If no graph G with $f(G) < 0$ is found this suggests $i_1(G) \leq i_2(G)$. If the extremal graphs found belong to a recognizable class, it may be possible to deduce a more precise inequality in $i_1(G)$, $i_2(G)$ and n. An example, is an inequality on the Randić index $Ra(G)$ (defined in [66]), the chromatic number $\chi(G)$ and the order n, obtained by observing that extremal graphs are cliques with pendant edges appended to one vertex [14].

(e) *Suggest a proof.* The way the extremal graphs are obtained, e.g. what transformations of G are used, may suggest strategies to prove conjectures for all graphs or for some classes of them. An example, is the proof that for all trees which are not even paths, the radius does not exceed the Randić index [14].

The *Graffiti* system of Fajtlowicz [23]-[27] has been used to generate hundreds of conjectures in graph theory which have been studied by many mathematicians and some of them proved or disproved. The following conjecture is a reformulation of one of them (i.e., # 834):

$$\delta.\bar{l}(G) \leq n \tag{30.12}$$

where δ denotes the minimum degree and $\bar{l}$ the average distance between pairs of distinct vertices of G. This conjecture has been refuted by AGX [14] in 153 sec of CPU time using an edge addition, 15 edge deletions, a split, two mergings, 6 moves, 3 2-Opts, another merging and split after VNS with neighborhoods of size up to 7, and a final edge deletion after another application of VNS with neighborhoods of size up to 7. Two other conjectures of *Graffiti* are also refuted in [14].

30.5 CONCLUSIONS

When evaluating metaheuristics, it appears that several, possibly conflicting, criteria should be taken into account. Desirable properties of metaheuristics include

(i) *Simplicity and coherence*: the metaheuristic should be based on a simple and clear principle, which should be widely applicable. Moreover, the various steps should follow naturally from the principle;

(ii) *Efficiency*: heuristics (obtained from the principle) for particular problem should be efficient, i.e. provide optimal or near-optimal solutions for all or at least most realistic instances. Preferably, the heuristics should solve optimally most problems of benchmarks, when available;

(iii) *Effectiveness*: heuristics for particular problems should provide good, perhaps optimal or near-optimal, solutions in moderate CPU time;

(iv) *Robustness*: heuristics for various problems should be effective and efficient, and for each of these problems should give good solutions for a variety of instances, i.e., not just be fine-tuned to some training sets and less good elsewhere;

(v) *User-friendliness*: heuristics should be well-defined, easy to understand and, most important, easy to use. This implies they should have few, ideally no, parameters.

While VNS is still in its infancy, results for a variety of combinatorial optimization problems described in this paper allow a first assessment. Clearly, VNS is based on a simple principle, i.e., systematic change of neighborhood within the search. This principle, explained in the basic schemes of Figures 30.1 and 30.2 is largely applicable and often in an easy way when a local search heuristic is available. (This simplicity implies that isolated steps of several heuristics or metaheuristics are particular cases or come close to some steps of VNS; their derivation often follows, when it is explicit, from different principles). Experiments with benchmarks of large p-median and multi-source Weber problems shows VNS is very efficient: it solves many instances exactly and otherwise gives solutions very close to the optimum. For several other problems VNS appears to perform better than recent heuristics developed by leading researchers. Moreover, VNS is effective: the best solutions are obtained in moderate computing time. This is particularly true when using VNDS. For instance, time to find the best solution of very large instances of p-median problem is less than that for a single descent with the fast interchange heuristic. Computational results on simulated and real data show VNS is robust i.e., about equally efficient and effective for both types of problems. Finally, VNS is very user-friendly. The basic steps are easy to understand and apply and versions without parameters (except for total CPU time allocated) are not hard to obtain, while remaining efficient and effective.

Developments of VNS have up to now focused on application of its general principle to combinatorial optimization problems. In this line, further developments of decomposition appear desirable, as the size of problem instances users would like to see solved increases with availability of ever larger data sets. Global optimization problems, in continuous variables, should be considered also (work on applying VNS to such problems is ongoing).

Moreover VNS could benefit from the numerous results obtained in the development of other metaheuristics, prominent among which is TS [31, 32, 33, 34, 35]. The basic principle of this metaheuristic, i.e. to exploit memory of the search through a series of increasingly sophisticated rules for forbidding moves (i.e., tabus) or for removing interdictions (i.e. aspiration levels) or for favoring moves, is almost orthogonal to the principle of VNS, i.e., to focus on efficient descent through selection of neighborhoods and systematic (but partly randomized) search for deeper valleys. Such a situation suggests hybrid heuristics could benefit from both principles and their study appears particularly promising.

New uses of heuristics appear to be very fruitful. This is true for application in graph theory, where VNS and AGX could be applied to many types of questions and to the study of numerous conjectures. It holds also for conjunction of VNS with advanced combinatorial optimization tools such as column-generation (with branch-and-bound). Initial success in solving exactly much larger instances of multi-source Weber, p-median and minimum sum-of-squares problem than done before [20, 21] suggest to try out this approach for many other combinatorial optimization problems.

Finally, an important and largely open theoretical problem is: *why* do metaheuristics perform as well as they do? An approach to this difficult question is to study the form of the objective function and particularly the distribution of local optima and shape of valleys. VNS, which does not alter the form of the valleys it explores, could be useful in this respect too.

Acknowledgment. Research of the authors supported by ONR grant N00014-95-1-0917, NSERC grant OGPOO 39682 and FCAR grant 32EQ 1048. We thank Jack Brimberg, Gilles Caporossi, Fred Glover, Anderson Parreira and Dionisio Perez-Brito for discussions of VNS and Dragoš Cvetković, Paul Seymour and Bjarne Toft for comments on and questions to AGX.

References

[1] K.H. Al-Sultan. A Tabu Search Approach to the Clustering Problem. *Pattern Recog.* **28**, 1443–1451 (1995).

[2] M.R. Anderberg. *Cluster Analysis for Application.* Academic Press, New York (1973).

[3] K. Appel and W. Haken. Every Planar Map is Four Colorable. Part I. Discharging. *Illinois J. Math.* **21**, 429–490 (1977).

[4] K. Appel and W. Haken. Every Planar Map is Four Colorable. Part II. Reducibility. *Illinois J. Math.* **21**, 491–567 (1977).

[5] K. Appel and W. Haken. Every Planar Map is Four Colorable. *A.M.S. Contemp. Math.* **98**, 1–741 (1989).

[6] G.P. Babu and M.N. Murty. A Near-Optimal Initial Seed Value Selection in k-Means Algorithm Using Genetic Algorithm. *Pattern Recog. Lett.* **14**, 763–769 (1993).

[7] J.E. Beasley. A Note on Solving Large p-Median Problems. *European J. Oper. Res.* **21**, 270–273 (1985).

[8] J.L. Bentley. Fast Algorithms for Geometric Traveling Salesman Problem. *ORSA J. Comput.* **4**, 387–411 (1992).

[9] K.D. Boese, A.B. Kahng and S. Muddu. A New Adaptive Multi-Start Technique for Combinatorial Global Optimizations. *Oper. Res. Lett.* **16**, 101–113 (1994).

[10] R.C. Brigham and R.D. Dutton. Ingrid: A Software Tool for Extremal Graph Research Theory. *Congressus Numerantium* **39**, 337–352 (1983).

[11] R.C. Brigham, R.D. Dutton and F. Gomez. Ingrid: A Graph Invariant Manipulator. *J. Symbolic Comput.* **7**, 163–177 (1989).

[12] J. Brimberg and N. Mladenović. A Variable Neighborhood Algorithm for Solving the Continuous Location-Allocation Problem. *Stud. in Locat. Analysis* **10**, 1–12 (1996).

[13] J. Brimberg, P. Hansen, N. Mladenović and É. Taillard. Improvements and Comparison of Heuristics for Solving the Multisource Weber Problem. *Les Cahiers du GERAD G-97-37*, Montréal, Canada (1997).

[14] G. Caporossi and P. Hansen. Variable Neighborhood Search for Extremal Graphs. *Les Cahiers du GERAD G-97-41*, Montréal, Canada (1997) (to appear in *Discrete Mathematics*).

[15] D. Cvetković and I. Pevac. Man-machine theorem proving in graph theory. *Artificial Intelligence* **35**, 1–23 (1988).

[16] D. Cvetković and S. Simić. Graph Theoretical Results Obtained by the Support of the Expert System GRAPH. *Bulletin de l'Académie Serbe des Sciences et des Arts T CVII*, 14–41 (1994).

[17] D. Cvetković, April 1997. Email communication.

[18] S. Cook. The Complexity of Theorem-Proving Procedures. in *Proc. of the 3rd annual ACM Symp. on Theory of Computing*, 151–158 (1971).

[19] L. Cooper. Location - Allocation Problems. *Oper. Res.* **11**, 331–343 (1963).

[20] O. du Merle, D. Villeneuve, J. Desrosiers and P. Hansen. Stabilisation dans le Cadre de la Génération de Colonnes. *Les Cahiers du GERAD G-97-08*, Montréal, Canada (1997).

[21] O. du Merle, P. Hansen, B. Jaumard and N. Mladenović. An Interior Point Algorithm for Minimum Sum-of-Squares Clustering. *Les Cahiers du GERAD G-97-53*, Montréal, Canada (1997).

[22] D. Erlenkotter. A Dual-Based Procedure for Uncapacitated Facility Location. *Oper. Res.* **26**, 992–1009 (1978).

[23] S. Fajtlowicz. On Conjectures of Graffiti. *Discr. Math.* **72**, 113–118 (1987).

[24] S. Fajtlowicz. On Conjectures of Graffiti - II. *Congressus Numerantium* **60**, 187–197 (1987).

[25] S. Fajtlowicz. On Conjectures of Graffiti - III. *Congressus Numerantium* **66**, 23–32 (1988).

[26] S. Fajtlowicz. On Conjectures of Graffiti - IV. *Congressus Numerantium* **70**, 231–240 (1990).

[27] S. Fajtlowicz. On Conjectures of Graffiti - V. *7th Intern. Quadrennial Conference on Graph Theory* **1**, 367–376 (1995).

[28] T. Feo and M. Resende. Greedy Randomized Adaptive Search. *J. Global Optim.* **6**, 109–133 (1995).

[29] M. Gendreau, A. Hertz and G. Laporte. New Insertion and Postoptimization Procedures for the Traveling Salesman Problem. *Oper. Res.* **40**, 1086–1094 (1992).

[30] M. Gendreau, A. Hertz and G. Laporte, The Traveling Salesman Problem with Back-Hauls, *Computers Oper. Res.* **23**, 501–508 (1996).

[31] F. Glover. Tabu Search - Part I, *ORSA J. Comput.* **1**, 190–206 (1989).

[32] F. Glover. Tabu Search - Part II, *ORSA J. Comput.* **2**, 4–32 (1990).

[33] F. Glover and M. Laguna. Tabu Search. In C. Reeves, editor, *Modern Heuristic Techniques for Combinatorial Optimization*, pages 70–150, Blackwell, Oxford (1993).

[34] F. Glover. Tabu Thresholding: Improved Search by Nonmonotonic Trajectories. *ORSA J. Comput.* **7**, 426–442 (1995).

[35] F. Glover and M. Laguna. *Tabu Search.* Kluwer, Boston (1997).

[36] A.D. Gordon. *Classification: Methods for the Exploratory Analysis of Multivariate Data*, Chapman and Hall, New York (1981).

[37] P. Hansen and B. Jaumard. Algorithms for the Maximum Satisfiability Problem. *Computing* **44**, 279–303 (1990).

[38] P. Hansen, B. Jaumard and N. Mladenović. How to Choose K Entities Among N. In I.J. Cox, P. Hansen and B. Julesz editors, *Partitioning Data Sets*, pages 105-116, DIMACS Series in Discrete Mathematics and Theoretical Computer Science **19**, American Mathematical Society, Providence, Rhode Island (1993).

[39] P. Hansen, B. Jaumard, N. Mladenović and A. Parreira. Variable Neighborhood Search for Weighted Maximum Satisfiability (in preparation).

[40] P. Hansen, N. Mladenović and É. Taillard. Heuristic Solution of the Multisource Weber Problem as a p-Median Problem. *Les Cahiers du GERAD*, G-96-10, Montréal, Canada (1996) (to appear in *Oper. Res. Lett.*).

[41] P. Hansen and N. Mladenović. Variable Neighborhood Search for the p-Median, *Les Cahiers du GERAD* G-97-39, Montréal, Canada, (1997) (to appear in *Location Science*).

[42] P. Hansen and N. Mladenović. Variable Neighborhood Search for Minimum Sum-of-Squares Clustering, *Les Cahiers du GERAD* (1998), Montréal, Canada, (forthcoming).

[43] P. Hansen, N. Mladenović and D. Perez-Brito. Variable Neighborhood Decomposition Search. *Les Cahiers du GERAD*, Montréal, Canada, (1998) (forthcoming).

[44] J.H. Holland. *Adaptation in Natural and Artificial Systems.* The University of Michigan Press, Ann Arbor, Michigan (1975).

[45] T.S. Hu, V. Klee and D. Larman. Optimization of globally convex functions. *SIAM J. Control Optim.* **27** 1026–1047 (1989).

[46] R.C. Jancey. Multidimensional Group Analysis. *Australian J. Botany* **14**, 127–130 (1966).

[47] D.S. Johnson and L.A. McGeoch, The Traveling Salesman Problem: a Case Study in Local Optimization. In E.H.L. Aarts and J.K. Lenstra, editors, *Local Search in Combinatorial Optimization*, pages 215–310, Wiley, London (1997).

[48] D.S. Johnson and M. Trick, eds., *Cliques, Coloring and Satisfiability: Second DIMACS Implementation Challenge*, DIMACS Series in Discrete Mathematics and Theoretical Computer Science **26** American Mathematical Society, Providence, Rhode Island (1996).

[49] O. Kariv and S.L. Hakimi. An Algorithmic Approach to Network Location Problems; part 2. The p-medians. *SIAM J. App. Math.* **37**, 539–560 (1969).

[50] S. Kirkpatrick, C.D. Gelatt and M. Vecchi. Optimization by Simulated Annealing. *Science* **220**, 671–680 (1983).

[51] R.W. Klein and R.C. Dubes. Experiments in Projection and Clustering by Simulated Annealing. *Pattern. Recogn.* **22**, 213–220 (1989).

[52] R.E. Kuenne, and R.M. Soland. Exact and Approximate Solutions to the Multisource Weber Problem. *Math. Program.* **3**, 193–209 (1972).

[53] S. Lin. Computer Solutions of the Traveling Salesman Problem. *Bell Syst. Tech. J.* **44**, 2245–2269 (1965).

[54] S. Lin and B.W. Kernighan. An Effective Heuristic Algorithm for the Traveling Salesman Problem. *Oper. Res.* **21**, 498–516 (1973).

[55] R.F. Love and H. Juel. Properties and Solution Methods for Large Location - Allocation Problems. *J. Oper. Res. Soc.* **33**, 443–452 (1982).

[56] J.B. MacQueen. Some Methods for Classification and Analysis of Multivariate Observations, *Proc. of the* 5th *Berkeley Symp. on Mathem. Statist. and Probab.* **1**, 281–297 (1967).

[57] K.-T. Mak and A. Morton. Distances Between Traveling Salesman Tours, *Discr. App. Math.* **58**, 281–291 (1995).

[58] F.E. Maranzana. On the Location of Supply Points to Minimize Transportation Costs. *Oper. Res. Quart.* **12**, 138–139 (1964).

[59] P. Mirchandani and R. Francis, eds., *Discrete Location Theory*, Wiley-Interscience (1990).

[60] N. Mladenović. A Variable Neighborhood Algorithm - a New Metaheuristic for Combinatorial Optimization. Abstracts of papers presented at Optimization Days, Montréal, p. 112 (1995).

[61] N. Mladenović, J.P. Moreno, and J. Moreno-Vega. A Chain-Interchange Heuristic Method, *Yugoslav J. Oper. Res.* **6**(1), 41–54 (1996).

[62] N. Mladenović and P. Hansen. Variable Neighborhood Search. *Computers Oper. Res.* **24**, 1097–1100 (1997).

[63] I.H. Osman and N. Christofides. Capacitated Clustering Problems by Hybrid Simulated Annealing and Tabu Search. *Int. Trans. Oper. Res.* **1**(3), 317–336 (1994).

[64] I.H. Osman and G. Laporte, Metaheuristics: A bibliography. *Ann. Oper. Res.* **63**, 513–628 (1996).

[65] C.H. Papadimitriou and K. Steiglitz. *Combinatorial Optimization, Algorithms and Complexity.* Prentice Hall, New Jersey (1982).

[66] M. Randić. On Characterization of Molecular Branching. *J. Americ. Chemical Soc.* **97**, 6609–6615 (1975).

[67] C. Reeves (eds.). *Modern Heuristic Techniques for Combinatorial Problems.* Blackwell, Oxford (1993).

[68] G. Reinelt. TSLIB-A Traveling Salesman Library. *ORSA J. Comput.* **3**, 376–384 (1991).

[69] M.G.C. Resende, L.S. Pitsoulis and P.M. Pardalos. Approximate Solution of Weighted MAX-SAT Problems Using GRASP. in Dingzhu Su, Jun Gu and Panos M. Pardalos (editors), *Satisfiability Problem: Theory and Applications*, DIMACS Series in Discrete Mathematics and Theoritical Computer Science **35**, American Mathematical Society, Providence, Rhode Island (1997).

[70] N. Robertson, D. Sanders, P. Seymour and R. Thomas. The Four-Color Theorem. *J. Combin. Theory, Ser. B* **70**, 2–44 (1997).

[71] P. Seymour, May 1997. Private Communication at the Workshop on Graph Coloring and Applications, Montréal (1997).

[72] H. Späth, *Cluster Dissection and Analysis (Theory, Fortran Programs, Examples)*, Ellis Horwood, Chichester, (1985).

[73] M.B. Teitz and P. Bart. Heuristic Methods for Estimating the Generalized Vertex Median of a Weighted Graph. *Oper. Res.* **16**(5), 955–961 (1968).

[74] S. Voß. A Reverse Elimination Approach for the p-Median Problem. *Studies in Locational Analysis* **8**, 49–58 (1996).

[75] R. Whitaker. A Fast Algorithm for the Greedy Interchange for Large-Scale Clustering and Median Location Problems. *INFOR* **21**, 95–108 (1983).

31 A VARIABLE DEPTH SEARCH ALGORITHM FOR THE GENERALIZED ASSIGNMENT PROBLEM

Mutsunori Yagiura, Takashi Yamaguchi and Toshihide Ibaraki

Department of Applied Mathematics and Physics,
Graduate School of Informatics, Kyoto University, Sakyo-ku, Kyoto 606-8501, Japan.
yagiura@kuamp.kyoto-u.ac.jp, ibaraki@kuamp.kyoto-u.ac.jp

Abstract: A *variable depth search* procedure (abbreviated as VDS) is a generalization of the local search method, which was first successfully applied by Lin and Kernighan to the traveling salesman problem and the graph partitioning problem. The main idea is to adaptively change the size of a neighborhood so that it can effectively traverse a larger search space while keeping the amount of computational time reasonable. In this paper, we propose a heuristic algorithm based on VDS for the generalized assignment problem, which is one of the representative combinatorial optimization problems known to be NP-hard. To the authors' knowledge, most of the previously proposed algorithms (with some exceptions) conduct the search within the feasible region; however, there are instances for which the search within the feasible region is not advantageous because the feasible region is very small or is combinatorially complicated to search. Therefore, we allow our algorithm to search within the infeasible region as well. We also incorporate an adaptive use of two different neighborhoods, shift and swap, within the sequence of neighborhood moves in VDS. Computational experiments exhibit good prospects of the proposed algorithm.

31.1 INTRODUCTION

To deal with computationally hard problems, approximate algorithms are used to provide reasonably good solutions in a practical amount of time. A *variable depth search* procedure (abbreviated as VDS) is a generalization of the local search method. The VDS was first successfully applied by Lin and Kernighan to the traveling salesman problem and the graph partitioning problem [4, 6]. The main idea is to adaptively change the size of neighborhood so that the algorithm can effectively traverse a larger search space while keeping the amount of computational time reasonable. In VDS, the neighborhood is usually defined to be the set of solutions obtained by a sequence of $\bar{d}$ simple neighborhood moves, where $\bar{d}$ is a parameter adaptively changed in the algorithm. It is important to include problem-specific characteristics into the VDS to make the adaptive change successful.

In this paper, we propose a heuristic algorithm based on VDS for the generalized assignment problem (abbreviated as GAP), which is one of the representative combinatorial optimization problems known to be NP-hard. In GAP, given n jobs and m agents, we are asked to determine a minimum cost assignment such that every job is assigned to exactly one agent and the resource constraint for each agent is satisfied. Among various heuristic algorithms developed for GAP are: a combination of the greedy method and the local search by Martello and Toth [7, 8]; a tabu search and simulated annealing by Osman [10]; a genetic algorithm by Chu and Beasley [2]; a VDS by Racer and Amini [11]; a tabu search based on ejection chain approach by Laguna et al. [5] (which is proposed for a generalization of GAP); and so on.

To the authors' knowledge, many of the previous algorithms conduct the search only within the feasible region. However, we often encounter instances for which the search within the feasible region is difficult because the feasible region is very small or is combinatorially complicated to search. In fact, just finding a feasible solution for GAP is already known to be NP-hard. Therefore, we allow our algorithm to search within the infeasible region as well. It should be noted that the tabu search by Laguna et al. [5], in which a strategic oscillation approach is employed, also allows the search within the infeasible region. It is noted as well that the VDS by Racer and Amini [11] also allow the search within the infeasible region by introducing a dummy agent with large cost; but in their case the search never returns to the infeasible region once a feasible solution is found.

We incorporate an adaptive use of two different neighborhoods, shift and swap. The shift neighborhood is defined to be the set of solutions obtainable by changing the assignment of one job, while the swap neighborhood is defined to be the set of solutions obtainable by exchanging the assignments of two jobs. We propose using two neighborhoods alternately and adaptively within the sequence of neighborhood moves in VDS.

Computational experiments have been conducted on benchmark problem instances with up to $n = 200$. The results indicate that the proposed algorithm produces solutions of comparable or better quality in reasonable amount of time for many of the tested problem instances, compared to the existing heuristic algorithms as cited above.

31.2 GENERALIZED ASSIGNMENT PROBLEM

Given n jobs $J = \{1, \dots, n\}$ and m agents $I = \{1, \dots, m\}$, we are asked to determine a minimum cost assignment such that every job is assigned to exactly one agent and the resource constraint for each agent is satisfied. We are also given the cost c_{ij} and the resource requirement a_{ij} if job j is assigned to agent i, and the amount b_i of the resource available to agent i. An assignment is a mapping σ: $J \to I$, where $\sigma(j) = i$ means that job j is assigned to agent i. The problem is formulated as follows:

$$\begin{array}{ll} \text{minimize} & cost(\sigma) := \sum_{j \in J} c_{\sigma(j),j} \\ \text{subject to} & \sum_{j \in J, \sigma(j)=i} a_{ij} \leq b_i, \forall i \in I. \end{array}$$

GAP is known to be NP-hard (e.g., [12]). It can be shown that just finding a feasible solution for GAP is NP-hard, since the partition problem [3] can be reduced to this problem with $m = 2$.

31.3 LOCAL SEARCH

VDS is a generalization of local search. The local search starts from an initial solution σ and repeats replacing σ with a better solution from its *neighborhood* $N(\sigma)$ until no better solution is found in $N(\sigma)$, where $N(\sigma)$ is a set of solutions obtained from σ by slight perturbations. A solution σ is called *locally optimal*, if no better solution exists in $N(\sigma)$. The local search is often applied to a number of randomly generated initial solutions, and the best among the obtained local optimal solutions is chosen. This is called the *random multi-start local search.*

We use two types of neighborhoods, shift neighborhood N_{shift} and swap neighborhood N_{swap}, in our algorithm, where

$$N_{shift}(\sigma) := \{\sigma' \mid \sigma' \text{ is obtained from } \sigma \text{ by changing the assignment of one job}\},$$
$$N_{swap}(\sigma) := \{\sigma' \mid \sigma' \text{ is obtained from } \sigma \text{ by exchanging the assignments of two jobs}\}.$$

As n is usually much larger than m, the size of shift neighborhood $|N_{shift}| = O(mn)$ is considered to be much smaller than that of swap $|N_{swap}| = O(n^2)$. In a shift move applied to an assignment σ, the resource requirement for the agent to which a job is added may increase greatly; while, in a swap move, resource requirements of the two relevant agents may not change much if the requirements of the two jobs are close. Note that, since the number of jobs assigned to each agent will not change if we use swap moves only, we should combine swap moves with shift moves in order to search different types of assignments. Based on these observations, we propose using two neighborhoods alternately and adaptively within the sequence of neighborhood moves in VDS.

It is often effective to allow the search to visit infeasible regions as well, in particular if the problems have tight constraints. In this case, it may be a good idea to evaluate solutions by the objective function penalized by infeasibility. Here we use the following function:

$$pcost(\sigma) := cost(\sigma) + \alpha \sum_{i \in I} p_i(\sigma), \tag{31.1}$$

where

$$p_i(\sigma) := \max\left\{0, \left(\sum_{j \in J, \sigma(j)=i} a_{ij}\right) - b_i\right\}$$

and α (> 0) is a prespecified parameter. The locally optimal solution under this function may not always be feasible; however, we can raise the probability of obtaining feasible solutions by: (1) keeping the best feasible solution obtained during the search as an incumbent solution, and (2) using sufficiently large α.

The local search, in which N is used for the neighborhood (e.g., $N = N_{shift}$ or N_{swap}), f is used to evaluate solutions (e.g., $f = cost$ or $pcost$), and an initial solution σ (not necessarily feasible) is given, is formally described as follows.

Algorithm $LS(N, f, \sigma)$

Step 1 If there is a solution $\sigma' \in N(\sigma)$ such that $f(\sigma') < f(\sigma)$, set $\sigma := \sigma'$ and return to Step 1. Otherwise go to Step 2.

Step 2 ($f(\sigma') \geq f(\sigma)$ holds for all $\sigma' \in N(\sigma)$.) Output σ and stop.

We implement Step 1 by the first admissible move strategy, in which solutions in $N(\sigma)$ are scanned according to a prespecified random order and the first improved solution is immediately accepted as the next solution.

31.4 VARIABLE DEPTH SEARCH

An important feature of our VDS is the alternate use of shift and swap neighborhoods, which will be explained later in this section. To understand this, consider first the behavior of local search in which $N_{shift} \cup N_{swap}$ is always used as the neighborhood. In this case, if the current solution σ is close to a local optimum, shift moves rarely occur, since the agent i to which a job is shifted usually becomes infeasible and its penalty $p_i(\sigma)$ increases a large amount. This means that as swap moves do not change the number of jobs assigned to each agent, the number of jobs of each agent is usually fixed at an early stage of the search and will never be changed until a locally optimal solution is reached. In other words, local search with neighborhood $N_{shift} \cup N_{swap}$ may not have enough search power even if the size $|N_{shift} \cup N_{swap}|$ is fairly large.

In order to vest diversifying power to our VDS, we systematically change the number of jobs assigned to each agent by employing the alternating use of shift and swap neighborhoods. Our algorithm forces to execute shift moves (which are usually degrading moves) as soon as a locally optimal solution is obtained using the swap neighborhood, and these two phases of local search are repeated until some stopping criterion is satisfied.

In VDS, an enlarged neighborhood is used, which is usually defined to be the set of solutions obtained by a sequence of $\bar{d}$ simple neighborhood moves, where $\bar{d}$ is a parameter adaptively changed in the algorithm. Our VDS starts from a solution $\sigma^{(0)} = \sigma_0$, which is already locally optimal with respect to N_{swap}. The first move from σ_0 to σ_1 is therefore a shift, followed by swap moves $\sigma_1 \to \sigma_2$,

$\sigma_2 \to \sigma_3, \ldots$ to a locally optimal solution $\sigma^{(1)} = \sigma_k$. This loop of a shift move and swap moves is iterated $\bar{d}$ times to generate a sequence of solutions $\sigma^{(0)}, \sigma^{(1)}, \ldots, \sigma^{(\bar{d})}$. Then the best one among $\sigma^{(0)}, \sigma^{(1)}, \ldots, \sigma^{(\bar{d})}$ is chosen as the initial solution for the next sequence of solutions. These sequences are organized into three nested loops in our VDS, as described below. Some part of the algorithm, such as the modified shift and swap neighborhoods, $N'_{shift}(\sigma)$ and $N^{dsj}_{swap}(\sigma_k, I_k)$, and the set I_k, will be explained later in more detail.

Algorithm VDS

Step 1 (initialization) Generate an initial solution $\sigma^{(0)} = \sigma_0$ (the details of the generation algorithm will be described later), set $d := 0$, $k := 0$ and $T := \emptyset$. Let σ' be the best feasible solution found during this stage of generating σ_0. Set $\sigma_{best} := \sigma'$ and $best := cost(\sigma')$; if no feasible solution is found, set $best := \infty$.

Step 2 (variable depth search) Repeat the following two steps, Step 2-a and Step 2-b, to generate a sequence of solutions $\sigma^{(0)}, \sigma^{(1)}, \ldots, \sigma^{(\bar{d})}$ until stopping criterion is fulfilled.

- **a** (shift move): If $d = 0$ and $N'_{shift}(\sigma^{(0)}) \backslash T = \emptyset$, or if $d \geq 1$ and $N'_{shift}(\sigma^{(d)}) = \emptyset$, then set $\bar{d} := d$ and exit to Step 3. Otherwise choose the solution σ with the minimum $pcost(\sigma)$ in $N'_{shift}(\sigma^{(0)}) \backslash T$ if $d = 0$, or in $N'_{shift}(\sigma^{(d)})$ if $d \geq 1$, and set $\sigma_{k+1} := \sigma$, and $k := k + 1$.
- **b** (swap move): If σ_k is feasible and $cost(\sigma_k) < best$, then set $\sigma_{best} := \sigma_k$, $best := cost(\sigma_k)$, $d^* := \bar{d} := d + 1$ and exit to Step 3. If $pcost(\sigma) \geq pcost(\sigma_k)$ for all σ in $N^{dsj}_{swap}(\sigma_k, I_k)$, then set $\sigma^{(d+1)} := \sigma_k$, $d := d + 1$ and return to Step 2-a. Otherwise choose a solution σ in $N^{dsj}_{swap}(\sigma_k, I_k)$, set $\sigma_{k+1} := \sigma$, $k := k + 1$ and return to Step 2-b. (The details of how to choose σ will be explained later.)

Step 3 (starting solution for the next variable depth search) If the incumbent solution σ_{best} has been updated during Step 2, set $\sigma^{(d^*)} := \sigma_{best}$; otherwise the solution $\sigma^{(d^*)}$ that minimizes $pcost(\sigma)$ among $\sigma^{(0)}, \sigma^{(1)}, \ldots, \sigma^{(\bar{d})}$ is chosen.

Step 4 (exit or go to the next variable depth search) If $d^* = 0$, then set $T := T \cup \{\sigma_1\}$; otherwise set $T := \emptyset$. If $N'_{shift}(\sigma^{(0)}) \backslash T = \emptyset$, go to Step 5; otherwise set $\sigma^{(0)} := \sigma_0 := \sigma^{(d^*)}$, $d := 0$, $k := 0$, and return to Step 2. (If $d^* = 0$ holds, it means that no improvement was attained in the previous variable depth search, and we try another sequence of solutions from the same $\sigma^{(0)}$. In this case, set T is augmented by σ_1 to choose a different solution for the new sequence. If $d^* \geq 1$, on the other hand, no restriction is imposed in the selection of the new sequence.)

Step 5 (halt or random restart) If Steps 1 - 4 are repeated r (prespecified integer) times, or the computational time exceeds τ (prespecified parameter) seconds, select σ_{best} and stop; otherwise return to Step 1.

Modification of the neighborhoods

In algorithm VDS, we use two neighborhoods, $N_{shift}(\sigma_k)$ and $N_{swap}(\sigma_k)$. To reduce computational effort and to improve performance, however, we modify these to $N'_{shift}(\sigma_k)$ or $N^{dsj}_{swap}(\sigma_k, I_k)$, respectively, defined as follows.

Modified shift neighborhood $N'_{shift}(\sigma_k)$: Let $J(\sigma, \sigma') := \{j \in J |\ \sigma(j) \neq \sigma'(j)\}$ be the set of jobs whose assignments were changed in the move from σ to σ', and let $J_k := J(\sigma_{k-1}, \sigma_k)$ $(k \geq 1)$. Note that $|J_k| = 1$ (resp., 2) if the move $\sigma_{k-1} \to \sigma_k$ is a shift move (resp., a swap move). Then $\tilde{J}_k := \cup_{1 \leq l \leq k, |J_l|=1} J_l$ is the set of jobs involved in the past shift moves in the sequence of $\sigma_0, \sigma_1, \ldots, \sigma_k$. If $k \geq 1$, in defining $N'_{shift}(\sigma_k)$, we forbid changing the assignment of jobs in $\tilde{J}_k$. This restriction is added to avoid cycling among the assignments in the sequence $\sigma_0, \sigma_1, \ldots, \sigma_{k+1}$. We also forbid a move from σ_k to a $\sigma \in N_{shift}(\sigma_k)$ unless

$$pcost(\sigma) - \{p_{\sigma(j)}(\sigma) - p_{\sigma(j)}(\sigma_k)\} < pcost(\sigma_{l^*}) \qquad (31.2)$$

with $J(\sigma_k, \sigma) = \{j\}$, and $pcost(\sigma_{l^*}) = \min_{0 \leq l \leq k} pcost(\sigma_l)$.

This is because the assignment σ not satisfying (31.2) has little chance of yielding an assignment better than σ_{l^*}, for the reason as described in Section 31.6. Let N'_{shift} denote the shift neighborhood resulting from the above rules.

Modified swap neighborhood $N^{dsj}_{swap}(\sigma_k, I_k)$: Let $I(\sigma, \sigma') := \{\sigma(j), \sigma'(j) |\ j \in J(\sigma, \sigma')\}$ be the set of agents which participate in the move from σ to σ', and let $I_k := I(\sigma_{k-1}, \sigma_k)$. We reduce $N_{swap}(\sigma_k)$ to $N'_{swap}(\sigma_k, I_k)$, where $N'_{swap}(\sigma_k, I_k)$ consists of those swaps in which at least one of the pertaining jobs is taken from an agent in I_k.

Swap neighborhood is further modified as follows. We call two swap moves from σ to σ' and from σ to σ'' are *disjoint* if $I(\sigma, \sigma') \cap I(\sigma, \sigma'') = \emptyset$. A set of swap moves is called *mutually disjoint* if every pair of swap moves are disjoint. The modified swap neighborhood $N^{dsj}_{swap}(\sigma_k, I_k)$ then consists of those solutions obtainable by simultaneously applying a set of mutually disjoint swap moves in $N'_{swap}(\sigma_k, I_k)$.

Details of Steps 1 and 2

Now we will explain the details of our algorithm VDS.

Generation of an initial solution in Step 1: First, generate randomly an assignment σ. Find a locally optimal solution σ'' by applying $LS(N_{shift}, pcost, \sigma)$, and then find another locally optimal solution σ_0 by $LS(N_{swap}, pcost, \sigma'')$. This σ_0 is used as the initial solution of Step 1.

Choice of the solution σ in Step 2-b: To choose σ_{k+1} from $N^{dsj}_{swap}(\sigma_k, I_k)$, we generate a sequence of solutions $\sigma'_1, \sigma'_2, \ldots, \sigma'_L$ realized by disjoint swap moves in $N'_{swap}(\sigma_k, I_k)$, by the following algorithm.

Algorithm DS

Step 1 Let $\sigma'_0 := \sigma_k$, $S_0 := \{\sigma \in N'_{swap}(\sigma'_0, I_k) |\ pcost(\sigma) < pcost(\sigma_k)\}$, and $l := 1$.

Step 2 Choose the $\sigma \in S_{l-1}$ that minimizes $pcost(\sigma)$, and set $\sigma'_l := \sigma$.

Step 3 Set $S_l := \{\sigma \in N'_{swap}(\sigma'_l, I_k) |\ pcost(\sigma) < pcost(\sigma'_l)$ and $I(\sigma'_l, \sigma) \cap (\cup_{1 \le h \le l} I(\sigma'_{h-1}, \sigma'_h)) = \emptyset\}$. If $S_l = \emptyset$, set $L := l$ and exit; otherwise let $l := l + 1$ and return to Step 2.

Then σ_{k+1} is defined to be the σ'_L when it exits from Step 3. In this case, we set $I_{k+1} := \cup_{1 \le l \le L} I(\sigma'_{l-1}, \sigma'_l)$, for which $|I_{k+1}| = 2L$ holds. See the appendix for discussions about the computational time of Algorithm DS.

31.5 COMPUTATIONAL RESULTS

Our algorithm in Section 31.4, and four of the known algorithms were run on a workstation Sun Ultra 2 Model 2200 (two 200MHz processors), where the computation was executed on only a single processor. We tested the benchmark instances called types C and D with up to $n = 200$ and $m = 20$, found in [2]. We also tested the instances called type E, found in [5]. Types C and D instances were taken from OR-Library [1], and type E instances were generated by ourselves [2].

We first examined the effect of parameter α of Algorithm VDS (α is included in *pcost* defined in (31.1)). Table 31.1 shows the average cost of 10 runs of Algorithm VDS with $r = 1$ and $\tau = \infty$. The column '#feasible' shows the number of runs in which feasible solutions are found. Average computational time for each run is also shown. We can observe that the performance of Algorithm VDS is quite sensitive on α.

We then tested (1) our VDS algorithm (denoted VDS), (2) random multi-start local search (denoted MLS), (3) VDS by Racer and Amini [11] (denoted R&A), (4) tabu search by Laguna et al. [5] (denoted TABU), (5) tabu search for the constraint satisfaction problem by Nonobe and Ibaraki [9] (denoted TABU-CSP). (An improved version of algorithm R&A is proposed in [1]; however, the new version is not tested here, since the idea of [1] is to start the search from good initial solutions generated by the heuristics of Martello and Toth [7, 8], which is also applicable to all other algorithms tested in this paper.) VDS, MLS and R&A are coded in C language by ourselves, while the codes of TABU and TABU-CSP are sent from the authors (which are also written in C). The parameter α in (31.1), used in VDS and MLS, was set to $\alpha = 2$ for types C and

[1]These are available at `http://mscmga.ms.ic.ac.uk/jeb/orlib/gapinfo.html`.
[2]These are available at `http://www.kuamp.kyoto-u.ac.jp/labs/or/members/yagiura/gap/`.

type	n	m	α	avr. cost	#feasible	avr. time (secs.)
C	100	10	0.25	—	0	0.3
C	100	10	0.50	—	0	1.2
C	100	10	1.00	1429.3	4	0.5
C	100	10	1.50	1423.9	10	0.4
C	100	10	2.00	1409.3	10	0.3
C	100	10	4.00	1414.5	10	0.2
C	200	10	0.25	—	0	10.5
C	200	10	0.50	—	0	7.5
C	200	10	1.00	2817.1	10	5.1
C	200	10	1.50	2816.2	10	2.9
C	200	10	2.00	2819.4	10	2.4
C	200	10	4.00	2821.1	10	2.5
D	100	10	0.25	—	0	2.6
D	100	10	0.50	—	0	9.6
D	100	10	1.00	—	0	57.7
D	100	10	1.50	6415.6	10	70.5
D	100	10	2.00	6394.6	10	47.8
D	100	10	4.00	6414.1	10	21.1
D	200	10	0.25	—	0	34.5
D	200	10	0.50	—	0	114.2
D	200	10	1.00	—	0	1010.1
D	200	10	1.50	12484.8	10	891.3
D	200	10	2.00	12495.7	10	584.4
D	200	10	4.00	12525.4	10	279.7
E	100	10	10.00	—	0	34.4
E	100	10	12.00	—	0	46.8
E	100	10	14.00	—	0	40.2
E	100	10	16.00	11600.1	10	74.9
E	100	10	18.00	11590.5	10	63.1
E	100	10	20.00	11600.8	10	33.8
E	100	10	40.00	11656.5	10	9.2
E	200	10	10.00	—	0	640.9
E	200	10	12.00	—	0	741.2
E	200	10	14.00	—	0	670.9
E	200	10	16.00	23319.6	10	1116.5
E	200	10	18.00	23334.4	10	452.9
E	200	10	20.00	23346.8	10	226.6
E	200	10	40.00	23428.3	10	87.7

Table 31.1 The effect of parameter α.

type	n	m	others[†]	GA[‡]
C	100	5	151.2	302.4
C	100	10	197.1	394.2
C	100	20	334.6	669.3
C	200	5	405.0	810.1
C	200	10	523.0	1046.0
C	200	20	896.1	1792.3
D	100	5	265.1	530.3
D	100	10	547.3	1094.7
D	100	20	1063.0	2126.1
D	200	5	971.4	1942.8
D	200	10	1594.8	3189.6
D	200	20	2782.5	5565.1
E	100	5	300.0	N.A.
E	100	10	600.0	N.A.
E	100	20	900.0	N.A.
E	200	5	1000.0	N.A.
E	200	10	2000.0	N.A.
E	200	20	3000.0	N.A.

Table 31.2 The computational times (secs.) of the six algorithms. Legend: † Computational time on Sun Ultra 2 Model 2200 (200MHz) allowed for algorithms VDS, MLS, R&A, TABU and TABU-CSP. ‡ Computational time on Silicon Graphics Indigo (100MHz) allowed for algorithm GA.

D instances, and to $\alpha = 20$ for type E instances. For MLS, $N_{shift} \cup N_{swap}$ is used for the neighborhood. R&A does not include any parameter. The parameters for TABU and TABU-CSP are set to the default values. All algorithms are iterated until their computational time for each problem instance exceeds the prespecified values in Table 31.2 (e.g., for VDS, parameter r is set to ∞ and τ is set to the values in the table).

For comparison purposes, we also include in Table 31.2 the results of the genetic algorithm cited in [2] by Chu and Beasley (denoted GA), where the computational time, reported in [2], is on a different workstation Silicon Graphics Indigo (R4000, 100MHz). The computational time allowed for algorithms VDS, MLS, R&A, TABU and TABU-CSP are set to the half of the time allowed for GA, taking into consideration rough estimation of the power of computers in use. The results of GA for type E instances are not available and denoted as 'N.A.' in the table.

Table 31.3 shows the best costs obtained by the tested algorithms, where '—' means that the algorithm could not obtain a feasible solution. There are five types of problems named types A, B, C, D and E. Among them, types A and B are very easy and there are not many differences among the tested algorithms; hence, only the results of types C, D and E are shown. It is known that types D and E are much more difficult than type C. In the table, '*' marks the best solution among those obtained by the six algorithms. From Table 31.3, we can observe that the solution quality of the proposed VDS is better than other five algorithms in most cases.

We have also compared our VDS with the tabu search by Osman [10] for type C instances with up to $n = 60$ [13]. The results indicate that VDS obtains

type	n	m	VDS	MLS	R&A	TABU	TABU-CSP	GA
C	100	5	*1931	*1931	1938	*1931	*1931	*1931
C	100	10	*1402	1412	1405	1403	1403	1403
C	100	20	1246	1263	1250	*1244	1245	*1244
C	200	5	*3457	3463	3469	*3457	3459	3458
C	200	10	*2809	2839	2835	2812	2817	2814
C	200	20	2400	2432	2413	*2396	2400	2397
D	100	5	*6365	6413	—	6386	6415	6373
D	100	10	6380	6476	6532	6398	6456	*6379
D	100	20	*6266	6396	6396	6283	6315	6269
D	200	5	*12778	12851	—	12788	12915	12796
D	200	10	*12496	12664	12707	12519	12777	12601
D	200	20	*12323	12628	12642	12416	12596	12452
E	100	5	*12685	12722	12917	12687††	12699	N.A.
E	100	10	*11582	11700	12047	11650††	11689	N.A.
E	100	20	*8483	8816	8883	8555††	8658	N.A.
E	200	5	*24946	24969	25649	25203††	24958	N.A.
E	200	10	*23334	23498	24710	23567††	23425	N.A.
E	200	20	*22439	22831	23947	22735††	22584	N.A.

Table 31.3 The best costs of the six algorithms. Legend: ††Results after 20,000 seconds on Sun Ultra 2 Model 2200 (200MHz).

solutions of comparable or even better quality within reasonable amount of computational time. It should be noted, however, that rigorous comparison is not possible, since the computers used for the two algorithms are different.

31.6 CONCLUSION AND DISCUSSION

We proposed a variable depth search algorithm for the generalized assignment problem. Our algorithm features: (1) the search within the infeasible region, and (2) an adaptive use of two different neighborhoods, shift and swap. Comparing our algorithm with other metaheuristic algorithms, it was observed that our algorithm was able to obtain solutions of better quality for types D and E instances, which is known to be very hard.

Let us call $\bar{d} - d^*$ ($\bar{d}$ and d^* are indices of the solutions in Step 3 of VDS) *overshoot*. This indicates in some sense the number of unnecessary moves. It is observed that the overshoot becomes quite large for types D and E instances as compared to other types. For example, the average number of overshoots for an instance of type D with $n = 200$ and $m = 20$ is 18.9, while that for an instance of type C with the same size is 3.0. Some additional rules to reduce this overshoot may be useful to improve the performance of VDS.

It is also observed that the performance of our VDS is sensitive to the parameter α (the weight on the penalty term). It may also be important to incorporate the ability of tuning α automatically in our algorithm.

Appendix

Remark on rule (31.2) The rule (31.2) is based on the following observation. Consider two solutions σ_k and σ' (σ' is not necessarily in $N'_{shift}(\sigma_k)$).

Then σ' can be obtained from σ_k by the set of shift moves $\sigma_k(j) \to \sigma'(j)$ for all jobs j in $J(\sigma_k, \sigma')$. Note that a swap move can be represented by two shift moves. Then

$$\sum_{j \in J(\sigma_k, \sigma')} \left\{ c_{\sigma'(j),j} - c_{\sigma_k(j),j} - \alpha \min \left\{ p_{\sigma_k(j)}(\sigma_k), a_{\sigma_k(j),j} \right\} \right\} \leq pcost(\sigma') - pcost(\sigma_k) \tag{31.3}$$

holds, since

$$\begin{aligned} p_i(\sigma') &\geq p_i(\sigma_k) - \min \left\{ p_i(\sigma_k), \sum_{j \in J(\sigma_k, \sigma'), \sigma_k(j) = i} a_{ij} \right\} \\ &\geq p_i(\sigma_k) - \sum_{j \in J(\sigma_k, \sigma'), \sigma_k(j) = i} \min \left\{ p_i(\sigma_k), a_{ij} \right\} \end{aligned}$$

for all $i \in I$. Each term in the left side of (31.3) denotes the increase of *pcost* by the shift $\sigma_k(j) \to \sigma'(j)$ of job j, in which the increase of $p_{\sigma'(j)}$ is neglected. Therefore, if an assignment σ' satisfying $pcost(\sigma') < pcost(\sigma_k)$ can be obtained by a sequence of moves, then there exists a first move $\sigma_k \to \sigma$ such that $\sigma \in N_{shift}(\sigma_k)$, $\sigma(j) = \sigma'(j)$ for $\{j\} = J(\sigma_k, \sigma)$ and

$$c_{\sigma(j),j} - c_{\sigma_k(j),j} - \alpha \min \left\{ p_{\sigma_k(j)}(\sigma_k), a_{\sigma_k(j),j} \right\} < 0.$$

The last condition is equivalent to

$$pcost(\sigma) - \left\{ p_{\sigma(j)}(\sigma) - p_{\sigma(j)}(\sigma_k) \right\} < pcost(\sigma_k), \tag{31.4}$$

since $pcost(\sigma) - pcost(\sigma_k) = c_{\sigma(j),j} - c_{\sigma_k(j),j} - \alpha \min \left\{ p_{\sigma_k(j)}(\sigma_k), a_{\sigma_k(j),j} \right\} + \left\{ p_{\sigma(j)}(\sigma) - p_{\sigma(j)}(\sigma_k) \right\}$. Therefore, in considering the first shift move from σ_k, we are restricted only to those σ satisfying (31.4), expecting that the subsequent swap moves from σ will eventually find an assignment σ' satisfying $pcost(\sigma') < pcost(\sigma_k)$ if such σ' exists.

In our rule of defining N'_{shift}, we further strengthen (31.4) to (31.2) by replacing the right-hand side by $pcost(\sigma_{l^*})$ $(\leq pcost(\sigma_k))$. This is because we want to find a solution σ' better than σ_{l^*} in our search in Step 2 of VDS.

Remark on algorithm DS Let us consider the computational time needed to choose the σ_{k+1} by Algorithm DS. We store S_l by the set $\{CH(\sigma'_l, \sigma) | \ \sigma \in S_l\}$, where $CH(\sigma'_l, \sigma) := \{\langle j; \sigma'_l(j) \to \sigma(j) \rangle | \ j \in J(\sigma'_l, \sigma)\}$. This is the set of changes of the assignments of the two jobs in $J(\sigma'_l, \sigma)$ involved in the swap moves $\sigma'_l \to \sigma$ of S_l. Then $\{CH(\sigma'_l, \sigma) | \ \sigma \in S_l\} \subset \{CH(\sigma'_{l-1}, \sigma) | \ \sigma \in S_{l-1}\}$ holds for all l. Let $\Delta pcost(CH(\sigma'_l, \sigma)) := pcost(\sigma) - pcost(\sigma'_l)$. Note that $\Delta pcost$ for the changes in $\{CH(\sigma'_l, \sigma) | \ \sigma \in S_l\}$ is independent of l during the execution of DS, since the swap moves remaining in S_l are disjoint to those previously used. Therefore, $\Delta pcost$ for all swap moves in S_l, $1 \leq l \leq L$, were already computed in Step 1 when S_0 was prepared in $O(|N'_{swap}(\sigma'_0, I_k)|)$

time. The time for Step 2 is clearly $O(|S_{l-1}|)$. The update of S_l in Step 3 is also possible in $O(|S_{l-1}|)$ time since it only requires the deletion from $\{CH(\sigma'_{l-1},\sigma)|\ \sigma \in S_{l-1}\}$ the changes $CH(\sigma'_{l-1},\sigma)$ of swap moves $\sigma'_{l-1} \rightarrow \sigma$ such that $I(\sigma'_{l-1},\sigma) \cap I(\sigma'_{l-1},\sigma'_l) \neq \emptyset$. Therefore, the computational time of DS is $O(|N'_{swap}(\sigma_k, I_k)| + \sum_{l=0}^{L-1} |S_l|) = O(|N'_{swap}(\sigma_k, I_k)| + L|S_0|)$, since $|S_l|$ is monotonically decreasing with l. On the other hand, the computational time would become $O(L|N'_{swap}(\sigma_k, I_k)|)$ if we just choose the best solution in $N'_{swap}(\sigma'_l, I_k)$ L times. Usually, $|S_0|$ is much smaller than $|N'_{swap}(\sigma_k, I_k)|$, and the algorithm DS thus reduces computational time.

Acknowledgements

The authors are grateful to Fred Glover and Manuel Laguna [5], and Koji Nonobe [9] for providing us their original codes.

References

[1] M.M. Amini and M. Racer, "A Hybrid Heuristic for the Generalized Assignment Problem," *European J. Oper. Res.*, 87 (1995) 343–348.

[2] P.C. Chu and J.E. Beasley, "A Genetic Algorithm for the Generalized Assignment Problem," *Computers Oper. Res.*, 24 (1997) 17–23.

[3] M.R. Garey and D.S. Johnson, *Computers and Intractability: A Guide to the Theory of NP-Completeness*, (Freeman, 1979).

[4] B.W. Kernighan and S. Lin, "An Efficient Heuristic Procedure for Partitioning Graphs," *Bell Syst. Tech. J.*, 49 (1970) 291–307.

[5] M. Laguna, J.P. Kelly, J.L. González-Velarde and F. Glover, "Tabu Search for the Multilevel Generalized Assignment Problem," *European J. Oper. Res.*, 82 (1995) 176–189.

[6] S. Lin and B.W. Kernighan, "An Effective Heuristic Algorithm for the Traveling Salesman Problem," *Oper. Res.*, 21 (1973) 498–516.

[7] S. Martello and P. Toth, "An Algorithm for the Generalized Assignment Problem," *Proc. IFORS*, (1981) 589–603.

[8] S. Martello and P. Toth, *Knapsack Problems: Algorithms and Computer Implementations*, (John Wiley & Sons, 1990).

[9] K. Nonobe and T. Ibaraki, "A Tabu Search Approach to the CSP (Constraint Satisfaction Problem) as a General Problem Solver," *European J. Oper. Res.*, 106 (1998) 599–623.

[10] I.H. Osman, "Heuristics for the Generalized Assignment Problem: Simulated Annealing and Tabu Search Approaches," *OR Spektrum*, 17 (1995) 211–225.

[11] M. Racer and M.M. Amini, "A Robust Heuristic for the Generalized Assignment Problem," *Ann. Oper. Res.*, 50 (1994) 487–503.

[12] S. Sahni and T. Gonzalez, "P-Complete Approximation Problems," *J. ACM*, 23 (1976) 555–565.

[13] T. Yamaguchi, "A Variable Depth Search for the Generalized Assignment Problem (in Japanese)," Graduate Thesis, Department of Applied Mathematics and Physics, Faculty of Engineering, Kyoto University, 1996.

32 GUIDED LOCAL SEARCH FOR THE VEHICLE ROUTING PROBLEM WITH TIME WINDOWS

Philip Kilby, Patrick Prosser and Paul Shaw

Department of Computer Science, University of Strathclyde
26 Richmond Street, Glasgow, Scotland. G1 1XH.
pjk@cs.strath.ac.uk, pat@cs.strath.ac.uk, ps@cs.strath.ac.uk

Abstract: We describe a heuristic solution method for the Vehicle Routing Problem with Time Windows. The method uses four improvement operators in a steepest descent search strategy. The Guided Local Search meta-heuristic is used to avoid local minima. The resulting algorithm is tested on Solomon's capacitated vehicle routing problems with time windows. We compare our results to the best heuristic approaches reported for the VRPTW. The new method performs significantly better than previous methods on classes where vehicle routes tend to be longer. The method is slightly worse on classes with shorter routes. We report 12 new best solutions for Solomon's problems.

32.1 INTRODUCTION

The Vehicle Routing Problem (VRP) is a practical problem that has been studied widely in the OR literature [7]. The problem is usually expressed as follows: given a set of customers requiring a visit, and a fleet of vehicles based at a depot that can perform the visits, construct a set of routes for the vehicles which minimizes the costs of operation. The objective function is usually expressed as costs related to the number of vehicles used and to distance travelled.

Vehicle routing problems generally come at least with capacity constraints on the vehicles (indicating that they can only carry a certain quantity of goods). However, other constraints are common. In this paper we examine the Vehicle Routing Problem with Time Windows (VRPTW). In this problem, as well as vehicles having limited capacity each customer can specify a time window for delivery.

Solution methods can be divided into two categories: exact and approximate. Recent exact approaches have used Integer Linear Programming, for example [2, 3, 4]. These methods are based on Lagrangian relaxation, which relaxes some of the constraints into the objective function. These methods have been progressing recently, and are now able to solve some problems with up to 150 customers [6]. However they can only reliably solve problems of up to 50 customers. Constraint Programming methods have been used on a related problem — the Travelling Salesman Problem with Time Windows [1, 10]. Results show that TSPs of up to 35 visits can be solved reliably. For practical problems, which may have thousands of customers, approximate methods, for instance local search, must be used.

In the recent past, great success has been achieved via the application of local search techniques and meta-heuristics to routing problems. Most popular has been the tabu search meta-heuristic [5, 12], genetic methods [11, 18, 19], simulated annealing [9], and hybrids, for instance [14, 17].

A new meta-heuristic, guided local search (GLS) [21, 23] is a memory-based method, and so shares similarities with tabu search. However, it operates by augmenting the cost function with a penalty term based on how near the search moves to previously visited local minima, thus encouraging diversification. This paper applies GLS to vehicle routing problems with time windows and capacity constraints. Results indicate that GLS can provide excellent results.

This paper is organised as follows. In Section 32.2, we introduce a local search algorithm for the vehicle routing problem. We begin first by describing the move operators and model, and then go on to describe the objective function and search itself. The GLS meta-heuristic is then presented in Section 32.3. The application of the meta-heuristic to a vehicle routing framework is discussed. Experiments are then performed using GLS on some standard benchmark problems from the literature that involve both capacity constraints on vehicles, and time windows at customers. Conclusions are drawn on the quality of the results in comparison to other methods.

32.2 LOCAL SEARCH

32.2.1 Improvement Heuristics

We describe the improvement heuristics we use for the vehicle routing problem. We are ignoring construction heuristics (those used to build a solution), since for the purposes of this paper, we consider the construction of a solution as part of the improvement. That is, in the initial solution no visits are allocated to any vehicle. A penalty is associated with not performing a visit, and so the search process constructs a solution in the process of minimizing cost. Section 32.2.4 explains this in detail.

Heuristics for routing plan improvement can be within a route, or between routes. Here, we will focus upon one heuristic that operates within routes (the two-opt procedure of [8]) and three that operate between routes: the relocate, exchange, and cross operators, described in [15].

The two-opt procedure is used for improving single vehicle routes, and operates by reversing part of the route. In Figure 32.1(a), the two-opt operator is shown. Here, the part of the route from A to B has been reversed, and in this instance, has resulted in a reduction in total length.

The relocate operator moves a customer visit from one route to another. Figure 32.1(b) shows that by moving visit A to another route between B and C, overall distance can be reduced. A visit can also be relocated to an "empty" route, resulting in the creation of a new vehicle itinerary with a single visit.

The exchange operator swaps customer visits in different vehicle routes. In Figure 32.1(c), the positions of visits A and B have been interchanged.

The cross operator swaps the end portions of two vehicle routes. Figure 32.1(d) shows the operation of the cross. The sections of the routes after visits A and B have been interchanged, resulting in a distance reduction. It is also possible to cross from the end of one route to the beginning of another, resulting in a concatenation of routes and an empty route. Moreover, if one of the routes involved in a cross operator is empty, the cross can split the non-empty route into two.

32.2.2 Model

We represent a solution to a vehicle routing problem as set of lists of customer visits, each list corresponding to a single vehicle route. A maximum number of lists are available, corresponding to the number of available vehicles. Lists left empty at the end of search correspond to vehicles that are not required to execute the routing plan. Associated with each vehicle is a capacity, which must always be greater than or equal to the sums of the loads of its deliveries. In the model, different vehicles can have different capacities.

Additionally, a *virtual* vehicle exists that notionally performs the visits not carried out by the real vehicles. The virtual vehicle is different from the real ones in three respects. First, the virtual vehicle can "make" any number of customer visits. Second, problem constraints (such as time and capacity) do not apply to the virtual vehicle, making all routes in the virtual vehicle legal. Third, the cost incurred by the virtual vehicle when it performs a customer visit is higher than that incurred by a real vehicle. This last point ensures that plans are created with work completed. The virtual vehicle is not used in the final routing plan, but indicates by its contents visits that could not be scheduled.

32.2.3 Objective

The objective function that we attempt to minimize is the total distance travelled by all vehicles. The cost of a customer visit being performed by the *virtual* vehicle is equal to $\alpha_1(d_{0i}+d_{i0})+\alpha_2$ where d_{ij} is the distance between customers i and j. We define customer 0 to be the depot. When $\alpha_1 = 1$ and $\alpha_2 = 0$, the

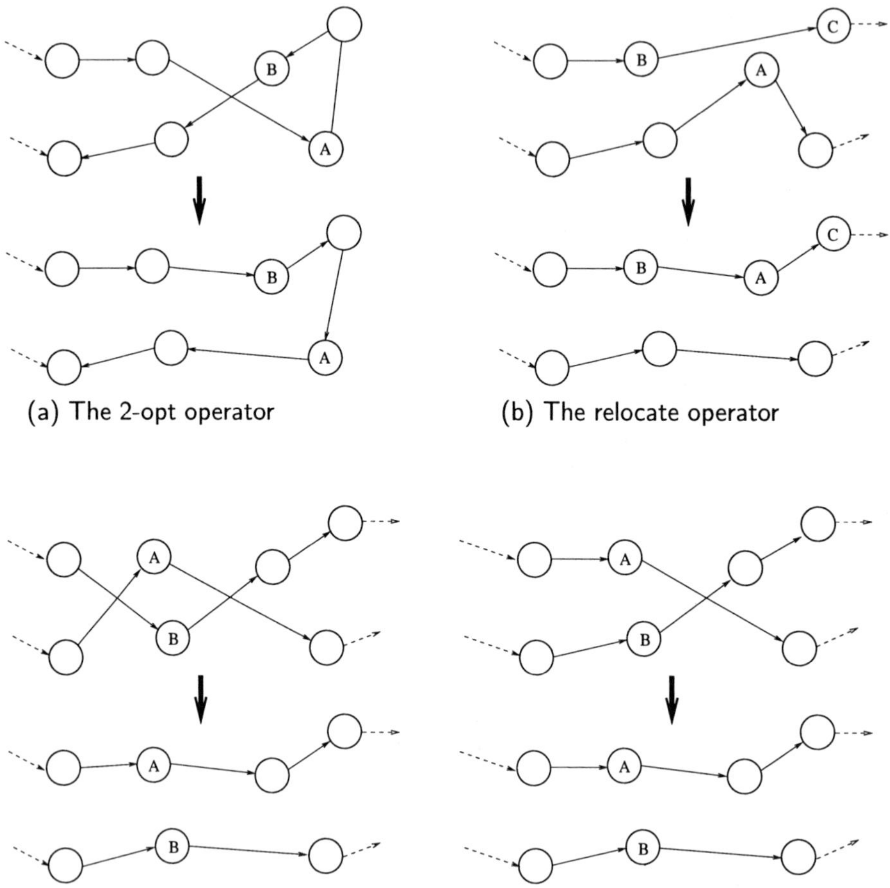

(a) The 2-opt operator
(b) The relocate operator
(c) The exchange operator
(d) The cross operator

Figure 32.1 Some Improvement Heuristics for the VRP/VRPTW

cost of the virtual vehicle making the visit is equal to that of a single vehicle being dispatched from the depot to make only that single visit. Higher values of α_1 and α_2 make it cheaper to perform visits by real vehicles.

Results in [13] suggest that good parameter settings are $\alpha_1 = 1.025$ and $\alpha_2 = 0.0005\Delta$, where Δ is the distance between the two most remote sites in the problem.

32.2.4 Search Method

The search method we adopt is an entirely greedy one. There are no random numbers involved, and the results are completely reproducible. We first of all begin the search by assuming all customer visits will be carried out by the virtual vehicle, and then use the four improvement heuristics presented to attempt to improve the cost of the solution. At each point in search, every possible 2-opt, relocate, exchange and cross operation is examined, and the cost and legality are calculated. The legal move that decreases the objective by the greatest amount is chosen, implementing the so-called "best acceptance" scheme. We stop the search when no moves will strictly reduce the objective. We make the restriction that the only heuristic that can be used in association with the virtual vehicle is the relocation of a visit from the virtual vehicle to a real vehicle.

32.2.5 Limitations

The local search method described above can quickly produce reasonable solutions to routing problems (typically within 10% of the best known solution [13]), but the variability of the results from problem to problem is high. We therefore now turn to meta-heuristic methods to stabilise the search process and increase the probability of attaining high quality solutions.

32.3 GUIDED LOCAL SEARCH

Guided local search (GLS) [21, 23] is a meta-heuristic based on penalties. The method works by adding a penalty factor to the objective function, based on the experience the search has gained. Broadly speaking, the method first identifies a local minimum. Aspects of the solution which are in some way "bad" are identified. The objective function is then changed by adding a term which penalises these aspects. The search iterates by finding a new local minimum and accumulating more penalties. In this way the search is discouraged from moving too close to previously visited local minima.

GLS has proved to be an effective meta-heuristic for a range of combinatorial optimisation problems such as the travelling salesman problem [23], quadratic assignment [23], function optimisation [22], partial constraint satisfaction [24], and resource allocation [20].

In what follows, we describe the general algorithm, and then its application to the vehicle routing problem.

32.3.1 The Guided Local Search Algorithm

Guided local search moves out of a local minimum by penalising particular solution features that it considers should not occur in a near-optimal solution. It defines a modified objective function, augmented with a set of penalty terms on these features. The usual local search method is then invoked to improve the augmented objective function. The cycle of local search and penalty term update can be repeated as often as required.

GLS requires the following components:

- A fixed set of features F. The only requirement for the feature set is that any solution to the problem must identify a subset of F. For $i \in F$, let the indicator function $f_i(\mathcal{S}) = 1$ if the feature i is in solution $\mathcal{S}$, and 0 otherwise.

- A cost vector c. c_i is the "cost" of feature i.

- A penalty factor λ.

GLS tracks penalties applied via a penalty vector p. p_i is the (integer) number of times feature i has been penalised so far. Assuming $O(\mathcal{S})$ is the original objective function for the problem, GLS defines an augmented objective function

$$O'(\mathcal{S}) = O(\mathcal{S}) + \lambda \sum_{i \in F} f_i(\mathcal{S}) p_i c_i$$

and requires a local search procedure that minimizes it. The user must therefore provide procedure LocalSearch($\mathcal{S}$,p) that performs a local search, starting at solution $\mathcal{S}$, and returns a new solution. *The improvement is carried out with respect to the augmented objective O'.*

GLS uses a function called ChoosePenaltyFeatures($\mathcal{S}$, p) which takes a solution and the current spread of penalties, and returns the set of features to be penalised. ChoosePenaltyFeatures penalises the most costly features in the current solution, weighted by the number of times the feature has already been penalised. This has the effect of concentrating penalties on "bad" features, while tending to accept bad features if they keep appearing in local minima (i.e., if they seem inevitable). ChoosePenaltyFeatures chooses the features $i \in \mathcal{S}$ for which $c_i/(p_i+1)$ is largest amongst the features in $\mathcal{S}$. Unless values in c are equivalent or multiples of each other, usually only one feature will be chosen.

Assuming the usual functions of InitialSolution() and StoppingCondition() exist, the GLS algorithm is described by Figure 32.2.

```
p := 0̄                                  // All elements zeroed
S := InitialSolution()                  // S is current solution
S := LocalSearch (S, p)                 // Make sure we start at a local min
S* := S                                 // S* is best solution
while not StoppingCondition() do
    f := ChoosePenaltyFeatures(S, p)
    for all i in f do
        p_i := p_i + 1
    S := LocalSearch (S, p)
    if O(S) < O(S*) then
        S* := S
S* := LocalSearch (S*, 0̄)               // Final local search with zero penalties,
                                        // to get to true local minimum
return S*
```

Figure 32.2 GLS algorithm

32.3.2 GLS for Vehicle Routing

We define our implementation of GLS for the VRPTW by instantiating each of the items described in Section 32.3.1.

- Feature set F: F is the set of arcs between any two customers. The identifier function $f(arc(a,b))$ is 1 when a directly precedes b in a vehicle route. This is similar to the approach of [21, 23] for the Travelling Salesman Problem (TSP), except that in the presence of time windows, we apply penalties to *directed* arcs.

- Feature costing: We assume that the cost c_a of directed arc feature a is the length of the arc d_a.

- Penalty factor λ: We have found that values of 0.1 to 0.3 work well, and have used 0.2 for the results in Section 32.4.

- LocalSearch($\mathcal{S}$,p): We use the "best acceptance" algorithm defined in Section 32.2.4. To implement the objective O', the local search is carried out using a modified distance matrix. In the modified matrix, each arc a has a penalised distance $d_a + \lambda p_a d_a$. The matrix is updated each time an element of p is changed.

InitialSolution() and StoppingCondition() are defined in Section 32.4.

32.4 COMPUTATIONAL RESULTS

Computational experiments were performed using GLS. The stopping condition used was 2000 iterations of the main loop. The initial solution used was that of all visits being performed by the virtual vehicle. The objective function was the total distance travelled by all vehicles. (There was no incentive to reduce the number of vehicles.) The penalty factor λ was set to 0.2.

The computer used was a DEC ALPHA 3000 model 300LX with a 21064 P3.0 processor, a clock speed of 125 MHz and 64Mb of memory. The DEC C++ compiler was used.

The problems studied are as used by Solomon [16]. These data are in 6 classes: C1, C2, R1, R2, RC1, RC2. All problems have 100 customers, a central depot, capacity constraints, time windows on the time of delivery, and a total route time constraint. The C1 and C2 classes have customers located in clusters. In the R1 and R2 classes the customers are at random positions. The RC1 and RC2 classes contain a mix of both random and clustered customers. Each class contains between 8 and 12 individual problem instances, numbered by appending two digits to the end of the class name (*e.g* RC104, R112, C207 *etc.*). There are a total of 56 instances in all. All problems in any one class have the same customer locations, and same vehicle capacities; only the time windows differ. The C1, R1 and RC1 problems have a short scheduling horizon, and require 9 to 19 vehicles. Classes C2, R2 and RC2 are more representative of "long-haul" delivery with longer scheduling horizons and fewer (2 – 4) vehicles.

Both travel time and distance are given by the Euclidean distance between points.

Since our objective function has no incentive to minimize vehicles, experiments for problem k were carried out with v_k, v_k+1, and v_k+2 vehicles, where v_k is the minimum number of vehicles reported in the best solution for problem k. For each problem, the solution where all customer visits could be scheduled in the lowest number of vehicles was kept.

In summary, given all 56 problems, 168 experiments (3 for each problem with differing numbers of vehicles) were carried out for 2000 iterations, with $\lambda = 0.2$. Since both our local search method and GLS contain no random element, only a single run on each problem was required.

The number of vehicles and total distance travelled for the problems within each class were averaged and are shown in Table 32.1. The data are compared against results reported in Rochat and Taillard [14], and Taillard et al. [17], which are the best two heuristic techniques reported for the VRPTW.

Comparisons between meta-heuristics are difficult because, in general, the longer they are left to run the better the solution they produce. Therefore, some cutoff time is required. However, this is very difficult as different researchers report their experience on different machines. We have attempted to be as fair as possible.

Rochat and Taillard use a 100 MHz Silicon Graphics Indigo. We believe that this is roughly equivalent to the Alpha machine used in our experiments. Taillard et al. use a 50 MHz Sparc-10 workstation. We estimate that this is about half the speed of the Alpha.

In both the above references, solution quality is reported at three checkpoints during search. We have compared against checkpoint 2 in both as the respective CPU times at this point corresponds most closely with the CPU resource used by our method. The mean CPU times per problem (rounded to the nearest 100 seconds) are: Rochat and Taillard–2800s, Taillard et al.–7400s, GLS–2900s.

	Rochat and Taillard [14]		Taillard et al. [17]		GLS	
Problem Class	Vehicles Used	Distance Travelled	Vehicles Used	Distance Travelled	Vehicles Used	Distance Travelled
C1	10.00	828.45	10.00	828.59	10.00	830.75
R1	12.58	1202.31	12.39	1230.48	12.67	1200.33
RC1	12.50	1368.03	12.00	1387.01	12.12	1388.15
C2	3.00	590.32	3.00	591.14	3.00	592.24
R2	3.09	969.29	3.00	1029.65	3.00	966.56
RC2	3.62	1155.47	3.38	1220.28	3.38	1133.42

Table 32.1 Average results by problem class

Examination of the averaged results indicates that for problem classes C1 and C2, there is little to choose between the three methods. In fact, these problem classes are easy, in that even a local search without a meta-heuristic converges close to the best known solutions. Typically, it is easy to get within 0.5% of the best known solutions quickly.

On class R1, GLS performs poorly, and is typically bad at reducing the numbers of vehicles used, using on average 0.7% more vehicles than Rochat and Taillard and 2.3% more vehicles than Taillard et al. Distances for GLS are similar to Rochat and Taillard, but are less than for Taillard et al. This is due to the fact that Taillard et al. use less vehicles, which (up to a point) typically means that more distance needs to be traversed. The distance impact is less apparent on a class such as R1 where many vehicles are needed. However, for the R2 and RC2 classes, reducing vehicles by one for a particular problem can have a large impact on distance travelled, since these problems only require 4 vehicles or less.

On class RC1, GLS sits between the other two methods in terms of performance. GLS uses 1.0% more vehicles than Taillard et al., but 3.1% less than Rochat and Taillard. Distance for GLS is roughly equivalent to Taillard et al., but is greater than for Rochat and Taillard, due to the latter using more vehicles.

On classes R2 and RC2, GLS easily outperforms the other two methods. On both classes, the number of vehicles used is the same as for Taillard et al. Comparing to Rochat and Taillard, GLS uses 3% and 7.4% less vehicles for classes R2 and RC2 respectively. With regard to travel distance, GLS reduces distance over Taillard et al. by 6.53% for R2 and 7.66% for RC2, while using the same number of vehicles. GLS reduces travel by 0.3% for R2 and 1.9% for RC2 over Rochat and Taillard, even though the latter uses more vehicles.

To summarise, GLS produces very similar results on classes C1 and C2 as the other methods. For class R1, GLS performs poorly, using more vehicles than the other two methods. GLS delivers intermediate performance on class RC1, but is much closer to the (better) results of Taillard et al. than those of Rochat and Taillard. For classes R2 and RC2, the results produced by GLS are significantly (around 7%) better.

It should be mentioned that both Rochat and Taillard and Taillard et al. use a tabu search approach coupled with a "route memory". The route memory holds good routes previously found by tabu search. The tabu search is applied to a solution constructed from routes drawn from the route memory. After tabu search is applied, the route memory is updated with the new routes. We conjecture that the route memory is responsible for the good performance of these methods on classes R1 and RC1. On classes R2 and RC2, each route is longer, and so the route memory is not as effective as there are potentially many more good routes and route decompositions than for R1 and RC1. Therefore, the route memory does not so quickly move to a state that contains the best, or near best routes. We expect GLS to benefit from a route memory in the same way as the tabu search method, delivering better performance on classes R1 and RC1.

Finally, we report on some new best solutions for Solomon's problems that we have found using GLS during our computational experiments. Some best solutions reported in the literature are based on integer valued distances. Due to ambiguity as to whether time windows are truly met using integer valued distances, solutions created using integer distances are open to dispute. Therefore, we quote as new best solutions those which are better than any produced by using *real* (not integer) distance values.

The cost values for the new best solutions are summarised in Table 32.2, and the complete solutions are listed in Appendix A.

Problem Name	Previous Best			New Best		Percentage Improvement
	Vehicles	Distance	Ref	Vehicles	Distance	
r109	11	1214.54	[17]	11	1205.96	0.7
r201	4	1254.80	[17]	4	1254.09	0.1
r204	2	869.29	[14]	2	867.33	0.2
r205	3	1038.72	[17]	3	998.72	4.0
r209	3	944.64	[14]	3	923.96	2.2
r210	3	967.50	[14]	3	963.37	0.4
rc201	4	1413.52	[10]	4	1406.94	0.5
rc203	3	1079.57	[14]	3	1068.07	1.1
rc204	3	806.75	[14]	3	803.90	0.4
rc205	4	1320.75	[10]	4	1302.42	1.4
rc206	3	1158.81	[17]	3	1156.26	0.2
rc207	3	1079.45	[10]	3	1074.10	0.5

Table 32.2 New Best Solutions to Solomon's Problems Found by GLS

32.5 CONCLUSION

We have described a local search method for vehicle routing problems based on multiple improvement heuristics, and the possibility of not scheduling some visits via the use of a *virtual* vehicle. Solutions are constructed via improvement from an initial solution where all visits are performed (at high cost) by the virtual vehicle.

A meta-heuristic, guided local search, was then used with the local search method, and was tested on Solomon's capacitated vehicle routing problems with time windows. We compared our results to the best heuristic approaches reported for the VRPTW. These approaches use tabu search combined with a route memory.

For problem classes with long routes, GLS easily outperforms these approaches by 7%. For classes with shorter routes, GLS was not as competitive, and was outperformed by up to 2.3%. We expect, however, that the performance of GLS on these instances would be greatly enhanced by the addition of a route memory.

Finally, GLS has provided 12 new best solutions to Solomon's problems. In particular, in class RC2, 75% of best solutions were improved and in class R2, 45% were improved.

As future work, we aim to adjust the cost function to take account of the number of vehicles, and to provide some guidance in the cost function or allowed moves to encourage at least one short route. (Having a short route increases the chances that this route will eventually be removed.) This will allow us to automatically reduce the number of vehicles as search progresses.

Other future work could include the investigation of different feature costing other than arc distance, looking at alternative methods of choosing features to penalise, and the actual penalty method itself.

Acknowledgement

The production of this paper was supported by the GreenTrip project, a research and development undertaking partially funded by the ESPRIT Programme of the Commission of the European Union as project number 20603. The partners in this project are Pirelli (I), ILOG (F), SINTEF (N), Tollpost-Globe (N), and University of Strathclyde (UK).

Appendix A: New Best Solutions

Name: **R109**
Distance: **1205.96**
Vehicles: **11**
1: 28–12–76–29–3–79–78–34–35
2: 92–14–44–16–86–38–43–58
3: 52–88–19–47–36–49–48
4: 39–67–23–75–22–41–74–56–4
5: 2–72–73–21–40–53–26–54–55–25
6: 5–98–59–99–84–61–85–91–100–37–93
7: 83–45–82–7–18–8–46–17–60–89
8: 27–69–31–30–51–81–50–77–68–24–80
9: 33–9–71–65–66–20–70–1
10: 62–11–64–63–90–32–10
11: 95–42–15–57–87–94–6–96–97–13

Name: **R201**
Distance: **1254.09**
Vehicles: **4**
1: 33–65–63–31–69–52–27–28–12–29–76–30–71–9–51–81–79–78–34–3–50–20–10–32–66–35–70–1
2: 2–72–39–75–23–67–21–73–40–53–87–57–41–22–56–26–68–54–74–4–55–25–24–80–77
3: 95–59–92–42–15–14–98–61–16–44–38–86–85–99–84–6–94–97–43–37–96–13–58
4: 5–83–45–82–47–36–64–11–19–62–88–7–8–18–90–49–46–48–60–17–91–100–93–89

Name: **R204**
Distance: **867.33**
Vehicles: **2**
1: 94–95–97–92–98–85–61–16–14–43–15–41–22–23–67–39–26–76–28–53–50–79–81–9–20–66–65–71–35–34–78–29–24–55–4–56–75–74–72–73–13–2–57–42–100–37–59–96–6–89
2: 27–52–18–82–7–88–31–1–69–10–62–11–19–46–17–86–44–38–91–99–87–93–8–45–84–5–60–83–48–47–36–49–64–63–90–32–70–30–51–33–3–77–68–80–12–54–25–21–40–58

Name: **R205**
Distance: **998.72**
Vehicles: **3**
1: 28–12–29–33–65–71–51–30–90–64–49–46–8–17–91–100–37–93–96–13–58
2: 5–45–83–59–95–92–98–14–42–15–2–21–72–39–67–23–75–22–41–73–40–53–76–50–3–79–81–9–78–34–35–66–20–32–10–70–48–60–89
3: 27–69–31–62–63–11–19–36–47–82–7–88–52–18–84–61–16–44–38–86–85–99–6–94–97–87–43–57–74–56–4–26–54–55–25–24–80–68–77–1

Name: **R209**
Distance: **923.96**
Vehicles: **3**
1: 28–12–29–76–69–31–88–62–63–90–70–30–71–9–51–81–33–50–3–79–78–34–35–65–66–20–32–10–48–60–91–93–89
2: 2–75–23–67–39–72–73–21–40–53–6–18–7–19–11–64–49–36–46–8–45–17–84–96–37–97–13–58–26
3: 27–52–82–47–83–5–99–95–92–98–100–44–14–38–86–16–61–85–59–94–87–42–57–15–43–41–22–74–56–4–25–55–54–24–80–68–77–1

Name: **R210**
Distance: **963.37**
Vehicles: **3**
1: 27–69–82–45–61–16–86–38–14–44–91–85–99–96–6–53–40–87–57–43–41–22–74–56–4–55–25–24–54–26–58
2: 28–33–65–71–51–30–88–62–11–64–36–46–8–83–84–18–7–48–47–49–19–63–90–32–10–31–52–94–13–97–37–98–100–93–17–5–60–89
3: 95–59–92–42–15–2–73–21–72–75–23–67–39–12–76–3–79–29–34–78–81–9–35–66–20–70–1–50–80–68–77

Name: **RC201**
Distance: **1406.94**
Vehicles: **4**
1: 65–14–47–59–75–16–15–11–12–73–78–79–7–6–8–46–3–4–1–55–68
2: 92–95–63–33–28–27–29–31–30–62–67–71–90–99–57–86–87–9–53–10–97–74–13–17–60–100–70
3: 42–36–39–72–45–5–2–98–88–61–44–40–38–41–81–94–50–34–32–26–89–48–24–25–77–58
4: 69–82–52–83–64–19–23–21–18–76–85–84–51–49–22–20–66–56–96–54–37–43–35–93–91–80

Name: **RC203**
Distance: **1068.07**
Vehicles: **3**
1: 90–65–92–50–34–31–29–27–26–28–33–32–30–62–67–85–63–89–76–49–22–20–51–84–95–56–66–82–10–13–16–17–86–74–59–97–75–58–77–25–24–83
2: 45–46–7–60–73–14–47–15–11–12–98–69–88–44–43–40–41–38–37–35–72–71–93–94–80
3: 81–61–42–36–39–54–96–91–64–19–18–48–21–23–57–52–99–87–9–53–78–79–8–6–1–3–5–4–2–55–68–70–100

Name: **RC204**
Distance: **803.90**
Vehicles: **3**
1: 69–98–82–10–11–15–16–17–47–14–12–53–60–78–73–79–7–8–46–45–5–3–1–4–6–2–88–55–100–70–61–68
2: 80–91–92–94–62–63–89–76–23–21–48–18–22–57–99–52–86–87–9–13–74–59–97–75–58–77–25–19–49–20–24–83–65–90
3: 81–96–54–41–39–42–44–43–40–36–35–37–38–72–71–93–67–84–51–85–33–32–30–28–26–27–29–31–34–50–95–56–64–66

Name: **RC205**
Distance: **1302.42**
Vehicles: **4**
1: 69–82–11–15–16–47–14–12–88–78–73–79–7–6–8–46–4–3–1–70–100
2: 92–95–33–28–27–29–31–30–63–76–85–67–84–51–49–22–20–66–56–50–34–32–26–89–48–24–25–77–58
3: 2–45–5–42–39–36–72–71–62–94–61–44–40–38–41–81–90–53–98–55–68–43–35–37–54–96–93–91–80
4: 65–83–64–19–23–21–18–57–86–52–99–9–87–59–75–97–10–74–13–17–60

Name: **RC206**
Distance: **1156.26**
Vehicles: **3**
1: 65–82–83–64–52–75–59–14–47–16–15–11–12–53–88–78–73–79–7–6–8–46–4–3–1–43–35–37–54–96–93
2: 69–98–2–45–5–44–42–39–38–36–40–41–61–81–71–67–94–90–99–66–56–84–50–34–32–26–89–20–48–25–77–58–74–24–91–80
3: 72–92–95–62–31–29–27–28–30–33–63–85–51–76–18–21–23–19–49–22–57–86–87–97–9–13–10–17–60–55–100–70–68

Name: **RC207**
Distance: **1074.10**
Vehicles: **3**
1: 69–98–88–2–5–70–61–42–44–40–38–41–72–71–81–94–92–84–85–51–49–20–22–74–58–97–13–10–17–60–55–100–68
2: 65–83–64–95–67–62–31–29–28–30–33–63–76–19–23–21–18–77–75–59–87–86–57–90–96–93–50–34–27–26–32–89–48–25–24–66–56–91–80
3: 82–99–52–9–11–15–16–47–14–12–53–78–73–79–7–6–8–46–4–45–3–1–43–36–35–37–39–54

References

[1] Y. Caseau and F. Laburthe. Solving small TSPs with constraints. In *Proceedings the 14th International Conference on Logic Programming*, 1997.

[2] M. Desrochers, J. Desrosiers, and M. Solomon. A new optimization algorithm for the vehicle routing problem with time windows. *Operations Research*, 40(2):342–354, 1992.

[3] M. L. Fisher. Optimal solution of vehicle routing problems using minimum K-trees. *Operations Research*, 42(4):626–642, 1994.

[4] M. L. Fisher and K. O. Jornsten. Vehicle routing with time windows: Two optimization algorithms. *Operations Research*, 45(3):488–492, 1997.

[5] M. Gendreau, A. Hertz, and G. Laporte. A tabu search heuristic for the vehicle routing problem. *Management Sci.*, 40(10):1276–1290, 1994.

[6] N. Kohl, J. Desrosiers, O. B. G. Madsen, M. M. Solomon, and F. Soumis. k-path cuts for the vehicle routing problem with time windows. Technical Report G-97-19, Les Cahiers du GERAD, 1997.

[7] G. Laporte and I. H. Osman. Routing problems: A bibliography. *Ann. Oper. Res.*, 61:227–262, 1995.

[8] S. Lin. Computer solutions of the traveling salesman problem. *Bell Syst. Tech. J.*, 44:2245–2269, 1965.

[9] I. H. Osman. Metastrategy simulated annealing and tabu search algorithms for the vehicle routing problem. *Ann. Oper. Res.*, 41:421–451, 1993.

[10] G. Pesant, M. Gendreau, J.-Y. Potvin, and J.-M. Rousseau. An exact constraint logic programming algorithm for the traveling salesman problem with time windows. to appear in Transportation Science, 1998.

[11] J.-Y. Potvin and S. Bengio. A genetic approach to the vehicle routing problem with time windows. Technical Report CRT-953, Centre de Recherche sur les Transports, University of Montreal, 1994.

[12] J.-Y. Potvin, T. Kervahut, B.-L. Garcia, and J.-M. Rousseau. A tabu search heuristic for the vehicle routing problem with time windows. Technical Report CRT-855, Centre de Recherche sur les Transports, University of Montreal, 1993.

[13] P. Prosser and P. Shaw. Study of greedy search with multiple improvement heuristics for vehicle routing problems. Technical Report RR/96/201, Department of Computer Science, University of Strathclyde, Glasgow, January 1997.

[14] Y. Rochat and E. D. Taillard. Probabilistic diversification and intensification in local search for vehicle routing. *Journal of Heuristics*, 1(1):147–167, 1995.

[15] M. W. P. Savelsbergh. Computer aided routing. Centrum voor Wiskunde en Informatica, Amsterdam, 1988.

[16] M. M. Solomon. Algorithms for the vehicle routing and scheduling problem with time window constraints. *Oper. Res.*, 35:254–265, 1987.

[17] E. Taillard, P. Badeau, M. Gendreau, F. Guertain, and J.-Y. Potvin. A new neighbourhood structure for the vehicle routing problem with time windows. Technical Report CRT-95-66, Centre de Recherche sur les Transports, University of Montreal, 1995.

[18] S. R. Thangiah, I. H. Osman, and T. Sun. Hybrid genetic algorithm, simulated annealing, and tabu search methods for vehicle routing problems with time windows. Working paper UKC/OR94/4, Institute of Mathematics and Statistics, University of Kent, Canterbury, 1994.

[19] S. R. Thangiah, R. Vinayagamoorthy, and T. Sun. Vehicle routing with time deadlines using genetic and local algorithms. In *5th International Conference on Genetic Algorithms*, 1993.

[20] E. Tsang and C. Voudouris. Fast local search and guided local search and their application to British Telecom's workforce scheduling problem.

Technical Report CSM-246, Department of Computer Science, University of Essex, August 1995.

[21] C. Voudouris. *Guided Local Search for Combinatorial Problems.* PhD thesis, University of Essex, April 1997.

[22] C. Voudouris and E. Tsang. Function optimization using guided local search. Technical Report CSM-249, Department of Computer Science, University of Essex, September 1995.

[23] C. Voudouris and E. Tsang. Guided local search. Technical Report CSM-247, Department of Computer Science, University of Essex, August 1995.

[24] C. Voudouris and E. Tsang. Partial constraint satisfaction problems and guided local search. Technical Report CSM-250, Department of Computer Science, University of Essex, September 1995.

33 MEMORY ADAPTIVE REASONING & GREEDY ASSIGNMENT TECHNIQUES FOR THE CAPACITATED MINIMUM SPANNING TREE PROBLEM

Erik Rolland[1], Raymond A. Patterson[2] and Hasan Pirkul[2]

[1]Department of Accounting & Management Information Systems
Fisher College of Business, The Ohio State University, Columbus, Ohio 43210, USA.
Rolland.1@osu.edu

[2]School of Management, The University of Texas at Dallas, Richardson
Texas 75083-0688, USA.
rpatters@utdallas.edu, hpirkul@utdallas.edu

Abstract: It is the purpose of this paper to investigate effects of adding randomization to a memory-based heuristic. The algorithms we propose are applied to the Capacitated Minimum Spanning Tree problem (CMST), and we study the combined effects of simultaneously applying a memory-based and a random-based heuristic to the CMST. This paper uses the Adaptive Reasoning Technique (ART) and concepts from the greedy randomized adaptive search procedure for solving the CMST. The resulting hybrid procedure is tested against the stand-alone Esau-Williams heuristic procedure, as well as the stand-alone greedy assignment technique. We find that randomization does not constructively add to the memory-based procedure, as ART alone typically outperforms all other approaches in terms of solution quality, while expending a modest amount of computational effort.

33.1 INTRODUCTION

The capacitated minimum spanning tree problem (CMST) plays an important role in the design of backbone telecommunications networks. This problem is NP-complete [18], and various solution methods have been proposed in the literature [4, 10, 13, 3, 20, 19]. In a recent paper, Adaptive Reasoning Technique

(ART: a new type of adaptive memory-based learning procedure) was described for this problem [19]. This paper extends this previous research by studying the combination of randomized greedy heuristics with the ART memory-based approach.

This paper is structured as follows: the next section outlines the description of the CMST and previous literature. The greedy and memory based solution procedures are described in Section 33.3. Section 33.4 demonstrates the computational results, while conclusions are summarized in Section 33.5.

33.2 PROBLEM DESCRIPTION AND PREVIOUS RESEARCH

The CMST is a problem of connecting a set of demand nodes to a central node through a minimum-cost tree network. Given a graph, $G(V, E)$, where V is the set of demand nodes (with the associated demand vector D_v), E is the set of possible edges in the graph (with the associated arc cost vector C_e, and an arc capacity K), and V^* is a designated goal node, the objective is to find a minimum-cost spanning tree that is rooted in node V^* where each sub-tree branch from node V^* does not contain more than K nodes. Several formulations have been proposed for this problem [7, 8, 9, 10].

One of the first, and best-known, heuristics for the CMST was proposed by Esau and Williams [4], denoted as EW. This procedure starts with an "empty" spanning tree (no arcs are used), and then adds arcs in a greedy fashion based on a modified cost structure without violating the capacity constraint, until it finds a feasible spanning tree. The EW modified cost structure for an arc (i, j) is the arc cost (c_{ij}) less the minimum distance directly to the goal node $(min\{c_{iV^*}, c_{jV^*}\})$. Altinkemer and Gavish [1, 2] presented the Parallel Savings Heuristic (PSH) method of solving the CMST. The PSH produces reasonably good solution values for the CMST. Hall [13] presents a cutting plane algorithm that often produces good solutions, but typically requires excessive CPU times. Gouveia and Martins [11] use a cutting plane algorithm, which may require extraordinarily long CPU time.

Given the limitations and shortcomings of the cutting-plane and heuristic solution procedures, a meta-heuristic algorithm is a viable solution approach. Sharaiha et al. [20] developed a tabu search procedure for solving the CMST. Their algorithm is designed specifically for the CMST, using techniques related to problem structure to improve on the solutions. Amberg et al. [3] proposed various tabu search-based procedures, as well as as a simulating annealing approach, which produce excellent results for the CMST.

Numerous heuristics for solving combinatorial optimization problems are based on greedy selection schemes. These heuristics greedily assign values to decision variables, but typically prohibit assignments that violate the problem constraints. The EW procedure discussed above is exactly such a technique. The problem with such greedy techniques is typically that they incorrectly assign values to decision variables early on in the selection process that eliminate later selections that may lead to obtaining a better objective function value for the problem. There are at least two approaches that can help circumvent this weakness in a greedy procedure:

1. Use of memory to inhibit poor selections by the greedy procedure. While iteratively applying a greedy procedure, memory can be used to keep track of recent choices, in order to prohibit poor selections by the greedy procedure. This assumes that the greedy heuristic is executed repeatedly, and memory is updated from one iteration to the next.

2. Avoid the selection of the best greedy option by randomly selecting from a first, second, third, or worse most greedy option.

Second-order heuristics [15, 16, 12] have previously been proposed for network optimization and the CMST. These approaches share some constructs similar to those found in ART, but without the dynamic memory functions of ART. Second-order heuristics were found to be successful for the CMST in [16].

Patterson et al. [19] demonstrated a different use of memory to manipulate the EW greedy heuristic procedure. Additional problem constraints are found through a learning and memory process for the CMST.

Hart and Shogan [13] proposed an algorithm for the vehicle routing problem based on randomized greedy selection and compared the cardinality implementation to the percentage implementation of randomized greedy selection for the Clarke and Wright heuristic. These ideas later were refined by Feo and Resende [5, 6], and form the basis for the greedy randomized adaptive search procedure (GRASP). With GRASP, the first-best greedy choice is not always made. Instead, a random choice is made between the first, second, third, or worse best greedy choice. Furthermore, a local search component is added. GRASP has been found to be a very efficient solution approach for numerous combinatorial optimization problems [6, 17]. It is the purpose of this paper to investigate the application of a modified and simplified version of GRASP to the CMST, and to study the combined effects of simultaneously using a memory-based heuristic and this greedy assignment technique to solve the CMST.

33.3 CONSTRAINT-BASED ADAPTIVE REASONING AND GRASP

For most problems there exists a greedy heuristic that produces solutions for the problem. We know that for Matroid problems (such as the un-capacitated version of the Minimum Spanning Tree problem), a greedy heuristic always produces the optimal solution. Many greedy heuristics are constructive; that is, they build up complete solutions from scratch, one solution component at a time. While studying the behavior of a greedy heuristic, such as the EW procedure [4] for the CMST, one will often see that the greedy heuristic really only makes a handful of crucial mistakes [15]. Actually, the mistakes can be divided into two categories: primary mistakes, that are due to the greedy selection criteria; and secondary mistakes, that are due to limited choices based on earlier primary mistakes. The primary mistakes are the crucial ones, and, given an opportunity, these are the mistakes that we would prevent.

ART is an adaptive reasoning technique that attempts to "learn" about the primary mistakes of a heuristic. In the ART framework, the greedy heuristic is executed repeatedly, and for each new execution we probabilistically introduce constraints that may prohibit certain solution elements from being considered by the greedy heuristic. That is, a prohibition of the edge (i, j) leads to the corresponding decision variable being set to zero ($x_{ij} = 0$ where $x_{ij} = 1$ if the edge from node i to node j is included in the solution or $x_{ij} = 0$ if the edge from node i to node j is not included in the solution). The prohibitions may last for more than one iteration, and as such, we may at any time have a collection of active constraints due to the prohibited solution elements. The length of the prohibition is determined probabilistically, but can often be a function of the cost of the solution element. The active constraints are held in a short-term memory. A long-term memory holds information regarding which constraints were in the active memory for the best set of solutions (for example, for the best ten solutions found so far). We can now augment these two memory functions with some basic principles of learning. For example, we can impose a certain degree of memory loss during the execution of the algorithm (some amount of memory loss is often important, even in human memory, since it enables us to explore "old territory" in light of newly discovered evidence). We can also impose a "propensity to learn" control variable, which can be varied over the execution time of ART (this is similar to a principle observed in humans: we are less "willing" to learn as we grow older). The end-result of the ART algorithm is a set of prohibitions that, when used in conjunction with a greedy algorithm, would enable us to find an optimal, or close to optimal, solution. The execution cycle of ART is as follows:

1. EW heuristic is used to solve the CMST, subject to any additional non-redundant problem constraints created by ART. The additional constraints are determined in the same manner which tabu restrictions are set in tabu search (the ART method as applied to the CMST sub-problem alone is fully described by Patterson et al. [19]).

2. The CMST solution found in Step 1 is evaluated:

   (a) Additional non-redundant constraints are added, and set to be in effect for a certain number of time periods (based on the parameters of the learning principles).

   (b) Additional non-redundant constraints whose time duration has expired are removed.

   (c) A long-term memory is updated to hold the 10 best found solutions.

   (d) Elements of learning are invoked as necessary.

3. Repeat from Step 1 until we have reached a pre-set maximum number of iterations.

A depiction of ART is given in Figure 33.1, whereas a more detailed flow chart is given in Figure 33.2. Figure 33.1 depicts the conceptual view of ART. The four main components of ART are as follows:

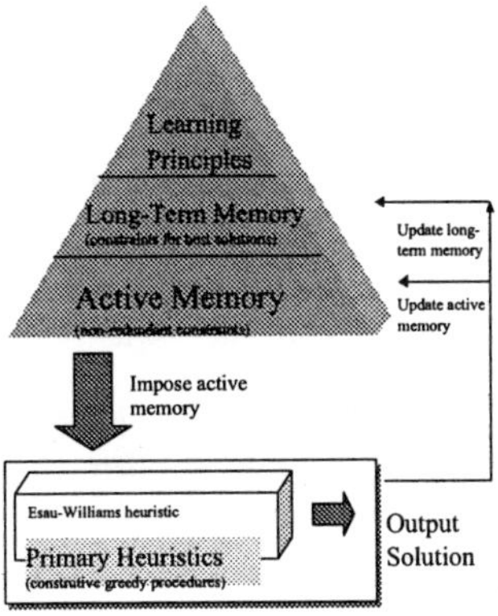

Figure 33.1 ART Components and Execution Cycle.

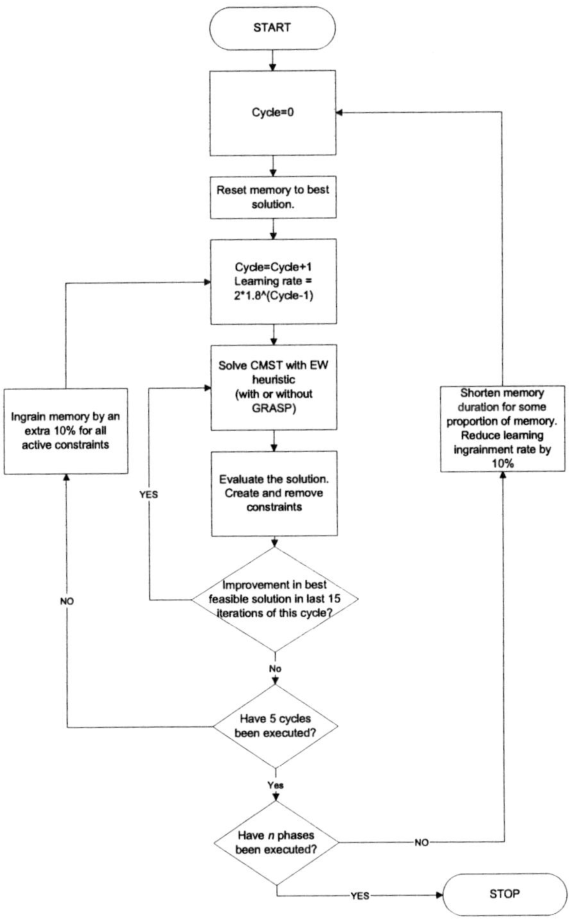

Figure 33.2 ART: A Detailed View.

1. Learning principles are applied to an EW solution to determine which decision variables are prohibited and for how long. These principles are loosely based on observations of human learning. The learning rules are based on common-sense notions of memory and learning such as: forgetting; reinforcing stronger memories through ingrainment; not trying to teach an old dog new tricks; variation in the willingness to learn over time; and variation in the time duration that learning is remembered.

2. A long-term memory is formed from the short-term memory associated with the set of best solutions. The long-term memory holds a summary of the non-redundant constraints found in the best solutions, and as such is a characterization of the combination of additional constraints that are fruitful for the problem at hand.

3. A short-term memory holds the iteration at which a constraint on a decision variable expires. This constraint is similar to a tabu restriction that eliminates a decision variable from consideration. This memory is set when a decision variable is selected by the primary heuristic. With a given probability, the short-term memory prohibits the use of the decision variable for a certain number of iterations. The time duration of the prohibition depends on both the cost associated with the arc and some degree of randomness. Since the EW primary heuristic makes few mistakes, the probability of actually prohibiting an arc is typically low for the CMST (the actual starting parameter used is 2%, and is increased by a factor of 1.8 for each phase of the algorithm (see Figure 33.2).

4. A constructive primary heuristic (EW) is used iteratively to solve the problem subject to the evolving set of additional problem constraints. The constraints that have been added prohibit particular decision variables from being included in the primary heuristic's selection process. The ART program alters the greedy selection sequence made by the primary heuristic algorithm. By altering a choice in the solution path of the greedy heuristic (through additional constraints), the subsequent choices made by the greedy heuristic are altered.

Greedy Assignment Technique According to Kontoravdis and Bard [17], "greedy randomized adaptive search procedures combine greedy heuristics, randomization, and local search. The computations are performed in two steps or phases. The first step includes the sequential construction of feasible solutions. At each iteration, all feasible moves are ranked according to an adaptive greedy function that takes into account the present state and one is randomly selected from a restricted candidate list. In the second step, local search is used to arrive at a local optimum." Our interpretation of GRASP for the CMST problem omits the local search procedure and only includes step one of traditional GRASP implementations. An alternative implementation would be to add a local search procedure to the top solutions found using both ART and GRASP. It is anticipated that in general, a starting position with lower cost solutions before the application of a local search procedure will result in lower cost solutions after the local search procedure is applied. Thus, GRASP as implemented in this paper does not include the local search procedure. That is, a detailed outline of our greedy assignment technique (yet under the acronym GRASP) is as follows.

With GRASP, the first-best greedy choice is not always made [14, 5, 6, 17]. Rather, a candidate list of the top greedy choices is created for each decision in the heuristic, from which the selection is randomly chosen. Thus, for the CMST, a simple GRASP implementation can be obtained by introducing the randomized selection into EW. Thus, one selects the arc to be chosen randomly among the n best alternatives (where n can be determined experimentally). This is referred to as n-GRASP. The basic EW procedure uses $n = 1$. When combining ART with GRASP, the modified GRASP capable EW heuristic replaces the traditional EW heuristic inside the ART framework (with n to be determined experimentally). No other modification to ART is necessary. A standalone GRASP procedure is then obtained by simply disabling the memory within ART. Thus, testing the GRASP procedure implies running the GRASP capable EW procedure numerous times within the ART looping framework. That is, the minimum number of repeated executions of the GRASP enabled EW is 350, where n is constant within this execution cycle. The maximum number of repeated executions of the GRASP enabled EW depends on how often a new "best solution" is found within the ART looping framework. Allowing executions of GRASP enabled EW to continue as long as improvements continue to be found facilitates a fair comparison with ART.

33.4 COMPUTATIONAL RESULTS

EW [4] is extremely fast, but does not produce high-quality solutions. The EW, ART, GRASP, and ART-GRASP hybrid heuristics were coded in Fortran 77, and executed on a Silicon Graphics INDY (R-4000 based MIPS) computer, with 32 MB of memory. All problems were solved with unit demand data. Twenty well know problem sets from the OR-Library[1] were obtained. Amberg et al. [3] and Sharaiha et al. [20] both obtained good results based on tabu search. The lower bounds on the optimal solutions (which in most cases are the optimal solutions) were kindly provided to us by Luis Gouveia ([10] and personal communication based on his research in progress). Some of these lower bounds are newer than the bounds reported in Sharaiha et al. [20]. The *tc* problem instances have the goal node in the center of the graph, whereas the *te* problem instances have the goal node at the end (corner) of the node scatter. The problems consist of 40 node and 80 node graphs. The 40 node problems were solved using arc capacities of 3, 5, and 10, whereas the 80 node problems were solved with arc capacities of 5, 10, and 20. Thus, a total of 60

[1] http://mscmga.ms.ic.ac.uk/info.html

known problem instances were solved. The 60 problem instances are all fully connected graphs.

Complete results for the *center* goal node problems *tc* are presented in Tables 33.1 and 33.2. The average gap between the bounds and the EW procedure was 4.44%, compared to 2.47% for size 2 GRASP-ART hybrid, 3.06% for size 2 GRASP, 2.73% for size 3 GRASP-ART hybrid, 2.74% for 3-GRASP, and 1.43% for ART.

Problem			Lower		Gap					
ID	Capacity	EW	Bound	Opt	EW	A2G	2G	A3G	3G	ART
tc40–1	3	774	742	Opt	4.31	1.21	0.67	0.40	0.40	0.00
tc40–2	3	749	717		4.46	1.53	1.53	1.39	1.53	0.98
tc40–3	3	728	716		1.68	1.12	1.12	0.98	1.12	0.42
tc40–4	3	804	775		3.74	1.42	0.52	0.65	0.90	0.65
tc40–5	3	760	741	Opt	2.56	0.54	0.54	0.81	0.54	0.00
tc40–1	5	595	586	Opt	1.54	0.34	1.19	1.02	1.37	0.34
tc40–2	5	588	578		1.73	0.87	0.87	0.87	0.87	0.87
tc40–3	5	602	577	Opt	4.33	1.39	1.39	1.39	1.39	0.69
tc40–4	5	645	617	Opt	4.54	0.00	1.30	0.16	1.13	0.00
tc40–5	5	615	600		2.50	1.50	1.50	1.83	1.50	0.83
tc40–1	10	516	498	Opt	3.61	3.21	2.81	2.81	1.61	0.00
tc40–2	10	505	490	Opt	3.06	2.65	2.24	1.63	2.24	0.41
tc40–3	10	517	500	Opt	3.40	0.00	1.60	1.60	1.60	0.00
tc40–4	10	524	512	Opt	2.34	0.78	1.37	1.56	1.17	0.39
tc40–5	10	540	504	Opt	7.14	2.78	4.76	2.38	1.19	0.00
Avg 40:					3.40	1.29	1.56	1.30	1.24	0.37
tc80–1	5	1182	1094		8.04	5.48	6.67	4.84	6.12	3.75
tc80–2	5	1170	1090		7.34	4.50	5.50	5.87	5.87	3.76
tc80–3	5	1131	1067		6.00	8.15	9.37	3.75	4.97	2.72
tc80–4	5	1151	1070		7.57	3.93	5.51	6.17	5.51	4.39
tc80–5	5	1338	1268		5.52	2.68	4.02	2.68	4.10	2.52
tc80–1	10	920	878		4.78	3.19	3.76	4.21	3.76	2.28
tc80–2	10	917	875		4.80	3.20	3.66	4.11	3.54	1.83
tc80–3	10	916	869		5.41	3.80	4.72	4.49	4.37	2.42
tc80–4	10	915	863		6.03	3.82	5.79	4.75	5.33	3.36
tc80–5	10	1069	998		7.11	4.71	5.51	5.71	5.71	4.01
tc80–1	20	856	834	Opt	2.64	1.44	1.92	3.36	1.92	0.48
tc80–2	20	836	820	Opt	1.95	1.95	1.46	2.93	1.71	0.73
tc80–3	20	856	828	Opt	3.38	1.93	2.66	2.90	2.66	1.45
tc80–4	20	866	820	Opt	5.61	1.71	3.17	1.95	3.17	0.98
tc80–5	20	971	916	Opt	6.00	4.26	4.69	4.59	5.02	2.73
Avg 80:					5.48	3.65	4.56	4.15	4.25	2.49
Avg 40–80:					4.44	2.47	3.06	2.73	2.74	1.43

Table 33.1 Results for the *tc* problems

Complete results for the *edge* goal node problems *te* are presented in Tables 33.3 and 33.4. The gaps were 5.57% for EW, compared to 3.44% for size 2 GRASP-ART hybrid, 3.21% for size 2 GRASP, 4.31% for size 3 GRASP-ART hybrid, 4.04% for 3-GRASP, and 2.48% for ART. Across both *te* and *tc*, the gap results were 5.01% for EW, compared to 2.95% for size 2 GRASP-ART hybrid, 3.14% for size 2 GRASP, 3.52% for size 3 GRASP-ART hybrid, 3.39% for 3-GRASP, and 1.95% for ART on average.

The results indicate that ART often improves on GRASP for a particular size candidate list, but that solution results are generally better when ART is used by itself. In other words, GRASP weakens ART by making less than first-best decisions. The ART memory function seems to be a more effective way of altering the deterministic heuristic process than GRASP alone or GRASP-ART hybrids.

Problem ID	Capacity	CPU time[1] A2G	2G	A3G	3G	ART
tc40–1	3	39	44	38	43	39
tc40–2	3	39	44	38	43	37
tc40–3	3	38	43	39	44	38
tc40–4	3	40	45	40	45	39
tc40–5	3	40	44	39	44	39
tc40–1	5	42	47	41	48	38
tc40–2	5	41	47	44	47	38
tc40–3	5	41	46	42	47	38
tc40–4	5	42	47	42	46	37
tc40–5	5	40	45	42	46	38
tc40–1	10	43	47	45	50	37
tc40–2	10	44	48	43	48	39
tc40–3	10	43	47	42	47	39
tc40–4	10	42	49	42	49	40
tc40–5	10	43	48	44	49	41
Avg:		41	46	41	46	38.5
tc80–1	5	276	331	284	324	264
tc80–2	5	303	325	263	310	272
tc80–3	5	277	332	273	320	271
tc80–4	5	280	319	268	325	261
tc80–5	5	296	327	287	332	258
tc80–1	10	289	329	288	344	274
tc80–2	10	289	334	281	337	273
tc80–3	10	287	335	289	338	272
tc80–4	10	278	333	289	345	284
tc80–5	10	292	351	285	339	284
tc80–1	20	292	344	283	349	270
tc80–2	20	294	343	288	339	266
tc80–3	20	286	334	288	347	281
tc80–4	20	291	345	299	349	278
tc80–5	20	297	339	284	345	270
Avg:		289	335	283	336	271.9
Avg tc:		165	190	162	191	155.2

[1] In seconds on a 133 MHz R4000 SGI Indigo

Table 33.2 CPU times for the tc problems

33.5 SUMMARY AND CONCLUSIONS

This paper examines the combination of memory-based and GRASP-based heuristics applied to the CMST. Our experimentation suggests that GRASP, and the combination of GRASP with ART, does not lead to overall improvement in solution quality as compared to ART alone. However, better solutions were found in some cases when using standalone GRASP, and GRASP with ART. This shows that the additional variance in the search through the solution space introduced by GRASP has some value. However, comparison of ART with and without GRASP, indicate that average solution quality generally deteriorates when GRASP is added to ART, and that stand-alone ART on the average does not benefit from GRASP.

Opportunities for future work include examining the effect of using the percentage version of GRASP instead of the cardinality version which was used in this paper. In addition, the number of GRASP choices available to choose from at each decision point in the EW heuristic can be varied during each iteration of the EW heuristic. For example, we could decrease the number of GRASP choices as the EW heuristic proceeds, resulting in more randomness at the front-end of the EW algorithm. Alternatively, the number of GRASP choices could be varied (e.g., randomly) over the many iterations of the EW heuristic.

Problem			Lower		Gap					
ID	Capacity	EW	Bound	Opt	EW	A2G	2G	A3G	3G	ART
te40–1	3	1215	1190		2.10	0.67	0.34	0.25	0.50	0.08
te40–2	3	1134	1103		2.81	0.73	2.45	1.63	1.63	1.00
te40–3	3	1146	1115	Opt	2.78	1.70	0.99	1.35	1.17	0.36
te40–4	3	1153	1132		1.86	0.71	0.18	1.50	0.62	0.62
te40–5	3	1147	1104		3.89	1.54	1.36	1.99	1.72	1.09
te40–1	5	857	830		3.25	2.29	2.53	2.77	3.73	2.05
te40–2	5	839	792		5.93	1.64	3.03	2.65	3.66	1.01
te40–3	5	820	797		2.89	1.63	1.00	1.38	2.26	2.76
te40–4	5	854	814		4.91	1.23	2.70	3.81	1.84	2.21
te40–5	5	816	784		4.08	2.42	2.42	3.57	1.79	1.15
te40–1	10	658	596		10.40	6.71	8.05	6.04	8.72	5.70
te40–2	10	632	573		10.30	4.19	3.49	6.28	5.06	0.70
te40–3	10	596	568		4.93	2.64	2.29	4.23	4.05	1.94
te40–4	10	638	596		7.05	1.68	0.34	3.36	2.35	1.34
te40–5	10	597	572	Opt	4.37	2.27	3.50	3.15	3.50	1.22
Avg 40:					4.77	2.14	2.31	2.93	2.84	1.55
te80–1	5	2604	2531		2.88	1.46	1.58	2.53	2.45	1.34
te80–2	5	2633	2522		4.40	3.41	2.58	3.81	3.77	1.94
te80–3	5	2723	2593		5.01	3.39	2.82	4.47	3.47	2.93
te80–4	5	2624	2539		3.35	2.72	2.80	4.33	3.31	1.89
te80–5	5	2593	2458		5.49	1.30	1.83	3.62	2.48	0.90
te80–1	10	1746	1631		7.05	4.90	5.15	6.38	7.42	3.92
te80–2	10	1748	1602		9.11	7.12	7.37	7.80	7.68	5.24
te80–3	10	1828	1660		10.12	5.66	5.72	6.33	6.81	4.58
te80–4	10	1685	1614		4.40	4.03	3.97	4.03	4.71	3.41
te80–5	10	1712	1586		7.94	5.42	4.98	5.30	6.62	3.91
te80–1	20	1330	1256		5.89	5.02	5.10	8.12	6.37	4.38
te80–2	20	1289	1201		7.33	7.74	4.91	7.91	5.58	4.41
te80–3	20	1340	1257		6.60	6.44	4.69	6.60	6.44	3.74
te80–4	20	1343	1247		7.70	8.34	5.37	9.78	8.02	6.26
te80–5	20	1334	1231		8.37	4.14	2.84	4.22	3.49	2.19
Avg 80:					6.38	4.74	4.11	5.68	5.24	3.40
Avg 40–80:					5.57	3.44	3.21	4.31	4.04	2.48
Avg tc–te:					5.01	2.95	3.14	3.52	3.39	1.95

Table 33.3 Results for the *te* problems

Acknowledgements

We are grateful to Luis Gouveia for providing us with the bounds for the *tc* and *te* test problems. Moreover, the helpful comments and clarifications of the referees as well as Stefan Voß are greatly appreciated.

References

[1] Altinkemer, K. and B. Gavish (1986) Parallel saving heuristics for the topological design of local access tree networks, Proc. Of IEEE Conference on Communications.

[2] Altinkemer, K. and B. Gavish (1988) Heuristics with constant error guarantees for the design of tree networks, Management Science, 34, 331–341.

[3] Amberg, A., W. Domschke, and S. Voß (1996) Capacitated minimum spanning trees: Algorithms using intelligent search, Combinatorial Optimization: Theory and Practice, 1, 9–39.

[4] Esau, L.R. and K.C. Williams (1966) On teleprocessing systems design, Part II – A method for approximating the optimal network, IBM Systems Journal, 5(3), 142–147.

Problem ID	Capacity	CPU time[1] A2G	2G	A3G	3G	ART
te40–1	3	39	43	39	44	36
te40–2	3	40	44	39	44	37
te40–3	3	39	44	39	44	36
te40–4	3	40	43	39	44	37
te40–5	3	39	44	39	43	38
te40–1	5	41	46	43	47	38
te40–2	5	43	47	43	49	40
te40–3	5	43	48	44	47	38
te40–4	5	45	48	43	47	38
te40–5	5	42	46	42	47	39
te40–1	10	43	47	45	50	37
te40–2	10	42	48	44	48	42
te40–3	10	42	47	43	49	41
te40–4	10	43	49	43	50	41
te40–5	10	44	48	45	51	40
Avg:		41.6	46.3	41.9	46.9	38.5
te80–1	5	277	321	287	330	260
te80–2	5	266	323	271	330	264
te80–3	5	267	315	268	335	270
te80–4	5	273	324	268	328	271
te80–5	5	289	318	269	332	255
te80–1	10	298	347	297	339	277
te80–2	10	286	326	278	350	274
te80–3	10	293	350	285	354	282
te80–4	10	286	339	294	349	264
te80–5	10	278	333	292	330	294
te80–1	20	288	354	285	338	276
te80–2	20	289	343	307	354	273
te80–3	20	291	350	303	345	292
te80–4	20	296	345	288	342	268
te80–5	20	296	348	295	367	299
Avg:		285	336	286	342	274.6
Avg te:		163	191	164	194	156.6
Avg te–tc:		189	220	188	223	179.5

[1] In seconds on a 133 MHz R4000 SGI Indigo

Table 33.4 CPU times for the *te* problems

[5] Feo, T.A. and M.G.C. Resende (1989) A probabilistic heuristic for a computationally difficult set covering problem, Operations Research Letters, 8, 67–71.

[6] Feo, T.A. and M.G.C. Resende (1995) Greedy randomized adaptive search procedures, Journal of Global Optimization, 6, 109–133.

[7] Gavish, B. (1982) Topological design of centralized computer networks – Formulations and algorithms, Networks, 12, 355–377.

[8] Gavish, B. (1983) Formulations and algorithms for the capacitated minimal directed tree problem, Journal of the Association of Computing Machinery, 30, 118–132.

[9] Gouveia, L. (1993) A Comparison of directed formulations for the capacitated minimum spanning tree problem, Telecommunications Systems, 1, 51–76.

[10] Gouveia, L. (1995) A 2^n Constraint formulation for the capacitated minimum spanning tree problem, Operations Research, 43, 130–141.

[11] Gouveia, L. and P. Martins (1995) An extended flow based formulation for the capacitated minimal spanning tree, presented at the Third ORSA Telecommunications Conference, Boca Raton, FL.

[12] Gouveia, L. and M.J. Lopes (1997) Using generalized capacitated trees for designing the topology of local access networks, Telecommunication Systems, 7, 315–337.

[13] Hall, L. (1996) Experience with a cutting plane algorithm for the capacitated spanning tree problem, INFORMS Journal on Computing, 8, 219–234.

[14] Hart, J.P. and A.W. Shogan (1987) Semi-greedy heuristics: An empirical study, Operations Research Letters, 6, 107–114.

[15] Karnaugh, M. (1976) A new class of algorithms for multipoint network optimization, IEEE Transactions on Communications, 24, 500–505.

[16] Kershenbaum, A., R. Boorstyn, and R. Oppenheim (1980) Second-order greedy algorithms for centralized network design, IEEE Transactions on Communications, 22, 1835–1838.

[17] Kontoravdis, G. and J.F. Bard (1995) A GRASP for the vehicle routing problem with time windows, ORSA Journal on Computing, 7, 10–23.

[18] Papadimitriou, C.H. (1978) The complexity of the capacitated tree problem, Networks, 4, 217–230.

[19] Patterson, R., E. Rolland, and H. Pirkul (1997) A memory adaptive reasoning technique for solving the capacitated minimum spanning tree problem, Working Paper, University of California, Riverside.

[20] Sharaiha, Y.M., M. Gendreau, G. Laporte, and I.H. Osman (1997) A tabu search algorithm for the capacitated minimum spanning tree problem, Networks, 29, 161–171.

34 A CHUNKING BASED SELECTION STRATEGY FOR INTEGRATING META-HEURISTICS WITH BRANCH AND BOUND

David L. Woodruff

GSM - University of California, Davis
Davis, CA 95616, USA.
dlwoodruff@ucdavis.edu

Abstract: Mathematical programming problems with some or all variables constrained to take on integer values are important in a large number of applications and are receiving a great deal of research attention. Meta-heuristics such as tabu search have been proposed for such problems and have been very successful in some settings but they suffer from the disadvantage that they cannot typically, by themselves, produce provably optimal solutions. Exact methods of solving these problems often include 'quick' heuristics to help solve sub-problems and/or improve bounds but incorporation of meta-heuristics has not received very much attention.

In this paper we examine an architecture for incorporating meta-heuristics in branch and bound algorithms. Since the meta-heuristics consume non-trivial computational resources, a major issue is deciding whether to launch a search from a node in the branch and bound tree. We propose methods for making this decision that make use of simple priorities, sampling and chunking to facilitate tests for novelty. Computational results are reported for 0-1 problems using XPRESS-MP for the branch and bound processing and a version of Reactive Tabu Search modified to support a novel two-level intensification/diversification strategy for those nodes selected as a starting point for heuristic search.

34.1 INTRODUCTION

Mathematical programming problems with some or all variables constrained to take on integer values are important in a large number of applications and are receiving a great deal of research attention. Although our methods are applicable in any setting where branch and bound is employed, in this paper all of our computational tests will be conducted on problems where the objective functions and constraints are linear. Linear formulations with integer variables are often appropriate in strategic planning, production planning with significant setup times, personnel scheduling and a host of other applications. The abstract formulation for linear problems with zero-one integers takes as data a row vector $\boldsymbol{c}$ of length n, a $m \times n$ matrix $\boldsymbol{A}$ and a column vector $\boldsymbol{b}$ of length m. Let D be the index set $1, \ldots, n$. The problem is to select a column vector, $\boldsymbol{x}$ of length n so as to

$$\min \boldsymbol{cx} \quad \text{(P)}$$

subject to

$$\boldsymbol{Ax} \geq \boldsymbol{b} \tag{34.1}$$

$$x_i \in \{0,1\} \; i \in I \tag{34.2}$$

$$x_i \geq 0 \; i \in D \setminus I \tag{34.3}$$

where the index set I gives the variables that must take on zero-one values.

Meta-heuristics, or heuristic search (HS) methods, such as tabu search [7] have also been proposed for such problems and have been very successful in some settings. A complete review is beyond the scope of our work here, but a few examples are [1, 8, 9, 10]. HS methods suffer from the disadvantage that they cannot typically, by themselves, prove that their solutions are optimal (exceptions include [6]). A related problem is that it is difficult to know when the search should be terminated.

Exact methods of solving these problems such as branch and bound or branch and cut operate by fixing some of the integer variables at an integer value; each set of fixed integer values defines a node in the search tree. The remaining integer values are left free to take on real or integer values; the resulting problem is referred to as the *relaxation* for the node. Exact methods progress from node to node to implicitly exhaust all possible combinations of integer values. The best integer feasible solution currently available can be used to bound out some combinations.

Exact methods sometimes include 'quick' heuristics to help solve sub-problems and/or improve bounds. For example, MINTO [11] uses what they call a 'diving' heuristic every ρ nodes to see if an integer feasible solution can be obtained using the solution for the node's relaxation as a starting point. The default value for ρ is 25; so the heuristic is employed for about 4% of the nodes. An algorithm that combines genetic algorithms with branch and bound for satisfiability problems has been proposed by researchers at Loughborough University [5].

In Section 34.2 we examine an architecture for incorporating HS in a branch and bound algorithm. Since these searches consume non-trivial computational resources, a major issue is deciding whether to launch a search from a node in the branch and bound tree. We propose methods for making this decision in Section 34.3. Computational results are reported in Section 34.4 for

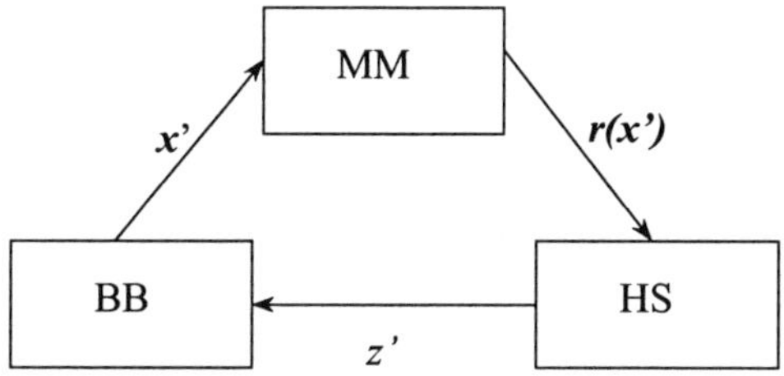

Figure 34.1 The Architecture

0-1 problems using XPRESS-MP for the branch and bound processing and Reactive Tabu Search (RTS) [3] for those nodes selected as a starting point for heuristic search. This section also describes a novel two-level intensification/diversification strategy. We modify RTS so that random diversification is no longer used; rather the chunking and branch and bound algorithm are relied on to produce new, high quality search starting points. The reaction mechanisms in RTS are used to adjust the tabu tenure parameter and to terminate the search. The paper ends with conclusions and a discussion of directions for further research.

34.2 THE ARCHITECTURE

In Figure 34.1 we display a simple architecture for incorporating meta-heuristics in a branch and bound algorithm. The branch and bound module is labeled BB. As each node is expanded, the solution to the resulting relaxation, $\boldsymbol{x}'$, is provided to the module labeled MM, which is the subject of our work here. This module examines $\boldsymbol{x}'$ and determines whether to launch a search using this solution after simple rounding, $r(\boldsymbol{x}')$ as a starting point. The search is conducted by a module named HS. If this module finds a feasible solution with a promising objective function value, z', it is provided to BB which will update the cutoff, if z' is better than the current cutoff. A key point is that the modules in this architecture can be assigned to separate computers and the communication requirements are modest and can be completely asynchronous.

This architecture itself is not particularly novel. For example, it is supported by the appl_primal (and also bypass_integral) calls in MINTO. Our main interests in this paper are on the nodes labeled MM and HS and their interaction.

34.3 SOLUTION SELECTION RULES

In order to give the rules used for selecting those nodes from which we will search, we need a little bit of notation. Let $\boldsymbol{x}^*$ be the best (integer) feasible solution seen so far by BB and HS. Let $z^* \equiv \boldsymbol{c}\boldsymbol{x}^*$ be the best objective value seen so far. Our interest here is entirely in solutions that are passed from BB to MM. Our algorithm does not pass solutions for nodes where the relaxation

Symbol	Parameter	Suggested Value
ρ	inverse of sampling frequency	$\|I\|/10$
p	chunking partition dimension	3
N	sample size	$3p$
α	threshhold for χ_p^2 inverse	0.7

Table 34.1 Parameters

is not LP feasible, and it does not pass solutions, $\boldsymbol{x}'$, if $\boldsymbol{c}\boldsymbol{x}' > z^*$ (i.e., if the node is *cutoff*).

The selection from among those solutions passed to MM proceeds in two phases. In the first phase, solutions are sampled and in the second phase distance measures based on chunking [12] are used to select solutions. The processing relies on the parameters given in Table 34.1, whose meanings will become clear. Before describing the two phases, we review chunking.

34.3.1 Chunking concepts

In this section we give the specific chunking application that we used. For more general discussion and development, see [12, 13]. We will refer to a non-empty subset of solution vector indexes as a *chunk*. An instance of a partial vector corresponding to a chunk is referred to as a *chunk instantiating value*. That is, a chunk instantiating value for a given chunk $C \subseteq D$ depends on the specific vector $\boldsymbol{x}$ whose values are under consideration. Let $\mathcal{P}$ be a partition of D with p cells. The mappings from chunk instantiating values to reals are referred to as *chunk valuation* functions, v. We will deal with a fairly specific valuation function. Let $\boldsymbol{v}_{\mathcal{P}}(\boldsymbol{x})$ be the (column) vector of chunk valuations for a given vector $\boldsymbol{x}$ under partition $\mathcal{P}$. Each element of the valuation vector is defined as

$$v_C(\boldsymbol{x}) = \frac{1}{\mid C \mid} \sum_{i \in C} x_i$$

where $\mid C \mid$ is the number of indexes in the chunk for each chunk $C \in \mathcal{P}$. If we index the chunks $C \in \mathcal{P}$, we can write v_i in the same way for $i = 1, \dots, p$.

We can use chunking to quantify the location and shape of a group of solution vectors. This characterization then allows us to quantify the distance between solution vectors using the shape to define a metric. In particular, we measure the distance to the location of the group used to define the shape.

To get the appropriate metric, we make use of a location vector (mean) $\boldsymbol{\mu}$ of length p and a $p \times p$ shape matrix (covariance matrix) $\boldsymbol{S}$. A p-vector $\boldsymbol{v}$ can be said to have a *Mahalanobis distance* from $\boldsymbol{\mu}$ using $\boldsymbol{S}$ that is given by

$$d^2(\boldsymbol{\mu}, \boldsymbol{S}; \boldsymbol{v}) = (\boldsymbol{v} - \boldsymbol{\mu})^T \boldsymbol{S}^{-1} (\boldsymbol{v} - \boldsymbol{\mu}) \tag{34.4}$$

Short distances are consistent with $\boldsymbol{\mu}$ and $\boldsymbol{S}$ and longer distances are not as consistent.

The motivation for this metric comes from multivariate probability theory. If $\boldsymbol{S}$ and $\boldsymbol{\mu}$ define a multivariate normal distribution, then for points governed by the distribution, the Mahalanobis distances are distributed χ^2_p (see e.g., Anderson [2], pages 72-75). As will be described later, we will make use of a matrix $\boldsymbol{S}$ which is the covariance estimate for the valuation vectors of the most recent N solutions passed to HS. We will make use of the χ^2_p distribution only as a method of characterizing distances in a way that depends on the data and p in a rational way that can be characterized by the single parameter α; hence we really are not relying at all heavily on normality.

34.3.2 Phase I

The first phase terminates when N solutions have been selected to be passed to HS. During this phase, we always select solutions, $\boldsymbol{x}'$, for which $r(\boldsymbol{x}')$ is the best integer feasible solution seen so far. We also select every ρ^{th} solution. For each solution selected, $\boldsymbol{x}'$, the valuation vector $\boldsymbol{v}(\boldsymbol{x}')$ is added to the valuation vector sample, $\boldsymbol{V}$.

34.3.3 Phase II

The second phase terminates when BB does. During this phase, we also always select solutions, $\boldsymbol{x}'$, for which $r(\boldsymbol{x}')$ is the best integer feasible solution seen so far. In addition, we make use of the mean, $\boldsymbol{\mu}_V$, and covariance matrix, $\boldsymbol{S}_V$, of the sample $\boldsymbol{V}$. For each solution selected, $\boldsymbol{x}$, the member of $\boldsymbol{V}$ that has been in the sample the longest is discarded and the valuation vector $\boldsymbol{v}(\boldsymbol{x})$ is added.

The mean and covariance matrix are computed in the usual way as

$$\boldsymbol{\mu}_V \leftarrow \frac{1}{N}\sum_{\boldsymbol{v}\in\boldsymbol{V}} \boldsymbol{v} \quad \text{and} \quad \boldsymbol{S}_V \leftarrow \frac{1}{N-1}\sum_{\boldsymbol{v}\in\boldsymbol{V}} (\boldsymbol{v}-\boldsymbol{\mu}_V)(\boldsymbol{v}-\boldsymbol{\mu}_V)^T . \tag{34.5}$$

We select solutions, $\boldsymbol{x}'$, if $d^2(\boldsymbol{\mu}_V, \boldsymbol{S}_V; \boldsymbol{v}(\boldsymbol{x}')) \geq (\chi^2_p)^{-1}(\alpha)$ where $(\chi^2_p)^{-1}(\cdot)$ is the inverse of the cumulative distribution function for the χ^2 distribution with p degrees of freedom.

In other words, solutions are selected if they are sufficiently 'distant' from the most recent N solutions used as a starting point. The idea is that computational effort should be expended only when the starting point is sufficiently different from others that have been explored or is sufficiently promising. The notion of 'sufficiently promising' is implemented by accepting BB solutions whose rounding results in a solution that is the best integer feasible solution seen so far. The notion of 'sufficiently different' is implemented using a metric that takes into account the characteristics and covariance structure of the solutions used to compute the valuation vectors in the population.

Furthermore, the metric changes over time in a very appealing way. Loosely speaking, the addition of a distant point tends to decrease the distance to the point and points near it. This is exactly the diversifying behavior we want for two reasons:

1. If a solution $\boldsymbol{x}'$ has a valuation vector that is sufficiently novel to have a distance that is beyond $(\chi^2_p)^{-1}(\alpha)$, we use it as a starting point. But then we add it to the population, which tends to reduce the Mahalanobis distance to the point and to others 'like it' (with equal or similar valuation vectors; where 'similar' is defined using the Mahalanobis metric that takes into account the covariance structure of the sample). This is good because we do not want to expend effort searching in the neighborhood of solutions whose neighborhoods have already been searched.

2. Since we bias the selection toward distant points, the covariance matrix tends to grow in size. Overall, this has the effect of making fewer and fewer points seem close. This is good because as the algorithm progresses, it becomes increasingly likely that we have seen the best solution that we will see and the goal should be to prove optimality. Hence, interruption of BB processing for HS should decrease over time and the test for sufficient novelty to warrant search should become more stringent. The use of Mahalanobis distances accomplishes this in a naturally adaptive way, without the addition of any parameters.

These concepts can be made rigorous, but readers can easily convince themselves of the effects on distance and matrix size by considering one dimensional valuation vectors (i.e., $p = 1$).

34.4 COMPUTATIONAL RESULTS

34.4.1 The hypothesis

Our main interest is in the processing done by MM and its interaction with HS and BB, so we conduct an experiment concerning the effects of its parameters while holding the BB and HS component characteristics fixed. Our hypothesis is as follows: the use of our *full architecture* with MM selection rules to start HS can improve the performance of a particular BB algorithm, at least on some classes of problems. Although our MM methods are general, our HS implementation was for pure integer problems only. Before giving the results of our experiments, we give the reader a summary of the HS and BB modules that we used.

34.4.2 Reactive tabu search

Although any type of search could, in principle, be used we use reactive tabu search (RTS) because it gives us an opportunity to exploit a two-level diversification strategy that will be explained after we review the characteristics of RTS. For a full description of the method, see the original paper by Battiti and Tecchiolli [3]. As with tabu search in general, RTS is defined relative to one or more neighborhood structures and evaluation functions when applied to the problem of minimizing a function $f(\cdot)$ subject to some constraints. For a set $B \subset \Re^n$, every solution $\boldsymbol{x} \in B$, let $\mathcal{N}(\boldsymbol{x}) \in B$ denote those solutions that can be reached from $\boldsymbol{x}$ with a single *move*. A move evaluation function $\hat{f}(\cdot)$, often resembles $f(\cdot)$, but may differ in order to take advantage of special knowledge of the problem.

For our particular application to problems with 0-1 decision vectors, we elect to make use of a simple "one-bit" neighborhood. Each solution has n neighbors with each neighbor formed by changing the value of one vector element. For $\hat{f}(\boldsymbol{x})$ we used $\boldsymbol{cx}$ plus M times the number of infeasible rows, where M is equal to the sum of the absolute values of the elements of $\boldsymbol{c}$.

The search proceeds from trial solution $\boldsymbol{x}^k$ to a trial solution

$$\boldsymbol{x}^{k+1} \in \mathcal{A} \subset \mathcal{N}(\boldsymbol{x}^k)$$

where $\mathcal{A}$ is the set of admissible moves. It is formed by removing from $\mathcal{N}(\boldsymbol{x}^k)$ any solutions that would reverse the attributes of the most recent κ moves and that do not satisfy any aspiration threshold. The moves that are in $\mathcal{N}(\boldsymbol{x})$ but not in $\mathcal{A}$ are said to be tabu. The value of κ is changed dynamically during reactive tabu search. It is increased if a repeated solution is detected and decreased slowly over time and/or if all moves are tabu. The initial value of κ for our experiments is $n/4$.

The distinguishing feature of RTS is its reliance on long term memory that is concerned primarily with repeated solutions. If too many solutions (given by the parameter REPT) are repeated too often (given by the parameter CHAOS) then the search is assumed to be trapped by a particular cup. We conjure the metaphor of a chaotic attractor. In addition to recording repeated solutions, the number of moves between repeated solutions is also stored (recorded in the variable Running_Average). This value is assumed to be proportional to the size of the domain of attraction of the chaotic attractor. When REPT solutions have been repeated CHAOS times, the search executes a large number of random moves (with the number proportional to Running_Average) in order to try to escape the domain of attraction of the chaotic attractor.

As a basis of comparison, note that other forms of tabu search likewise make use of frequency-based memory as a foundation for diversification strategies (to drive the search into new regions), but the frequencies are generally recorded for attributes of solutions, such as values of variables and functions, rather than for complete solutions. In addition, the trigger for diversification in other forms of TS typically consists of either an iteration count or a limit on the number of iterations without finding a new best solution.

For our particular application, we make use of the parameter values used in [13] (i.e., REPT = CHAOS = 2), but rather than making a series of random moves when diversification is indicated, we simply terminate the search. The effect is to let BB and MM diversify the search in a systematic way rather than using randomization. To use the language of tabu search, we use BB and MM for intensification (to find good starting points) and for diversification (to find diverse starting points). The reactive mechanisms in RTS are used to make internal parameters such as κ adaptive and to signal the need for diversification. So we let the RTS algorithm at a lower level signal the need for diversification by the branch and bound algorithm at a higher level. This multilevel diversification/intensification scheme is somewhat novel and proves to be effective.

34.4.3 XPRESS MP

For the BB module, we made use of XPRESS-MP version 9.3 [4]. Default parameter settings were used. We chose XPRESS-MP because the package provides excellent low-level support for HS. The searches were launched using the callback facility provided by XPRESS-MP.

34.4.4 Instances tested

All experiments were done on a 120MHz Pentium with 8MB of main memory allocated to the solver. The MM and HS coding was done in C++.

The chunk partitions were simply a division of D into p roughly equal cells. The result was that the chunks were sets of contiguous indexes. More sophisticated chunking schemes are discussed in [12], but are left as future research for this application. We tested values of 2, 3, and 4 for p. Clearly, $N > p$ or the covariance matrix will be singular. We tried values of $3p$ and $5p$ for N. The search is fairly insensitive to these parameters. Because of the adaptive nature of the distances (as we included distant points in the sample, the distances shrink), the search is very insensitive to the parameter α; a good value seems to be 0.7.

Not very many instances are required to support our hypothesis that the architecture proposed here can be beneficial. It certainly is not always beneficial. In particular, it is of no help (and wastes some CPU time) when the instance can be solved quickly by BB. We now describe experiments with some instances that support our hypothesis and illustrate some key observations about the performance of the algorithms.

34.4.4.1 ex_dash. This instance has 575 columns and 10 rows. All of the rows have both upper and lower bounds. Figure 34.2 shows time versus solution quality for the suggested parameter configurations given in Table 34.1. If only a few seconds are allocated, then BB running alone gets better results. For longer periods of time, the full architecture is superior until after a very long period, when BB alone can find a solution of equal quality. Since the CPU times are determined by the particular software and the hardware configuration, to support reproducibility the objective function value versus the node count are also shown in Figure 34.3. As one would expect, the full architecture provides even more improvement when effort is measured in nodes.

A larger set of runs are detailed in Appendix A to support the contention that the use of MM and HS to augment BB results in improved performance over a broad range of parameter settings for this problem instance. Close examination of this table reveals that performance is not very sensitive to the chunking parameter p in the ranges tested (between 2 and 4).

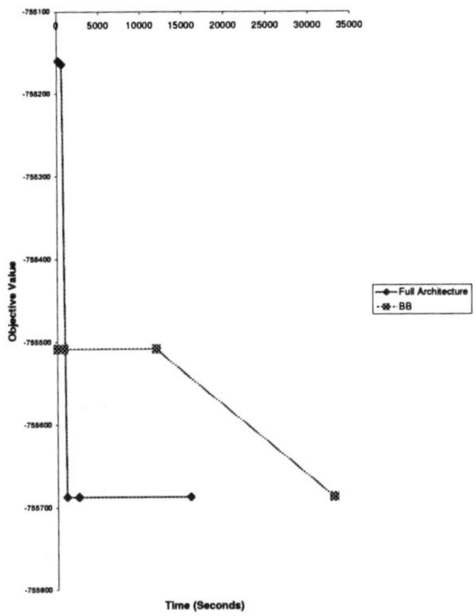

Figure 34.2 Comparison of CPU Times for the ex_dash instance.

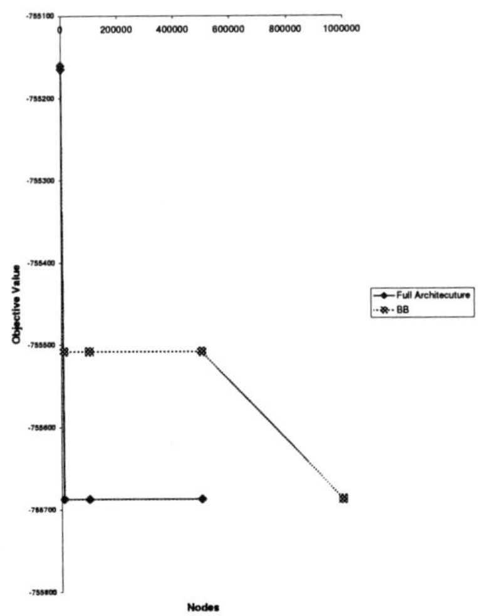

Figure 34.3 Comparison of Nodes Needed for the ex_dash instance.

34.4.4.2 P0548. The exponential growth in the branch and bound tree size implies that for every computer there are instances that will exceed the memory of the computer. This is one of the reasons to use HS in conjunction with BB; by finding good bounds early, the tree size can be reduced and/or good results can be found before the tree overruns available memory.

This instance allows us to "simulate" an instance that is too big for our technology. We put the word in quotes because this instance really is too big for the technology we were using (although it can be solved by similar computers with more memory). It has 176 rows and 548 columns. The instance causes memory and disk overflows if BB is allowed to run for very long on our PC. Also, clique and cover cuts are a tremendous boost for this instance, but we were not using cutting planes.

When running alone, BB quickly finds a solution with objective value 29008 and can find one with value 28926 by the time it runs out of memory resources after about 1200 seconds. With the full architecture, it takes over one minute to find a feasible solution, but then it finds one with objective value 28434, which it cannot improve upon before exceeding available memory. The full architecture is able to get this improvement regardless of the parameter settings.

34.4.4.3 Stein45. This Steiner tree problem instance has 331 rows and 45 integer variables. The instance is small, so that we prove optimality using only our small PC. Regardless of the parameter settings, the full architecture algorithm can find the optimal solution in about half the time required for BB alone to find it (about 60 seconds). On the other hand, BB alone can prove optimality about 30% faster than the full architecture. This is pro-intuitive and borne out by experiments with other instances. The use of MM and HS can help find good solutions faster, but the extra pruning that this enables does not compensate for the extra time spent searching so that if optimality is to be proven one can do it faster without the extra modules.

34.4.4.4 Mod008. This small instance (6 rows and 319 columns) is representative of the large class of instances that can be solved quickly by BB and for which the full architecture offers no improvement. The CPU time degradation for proving optimality is about 30% if one uses the full architecture described in this paper.

34.5 CONCLUSIONS AND DIRECTIONS FOR FURTHER RESEARCH

In this paper we described an architecture for incorporating meta-heuristics in branch and bound algorithms. We then proposed and tested methods for making the decision to search for better solutions using the solution to the relaxation at a node as a starting point. The selection methods make use of simple priorities, sampling and chunking to facilitate tests for novelty. We modify RTS so that random diversification is no longer used; rather the chunking and branch and bound algorithm are relied on to produce new, high quality search starting points. The reaction mechanisms in RTS are used to adjust the tabu tenure parameter and to terminate the search.

Computational results are reported for 0-1 problems using XPRESS-MP for the branch and bound processing and modified version of Reactive Tabu Search for those nodes selected as a starting point for heuristic search. Our results suggest that the approach can be helpful for some, but not all instances.

We argue that this makes the approach worthy of further research. Like many options in integer programming, it is not always helpful on every instance, but when it is effective it is very effective. When it is not effective, the computational cost is modest. Hence, for the many applications where similar instances are solved day after day, it is worth trying the method to see if it helps on the family of instances at hand.

The main areas of future research are improvement of the search used in HS, improved control of the search by MM using information obtained during the process, and perhaps some use of information obtained by MM to alter branching and variable selection priorities in BB. Integer programming problems are very important and large amounts of research continue both in the area of exact methods and in the area of meta-heuristics. We hope that this paper has taken steps toward application of the two techniques in unison.

Acknowledgments

This work was supported in part by the National Science Foundation grant DMS -94.06193.

Appendix A — Results for ex_dash

Table 34.2 lists results for various parameter settings for the instance ex_dash. The column labeled "Max Nodes" indicates the termination criteria. The times are inflated a great deal by time spent paging and are somewhat imprecise because they were collected on a machine using Windows95, but they are adequate for assessing parameter sensitivities. The messages here are that the value for ρ is somewhat important but the search seems to be fairly insensitive to most parameters and there is no discernable pattern of response.

References

[1] Aboudi R. and Jörnsten, K., "Tabu Search for General Zero-One Integer Programs using the Pivot and Complement Heuristic," *ORSA J. Comput.*, 6, 82-93, 1994.

[2] T.W. Anderson, *An Introduction to Multivariate Statistical Analysis*, Wiley, New York, 1984.

[3] R. Battiti and G. Tecchiolli, "The Reactive Tabu Search," *ORSA J. Comput.*, 6, 126-140, 1994.

p	Max Nodes	α	N	ρ	Time (sec)	Obj
3	100	0.8	6	287	99	-755160
3	100	0.7	9	575	100	-755160
3	1000	0.7	9	575	174	-755160
3	10000	0.7	9	575	893	-755541
3	100000	0.7	9	575	2436	-755567
3	500000	0.7	9	575	16490	-755567
2	100000	0.7	6	575	2369	-755567
2	500000	0.7	6	575	16303	-755567
4	100000	0.7	12	575	2570	-755567
4	500000	0.7	12	575	16620	-755567
3	1000000	0.7	9	575	41494	-755687
3	100	0.7	9	57	175	-755160
3	1000	0.7	9	57	589	-755164
3	10000	0.7	9	57	1122	-755687
3	100000	0.7	9	57	2561	-755687
3	500000	0.7	9	57	15947	-755687
2	100000	0.7	6	57	2311	-755567
2	500000	0.7	6	57	16294	-755567
4	100000	0.7	12	57	14085	-755567
3	100	0.7	15	57	174	-755160
3	1000	0.7	15	57	1052	-755353
3	10000	0.7	15	57	1462	-755541
3	100000	0.7	15	57	2936	-755567
3	500000	0.7	15	57	16943	-755567
2	100000	0.7	10	57	2712	-755567
2	500000	0.7	10	57	16674	-755567
4	100000	0.7	20	57	3019	-755567
4	500000	0.7	20	57	16979	-755567
3	1000000	0.7	15	57	40932	-755687
3	100	0.3	9	57	173	-755160
3	1000	0.3	9	57	588	-755164
3	10000	0.3	9	57	1131	-755687
3	100000	0.3	9	57	2532	-755687
3	100	0.3	9	57	194	-755160
3	1000	0.3	9	57	591	-755164
3	10000	0.3	9	57	1121	-755687
3	100000	0.3	9	57	2607	-755687
3	500000	0.3	9	57	15968	-755687
2	100000	0.3	6	57	2314	-755567
2	500000	0.3	6	57	16410	-755567
4	100000	0.3	12	57	14126	-755567
2	100000	0.7	6	57	2281	-755567
2	500000	0.7	6	57	16234	-755567
4	100000	0.7	12	57	14326	-755567
4	500000	0.7	12	57	28178	-755567

Table 34.2 Parameter Variations for the Instance ex_dash

[4] Dash Assoc., *XPRESS-MP Reference Manual*, Version 9.32, Binswood Ave, Leamington Spa, Warwickshire, CV32 5TH, UK, 1996.

[5] French, A.P., A.C. Robinson, and J.M. Wilson, "Solving Satisfiability Problems Using a Hybrid Genetic-Algorithm/Branch-and-Bound Approach," Working Paper, Loughborough University, Loughborough, LE11 3TU, England, 1997.

[6] Glover, F., "Ghost Image Methods for Integer Programming," working paper, University of Colorado, Boulder, CO, USA, 1996.

[7] Glover, F. and Laguna, M. "Tabu Search," *Modern Heuristic Techniques for Combinatorial Problems*, C. Reeves, (ed.), Blackwell Scientific Publishing, 70-150, 1993.

[8] Glover, F. and Løkketangen, A. "Solving Zero-One Mixed Integer Programming Problems Using Tabu Search," *European J. Oper. Res.*, to appear.

[9] Løkketangen, A. and Glover, F. "Probabilistic Move Selection in Tabu Search for Zero-One Mixed Integer Programming Problems," in *Metaheuristics: Theory & Applications*, I.H. Osman and J.P. Kelly, eds., Kluwer, 467-487, 1996.

[10] Løkketangen, A., Jörnsten, K. and Storøy, S. "Tabu Search Within a Pivot and Complement Framework," *Int. Trans. Oper. Res.*, 1, 305-316, 1994.

[11] Savelsbergh, M.W.P. and G.L. Nemhauser, "Functional Description of MINTO, a Mixed Integer Optimizer, Version 2.3, ISYE, Georgia Institute of Technology, Atlanta, GA, 30332, USA, 1996.

[12] Woodruff, D.L., "Proposals for Chunking and Tabu Search," *European J. Oper. Res.*, 106, 585-598, 1998.

[13] Woodruff, D.L., "Chunking Applied to Reactive Tabu Search," in *Metaheuristics: Theory & Applications*, I.H. Osman and J.P. Kelly, eds., Kluwer, 555-569, 1996.